aviation
fundamentals

SANDERSON

S5336C

FOREWORD

In the early days of aviation, following the Wright brothers' historic flight, a local citizen and his son approached an aviation pioneer while he was working on his frail craft in a cow pasture. The citizen asked, "Really, what good is the airplane?" The aviation pioneer answered by posing the question, "What good is your little boy?" The citizen answered, "Why, some day he'll grow up to be a man." The aviation pioneer replied, "So will the airplane."

History has witnessed the "coming of age" of aviation and when the first Earth satellite was launched in 1957, the space age took its first tottering steps. We might view aviation as a young adult and space travel as a pre-adolescent, with both of these dynamic areas of human skill, technology, and knowledge still having their greatest potential lying ahead of them.

Communicating the knowledge and potential of this fascinating segment of human endeavor, as well as the present impact of air vehicles upon society, is the goal of aviation/education.

Sanderson believes the most direct and interesting approach to achieve these goals is by gaining a foundation in fundamental subjects. The future study of other subjects will be based on these fundamentals. For example, one does not approach the study of mathematics without first starting with basic arithmetic. In the same way, the student of aviation starts with such fundamentals as the science of flight. He broadens his knowledge through the study of the history of aviation and adds depth to his knowledge through the study of such subjects as meteorology, navigation, and communications.

These subjects are some of the ABC's of aviation. They capture interest and provide a foundation on which a student can build for a specific area of study. These fundamentals will expand the knowledge and usefulness of the citizen in a society where aviation plays an increasingly important role and they provide for the exploration of career opportunities. Even if the student elects to progress no farther than the completion of the *Aviation Fundamentals* course, many limitations will be overcome and new insights gained.

In addition to receiving a foundation for future career opportunities, the student of aviation is challenged by a subject as modern as the age in which he lives. Meaning and direct application are found for the principles of math, physics, biology, chemistry, sociology, geography, and other subjects. An appreciation of the ease with which long distances are traveled will develop.

The progress of aviation in the future will be dependent upon the knowledge and awareness of the great impact of aviation upon the individual citizen. Imparting this knowledge and awareness is the objective of *Aviation Fundamentals*.

ACKNOWLEDGEMENTS

Sanderson wishes to express appreciation to the following companies and organizations for their contributions of illustrations.

Aviation Data/Advisory Service
Avco Lycoming Division
Beech Aircraft Company
Boeing Aircraft Company
British Aircraft Company
Cessna Aircraft Company
Convair Division of General Dynamics
Thomas Y. Crowell Company
Echelon Publishing Co.
Federal Aviation Administration
Gates-Lear Jet
General Aviation Manufacturers Association
Hughes Aircraft Company
Kansas Commission on Aerospace Education
King Radio Corporation
Lockheed Aircraft Corporation
Martin Marietta Corporation
McDonnel Douglas Corporation
National Aerospace Education Council
North American Aviation
Piper Aircraft Company
Sikorsky Aircraft Division of United Aircraft
  Corporation
Smithsonian Institution
United Air Lines

Gratitude is also expressed to Mr. James H. Connett for his work in researching the Aviation History Chapter.

TABLE OF CONTENTS

chapter 1

INTRODUCTION TO THE AIRPLANE

INTRODUCTION

This chapter will acquaint the student with the basic concepts of airplane design variation, structure, and the function of each major component. He will also become familiar with the dynamic nature of airflow and the dynamic and static forces acting on an airplane in flight. Additionally, this chapter will provide basic knowledge of the factors affecting aircraft performance.

SECTION A - THE AIRPLANE

VARIETIES OF AIRCRAFT

The FAA has developed guidelines which assist in differentiating between the vast array of aircraft. One method of classification is based on the manner in which an aircraft sustains itself while airborne. From this standpoint, aircraft are divided into the four following categories:

1. Lighter-than-air.
2. Gliders.
3. Rotorcraft.
4. Airplane.

LIGHTER-THAN-AIR

Historically, *balloons* were the first aircraft that enabled man to break the bond that had held him to earth. Today, the balloon survives principally for sport, and the dangerously flammable hydrogen, used previously to provide lifting power, has been replaced by hot air. (See Fig. 1-1.)

Another variety of lighter-than-air craft is the *blimp.* (See Fig. 1-2.) A rigid keel runs along the bottom of the gas envelope, and to this keel a control car or cabin is attached. Generally, twin

Fig. 1-1. Balloon

Fig. 1-2. Blimp

powerplants are used, one on each side of the control car.

A third variation of lighter-than-air craft was the *dirigible* or Zeppelin. (See Fig. 1-3.) These giant airships differed from the blimp in that they had a lightweight aluminum framework.

Fig. 1-4. Sailplane

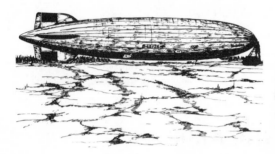

Fig. 1-3. Dirigible with Lightweight Internal Framework

GLIDERS

The type of *glider* most prominent today is the high-performance sailplane. (See Fig. 1-4.) This type of craft possesses excellent aerodynamic characteristics and has enjoyed great popularity in a sport which allows the pilot to pit his knowledge of air currents and skill in controlling the sailplane to achieve great altitudes and long duration flights.

ROTORCRAFT

Rotorcraft may be subdivided into *gyroplanes* which attain thrust with a pusher-type propeller and lift with a free-wheeling rotor, and *helicopters* on which the rotor is powered in order to achieve lift. (See Fig. 1-5.)

AIRPLANES

By far, the most important type of aircraft in terms of its numbers, economic contributions, and sociological impact is the airplane.

As with other types of aircraft, airplanes may be classified on the basis of certain characteristics. In the case of airplanes, this breakdown is made on the basis

GYROPLANE

HELICOPTER

Fig. 1-5. Gyroplane and Helicopter

of their intended use, number of engines, type of landing gear, and the location and configuration of their wings. As shown in figure 1-6, single-engine land, single-engine sea, multi-engine land, and multi-engine sea airplanes denote a more detailed classification.

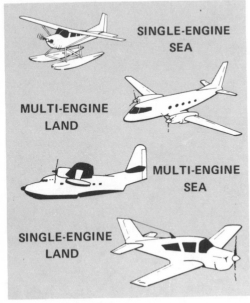

Fig. 1-6. Airplanes

AIRPLANE COMPONENTS

The major component parts of an airplane are: (See Fig. 1-7.)

1. Wings.
2. Fuselage or body.
3. Tail assembly or empennage.
4. Landing gear.
5. Powerplant.

WINGS

WING CONFIGURATION AND LOCATION

A casual walk around any large airport will show that most airplanes are *monoplanes*, or airplanes with one wing. (See Figs. 1-8 and 1-9.) Airplanes with two wings, called *biplanes*, are likely to be agricultural aircraft, restored antiques, or sport airplanes designed and built by the homebuilder. As shown in figure 1-10, wings can take various shapes and may be located on the fuselage in different positions.

WING CONSTRUCTION

Many high-wing airplanes have external braces or wing struts which transmit the flight and landing loads through the struts to the main fuselage structure. A few high-wing and most low-wing airplanes, on the other hand, have a *full-cantilever* wing designed to carry the loads without external bracing or struts.

Wings are designed to be extremely strong and, at the same time, lightweight. *Spars* run the length of the wing from wing root to wingtip and bear the major portion of bending loads. Bending or twisting loads are transferred to the metal skin and absorbed by tension or compression of the skin. *Ribs* maintain the shape of the wing and stiffen the skin. Additional resistance to buckling-under loads is provided by parts known as *stringers* and *formers*. In almost all aircraft, some of the space inside the wing is occupied by fuel tanks. (See Fig. 1-11.)

COMPONENT PARTS

Flaps and *ailerons* are hinged along the trailing edge of most airplane wings. (See Fig. 1-11.) The ailerons make up the outboard segment of the wings' trailing edges and move in opposite directions in order to make the aircraft roll or bank. The flaps move together and are used primarily during the takeoff and landing phase of a flight.

FUSELAGE

The *fuselage*, or body, of the airplane provides an attachment point for the wing, tail assembly, landing gear, and powerplant. In addition, it provides space for the crew and passengers and houses the various controls and instruments required for flight.

Fuselages are sometimes categorized according to the type of construction employed. In the *truss-type* construction, the forces of compression, tension, and shear are absorbed by a framework of steel tubing. (See Fig. 1-12.) The *semi-monocoque* type is an alternate method

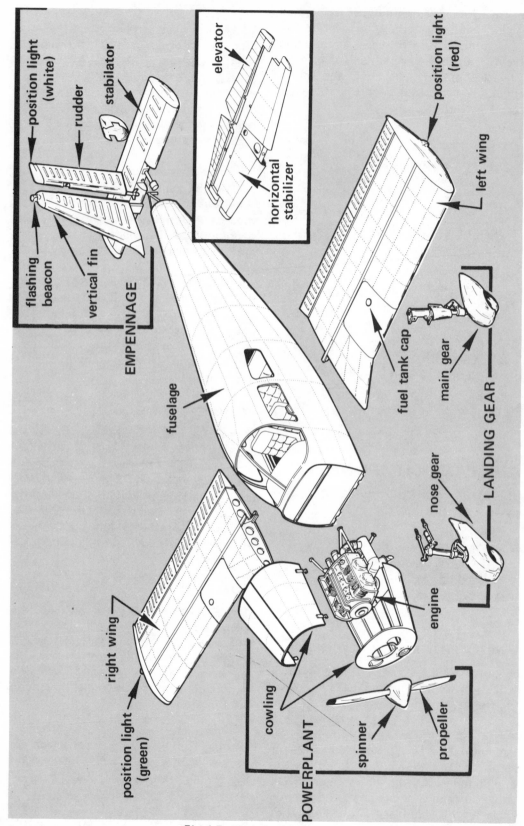

Fig. 1-7. Airplane Components

Fig. 1-8. Typical Monoplane in Low-
Wing Configuration

Fig. 1-9. Typical Monoplane in High-
Wing Configuration

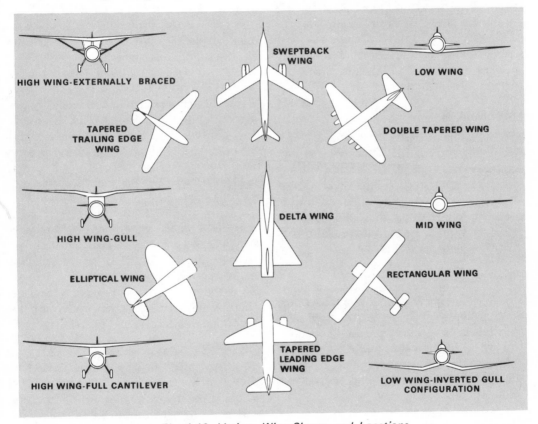

Fig. 1-10. Various Wing Shapes and Locations

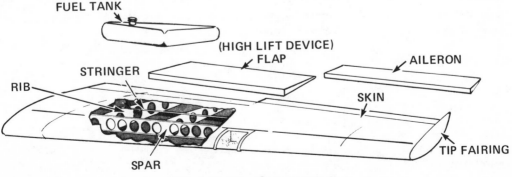

Fig. 1-11. Wing Components

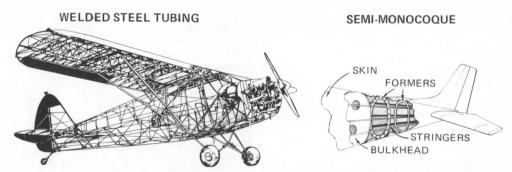

WELDED STEEL TUBING

SEMI-MONOCOQUE

SKIN
FORMERS
STRINGERS
BULKHEAD

Fig. 1-12. Two Methods Employed in Fuselage Construction

of construction. This means that the skin, made of stamped and formed sheets of aluminum alloy, carries a major portion of the stresses. A part of the structural strength also lies in internal members such as the floor, bulkheads, stringers, formers, and stiffeners.

EMPENNAGE

The *empennage* most often consists of a fixed vertical stabilizer, attached movable rudder, horizontal stabilizer, and hinged movable elevator. (See Fig. 1-13.) Some airplanes have a pivoting one-piece horizontal stabilizer which needs no elevator. A horizontal surface designed in this manner is called a stabilator.

TRIM TABS

Most airplanes have a small hinged section on the elevator called the *trim tab*. The trim tab is connected by cables to a control wheel in the cabin and is used to relieve control pressures at any desired flight attitude, thus helping to prevent pilot fatigue.

The direct effect of a trim tab is to exert pressure on the control surface to which it is attached. Figure 1-14 shows an elevator trim tab in the down position, thus

exposing it to the pressures of the slipstream. This forces the elevator up, and the nose of the airplane rises. More complex airplanes may also have trim tabs on the rudder and the ailerons.

LANDING GEAR

Even though the airplane is basically designed to fly, provisions must be made for ground operations such as taxiing, takeoff, and landing. The landing gear of an airplane may be of the so-called *conventional* type which incorporates two main wheels and a tailwheel (see Fig. 1-15), or it may be of the *tricycle* type with two main wheels and a steerable nosewheel (see Fig. 1-16). The wheels may be fixed in position, or by utilizing a retraction mechanism, they may be withdrawn into the wings or fuselage thereby effecting a great reduction in aerodynamic drag. (See Fig. 1-17.)

In order to absorb landing shocks, the wheels of most airplanes are attached to *oleo struts*. (See Fig. 1-18.) The oleo struts make use of confined air and oil in order to absorb impacts. Other aircraft may have a spring steel landing gear, and a few aircraft use shock-absorbing biscuits, shock cords, and torsion bar systems.

VERTICAL STABILIZER (FIN)
RUDDER
ELEVATOR
HORIZONTAL STABILIZER

HORIZONTAL STABILIZER AND ELEVATOR

STABILATOR

Fig. 1-13. Components of the Tail Assembly or Empennage

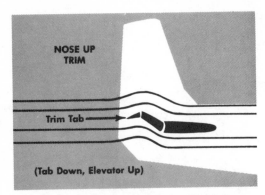

Fig. 1-14. Function of the Elevator Trim Tab

Fig. 1-17. Airplane with Landing Gear Retracted

Fig. 1-15. Typical Conventional Landing Gear Installation

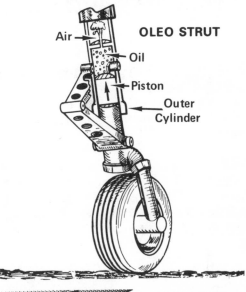

Fig. 1-16. Modern Tricycle-Geared Airplane

The utility of the airplane has been greatly extended through the adaptation of specialized types of landing gears. Figure 1-19 illustrates the distinctive landing gears provided for skiplanes, floatplanes, and amphibious airplanes. Pilots utilizing ski-equipped airplanes can operate from frozen lakes, glaciers, and snowbound airports.

Fig. 1-18. Airplane Landing Gear Shock Struts

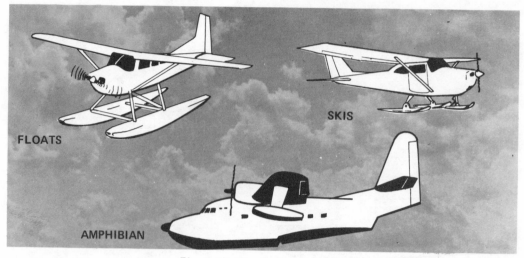

Fig. 1-19. Special Landing Gear

BRAKES

The brakes may be considered as components of the landing gear. Small general aviation aircraft utilize a simple self-contained hydraulic system to apply braking action.

On some aircraft, the brakes on the left and right wheels are equipped with individual hydraulic brakes which are operated by applying toe pressure to the rudder pedals. The rotation of the pedals actuates the brake cylinders, resulting in brake fluid being forced through the brake lines to the brakes, where a braking action is applied to the main landing gear wheels. A separate handle is used to set the brakes for parking. (See Fig. 1-20.) Steering is accomplished by a mechanical linkage from the rudder pedals to the nose- or tailwheels.

On some aircraft, the brakes on the left and right main wheels are applied simul-

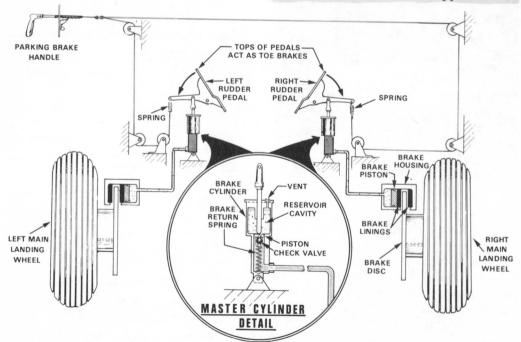

Fig. 1-20. Typical Individual Hydraulic Brake System

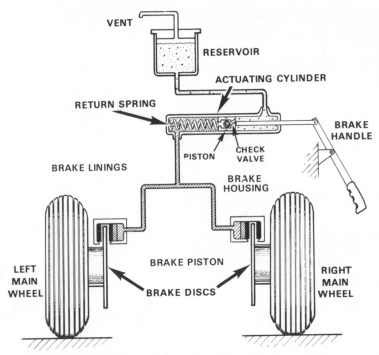

Fig. 1-21. Typical Hand Brake System

taneously by means of a hand brake. (See Fig. 1-21.) This type of system is primarily designed for deceleration and parking operations. Steering is accomplished by a mechanical linkage from the rudder pedals to the nose- or tailwheels.

THE POWERPLANT

The powerplant may consist of an engine-propeller combination or, in the case of jet propulsion, the engine itself is considered the powerplant. A commonly used powerplant is the gasoline-powered internal combustion engine.

In single-engine aircraft, an engine mount holds the powerplant to a sealed firewall at the front of the cabin section. (See Fig. 1-22.) The cowling is used to enclose the engine and serves several additional purposes. The cowling:

1. Shrouds the engine to provide engine cooling by ducting the cooling air around the engine cylinders.

2. Streamlines the front of the airplane so that the on-rushing air will pass the nose of the airplane with a minimum of drag and turbulence.

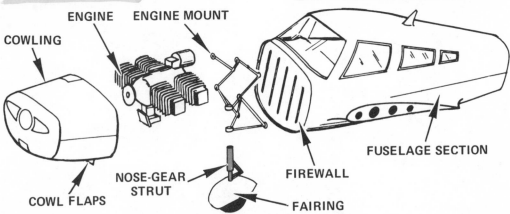

Fig. 1-22. Details of Engine Compartment

3. Protects the engine and accessories from rain and snow when the aircraft is parked.

4. Enhances the appearance of the airplane.

PROPELLERS

In most cases, the propeller is mounted directly on the crankshaft which is driven by the moving pistons as they go through their power cycle. Each blade of the propeller is shaped like a small airfoil or wing. (See Fig. 1-23.) These blades are "twisted" to insure that the blades of the propeller "bite" into the air at the proper angle and provide forward lift, much like the wing provides vertical lift. In a sense, then, the propeller becomes a revolving wing. The path of the propeller on a cruising airplane is shown in figure 1-24.

Details of powerplants and propellers are discussed in Chapter 2.

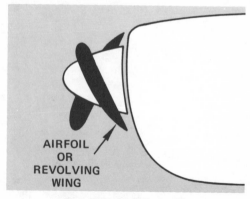

AIRFOIL OR REVOLVING WING

Fig. 1-23. The Propeller

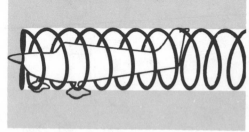

Fig. 1-24. The Path of the Propeller

SECTION B - AERODYNAMICS

Once the aircraft leaves the ground, it is supported in the air by aerodynamic forces. An examination of the forces acting upon an airplane and the application of the principles upon which these forces are based follows.

BERNOULLI'S PRINCIPLE

One of the most significant physical laws that led to the creation of the airfoil is the *Bernoulli Principle,* which states the relationship between pressure, fluid flow velocity, and the potential energy of fluids (liquids and gases). This principle, discovered by Daniel Bernoulli (1700-1782), a Swiss scientist, states that: "As the velocity of the fluid increases, the pressure in the fluid decreases;" and conversely, "as the velocity of the fluid decreases, the pressure in the fluid increases."

An operating example of Bernoulli's Principle is the *venturi tube.* (See Fig. 1-25.) This device is a tube which is narrower in the middle than at the ends. As air passes through the tube, it speeds up as it reaches the narrow portion and slows down again as it passes the restriction.

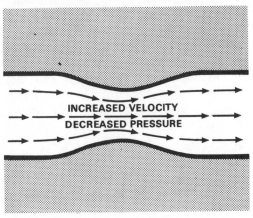

INCREASED VELOCITY
DECREASED PRESSURE

Fig. 1-25. Flow of Air Through a Venturi Tube

The previous statement can be explained as follows: if the air flowing through the entrance to the tube has 1,000 molecules per inch in the tube and is moving at a rate of one inch per second, there are 1,000 molecules flowing through the tube each second. Assuming that the area of the narrow part of the tubing is one-half that of the rest of the tube, the molecules in the air must speed up to get 1,000 molecules per second through this restricted area. With one-half the area, the speed must be twice that in the rest of the tube to permit the same volume of air to pass through the restriction.

More energy is imparted to the molecules as they accelerate. This leaves less energy to exert pressure and the pressure thus decreases. The throat in a carburetor, as described in Chapter 2, is an application of the venturi principle.

AIRFOILS

The principles applied to the venturi are also utilized to create lift for an airplane. However, in lieu of a venturi, a wing with an airfoil shape is used to create a pressure differential in the air.

An *airfoil* is any shape which is designed to produce lift. Although the wing is the primary part of the airplane that produces lift, other airfoils find application as propeller blades and tail surfaces. An airfoil has a leading edge, a trailing edge, a chord, and camber as shown in figure 1-26.

The *leading edge* is the part of the airfoil that first meets the oncoming air. The *trailing edge* is the aft end of the airfoil where the airflow over the upper surface joins the airflow over the lower surface.

The *chord line* is an imaginary straight line drawn from the leading edge to the trailing edge. This line has significance only in determining the angle of attack of an airfoil and in determining wing area.

The *camber* of an airfoil is the curvature of its upper (upper camber) and lower (lower camber) surfaces.

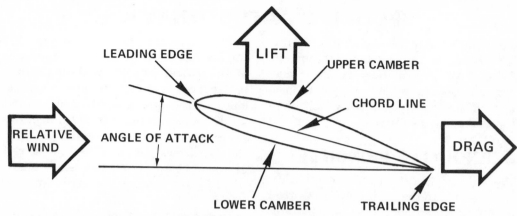

Fig. 1-26. Airfoil Terminology

The *relative wind* is the wind moving past the airfoil. The direction of this wind is relative to the attitude or position of the airfoil and it is always parallel to the flight path of the aircraft. (See Fig. 1-27.) The velocity of the relative wind is the speed of the airfoil through the air.

The *angle of attack* is the angle formed by the chord of the airfoil and the direction of the relative wind. The pitch attitude of the airfoil and the angle of attack are the same only in level flight. In other flight conditions, they are different values. (See Fig. 1-27.)

THE FOUR FORCES

An aircraft in straight-and-level flight is acted upon by four forces: lift, gravity, thrust, and drag. *Lift* is the upward acting force; *gravity*, or weight, is the downward acting force; *thrust* acts in a forward direction; and *drag* is the backward, or retarding force produced by air resistance. (See Fig. 1-28.)

Lift opposes weight and thrust opposes drag. When an aircraft is in straight-and-level flight, the opposing forces balance each other; lift equals weight and thrust equals drag. Any inequality between thrust and drag, while maintaining straight-and-level flight, will result in acceleration or deceleration until the two forces again become balanced.

LIFT

According to the Bernoulli Principle, there is an acceleration or increase in the velocity of air as the air flows around an airfoil shape; therefore, there is an acceleration of the relative wind as it flows above and below the surface of the airplane wing. Because the camber of the upper wing surface is greater than that of the lower surface, air flowing above the wing will be accelerated more than air flowing beneath the wing. The Bernoulli Principle also states that an increase in the velocity of a fluid, such as air, results in a decrease of pressure with-

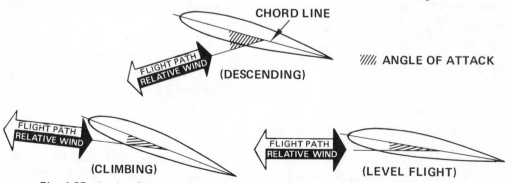

Fig. 1-27. Angle of Attack Depends Only on Relative Wind; Not on Pitch Angle

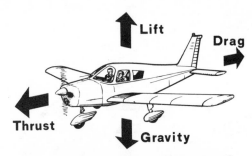

Fig. 1-28. The Four Forces Acting on an Aircraft in Straight-and-Level Flight

in that fluid. As a result, the reduction in air pressure above the wing will be greater than the pressure reduction along the lower wing surface. This difference of pressure accounts for the upward force called lift. (See Fig. 1-29.)

At *high angles of attack*, an additional force is derived from the impact of air against the lower surface of the wing because it is inclined to the relative wind. This principle may be observed by putting a hand or flat surface out of the window of a fast-moving car. The impact of the air on the bottom surface when it is inclined sharply upward creates a lifting force that is easily sensed. (See Fig. 1-30.)

Even at high angles of attack, the lift generated by the impact of air on the bottom surface of the wing amounts to only a fraction of the lifting force needed to sustain the aircraft in flight. More than 75% of the lift is caused by the lower pressure above the airfoil. The wing is

not so much pushed up from below by excess air pressure as pulled up from above by a suction force.

Lift can be increased in two ways; by increasing the forward speed of the airplane or by increasing the angle of attack. The pilot can increase the forward speed of the aircraft by applying more power. This increases the speed of the relative wind over the airfoil.

THE STALL AND ITS CAUSE

Increasing the angle of attack will increase lift up to a point. As the airfoil is inclined, the air flowing over the top of the airfoil is diverted over a greater distance resulting in an even greater increase in air velocity and more lift.

However, as the airfoil is given a greater angle of attack relative to oncoming air, it becomes more difficult for the air to flow smoothly across the top of the wing. Thus, it starts to separate from the wing and enters a burbling or turbulent pattern. The angle at which airflow separation and turbulence occurs on the upper wing surface is called the *critical angle of attack*. This turbulence results in a loss of lift in the area of the wing where it is taking place.

The separation point starts near the trailing edge of the wing and progresses forward as the angle of attack is increased. (See Fig. 1-31.) Finally, the separation

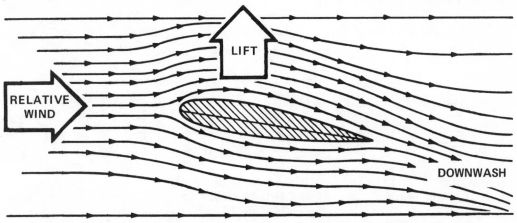

Fig. 1-29. Lift is Generated by Air Traveling Faster Above Airfoil Than Beneath

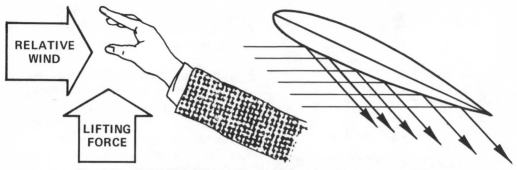

Fig. 1-30. At High Angles of Attack, Part of Lifting Force is Obtained From Air Deflected Downward

point moves so far forward that most of the wing loses its lift and a *stall condition* occurs.

FACTORS AFFECTING THE STALL SPEED

Gross Weight

As the gross weight of the airplane is increased, the stall speed increases. Due to the greater weight, a higher angle of attack must be maintained to produce the extra lift required to support the additional weight; therefore, the critical angle of attack will be reached at a higher airspeed when the aircraft is loaded to maximum gross weight than when the pilot is flying solo.

Flaps

The use of flaps results in a reduction in stall speed. The stall speed chart shown in figure 1-32 illustrates this fact. Notice that there is a progressive reduction in stall speed from 63 m.p.h. with the flaps up, to 53 m.p.h. with the flaps extended 30°.

Angle of Bank

Figure 1-33A illustrates the relationship between angle of bank and stall speed. Note that there is a progressive increase in stalling speed as the angle of bank is increased from zero. The stall speed increases due to the additional load factor or weight supported by the wings. As load factor increases, the total lift must also increase to maintain level flight. This additional lift is generated by increasing the angle of attack. Therefore, in a turn, the critical or stall angle of attack is reached at a higher airspeed than in level flight. For example, the stall speed is approximately 40% greater in a 60° bank turn than when the aircraft is in straight-and-level flight. (See Fig. 1-33A, item 1.)

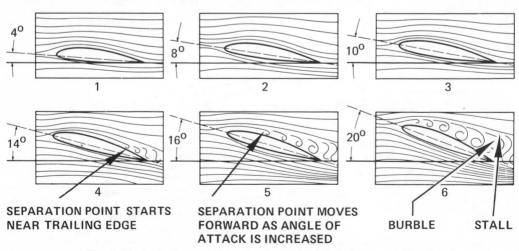

SEPARATION POINT STARTS NEAR TRAILING EDGE

SEPARATION POINT MOVES FORWARD AS ANGLE OF ATTACK IS INCREASED

BURBLE　　**STALL**

Fig. 1-31. Stall Occurs When Angle of Attack is Increased Too Much

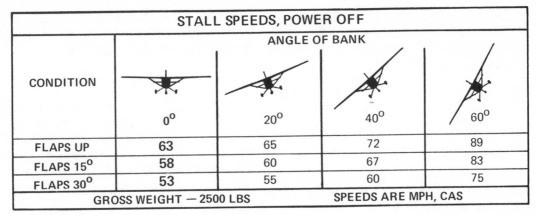

STALL SPEEDS, POWER OFF				
CONDITION	ANGLE OF BANK			
	0°	20°	40°	60°
FLAPS UP	63	65	72	89
FLAPS 15°	58	60	67	83
FLAPS 30°	53	55	60	75
GROSS WEIGHT — 2500 LBS			SPEEDS ARE MPH, CAS	

Fig. 1-32. Effect of Flaps

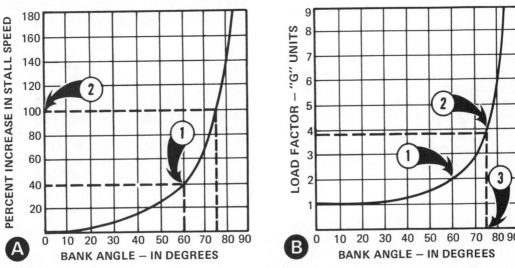

Fig. 1-33. Stall Speed and Load

Center of Gravity Location (Weight Distribution)

If the weight distribution of the aircraft is located farther forward (forward CG), additional down force on the tail must be developed through elevator back pressure in order to produce a critical angle of attack. If the elevators are already in the full-up position, the additional aerodynamic force must be generated by increased airflow over the horizontal tail surfaces. In order to do this, the airspeed must be greater; therefore, the *stall speed increases.*

Conversely, the farther aft the weight distribution, the *lower* the stalling speed. The pilot should not interpret this to mean that a far aft weight distribution is desirable as the problems involved in stability and stall recovery will far outweigh any advantages obtained by the decrease in stall speed.

Load Factor

Load factor is the ratio of the load supported by the wings to the actual weight of the airplane and its contents. With the bank required to produce a load factor of two G's (twice the force of gravity), the wings support twice the weight of the airplane; at a load factor of four G's, they support four times the weight of the airplane. (See Fig. 1-33B, items 1 and 2.)

Airplanes certificated by the FAA in the *normal category* are required to have a minimum limit load factor of 3.8. The limit load factor is that load factor that an airplane can sustain without incurring permanent structural damage. Note, on load factor charts, that this limit load factor is attained in a level, banked turn of approximately 75° (item 3). At this angle of bank, the stall speed is also approximately 100% greater than the value for straight-and-level flight. (See Fig. 1-33A, item 2.)

Frost, Snow, and Ice

Even a light accumulation of frost, snow, or ice on the wings can cause a significant increase in stall speed. The accumulation disrupts the smooth flow of air over the wing, thus decreasing the lift it produces and increasing the stall speed. The stall speed may be increased so much that an airplane may be unable to attain the necessary speed to take off or, even though the airplane may become airborne, it could be so close to the stall speed that it would not be possible to maintain flight. Frost, snow, and ice should always be removed from the wings before flight is attempted.

Turbulence

Air turbulence can cause large, momentary increases in stall speed. An upward vertical gust causes an abrupt change in the direction of the relative wind and the angle of attack. It is this abrupt increase in angle of attack which can produce a stall. Pilots making landing approaches under turbulent conditions generally maintain an approach speed somewhat higher than the normal approach speed in order to guard against a stall.

MISCONCEPTIONS REGARDING STALLS

Some persons believe that stalls occur only at relatively low airspeeds; however, this is not true. An airplane can be stalled *at any airspeed*. All that is necessary is to exceed the critical angle of attack. This can be done at any airspeed if the pilot applies abrupt or excessive back pressure on the elevator control. The stall that occurs at a relatively high speed is referred to as an *accelerated* or *high-speed stall*.

Another misconception is that it is necessary for the airplane to have a relatively high pitch attitude in order for it to stall. An airplane can be stalled *in any attitude;* level, banked, or even inverted. Again, all that is necessary is to exceed the critical angle of attack. This could occur, for example, in a level attitude during landing approach if the airspeed is allowed to become too low and the angle of attack becomes too great. (See Fig. 1-34.)

Another misconception is that the speed of the airplane over the ground, or ground speed, is related to the stall speed. For example, this question was posed to a group of student pilots: "If an airplane's stalling speed was 60 m.p.h. and you are

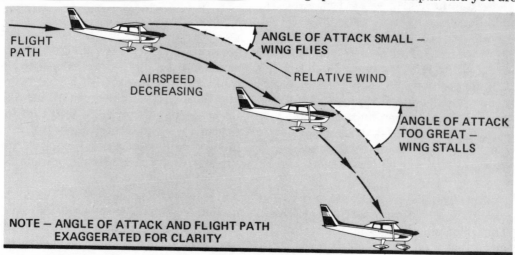

FLIGHT PATH

ANGLE OF ATTACK SMALL — WING FLIES

AIRSPEED DECREASING

RELATIVE WIND

ANGLE OF ATTACK TOO GREAT — WING STALLS

NOTE — ANGLE OF ATTACK AND FLIGHT PATH EXAGGERATED FOR CLARITY

Fig. 1-34. Stall Can Occur in Level Attitude

flying at an airspeed of 70 m.p.h. into a 30 m.p.h. wind, what would happen if you maintained this airspeed of 70 m.p.h., but turned downwind?" A number of the students said the airplane would stall; however, this is not correct. *Airspeed* is the only speed which is of any significance in aerodynamics. Once it is off the ground, an airplane "feels" nothing but its own speed through the air. Its stall speed is not affected by what the airplane's speed happens to be in relation to the ground. The airplane in flight "feels" no wind; it simply proceeds through the air mass operating with the same mechanical efficiency upwind, downwind, crosswind, or in no wind at all.

STALL RECOVERY

When the wing stalls, the airplane descends since insufficient lift is being generated to hold the aircraft aloft. The pilot can recover from a stall condition by one of two methods:

1. By lowering the nose of the aircraft. This decreases the angle of attack of the wings and allows the airflow over the wings to smooth out again.
2. By accelerating the aircraft. This is done by the application of more engine power. However, if the aircraft is moving under full power when the stall occurs, the only way to recover is to lower the nose.

As noted above, lift is affected by the angle of attack of the airfoil with respect to the relative wind. To hold the airfoil in a particular position requires an external aerodynamic force. Through the use of elevators or a stabilator, the pilot can control the angle of attack. The pilot holds the surface in the position that causes the airflow over the tail surfaces to exert the required amount of force to hold the wings in the desired attitude.

DESIGN OF THE WING

The shape of the wing is very important for the type of operation that the aircraft is designed to accomplish. The shape of the airfoil selected is determined by the primary function of the airplane. If the aircraft is designed for slow speed, a thick airfoil is used; for high speed aircraft, a thin airfoil is utilized.

Planform

Planform is the shape of the wing as seen from directly above or below. The relationship between the length and the width of a wing is called the *aspect ratio.* It is computed by dividing the span (distance from wingtip to wingtip) by the average chord of the wing. (See Fig. 1-35.) The average chord is called the mean aerodynamic chord (MAC).

In general, the higher the aspect ratio, the more efficient the wing. A long narrow wing will create more lift per square foot of area than a short wide wing. The aspect ratios on gliders can be as great as 20 to one. The average aspect ratio of general aviation aircraft is approximately six to one.

The wingtip is the least efficient portion of the wing because the air under the wing rolls from beneath the wingtip into the low pressure area above the upper surface of the wing. This rolling air causes a swirl, or vortex, at the tip of the wing and decreases the amount of smooth air which creates lift. (See Fig. 1-36.) Thus, a narrow chord wing has less wing surface in the disturbed vortex area and has less loss of lift than a wide chord wing.

The swirling air trailing behind the wingtips is called wake turbulence. A discussion of the operational aspects of flight in areas where wake turbulence from large aircraft is most often encountered is included at the end of Section C of this Chapter.

The advantages and disadvantages of the various wing planform shapes shown previously in figure 1-10 are noted below:

1. STRAIGHT WING — excellent stall characteristics. Most economical to build. Inefficient from structural, weight, and drag standpoints.
2. TAPERED WING — efficient from structural, weight, and drag standpoints. Stall characteristics are not as good as straight wing.

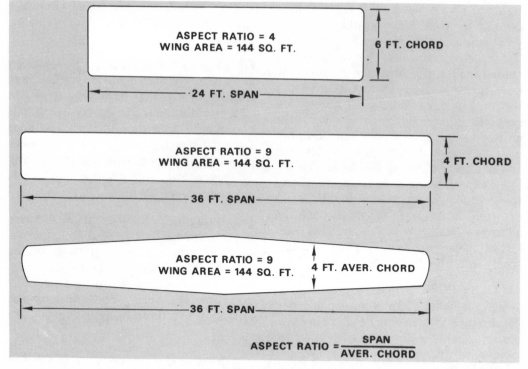

Fig. 1-35. Aspect Ratio

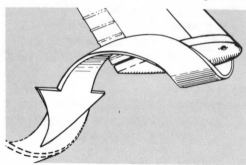

Fig. 1-36. Wingtip Vortex

3. ELLIPTICAL WING — most efficient from structural, weight, and drag standpoints. Stall characteristics are not as good as straight wing and wing is more expensive to build than tapered wing.

4. SWEPTBACK AND DELTA WINGS — efficient at high speeds near the speed of sound, but inefficient at approach and landing speeds.

Incidence

The angle of incidence is the name given to the angle between the wing chord line and the airplane's longitudinal axis. (See Fig. 1-37.) Choosing the right angle of in-

cidence in designing the airplane can improve flight visibility over the nose and reduce drag in cruising flight. Most airplanes have a slight positive angle of incidence so that the wing has a positive angle of attack when the fuselage is perfectly level, as in cruising flight.

Washout or Twist

To provide positive aileron control throughout the stall, wing twist is often built into the wing. (See Fig. 1-38.) In this configuration, the wingtip has a lower angle of incidence than the wing root (usually two or three degrees). This results in the wingtips having a lower angle of attack than the root during the approach to a stall. Thus, the ailerons and the wingtip will still be flying and providing control when the wing root has stalled. This also produces a more gentle stall since the wing does not stall all at once.

Stall Strips

Stall strips are sometimes employed to make the wing root stall first. As the

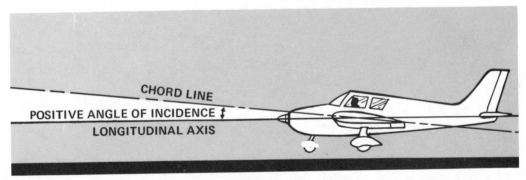

Fig. 1-37. Angle of Incidence

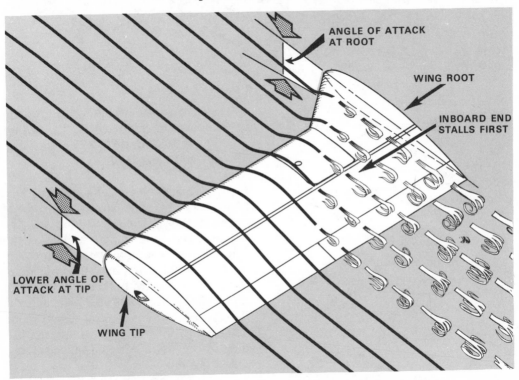

Fig. 1-38. Wing Twist Maintains Aileron Control During the Stall

angle of attack increases, these strips break up the airflow and start the stall at the root. (See Fig. 1-39.)

Spanwise Airfoil Variation

Some wings employ a high-speed airfoil at the root and a low-speed type at the tip to keep the wingtips from stalling. The low-speed airfoil can fly at higher angles of attack than the high-speed root section. Thus, the root stalls first.

WING FLAPS

As noted previously, increasing the camber of the wing increases lift. A de-vice used to change the camber is the *flap*. It is a movable device which results in the turning down of the trailing edge of the wing. Some of the more common types of flaps are: (See Fig. 1-40.)

1. PLAIN FLAP — This device is hinged to the trailing edge of the wing and is the simplest method of changing the camber of an airfoil. In the re-tracted position, the flap stream-lines and adds to the area of the wing surface.

2. SLOTTED FLAP — This type of de-vice moves aft and down when ex-tended. A slot is formed between the

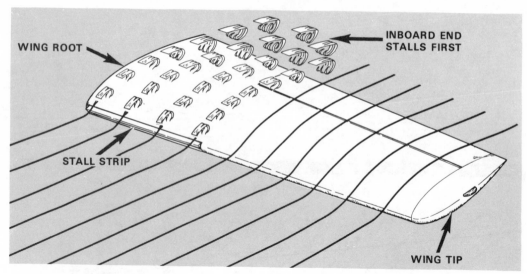

Fig. 1-39. Stall Strips Cause Stall to Begin at Wing Root

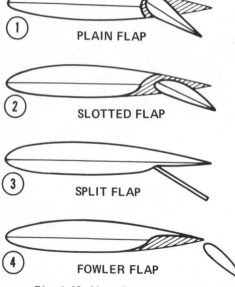

Fig. 1-40. Most Common Types of Wing Flaps

wing and flap permitting a smooth flow of air between the two components.

3. SPLIT FLAP — This is a flat surface hinged to, or set into, the lower side of the trailing edge of the wing. The shape of the upper surface of the wing is unchanged when the split flap is lowered. This type of device creates more drag than the plain flap because of its turbulent wake.

4. THE FOWLER FLAP — This mechanism moves backward on rollers in a track and then is rotated downward to provide high lift capability. When down, this device increases lift and wing area.

Flaps normally have a deflection range of from $0°$ to $40°$ from the chord of the wing. The deflection of the flap increases the upper camber of the wing, thus increasing the negative pressure on top of the wing and allowing more positive pressure to build up underneath. Thus, lift is *increased* by lowering the flaps. However, as the wing flaps are extended into the airstream to increase lift, they also cause additional induced drag. This drag increases with the position of the flaps — the greater the flap deflection, the greater the drag.

Pilots make use of the lift-to-drag ratio generated by wing flaps in several flight operations. For example, during takeoff or climb over obstacles, the pilot desires high lift with as little drag as possible. To accomplish this, wing flaps are usually extended to one-fourth or one-half the flap travel. This permits additional lift to be obtained with a minimum of additional drag.

Beyond the half-flap position, although it is true that wing flaps produce more lift with deflection, the drag increases to the extent that the flaps are acting more as air brakes than lifting devices. It is because of this braking effect that more than one-half flap travel is not used in takeoff or in operations where high lift with low drag is desired.

The braking effect of the flaps in the full-flap position is very advantageous for landing, however. The landing flight path, without flaps, is very shallow. When landing over obstacles, this causes the aircraft to land further down the field. With the air-brake effect of full flaps, the aircraft descends at a much steeper angle. This permits landing closer to the approach end of the runway. The landing speed, with flaps, is slower during the touchdown, and the aircraft is easier to stop.

GRAVITY

Gravity is the force that lift has to overcome for the airplane to fly and amounts to nothing more than the weight of the loaded airplane.

THRUST

Forward motion is essential for the flight of all fixed-wing airplanes. This force, called *thrust*, is obtained through the use of a propeller or jet engine. During straight-and-level flight at a constant airspeed, thrust and drag are equal. When the thrust is increased by increasing engine power, thrust momentarily exceeds drag, and the airspeed increases. The increase in airspeed causes a corresponding increase in drag. Airspeed, however, does not increase at the same rate as drag. As airspeed is increased, drag increases at a much *faster* rate. At a new higher airspeed, thrust and drag forces again become equalized and speed again becomes constant.

For the airplane to remain in steady flight, equilibrium must be maintained; lift must be equal to airplane weight, and powerplant thrust must be equal to airplane drag. Thus, airplane drag determines the thrust required to maintain level flight.

DRAG

Drag may be subdivided into induced drag and parasite drag. *Induced drag* is simply that drag which is created in the process of developing lift. Increasing the angle of attack or applying flaps to increase lift are two actions which will increase induced drag. (See Fig. 1-41.)

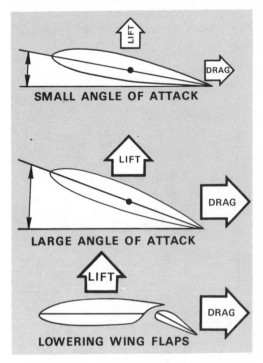

SMALL ANGLE OF ATTACK

LARGE ANGLE OF ATTACK

LOWERING WING FLAPS

Fig. 1-41. Increasing Induced Drag

In addition to the induced drag caused by the development of lift, there is *parasite drag* due to skin friction and form. Parasite drag is present any time the airplane is moving through the atmosphere, even in a zero lift condition. Components of the airplane, such as the wings, fuselage, tail, and landing gear, contribute to the drag because of their own form. Additionally, any loss of momentum of the airstream due to cowl openings for powerplant cooling, or gaps such as between the wing and aileron, creates additional parasite drag. The sum of all the drag due to form, friction, leakage, momentum losses, and interference is termed *total parasite drag,* since none of these factors are directly associated with the development of lift.

The total drag of the airplane in flight is the sum of the *induced* and *parasite* drags. Figure 1-42 illustrates the variation of drag with speed for a given airplane in level flight at a particular weight, configuration, and altitude. Note that the parasite drag increases with speed, while induced drag decreases with speed. The total drag of the airplane shows the predominant influence of *induced drag at low speeds* and *parasite drag at high speeds*.

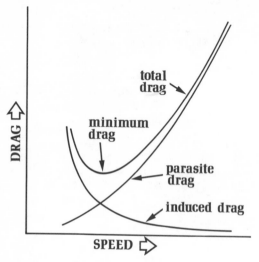

Fig. 1-42. Effect of Speed on Drag

GROUND EFFECT

When the airplane wing operates within a wingspan distance above the ground, induced drag is significantly reduced. Within this height above the ground, the strength of wing vortices and the wing downwash is substantially reduced since these forces are restricted in downward movement.

Therefore, as the aircraft nears the ground, drag effects are reduced rapidly, until just before touchdown drag can be decreased as much as 40%. The low-wing airplane is more subject to ground effect than mid- or high-wing airplanes because the wing is closer to the ground.

LIFT-TO-DRAG RATIO

The most important parameter of airplane performance that is obtained at the best lift-to-drag ratio (greatest amount of lift with the least amount of drag) is the power-off glide ratio. For example, if the airplane in a glide has a lift-to-drag ratio of 15:1 (15 times as much lift as drag), each mile of altitude can be traded for 15 miles of horizontal distance. To obtain this glide performance, the airplane should be flown at the angle of attack which yields the best lift-to-drag ratio. This is usually expressed in terms of a *best angle-of-glide* airspeed which can be found in the owner's manual for the airplane.

THE THREE AXES

There is one point in the airplane, regardless of the airplane's attitude, about which it can be balanced perfectly. This point is the center of the airplane's total weight and is called *center of gravity*, or "CG." The location of this point varies with the loading of the airplane. For example, it is located further forward in an airplane loaded with two persons in the front seats than it would be with both front and rear seats occupied.

All movements of the airplane in flight revolve around the center of gravity. These movements can further be classified as being made about one or more of three axes of rotation. These axes are called vertical, lateral, and longitudinal, and all pass through the airplane's center of gravity. (See Fig. 1-43.)

An axis is defined as a line passing through a body about which the body revolves. A turning maneuver revolves the aircraft about the *vertical axis;* climbing and diving produce movement around the *lateral axis;* and banking right or left rotates the aircraft around the *longitudinal axis.*

The control surfaces produce motion about each of these axes. Although rotation about the lateral, or pitch axis, is affected by several factors, the control of pitch is primarily the function of the *elevators* or *stabilator.* This surface is con-

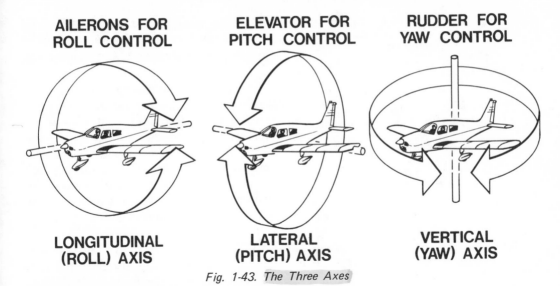

AILERONS FOR ROLL CONTROL	ELEVATOR FOR PITCH CONTROL	RUDDER FOR YAW CONTROL
LONGITUDINAL (ROLL) AXIS	LATERAL (PITCH) AXIS	VERTICAL (YAW) AXIS

Fig. 1-43. The Three Axes

trolled by fore and aft movement of the control wheel or stick. When forward pressure is applied to the control wheel, the elevators or stabilator move downward, producing an upward force on the horizontal tail. This raises the tail, and therefore lowers the nose. When back pressure is applied to the wheel, the elevators move upward, resulting in a downward tail movement. (See Fig. 1-44.)

The *ailerons* create movement known as *roll* about the longitudinal axis. The ailerons are moved by rotating the control wheel clockwise or counterclockwise. A downward deflected aileron produces additional lift, and the wing to which it is attached rises. The wing with the aileron deflected upward produces less lift and descends. (See Fig. 1-45.)

Once the ailerons have been properly positioned, the pilot returns the control wheel to a neutral position. The aircraft then continues to bank until the pilot applies opposite control pressure to take the aircraft out of the bank.

Increasing the lift, though, also increases the drag. The wing with the lowered aileron will generate *both* greater lift and greater drag. The aircraft will yaw in the direction of the wing with the lowered aileron, which is the direction opposite

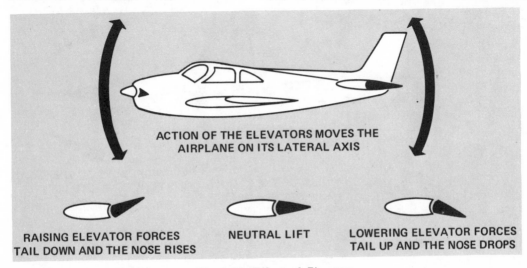

ACTION OF THE ELEVATORS MOVES THE AIRPLANE ON ITS LATERAL AXIS

RAISING ELEVATOR FORCES TAIL DOWN AND THE NOSE RISES

NEUTRAL LIFT

LOWERING ELEVATOR FORCES TAIL UP AND THE NOSE DROPS

Fig. 1-44. Effect of Elevators

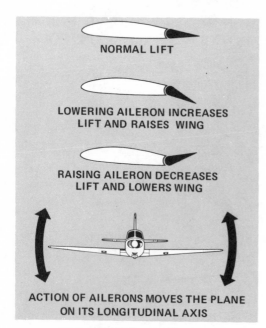

ACTION OF AILERONS MOVES THE PLANE
ON ITS LONGITUDINAL AXIS

Fig. 1-45. Effect of Ailerons

der is to the left and to the right and is controlled by pressure on the rudder pedals mounted on the floor of the cabin.

Pressure applied to the right rudder pedal displaces the rudder to the right. This increases the camber on the left side of the vertical tail components, creating a low-pressure area on the left side and a high-pressure area on the right side of the tail. (See Fig. 1-46.) The resulting pressures cause the tail to move left and the nose of the aircraft to move right.

Applying pressure to the left rudder pedal provides an opposite reaction. Left pedal movement displaces the rudder to the left. A low-pressure area forms on the right side of the tail and a high-pressure area on the left side. The tail moves to the right and the nose of the airplane moves left.

the turn being performed. This tendency is called *adverse yaw.*

Movement about the vertical axis is referred to as *yaw.* The *rudder* is a movable control surface on the fin and is responsible for rotation or control around the vertical axis. Movement of the rud-

The rudder is used initially in a turn to overcome adverse yaw. During any flight training maneuver, the student learns that, after a turn is established, the rudder is neutralized to maintain the desired turn.

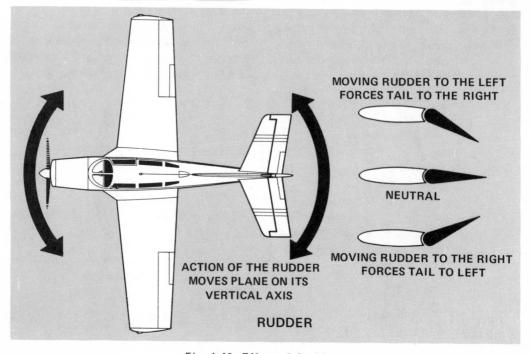

ACTION OF THE RUDDER
MOVES PLANE ON ITS
VERTICAL AXIS

MOVING RUDDER TO THE LEFT
FORCES TAIL TO THE RIGHT

NEUTRAL

MOVING RUDDER TO THE RIGHT
FORCES TAIL TO LEFT

RUDDER

Fig. 1-46. Effect of Rudders

AIRCRAFT STABILITY

An airplane in flight is constantly affected by minor control pressures and outside forces that move it from the desired flight attitude. The tendencies of an airplane, once it is disturbed from an attitude, depend upon the type of stability that is designed into the airplane.

Stability refers to the characteristic that causes an airplane to tend to return to, stay at, or move farther from its original attitude after it has been displaced. *Positive static stability* refers to the initial tendency to return to the original attitude. An airplane that has *positive dynamic stability*, on the other hand, will have the tendency to return to the original attitude through a series of decreasing oscillations.

An airplane that tends to stay in an attitude in which it is placed has *neutral static stability*. On the other hand, an airplane that tends to return to the original attitude (responds to positive static stability) but overshoots and continues to perform oscillations of equal magnitude has *neutral dynamic stability*.

An airplane that has *negative static stability* will constantly change from one attitude to another and will require the continuous use of controls to maintain a particular attitude. *Negative dynamic stability* is the tendency of an airplane to deviate farther and farther from the original attitude through a series of oscillations of increasing magnitude. Of course, no airplane may be certified that possesses this characteristic.

An example of the different types of stability can be seen by raising the wing of an aircraft while in flight. In a modern aircraft, the resulting tendency of the wing will be to return to a level attitude, thereby exhibiting positive static and positive dynamic stability. Neutral stability would have caused the airplane to stay in a wing-high attitude. Conversely, negative stability would have caused the wing to progress farther from a wings-level attitude through a series of oscillations of increasing magnitude.

LONGITUDINAL STABILITY

If an aircraft is stable along its longitudinal axis, it will not pitch up or down unless some force raises or lowers the nose of the aircraft. This movement is sometimes called *nosing up* or *nosing down.* If the aircraft is statically stable along its longitudinal axis, it will resist any force which might cause it to pitch, and it will return to level flight when the force is removed. Of the three types of stability of an aircraft in flight, this type is the most important.

To obtain longitudinal stability, the aircraft is designed so that it is slightly nose-heavy. During normal flight, the aircraft has a continuous tendency to dive. However, this tendency is offset by the horizontal tail surfaces (stabilizer) being set at a negative angle of attack. This position produces a downward or negative lift so that the downward force on the tail exactly counteracts the nose heaviness at a predetermined aircraft speed. (See Fig. 1-47.)

When power is reduced to idle, the airplane loses speed due to lack of thrust. Thus, the speed of the air over the tail decreases, resulting in less downward force on the tail. This permits the nose to drop, allowing gravity to act as a thrust force.

As the aircraft dives, its airspeed increases, which in turn provides more downward force on the tail. This pushes the tail down and the nose up, causing the aircraft to go into a climb. As the climb continues, the speed again decreases, and the downward force on the tail becomes gradually less until the nose drops once more. This time, if the aircraft is dynamically stable, the nose does not drop as far as it did the first time, and the aircraft enters a much shallower dive. The speed then increases until the aircraft again

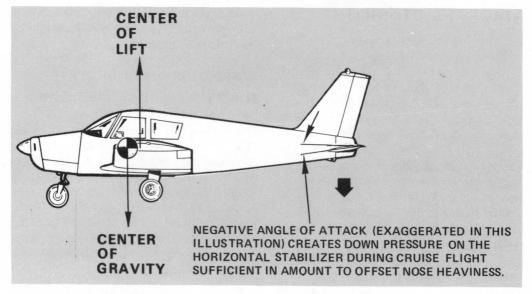

CENTER
OF
LIFT

CENTER
OF
GRAVITY

NEGATIVE ANGLE OF ATTACK (EXAGGERATED IN THIS
ILLUSTRATION) CREATES DOWN PRESSURE ON THE
HORIZONTAL STABILIZER DURING CRUISE FLIGHT
SUFFICIENT IN AMOUNT TO OFFSET NOSE HEAVINESS.

Fig. 1-47. Longitudinally Stable Aircraft is Slightly Nose Heavy

goes into a shallower climb, as before. After several such oscillations, the aircraft will finally settle down to a speed at which the downward force on the tail exactly offsets the tendency of the aircraft to dive. The aircraft can then make a smooth glide down, regardless of whether the power is on or off.

When an aircraft is balanced so that the center of gravity is *behind* the center of lift, the aircraft is tail-heavy and has a tendency to climb. This climbing tendency may be offset only by increasing the lift on the horizontal tail surfaces to produce a positive lift or upward force.

If the throttle is retarded when the airplane is in a tail-heavy condition, the lifting force developed by the tail is decreased because of the loss of thrust. As a result, the aircraft becomes tail-heavy and starts to climb. Since there is nothing to check this climbing tendency, the aircraft continues to climb until it stalls.

If the same aircraft is put into a dive with the controls released, the lift on the tail becomes greater and greater as the speed increases. This, in turn, forces the nose of the aircraft down and causes the aircraft to dive more

and more steeply, until it finally may go partly on its back.

Most airplanes in use today have good longitudinal stability and have sufficient trim tab control to correct for almost any condition of loading.

LATERAL STABILITY

If an aircraft is stable about the lateral axis, the wingtips will hold their positions in level flight unless some force is applied to displace them. Any force causing the wings to roll will be resisted, and the wings will return to straight-and-level flight once the force is removed. The most common design factor for providing lateral stability is *dihedral.* Dihedral is the upward angle of an airplane's wings with respect to the horizontal. (See Fig. 1-48.) Low-wing airplanes have more dihedral than high-wing airplanes because the center of gravity is above the wing in a low-wing airplane. When the center of gravity is below the wing, such as in a high-wing airplane, it has a lateral stabilizing effect and very little dihedral is required.

When an aircraft with dihedral rolls so that one wing is lower than the other, the lower wing will have more effective lift

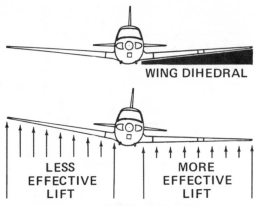

Fig. 1-48. Dihedral Assists Lateral Stability

than the raised wing because it is not tilted from the horizontal as much. The imbalance in lift tends to raise the lower wing and restore level flight.

Sweepback (see Fig. 1-49) has a lateral stabilizing effect also. When an aircraft with sweepback wings begins to slip, the leading edge of the low wing meets the relative wind more nearly perpendicular than the up wing. This results in more lift on the low wing which causes a restoring force to return the wing to a level position.

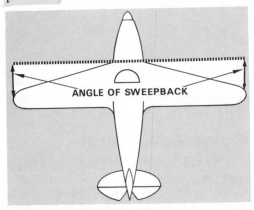

Fig. 1-49. Sweepback

Sweepback also contributes to directional stability. If a sweepback wing momentarily yaws because of turbulence or rudder application, the wing which moves forward has the relative wind more perpendicular to the leading edge. This increases the airspeed over the forward wing and consequently decreases the airspeed over the wing which yawed back-

ward. Increased airspeed supplies additional drag while decreased airspeed lessens drag. The increased drag on the forward wing pulls it back while the opposite reaction occurs to the trailing wing. For this reason, both wings will tend to return to their original position automatically correcting the airplane's direction.

DIRECTIONAL STABILITY

Directional stability tends to keep the aircraft flying in a straight direction. An aircraft acts something like a weathervane. If it swings away from its course by rotating about its vertical axis (yawing), the force of the air on the vertical tail surfaces tends to swing it back to its original line of flight. Here again, the location of the center of gravity is important. The area of the side of the airplane behind the center of gravity must be greater than the side area in front of the center of gravity in order to prevent yaw, because, if there were more side area in front of the center of gravity, any tendency of the aircraft to rotate about its vertical axis would then tend to turn the aircraft around. Directional stability, again, is the result of aircraft design. Adequate vertical stabilizer surface provides the necessary directional stability. (See Fig. 1-50.)

TURNING EFFECTS

There are several factors inherent in airplane design that result in a turning tendency. These include the *torque reaction* of the engine, *gyroscopic precession*, *asymmetrical thrust* from the propeller, and *slipstream rotation*.

TORQUE

Except for some twin-engine aircraft, propellers rotate clockwise as seen from the pilot's seat. The torque reaction from the spinning propeller causes the airplane to tend to rotate counterclockwise. (See Fig. 1-51.)

To minimize pilot effort, aircraft are adjusted, or rigged, to compensate for torque in cruising flight. In addition,

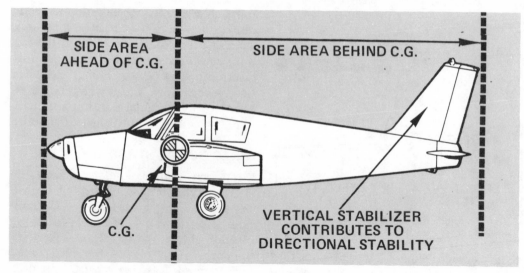

Fig. 1-50. Directional Stability

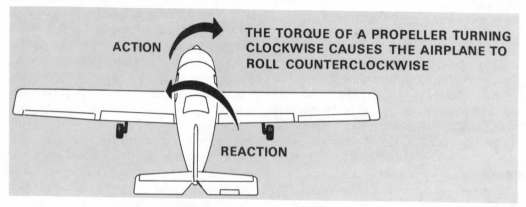

Fig. 1-51. Torque Effect

pilot-adjustable aileron trim tabs often are utilized in high performance aircraft, to correct the torque effects of various power settings.

PRECESSION

The turning propeller on an airplane possesses characteristics of a gyroscope. One of the properties of gyroscopic action is precession. *Gyroscopic precession* can be explained as the resultant action or deflection of the spinning object when a force is applied to this object. The reaction to a force applied to a gyro acts approximately 90° in the direction of rotation from the point where the force is applied. Therefore, if an airplane is rapidly moved from a nose-high pitch attitude to a nose-low pitch attitude, gyroscopic precession will create a tendency

for the nose to yaw to the left. The left-turning force is quite apparent in tail-wheel type airplanes as the tail is raised on the takeoff roll. (See Fig. 1-52.)

ASYMMETRICAL THRUST

In a propeller-driven airplane at a high angle of attack, asymmetrical thrust is created because the descending propeller blade on the right side of the engine has a greater angle of attack than the ascending blade on the left. This produces greater thrust from the right side of the propeller resulting in a left yaw known as *P-factor*. (See Fig. 1-53.) It should be remembered that P-factor creates a left-turning tendency *only* when the airplane is flying with a positive angle of attack. P-factor is not prevalent in level cruise flight since both propeller blades

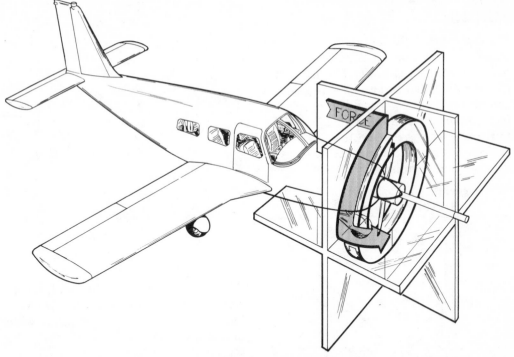

Fig. 1-52. Gyroscopic Effect

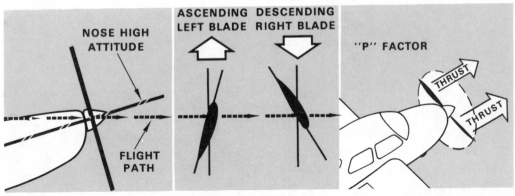

Fig. 1-53. "P"-Factor

are at the same angle of attack, producing *equal* thrust.

SPIRALLING SLIPSTREAM

The propeller also induces a slipstream rotation which causes a change in flow direction across the vertical stabilizer. Due to the direction of propeller rotation, the slipstream strikes the left surface of the vertical fin causing the airplane to yaw left. (See Fig. 1-54.) Some airplane manufacturers correct this problem by offsetting the rudder, thereby allowing the airplane to fly a straight path.

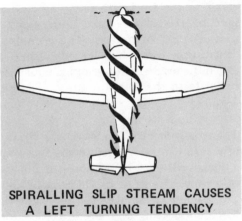

SPIRALLING SLIP STREAM CAUSES A LEFT TURNING TENDENCY

Fig. 1-54. Slipstream Effect

WHY THE AIRCRAFT TURNS

Turns are made by inclining the lift of the wings to one side to produce a force which turns the aircraft. The lift force can be subdivided and represented as two forces: one acting vertically and one acting horizontally. (See Fig. 1-55.) The horizontal force is the part that makes the aircraft turn, while the vertical force is the part that overcomes gravity.

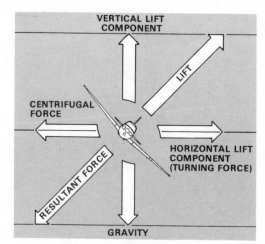

Fig. 1-56. Forces Balance in a Coordinated Turn

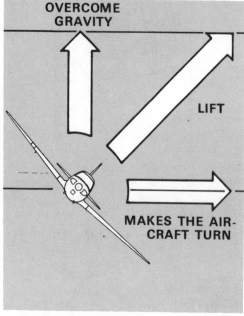

Fig. 1-55. Lift has Vertical and Horizontal Components in a Turn

In a level turn, there are pairs of opposing forces in balance. The centrifugal force acting on the aircraft is opposite and equal to the turning force. The vertical part of the lift is equal and opposite to the force of gravity. The combined centrifugal force and gravity force is called the resultant force and is equal and opposite to the total lift force. (See Fig. 1-56.)

The arrangement of forces on the aircraft in a turn can be likened to the forces on a bucket of water being swung at the end of a rope. The strain on the rope holding the bucket is similar to the total lift force supporting the aircraft. (See Fig. 1-57.)

Centrifugal force, which tends to swing the bucket away from the hand, and the gravity force combine to form the resultant force or load-factor force which makes the bucket feel heavier than when the bucket is at rest. It is this same resultant force that holds the water in the

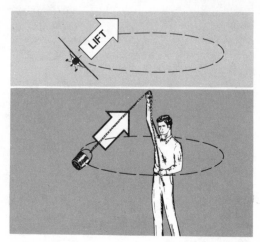

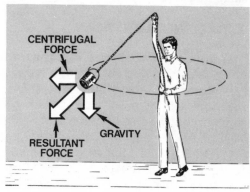

Fig. 1-57. Forces on an Aircraft

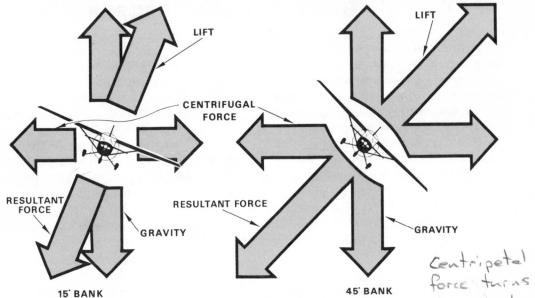

Fig. 1-58. The Steeper the Bank, the Greater the Load Factor

Centripetal force turns the airplane.

bucket with no tendency for the water to spill over the sides.

In a properly executed turn, the pilot feels heavier in a manner similar to the water in the bucket. The pilot does not experience any side forces like those felt when an automobile follows a curve on a highway.

The steeper the bank, the greater the total lift needed and the greater the force required to make the airplane turn. This also means that the centrifugal force and the gravitational force combine to make a greater resultant force and higher load factor and to make the pilot feel heavier in the seat. (See Fig. 1-58.)

Figure 1-33B illustrates the load factors imposed for various degrees of bank. For example, 60° of bank would produce a load factor of two "G's," thus making the effective weight of the aircraft twice its normal weight. If the aircraft weighs 2,000 pounds in straight-and-level flight, a level 60° bank turn would make the effective weight of the aircraft 4,000 pounds.

In level flight, the total lift force is equal to gravity, as shown in figure 1-59. When the aircraft is banked, the total lift is diverted. Since the total lift is still equal to gravity, not enough lift is acting vertically to counteract gravity and the aircraft descends. In order to maintain altitude in a turn, the total lift must be increased until enough lift will act vertically to counteract gravity. This is done by increasing the angle of attack by raising the nose with back pressure on the control wheel.

In summary, the aircraft in flight is subjected to many forces, some of which act together, while others act in opposition. In order to master the art and science of flight, the pilot must develop a familiarity and working knowledge of the interplay of forces acting on the airplane.

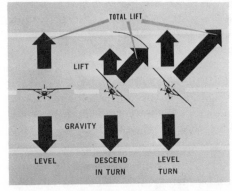

Fig. 1-59. Lift Must be Increased to Keep Aircraft From Descending in a Turn

SECTION C - BASIC AIRPLANE PERFORMANCE

The *performance* of an airplane basically means how well it accomplishes the job for which it was designed—whether it was designed for speed, utility, or maneuverability. Generally, when pilots speak of performance, they talk about acceleration and speed. A high-performance airplane can accelerate and climb quickly and fly faster than an airplane with relatively low performance. However, an airplane specifically designed for short field landings and takeoffs may have excellent takeoff and landing performance, but may not have exceptional acceleration or speed.

Atmospheric conditions greatly influence the performance of an airplane. A well-informed pilot knows how to obtain and interpret the published information about his airplane and how to operate the aircraft within the performance capabilities imposed by airplane design and atmospheric conditions.

EFFECTS OF TEMPERATURE AND ALTITUDE

The density of the air on a given day largely determines aircraft performance, since lift and drag are directly affected. As the air density increases, lift and drag increase; as the air density decreases, lift and drag decrease.

Air density is affected by three major factors: pressure, temperature, and humidity. For example, at an altitude of 18,000 feet, the air is approximately half as dense as it is at sea level and, likewise, the pressure is approximately half. (See Fig. 1-60.) Due to this decrease in air density, an airplane requires a longer takeoff distance at higher elevation airports than under the same conditions at sea level fields.

Because air expands when heated, warm air is less dense than cool air. A cubic foot of airspace on a hot day will contain fewer air molecules than that same cubic foot on a cooler day. Therefore, when other conditions (such as pressure and humidity) remain the same, an airplane will require a longer takeoff run on a hot day than on a cool day. Likewise, because water vapor weighs less than an equal amount of dry air, moist air (high relative humidity) is less dense than dry air (low relative humidity). Therefore, other conditions remaining the same, an airplane will require a longer takeoff run on a humid day than on a dry day.

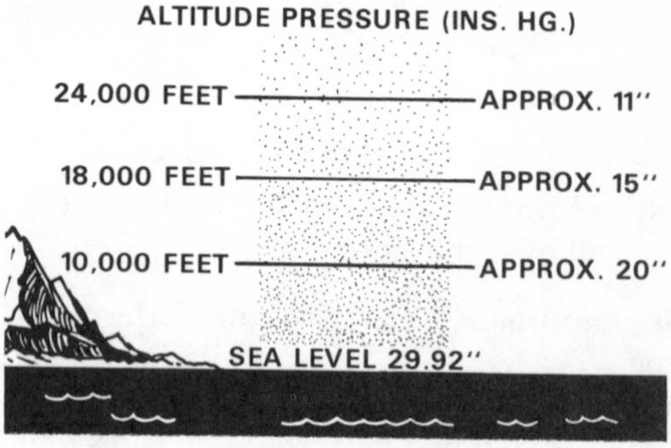

ALTITUDE PRESSURE (INS. HG.)

Fig. 1-60. Pressure Decreases
With Altitude

DENSITY ALTITUDE AND AIRCRAFT PERFORMANCE

Density altitude is a measure of air density and, therefore, is very important in determining airplane performance. Under non-standard conditions of temperature and pressure at a given airport, density altitude will differ from the airport elevation. As air density decreases (i.e., air becomes thinner), density altitude increases and vice versa. Low atmospheric pressure, high temperature, and high humidity all result in a decrease in air density and a corresponding increase in density altitude.

In computing density altitude, the small effect of humidity is usually disregarded. Therefore, temperature and barometric pressure are the two variables considered in most aircraft performance computations.

Density altitude, as it relates to changing temperatures, can be easily misunderstood. High density altitude does not mean that the air has a high density, but rather that the air has a low density associated with high altitude in the standard atmosphere. On a hot day, the density altitude at a particular airport may be 2,000 or 3,000 feet higher than the elevation of that field. As a direct consequence, an aircraft operating near the airport will perform as though it were in air at the higher altitude.

With an increase in density altitude, the engine horsepower decreases (unless it is a supercharged engine) since the actual amount of air to support combustion in the engine has decreased. In addition, both the wing and the propeller lose lifting efficiency in the thinner air.

DENSITY ALTITUDE GRAPH

One way of determining density altitude for a certain condition of temperature and pressure is through the use of a density altitude graph as shown in figure 1-61.

To understand how to use this graph, a sample problem should be worked. For example: pressure altitude: 6,000 feet; outside air temperature: +30° C. (See Fig. 1-61.)

1. First, locate the vertical line which represents the outside air temperature (+30° C) by referring to the scale at the bottom of the graph. (See Point A.)
2. Then, proceed vertically along the +30° C temperature line until the 6,000 foot pressure altitude line is intersected. (See Point B.)

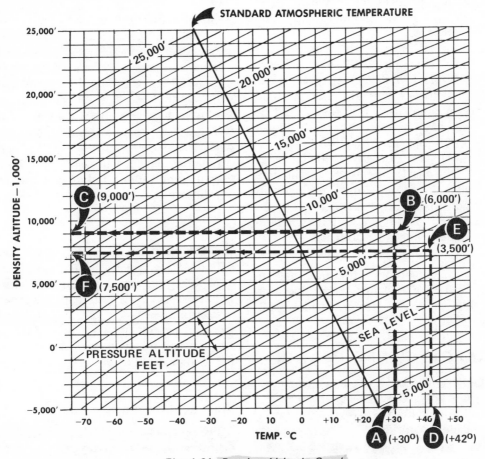

Fig. 1-61. Density Altitude Graph

NOTE: Pressure altitude is determined by setting the altimeter for 29.92 in. Hg. and reading the indicated altitude.

3. Proceed horizontally from Point B to the left edge of the graph and read density altitude at Point C. The solution to this problem is 9,000 feet density altitude.

To determine density altitude when the pressure altitude and outside air temperature values do not fall on the printed grid lines, it is necessary to interpolate between lines. For example: pressure altitude: 3,500 feet; outside air temperature: +42° C. This problem is solved as follows:

1. Locate +42° C at Point D. Since each vertical line on this graph is equal to 5°, this point is two-fifths the distance between +40° and the next graduation to the right, 45°.
2. Proceed vertically up from Point D until reaching 3,500 feet pressure altitude (Point E). Since the pressure altitude lines are drawn at 1,000 foot intervals, 3,500 feet would be midway between the 3,000 and 4,000 foot pressure altitude lines.
3. Moving horizontally to the left edge of the chart, the pilot reads the density altitude at Point F. Since this point is halfway between the 7,000 and 8,000 foot graduations, the density altitude is 7,500 feet.

An alternate method of determining density altitude is to solve the problem on the flight computer. (Refer to Chapter 10, Page 10-6.)

TAKEOFF PERFORMANCE

Takeoff is probably the maneuver most critically affected by density altitude. The pilot should always take the following factors into consideration before beginning a takeoff on a runway of marginal length.

1. Density altitude (pressure altitude corrected for temperature).
2. Headwind.
3. Airplane gross weight.
4. Runway gradient and type of surface.
5. Air turbulence.
6. Humidity.

DENSITY ALTITUDE AND TAKEOFF PERFORMANCE

Any time marginal or questionable takeoff conditions exist, the pilot must compute density altitude. By consulting the takeoff performance charts or graphs in the airplane owner's manual, the pilot can determine the adverse effect of high density altitude on his proposed takeoff. It is possible, under extreme conditions, that he may find that a takeoff is impossible within the limits of the available runway.

WIND

Wind direction and velocity will greatly affect takeoff performance. A headwind of considerable velocity can shorten the takeoff run and help compensate for the adverse effects of high density altitude. Crosswinds, and especially tailwinds, on the other hand, can lengthen the takeoff roll and reduce the subsequent angle of climb. Decreased controllability and takeoff performance can make crosswind takeoffs undesirable and downwind takeoffs hazardous.

GROSS WEIGHT

The gross weight of an airplane on takeoff also affects length of the takeoff roll and climb performance. Takeoff performance information, found on most charts and graphs, is based on maximum gross weight. Operations at lower takeoff weights will normally yield shorter takeoff distances and better climb performance.

RUNWAY GRADIENT AND SURFACE

Runway gradient refers to the uphill or downhill slope of a runway. An uphill runway decreases airplane acceleration, thereby increasing the takeoff distance. A downhill slope, on the other hand, will help decrease takeoff distance. In addition, the surface covering the runway can vary runway length requirements. Soft, muddy, or rough fields can increase runway requirements, and in extreme cases, may make a takeoff impossible.

TURBULENCE

Turbulence caused by high winds or other atmospheric conditions in the vicinity of an airport, can decrease takeoff maneuverability and climb performance. Essentially, this is because turbulence can increase the possibility of a stall. Therefore, increased climb speeds must be maintained in rough air.

HUMIDITY

Although not a factor normally included in takeoff performance computations, high humidity will increase takeoff distance and decrease climb performance. An unusually hot and humid day is cause for the pilot to exercise additional caution prior to attempting a marginal takeoff.

TAKEOFF PERFORMANCE CHARTS

The performance charts published for specific aircraft describe or predict the way in which the airplane should perform under a given set of conditions. These charts are published in either tabular or graph form.

Pilots should consult performance charts any time there is doubt about the airplane's capability, whether it be due to the length and/or condition of the runway, the high density altitude, or a lack of familiarity with the equipment being flown. In addition, Federal Aviation Regulations require that a pilot know the takeoff and landing distances his aircraft requires for each operation he conducts.

As an added safety measure, he should consider the performance of his airplane less than that predicted by the performance charts. For example, a pilot flying an older airplane, slightly out of rig and equipped with several protruding radio antennas, will find it difficult to meet

TAKEOFF DISTANCE

CONDITIONS:
Flaps Up
Full Throttle Prior to Brake Release
Paved, Level, Dry Runway
Zero Wind

NOTES:
1. Maximum performance technique.
2. Prior to takeoff from fields above 5000 feet elevation, the mixture should be leaned to give maximum RPM in a full throttle, static runup.
3. Decrease distances 10% for each 9 knots headwind. For operation with tailwinds up to 10 knots, increase distances by 10% for each 2 knots.
4. Where distance value has been deleted, climb performance after lift-off is less than 150 fpm at takeoff speed.
5. For operation on a dry, grass runway, increase distances by 15% of the "ground roll" figure.

WEIGHT LBS	TAKEOFF SPEED KIAS		PRESS ALT FT	0°C		10°C		20°C		30°C		40°C	
	LIFT OFF	AT 50 FT		GRND ROLL	TOTAL TO CLEAR 50 FT OBS	GRND ROLL	TOTAL TO CLEAR 50 FT OBS	GRND ROLL	TOTAL TO CLEAR 50 FT OBS	GRND ROLL	TOTAL TO CLEAR 50 FT OBS	GRND ROLL	TOTAL TO CLEAR 50 FT OBS
1600	53	60	S.L.	655	1245	710	1335	765	1435	820	1540	880	1650
			1000	720	1365	775	1465	835	1575	900	1690	970	1815
			2000	790	1500	855	1615	920	1735	990	1865	1065	2005
			3000	870	1650	935	1780	1010	1915	1090	2065	1170	2225
			4000	955	1820	1030	1965	1115	2125	1200	2290	1290	2475
			5000	1050	2015	1140	2185	1230	2360	1325	2555	1430	2770
			6000	1160	2245	1255	2435	1360	2640	1465	2870	1580	3120
			7000	1285	2510	1390	2730	1505	2970	1625	3240	- - -	- - -
			8000	1420	2820	1540	3080	1670	3370	- - -	- - -	- - -	- - -

Fig. 1-62. Takeoff Distance Chart

the performance specifications indicated on the charts. A careful pilot always leaves a respectable margin of safety between expected and required performance.

TAKEOFF PERFORMANCE TABLES

The takeoff performance table should be used for flight planning purposes, especially when a takeoff is contemplated at a high altitude airport or on an airstrip of marginal length. For example, referring to the chart shown in figure 1-62, the pilot can determine the ground roll and distance to clear a 50-foot obstacle under the following conditions:

```
Weight . . . . . . . . . . . .   1,600 lbs.
Wind . . . . . . . . . . . . . . .   Calm
Pressure altitude  . . . . . .   Sea level
Temperature . . . . . . . . . .   20°C
Takeoff speed @ 50 ft.  . . .   60 KIAS
```

By looking at the sea level data at 20° Celsius, the pilot finds that the ground roll required is 765 feet and the distance required to clear a 50-foot obstacle is 1,435 feet. It should be noted that, as the pressure altitude increases, the runway length requirements also increase until the takeoff roll has almost doubled at the 7,000-foot level when compared to the sea level value.

The optimum use of performance tables requires a pilot to take into account the various factors used in determining performance data. For instance, as noted at the top of the chart, the values in figure 1-62 were calculated with the flaps up and on a paved, level, dry runway. In addition, note 3 states that distance must be decreased by 10 percent for each nine knots of headwind. Note 5 shows that distance increases by 15 percent of the ground roll figure for operations on dry grass runways.

Factors affecting takeoff performance that are not found on this chart include humidity, runway gradient, and turbulence. In addition, pilot technique plus the age and condition of the airplane can greatly affect these figures.

INTERPOLATION

Interpolation is a process by which the pilot can determine values for conditions which may not be shown directly on a performance chart. In order to help the pilot understand the process of interpolation, a sample problem is presented using the performance table in figure 1-62.

For the sample problem, a takeoff is planned from a paved, level, dry runway at an airport with a pressure altitude of 5,500 feet and a temperature of 20° Celsius. Since the table shows the takeoff ground roll for 5,000 feet and 6,000 feet, interpolation must be used to determine the ground roll at 5,500 feet. In the problem, the pilot determines that 5,500 feet is one-half the distance between 5,000 and 6,000 feet. Therefore, the difference between the two ground rolls can be divided by one-half and added to the 5,000-foot value. This procedure provides the ground roll at 5,500 feet. The following calculations explain this interpolation, which results in a ground roll of 1,295 feet.

```
Ground roll @ 5,000 ft. . . . .   1,230 ft.
Ground roll @ 6,000 ft. . . . .   1,360 ft.
1,360 ft. - 1,230 ft.  . . . . . .   130 ft.
130 ft. x ½ . . . . . . . . . . . . .   65 ft.
1,230 ft. + 65 ft.  . . . . . . .   1,295 ft.
```

TAKEOFF PERFORMANCE GRAPHS

Some aircraft manufacturers publish graphs in their owner's manuals. (See Fig. 1-63.) The information derived from graphs is usually similar in content to that found in a table and is based on an aircraft loaded to maximum weight.

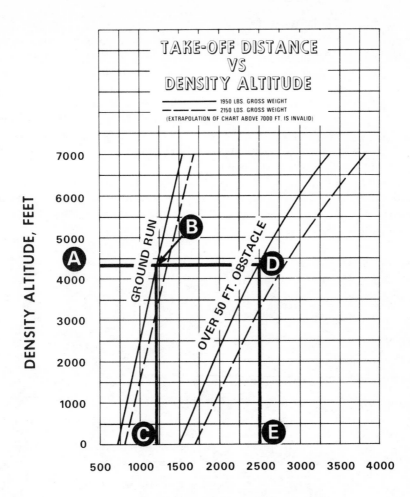

Fig. 1-63. Takeoff Distance

The sample takeoff distance graph in figure 1-63 shows that distances can be figured for takeoff with and without a 50-foot obstacle ahead. To determine takeoff distance, it is first necessary to compute density altitude. (See Fig. 1-61.)

Assume that the density altitude for an airport is 4,300 feet and the pilot must take off over a 50-foot obstacle. To find the ground run expected on this takeoff, the pilot locates 4,300 feet at the left edge of the graph (see Fig. 1-63, Point A),

and moves horizontally to the right until intersecting the *ground run* line (Point B). Dropping vertically to the bottom edge of the graph (Point C), the pilot finds that his aircraft will break ground 1,200 feet from the start of the takeoff roll.

To determine the distance required to clear a 50-foot obstacle, the pilot continues horizontally along the 4,300 foot level until he intersects the *over 50-foot obstacle* line (Point D). Moving vertically downward to the bottom line of the graph (Point E), he finds that it will take 2,500 feet in order to clear a 50-foot obstacle under these conditions.

THE DENALT COMPUTER

The DENALT (density altitude) computer comprises a very handy additional method to compute takeoff distance and rate of climb. This computer is produced by the U.S. Government and can be purchased from the U.S. Government Printing Office, Washington, D.C., 20402. The computer shown in figure 1-64 is designed specifically for use with fixed-pitch propeller airplanes. Another computer, very similar in format and use, is produced and is available for variable-pitch propeller airplanes. Should the student wish to order a DENALT computer, he should specify which type he desires — fixed-pitch or variable-pitch.

As an example of the use of the DENALT computer, assume the following conditions:

Fig. 1-64. The Denalt Computer

Given:
1. Surface temperature — +80°F
2. Airport pressure altitude — 4,000 feet
3. Takeoff distance under sea level standard conditions — 1,700 feet

Find:
Corrected takeoff distances.

Solution:
1. Set 80° F in the AIR TEMP window as shown in figure 1-64, item 1.
2. Note that opposite the pressure altitude of 4,000 feet, the takeoff factor of 2.0 is located. (See Fig. 1-64, item 2.)
3. Multiply the takeoff distance on a standard day at sea level by the takeoff factor of 2.0 to find the correct takeoff distance of 3,400 feet.

Note that the lower portion of the DENALT computer is used to calculate the corrected rate of climb using procedures similar to those just outlined.

CLIMB PERFORMANCE

Most of the factors affecting takeoff performance also affect the subsequent climb capability of an airplane. A high density altitude, along with increased relative humidity, reduces the climb capacity of an airplane. In addition, turbulent air, variable piloting techniques, plus the condition and age of the aircraft, can all result in a different rate of climb than is specified in the airplane flight manual or the owner's handbook.

There are three climb airspeeds with which pilots should become familiar in order to obtain the best performance from their aircraft; best rate-of-climb airspeed, best angle-of-climb airspeed, and normal-climb airspeed.

BEST RATE-OF-CLIMB AIRSPEED

An airplane operated at the *best rate-of-climb speed* will obtain the *greatest gain in altitude over a given time period.* (See Fig. 1-65.) This speed is normally used during the climb after all obstacles have been cleared and is usually maintained until leaving the traffic pattern. Best rate-of-climb speed is also useful when a climb over higher terrain is required since this speed will allow the pilot to climb to a terrain clearance altitude in the shortest period of time.

The best rate-of-climb airspeed will decrease with an increase in altitude. Prolonged operation at this speed in some airplane types can produce high engine temperatures, especially on days when the outside air temperature is high. Pilots must closely monitor the oil temperature and cylinder head temperature gauges any time a steep climb is used for a long period of time.

BEST **RATE-OF-CLIMB** AIRSPEED

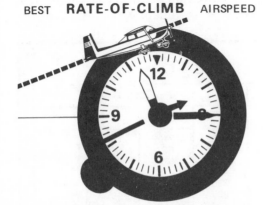

GREATEST ALTITUDE GAINED
OVER A GIVEN TIME PERIOD

BEST **ANGLE-OF-CLIMB** AIRSPEED

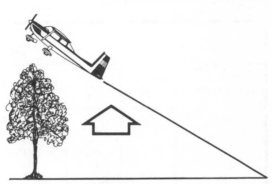

GREATEST GAIN IN ALTITUDE
OVER DISTANCE TRAVELLED

Fig. 1-65. Best Rate-of-Climb and Angle-of-Climb Airspeeds

BEST ANGLE-OF-CLIMB AIRSPEED

The *best angle-of-climb airspeed* yields the *greatest gain in altitude over a given distance.* To help distinguish between best rate and best angle airspeed, one can think of best rate-of-climb performance expressed in altitude gained per minute and best angle-of-climb performance as expressed in altitude gained per horizontal distance traveled. (See Fig. 1-65.)

The best angle-of-climb airspeed is used after liftoff to clear obstacles that may be at the end of the runway. Since this airspeed is relatively low and full power is used for obstacle clearance, it is recommended that the best angle-of-climb speed be maintained only for short time intervals in order to prevent engine overheating. Once an obstacle is cleared, the pilot should lower the nose of the airplane to increase the airspeed to at least the best rate-of-climb airspeed. The airspeed recommended for best angle-of-climb *increases* with altitude.

NORMAL CLIMB SPEED

The airplane owner's handbook will usually specify a *normal* or *cruise climb airspeed* to be used for prolonged time periods. While the normal climb airspeed certainly does not produce the climb rate associated with the best rate-of-climb airspeed, it does provide adequate performance for most operations while insuring adequate engine cooling and improved "over the nose" visibility.

After departing the airport traffic pattern, and if no high terrain must immediately be crossed, the pilot should climb at the normal climb speed. Not only will this enhance engine cooling, but it also will provide better cross-country speeds, cutting down the total time en route to destination.

CLIMB PERFORMANCE TABLES

Climb performance tables are included in the airplane owner's handbook to give the pilot an idea of the approximate performance that can be expected under various conditions. It should be remembered, however, that several conditions affecting climb performance are not taken into consideration on these charts and the pilot must adjust his climb requirements and expectations accordingly.

The climb performance table, shown in figure 1-66, shows that, at sea level, with an outside air temperature of 0° Celsius, a gross weight of 1,600 pounds, and the best rate-of-climb airspeed of 68 knots indicated airspeed, the pilot can expect a rate of climb of 710 feet per minute. In addition, it can be seen that the climb airspeed decreases as the altitude increases and that the rate of climb decreases with an increase in altitude or temperature. As shown to the left of the table, the climb performance values were calculated with flaps up and full throttle.

RATE OF CLIMB

CONDITIONS: Flaps Up Full Throttle	WEIGHT LBS	PRESS ALT FT	CLIMB SPEED KIAS	RATE OF CLIMB - FPM			
				-20°C	0°C	20°C	40°C
	1600	S.L.	68	770	710	655	595
		2000	67	675	615	560	500
		4000	65	580	520	465	405
		6000	64	485	430	375	310
		8000	63	390	335	280	215
		10,000	62	295	240	185	- - -
		12,000	61	200	150	- - -	- - -

Fig. 1-66. Rate-of-Climb Chart

CLIMB PERFORMANCE GRAPHS

The climb performance graphs, shown in some flight manuals, yield similar information to that found on the climb table. On the sample graph in figure 1-67, the rate of climb is computed for an airplane operating at maximum certificated weight. To determine rate-of-climb for an airplane operating from an airport with a density altitude of 5,500 feet, the pilot moves horizontally from 5,500 feet, on the left side of the graph (Point A), to the solid black rate-of-climb line (Point B). From this point, he proceeds straight down and reads the rate of climb at the bottom edge of the graph (Point C). The pilot finds that the rate of climb at a density altitude of 5,500 feet is 440 feet per minute.

NOTE: Remember that the altitudes listed on the graph are density altitudes; therefore, a density altitude computation (see Fig. 1-61) must be performed to make the graph data accurate.

MISCONCEPTIONS REGARDING WIND AND CLIMB PERFORMANCE

One of the least understood areas of aeronautical knowledge is the effect of wind on airspeed and climb performance. One pilot who stalled an aircraft shortly after

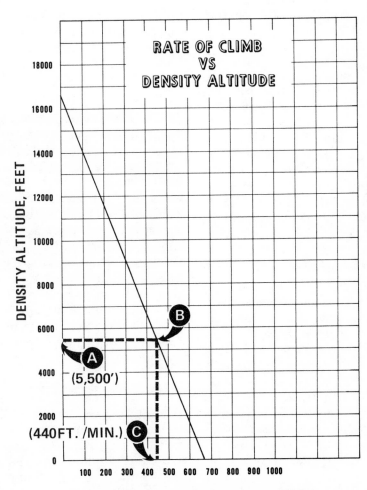

Fig. 1-67. Rate-of-Climb Graph

takeoff attributed the stall to a turn downwind which decreased the airspeed over the wings resulting in a loss of lift. This, of course, could not have been the cause since airspeed is not determined by atmospheric wind direction or velocity. Airspeed, rather, is the speed at which the airplane is moving through the atmosphere and is unaffected by the wind. Atmospheric wind affects only the ground speed, or speed at which the airplane is traveling over the earth's surface. Because of this, the only climb performance affected by atmospheric wind is the *angle of climb*, or gain in altitude over a given distance.

A pilot making a downwind turn will retain the *same* rate of climb; however, the ground speed will increase, resulting in less altitude gained for the distance traveled.

During turns close to the ground where the pilot can visually judge altitude, he may believe that a downwind turn has actually resulted in a loss of lift and a decrease in climb rate. This is just an illusion, however, due to the decrease in angle of climb over the ground. Although rate of climb remains the same, the decrease in climb angle could, in some cases, be hazardous, not because it produces a stall, but because of the decrease in obstacle clearance capability.

CRUISE PERFORMANCE

The manufacturers of today's light airplanes provide cruise performance charts or graphs to indicate rate of fuel consumption, true airspeed, range, and endurance at different altitudes and power settings. These charts may be relatively simple or quite sophisticated, but in either case, it is important to remember that the performance data is based upon specific conditions of fuel/air mixture, atmospheric temperature and pressure, gross weight, and wind.

Any deviation from the specific information upon which the chart computations are based will affect the accuracy of

some or all of the results. For example, most cruise performance charts are based on zero wind and standard atmospheric conditions. Yet, during actual operations, these conditions seldom prevail. It should be noted that wind has a very significant effect on the distance an airplane can fly, but no effect on its rate of fuel consumption or the total time it can remain aloft.

CRUISE PERFORMANCE TABLES

A table like the one shown in figure 1-68 is used in many aircraft flight manuals. This table is for an airplane with a fixed-pitch propeller. Therefore, for every pressure altitude shown, the power settings are expressed in r.p.m. The percent of brake horsepower, true airspeed, and fuel consumption correspond to a specific pressure altitude, r.p.m. setting, and temperature range.

The fuel consumption for any percent of brake horsepower is always about the same regardless of altitude. Endurance and range, however, improve somewhat with altitude. It can be seen, too, that any given percentage of brake horsepower is achieved with a different r.p.m. to maintain a particular percent of brake horsepower with increasing altitude. This is due to the fact that air becomes less dense as altitude increases. The engine must turn more revolutions (combustion cycles) per minute at higher altitudes to produce the same power that it would produce with fewer r.p.m.'s at lower altitudes.

As can be seen from the table, a pilot flying at a pressure altitude of 6,000 feet with a standard temperature, a power setting of 2,600 r.p.m. and 64% brake horsepower, should obtain approximately 99 knots true airspeed and a fuel consumption rate of 4.8 gallons per hour.

CRUISE PERFORMANCE

CONDITIONS:
Recommended Lean Mixture
1600 Pounds
Flaps Up

ALTITUDE	RPM	20°C BELOW STANDARD TEMP			STANDARD TEMPERATURE			20°C ABOVE STANDARD TEMP		
		% BHP	KTAS	GPH	% BHP	KTAS	GPH	% BHP	KTAS	GPH
2000	2650	- - -	- - -	- - -	78	103	5.9	72	102	5.4
	2600	80	102	6.0	73	101	5.5	68	100	5.1
	2500	70	97	5.3	65	96	4.9	60	95	4.6
	2400	62	92	4.7	57	91	4.3	53	91	4.1
	2300	54	87	4.1	50	87	3.9	47	86	3.7
	2200	47	83	3.7	44	82	3.5	42	81	3.3
4000	2700	- - -	- - -	- - -	78	105	5.8	72	104	5.4
	2600	75	101	5.6	69	100	5.2	64	99	4.8
	2500	66	96	5.0	61	95	4.6	57	95	4.3
	2400	58	91	4.4	54	91	4.1	50	90	3.9
	2300	51	87	3.9	48	86	3.7	45	85	3.5
	2200	45	82	3.5	42	81	3.3	40	80	3.2
6000	2750	- - -	- - -	- - -	77	107	5.8	71	105	5.3
	2700	79	105	5.9	73	104	5.4	67	103	5.1
	2600	70	100	5.2	64	99	4.8	60	98	4.5
	2500	62	95	4.7	57	95	4.3	53	94	4.1
	2400	54	91	4.2	51	90	3.9	48	89	3.7
	2300	48	86	3.7	45	85	3.5	42	84	3.4

Fig. 1-68. Cruise Performance Table

CRUISE PERFORMANCE GRAPHS

Some owner's handbooks use cruise performance graphs instead of tables. In the Range vs. Density Altitude graph shown in figure 1-69, the pilot can determine the percent of brake horsepower combinations and expected range in statute miles for various density altitudes. For instance, at a density altitude of 7,500 feet, utilizing 65% rated horsepower, in an aircraft with 50 gallons of fuel, the pilot can expect a range of 830 miles.

To determine the percent of brake horsepower and fuel consumption for various altitude/r.p.m. combinations, the pilot should consult the Power vs. Density Altitude performance chart (see Fig. 1-70). Using the example from the previous paragraph, the pilot finds that he must use 2,430 r.p.m. to obtain 65% power. Also, the fuel consumption rate is 7.3 gallons per hour at this power setting.

LANDING PERFORMANCE

Performance information for the landing phase of flight is also published in owner's handbooks and flight manuals. Landing techniques used by pilots are more variable than for most other maneuvers, making published landing performance information less accurate than other types of performance data.

APPROACH SPEED

The owner's handbook will recommend an approach airspeed for various flap settings. This recommended indicated airspeed should be used regardless of temperature and altitude combinations. Although temperature and pressure changes will change air density and, therefore, lift, a given indicated airspeed will continue to insure the same amount of lift. With an increase in temperature and a decrease in pressure, however, the true airspeed and resultant ground speed will increase for a given indicated airspeed. In less dense air, the airplane must move through the atmosphere faster to obtain a desired indicated airspeed.

This fact is important for pilots who attempt landing approaches at airports of varying elevations. A pilot who has made the majority of his landing approaches at

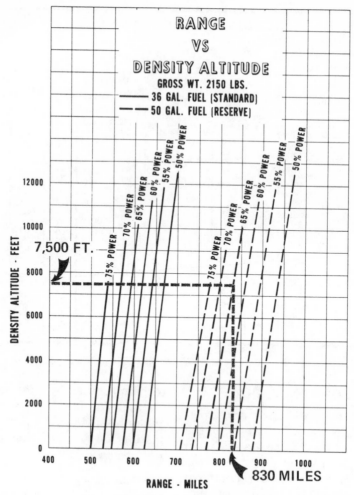

Fig. 1-69. Range Graph

low elevation airports soon becomes accustomed to the speed over the ground yielded by his normal approach speed, and therefore, begins to monitor the airspeed indicator less and less. If this pilot attempts to use ground speed as the approach speed determinant at high elevation airports, he will find that the indicated airspeed will be significantly lower than the normal approach speed and may even approach a stall. This is why maintaining the proper pitch attitude in order to stabilize the airspeed on landing approaches is so important.

WEIGHT AND CONFIGURATION

Landing weight is another factor that must be considered during approaches. It is usually a good idea to approach with a slightly higher airspeed when the weight

is quite high to increase maneuverability at the greater weight and to make a go-around from an aborted landing less critical.

The use of flaps on a landing approach will vary with field conditions and length. Normally, a short or soft field calls for the use of full flaps and a lower indicated approach speed.

LANDING PERFORMANCE TABLES

As stated previously, some handbooks supply landing performance tables instead of graphs as shown in figure 1-71. The table shown is used to compute the landing capability using 40° of flaps and an approach speed at 50 feet of 52 knots indicated airspeed. The table indicates

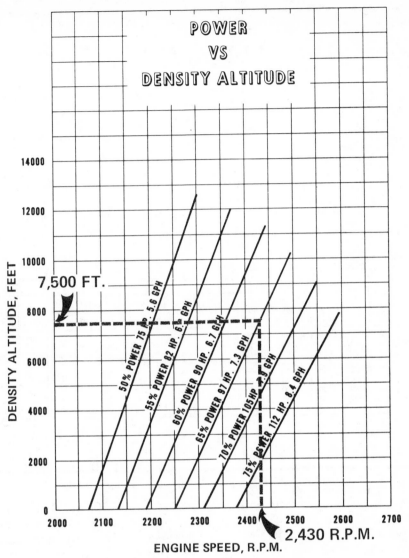

Fig. 1-70. Power vs. Density Altitude Graph

that both the ground roll and the distance to clear a 50-foot obstacle increase with altitude. This is because of the higher ground speed resulting from a given indicated airspeed at higher altitudes.

In addition, values are decreased 10% for each 9 knots of headwind. For example, at a pressure altitude of 6,000 feet with an outside air temperature of 30°C and a 9-knot headwind, 526 feet of actual ground roll and 1,156 feet total distance

required to clear a 50-foot obstacle are required for landing.

LANDING PERFORMANCE GRAPHS

Landing performance graphs are often used to give the pilot information on landing distances. (See Fig. 1-72.) In this graph, landing distances are computed at various density altitudes for one weight condition.

Referring to the graph in figure 1-72, the pilot finds that he will require 1,180 feet

LANDING DISTANCE

CONDITIONS:
Flaps 40°
Power Off
Maximum Braking
Paved, Level, Dry Runway
Zero Wind

NOTES:
1. Maximum performance technique.
2. Decrease distances 10% for each 9 knots headwind. For operation with tailwinds up to 10 knots, increase distances by 10% for each 2 knots.
3. For operation on a dry, grass runway, increase distances by 45% of the "ground roll" figure.

WEIGHT LBS	SPEED AT 50 FT KIAS	PRESS ALT FT	0°C		10°C		20°C		30°C		40°C	
			GRND ROLL	TOTAL TO CLEAR 50 FT OBS	GRND ROLL	TOTAL TO CLEAR 50 FT OBS	GRND ROLL	TOTAL TO CLEAR 50 FT OBS	GRND ROLL	TOTAL TO CLEAR 50 FT OBS	GRND ROLL	TOTAL TO CLEAR 50 FT OBS
1600	52	S.L.	425	1045	440	1065	455	1090	470	1110	485	1135
		1000	440	1065	455	1090	470	1110	485	1135	505	1165
		2000	455	1090	470	1115	490	1140	505	1165	520	1185
		3000	470	1115	490	1140	505	1165	525	1195	540	1215
		4000	490	1140	505	1165	525	1195	545	1225	560	1245
		5000	510	1170	525	1195	545	1225	565	1255	585	1285
		6000	530	1200	545	1225	565	1255	585	1285	605	1315
		7000	550	1230	570	1260	590	1290	610	1320	630	1350
		8000	570	1260	590	1290	610	1320	630	1350	655	1385

Fig. 1-71. Landing Performance Table

to land over a 50-foot obstacle with no wind at a density altitude of 4,300 feet. Also, his ground roll will be 600 feet.

The landing surface determines braking effectiveness and, therefore, affects the length of the landing roll. Hard-surfaced runways provide the best braking while icy, grassy, or wet runways can significantly increase stopping distance.

WIND COMPONENT CHART

Since the amount of headwind is a determining factor in computing the takeoff and landing distance, the thorough pilot should never assume that the wind velocity is a direct headwind. The magnitude of crosswind and headwind components can be determined by using the wind component chart shown in figure 1-73. For example, assume that the wind is re-

ported from 350° at 20 knots for a takeoff on runway 32. The pilot notes that the angle between the wind and the runway direction is 30°.

By entering the chart at the 30° angle off the runway line (Point A) and proceeding downward until intersecting the 20-knot speed arc (Point B), the headwind component of 17 knots can be read directly left of the speed arc intersection (Point C), and a crosswind component of 10 knots can be read directly below the speed arc intersection Point D. Therefore, the pilot would base his takeoff performance on a 17-knot headwind component rather than a headwind of 20 knots.

The understanding and application of weight and balance data and performance tables, graphs, and charts will not only benefit the pilot with safer operation of

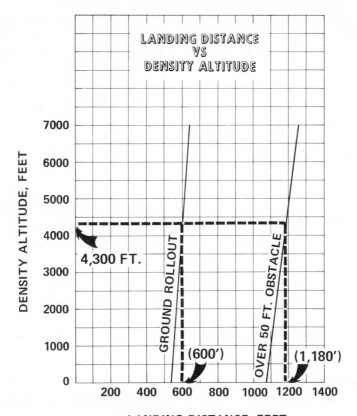

Fig. 1-72. Landing Performance Graph

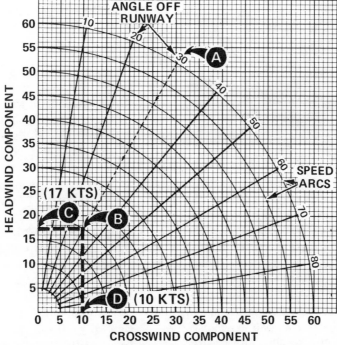

Fig. 1-73. Wind Component Chart

his aircraft, but will also result in maximum utility and performance from his aircraft.

WAKE TURBULENCE

One additional factor that can severely limit aircraft performance is wake turbulence. Of special concern to pilots of light aircraft is the turbulence which is created by large aircraft operating in the vicinity of the airport.

The greatest danger from heavy aircraft turbulence is the effect of *wingtip vortices.* These vortices are the result of high pressure air under the wings, spilling around and over the wingtips, to equalize the low pressure area above the upper wing surface. If one could actually see the vortices generated by the wings, they would appear as twin horizontal tornadoes, originating at the wingtips. This aerodynamic phenomenon was discussed briefly in the aerodynamics section. (See Fig. 1-74.)

Fig. 1-74. Wingtip Vortices

These rotating cones of air are created as a result of the generation of lift by the wings of the aircraft. The wing transmits energy, proportional to the weight of the aircraft, to the air through which it flies. The intensity of this turbulence is directly related to the wing span, weight, forward speed of the aircraft, and density of the air. In general, the slower the speed of the aircraft, the shorter the wing span, and the greater the weight, the greater will be the velocity of the air in the

vortex cores generated by the aircraft. Therefore, the greatest turbulence is generated by a large aircraft when it is flying slowly at a high angle of attack. This condition primarily exists during takeoff, climbout, and landing.

Actual tests have demonstrated that vortex core velocities may exceed 90 knots immediately behind a heavy jet which is flying very slowly. In addition, experience has shown that the vortex cores tend to descend at 400 to 500 feet per minute to a level-off point 800 to 900 feet below the flight altitude of the jet.

At cruise, the airspeed is greatest; therefore, the aircraft weight is spread over the greatest amount of air. The turbulence thus created is at a minimum for that aircraft. Conversely, when the aircraft is on the ground, the wings are not supporting the weight of the aircraft; thus, vortex cores are not generated. Prior to liftoff, or after touchdown, wingtip vortices are almost non-existent.

PROCEDURES FOR OPERATING NEAR WAKE TURBULENCE

Some commonsense rules for takeoff behind a large aircraft include: (See Fig. 1-75.)

1. Start the takeoff roll from the same point as the large aircraft to insure liftoff ahead of the large aircraft's point of takeoff. Climb out above or to one side of the large aircraft's flight path to avoid the turbulence.
2. When taking off after the landing of a large aircraft, be sure to plan the liftoff point beyond the touchdown area of the large aircraft.
3. Climb out parallel to the runway but on the upwind side since wind tends to blow turbulence to the downwind side of the runway.
4. If there is no wind, wake turbulence has a tendency to move outward (laterally) from the center of the flight path, thereby remaining over the runway. Therefore, the situation

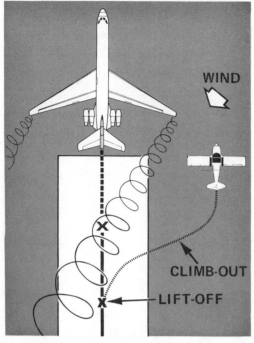

Fig. 1-75. Takeoff After a Large Aircraft

may dictate a delayed takeoff to allow the turbulence to dissipate.

In flight, if it is necessary to cross behind a large aircraft, pass slightly above its flight path to avoid the vortex.

Here are some good rules for landing behind a large aircraft: (See Fig. 1-76.)

1. If cleared to land close behind a large aircraft, it is the pilot's prerogative to refuse the landing clearance if the distance between aircraft is too short.
2. Avoid crossing the path of a large aircraft if possible. This can be accomplished by keeping the traffic pattern inside or above that of the aircraft ahead.
3. On approach, maintain the descent path at or above the descent path of the large aircraft and touch down beyond the place where the large aircraft is landing. It is well to maintain greater airspeed than normal to facilitate control if turbulence is encountered.
4. If a large aircraft has just taken off as the pilot approaches, he should plan his touchdown to take place well in advance of the large aircraft's liftoff point.

JET BLAST

The number of ground accidents involving loss of life, serious injury, and aircraft damage due to jet blast indicates that some pilots are not aware of this serious hazard. No pilot would deliberately taxi an aircraft in a hurricane, yet some have taxied behind a jet generating winds of hurricane force. Here, for example, are excerpts from an NTSB special study on jet blast: "aircraft flipped over from jet blast of DC-9"; "707 cleared for takeoff flipped aircraft on back."

The reasons for these accidents are clearly illustrated in figure 1-77. At idle thrust, a four-engine, wide-bodied jet generates winds of 50 m.p.h. at 50 feet behind it. However, at breakaway and takeoff thrust, hurricane-like winds are generated.

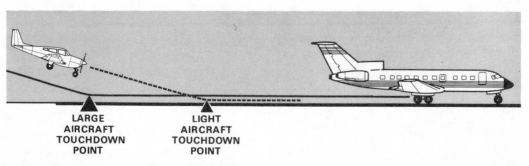

Fig. 1-76. Landing Behind a Large Aircraft

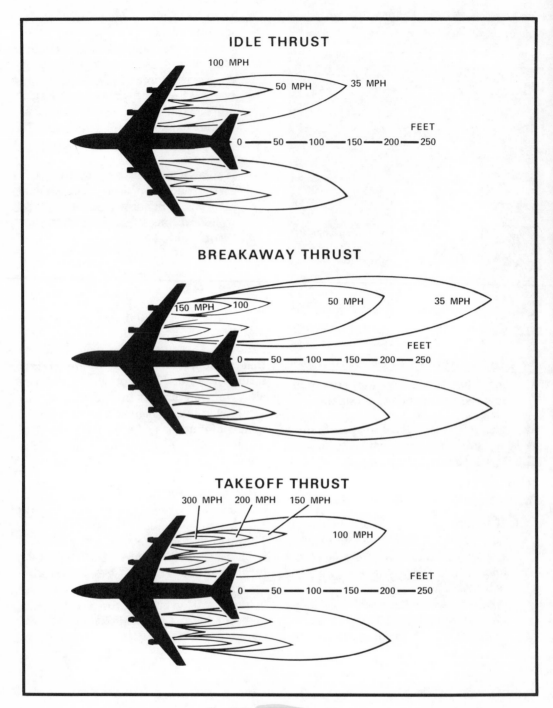

Fig. 1-77. Jet Blast Profiles

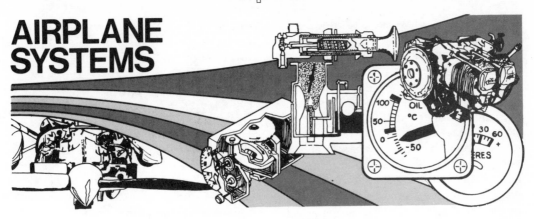

chapter 2

AIRPLANE SYSTEMS

INTRODUCTION

In order for a pilot to be fully competent in an airplane, it is necessary that he understand the aircraft systems. Obtaining the maximum performance from an aircraft depends upon the pilot being familiar with the operation of the systems which provide fuel, electricity, power, and control of the aircraft on the ground and in flight. An understanding of the airplane systems provides a background which enables the pilot to perform normal duties and handle unexpected situations smoothly and safely.

SECTION A - FLIGHT CONTROL SYSTEMS

Aircraft control systems are classified as primary and secondary. The primary control systems consist of those which are necessary for the safe flight of an aircraft, such as ailerons, elevator (or stabilator), and rudder. Secondary control systems improve the performance characteristics of the aircraft, or relieve the pilot of excessive control forces. Examples of secondary control systems are wing flaps and trim systems.

AILERONS

The ailerons are located on the outboard trailing edges of both wings and are used to rotate the aircraft around the *longitudinal* or *roll axis*. These surfaces move in opposite directions from each other and are controlled by rotation of the pilot's control wheel. A series of chains, cables, bellcranks, and pulleys connect the ailerons to each other and to the control wheel in the cockpit. (See Fig. 2-1.)

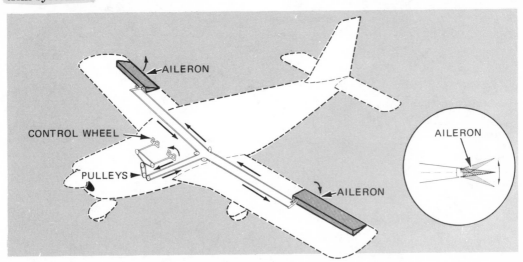

Fig. 2-1. Aileron Control System

When the control wheel is rotated to the left, the right aileron moves down and the left aileron moves up. The downward deflected right aileron provides the outer panel of the right wing with increased camber on the upper surface. This creates a higher velocity airflow over this section of the wing, resulting in an area of lower pressure which causes the right wing to move up. Conversely, the left aileron, being deflected up, forces the left wing down by decreasing the lift produced by the left outer wing panel. The upward movement of the right wing and downward movement of the left wing causes the airplane to roll to the left.

When the control wheel is turned to the right, an opposite aileron movement results and the airplane rolls to the right. The airplane will continue to roll in the direction that the control wheel is turned until the controls are neutralized to establish the desired angle of bank.

ELEVATORS AND STABILATORS

The pitch motion around the *lateral* or *pitch axis* is controlled by a movable horizontal tail surface actuated by fore and aft movement of the control wheel. Two types of movable horizontal tail surfaces are used in today's aircraft to control the motion about the lateral axis: *elevators* and *stabilators*. (See Fig. 2-2.)

Elevators are movable control surfaces that are hinged to the horizontal stabilizer. Like ailerons, the elevators are connected to the control wheel by a system of control cables, pulleys, and bellcranks, while the horizontal stabilizer is fixed

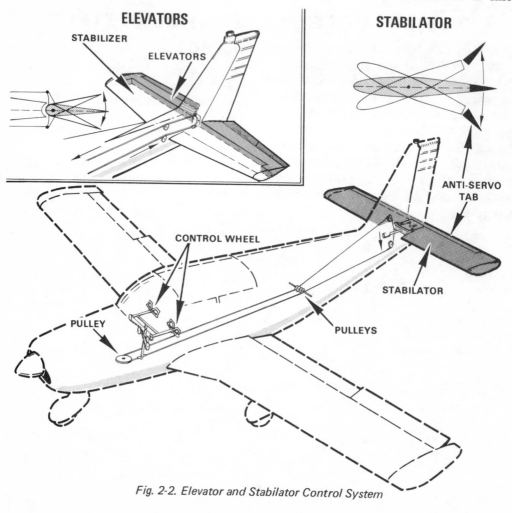

ELEVATORS

STABILIZER

ELEVATORS

STABILATOR

ANTI-SERVO TAB

CONTROL WHEEL

STABILATOR

PULLEY

PULLEYS

Fig. 2-2. Elevator and Stabilator Control System

and provides horizontal stability for the airplane. A change in position of the elevators modifies the *camber* of the airfoil which increases or decreases lift.

The stabilator is a one-piece, horizontal tail surface that pivots up and down. A change in position of the stabilator changes the angle of attack resulting in an increase or decrease in lift on this surface. When the pilot wants to raise the aircraft nose, he applies back pressure on the control wheel. This moves the trailing edge of the stabilator tail surface up resulting in a negative angle of attack. This causes the tail to move down and the nose of the aircraft to move up.

An *anti-servo* tab, mounted on the trailing edge of the stabilator, moves in the same direction as the trailing edge of the stabilator (see Fig. 2-2) to provide the pilot with a control feel similar to that experienced with conventional elevators. Without this anti-servo tab, control forces from a stabilator would be so light that the pilot would have a tendency to overcontrol.

The function of the elevators or stabilator is to control the angle of attack of the wings. Climbs and descents are established by a change in the throttle setting with the elevators or stabilator adjusting the angle of attack of the wings.

RUDDERS

The *rudder* moves the aircraft about the *vertical* or *yaw axis* resulting in a motion known as *yaw*. The movement of the rudder is to the left or to the right and is controlled by movement of the rudder pedals connected to the rudder through a series of cables, pulleys, and bellcranks. These pedals are mounted under the instrument panel near the floor of the pilot's compartment. A *vertical stabilizer*, often called the *fin*, provides directional stability and a streamlined structure for mounting the rudder. (See Fig. 2-3.)

Pressure applied to the right rudder pedal displaces the rudder to the right. This increases the camber on the left side of the vertical tail surface, creating a low pressure area on the left side and a high pres-

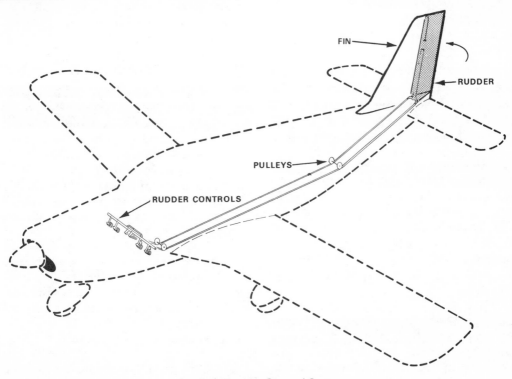

Fig. 2-3. Rudder Control System

Fig. 2-4. Elevator Trim Controls

sure area on the right side. The resulting pressures cause the tail of the aircraft to move left and the nose to move right. Applying pressure to the left rudder pedal provides an opposite reaction. The rudder is used in conjunction with the ailerons to coordinate turns.

TRIM SYSTEMS

Several types of trim systems are provided to minimize elevator or stabilator control pressures. These systems are:

1. Trim tab.
2. Adjustable stabilizer.
3. Movable tail.

TRIM TAB

The trim tab system incorporates an adjustable trim tab which is mounted on the trailing edge of the elevator or stabilator. In the case of the stabilator, the anti-servo tab also serves as a trim tab.

The trim tab is controlled by a trim wheel or crank in the cabin. (See Fig. 2-4.) By adjusting the tab control, the pilot can reduce control wheel pressure for various elevator or stabilator settings. A tab-position indicator is incorporated in the tab-control mechanism to show the nose-up or nose-down position of the tab setting.

Placing the tab control to its full nose-down position moves the tab to its full-up position. (See left side of Fig. 2-5.) The tab, being up into the airflow over the tail, forces the trailing edge of the elevator or

stabilator down. This brings the tail up and results in a nose-down movement of the aircraft.

Setting the tab control in the full nose-up position moves the tab to its full-down position. (See right side of Fig. 2-5.) With the tab in the down position, the air flowing under the horizontal tail surface hits the tab and forces the trailing edge of the elevator or stabilator up, giving the aircraft a nose-up movement. Trim tabs can be set at any desired position to provide various elevator or stabilator settings. Some aircraft have trim tabs on all movable primary control surfaces (aileron, elevator, and rudder).

ADJUSTABLE STABILIZER

The adjustable stabilizer trim system enables the entire stabilizer to be rotated by a control wheel or crank in the cabin. Hinges are provided at the aft end of the stabilizer so that the surface rotates when the screw jacks are turned. A tab-position indicator is incorporated in the stabilizer control mechanism to show the nose-up or nose-down position of the stabilizer setting.

Moving the trim control to the nose-down setting rotates the leading edge of the stabilizer up. In this position, the stabilizer has a higher angle of attack. This causes the tail to lift and results in a nose-down movement of the aircraft.

Setting the trim control in the nose-up position reduces the angle of attack of the

NOSE-DOWN TRIM

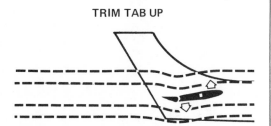

TAB UP; ELEVATOR DOWN

NOSE-UP TRIM

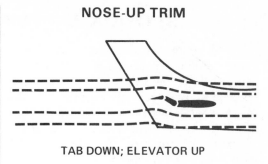

TAB DOWN; ELEVATOR UP

ELEVATOR

STABILATOR

TRIM TAB UP

TRIM TAB DOWN

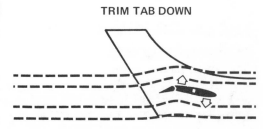

Fig. 2-5. Elevator and Stabilator Trim Action

stabilizer. This results in an upward movement of the aircraft nose. The stabilizer can be set at any position between full-up and full-down to provide the amount of trim desired.

MOVABLE TAIL

The movable tail trim system permits the pilot to adjust the position of the tail to minimize elevator control pressures. There are no elevator trim tabs in this system. The complete empennage, or tail section, pivots to provide elevator trim.

Movement of the tail trim system is controlled by a control wheel in the cabin.

A tab-position indicator shows the nose-up or nose-down setting of the tail.

When the trim control wheel is turned full forward, the complete tail rotates aft. This places the stabilizer in a higher angle of attack position, resulting in a nose-down movement of the aircraft.

Placing the trim control wheel in the nose-up position rotates the complete tail forward and reduces the angle of attack of the stabilizer. This results in the aircraft nose moving upward. The tail can be set in any position between full-up and full-down to provide the amount of trim desired.

SECTION B - FLIGHT INSTRUMENTS

One of the most important areas of aeronautical knowledge is flight instruments and their respective systems. Acquiring such a knowledge enables pilots to visualize what takes place behind an instrument panel and what makes each of the instruments function. In addition, with this knowledge, pilots are much more capable of interpreting the different types of information displayed on the instruments during the various conditions of flight.

THE MAGNETIC COMPASS

The magnetic compass was one of the first instruments to be installed in an aircraft and is still the *only direction-seeking instrument* found in most general aviation aircraft. (See Fig. 2-6.) The compass is a very reliable, self-contained unit and is independent of external vacuum or electrical power. To use it satisfactorily, however, the pilot must understand principles of magnetism and the various errors common to compass operation.

MAGNETIC PRINCIPLES

A magnet is a piece of metal that has the property of attracting another metal. The force of attraction is greatest at the poles or points near each end of the mag-

Fig. 2-6. The Magnetic Compass

net and least in the area halfway between the two poles. Lines of force flow from each of these poles in all directions, bending around and flowing toward the other pole to form a *magnetic field*. (See Fig. 2-7.)

Fig. 2-7. Lines of Force Around a Bar Magnet

A magnetic field surrounds the earth with lines of force oriented approximately to the North and South Magnetic Poles. Because the aircraft compass is oriented to the magnetic pole, in-flight allowance must be made for the *difference in location* of the geographic and magnetic poles if the heading reference is to be the geographic true pole.

Lines of equal magnetic declination or variation are called *isogonic lines* and are measured in degrees east or west of the earth's geographic North Pole. (See Fig. 2-8.) These lines are plotted in degrees of east or west variation on aeronautical charts. A line connecting the points of zero degrees variation is called the *agonic line*.

COMPASS CONSTRUCTION

The magnetic compass is located away from radios and electrical equipment to limit the effect of any magnetic disturbance in the aircraft. It contains a small internal light bulb for night illumination

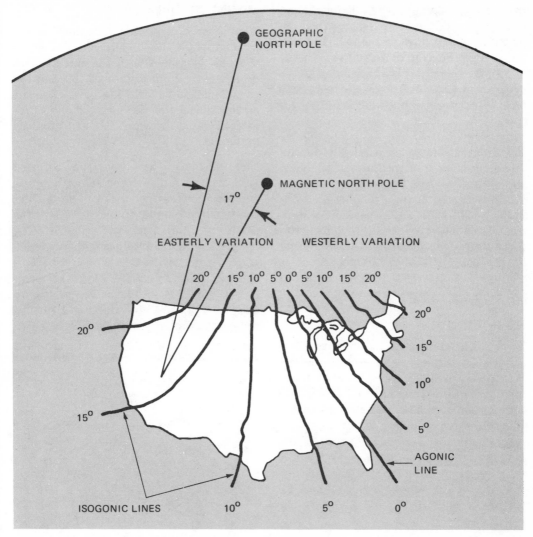

Fig. 2-8. Lines of Variation

which is mounted in the front of the instrument case.

The compass card is graduated in five-degree increments, and the magnetic headings (without the last zero) are printed every 30°. In addition, the compass card is equipped with a float to keep it in a horizontal position in the compass fluid. (See Fig. 2-9.) The pivot underneath the card rests on a jeweled bearing which permits the card to rotate freely, while two long magnets mounted underneath give the compass its directional quality. These magnets always attempt to align themselves with the magnetic field of the earth.

The float assembly is sealed in a chamber filled with acid-free, white kerosene. This liquid serves several purposes. First, due to buoyancy, part of the weight of the card is taken off the pivot which supports

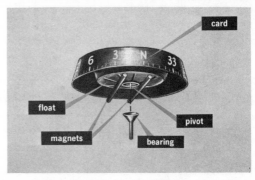

Fig. 2-9. Magnetic Compass Card Assembly

the card. Second, the fluid decreases the erratic swing of the compass, known as oscillation, which may be the result of turbulence or rough pilot technique. In addition, the fluid lubricates the pivot point on the pedestal which serves as the mount for the float assembly and the compass card.

The float assembly is balanced on the pivot allowing free rotation of the card and a tilt of up to 18 degrees. At the rear of the compass bowl, a diaphragm is installed to allow for any expansion or contraction of the liquid, thus preventing the formation of bubbles or possible bursting of the case. Mounted behind the glass window is a *lubber* or reference line to designate the compass indication.

DEVIATION

Two small compensating magnets, adjustable by screws, are located in the top or bottom of the compass case to minimize *deviation*. Deviation is the deflection of the compass needle from a position of magnetic north as a result of local magnetic disturbance in the aircraft. Adjusting the compensating screws to minimize deviation is known as *swinging the compass*, a procedure which should be performed periodically by maintenance personnel.

The deviation noted should be recorded on a compass correction card. (See Fig. 2-10.) When flying compass headings, the pilot must refer to this card and make the appropriate adjustment for the particular heading he wishes to fly. It is important for the pilot to avoid placing metallic objects near the compass during flight, since this practice may induce large errors into the indications.

MAGNETIC DIP

A number of compass errors are caused by a deflection of the aircraft compass as it seeks alignment with the earth's lines of magnetic force. Lines of force in the earth's magnetic field are parallel to the earth's surface at the equator and curve increasingly downward closer to the magnetic poles.

The magnetic bars on the compass card tend to assume the same direction and position as the lines of force, thus, the bars are parallel to the surface of the earth at the magnetic equator. The magnetic bars point increasingly downward as the aircraft is moved closer to the magnetic pole. (See Fig. 2-11.) This characteristic is known as *magnetic dip* and is responsible for northerly turning error and acceleration/deceleration error.

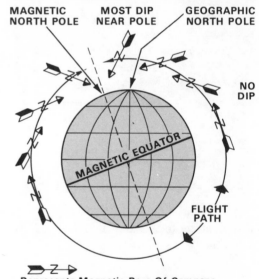

MAGNETIC NORTH POLE   MOST DIP NEAR POLE   GEOGRAPHIC NORTH POLE

NO DIP

MAGNETIC EQUATOR

FLIGHT PATH

Represents Magnetic Bars Of Compass

Fig. 2-11. Magnetic Dip

TURNING ERRORS

Northerly turning error is the most pronounced of the in-flight errors and is

FOR (MH) magnetic heading	0°	30°	60°	90°	120°	150°	180°	210°	240°	270°	300°	330°
STEER (CH) compass heading	359°	30°	60°	88°	120°	152°	183°	212°	240°	268°	300°	329°
RADIO ON [X]					RADIO OFF [ ]							

Fig. 2-10. Compass Correction Card

most apparent when turning to or from headings of north and south. When the airplane is banked, the compass card also banks because of the centrifugal force acting upon it. While the compass card is in this banked attitude, the vertical components of the earth's magnetic field cause the earth-seeking ends of the compass to dip to the low side of the turn, giving the pilot an erroneous turn indication. (See Fig. 2-12.)

When making a turn *from a heading of north* in the Northern Hemisphere, the compass briefly gives an indication of a turn in the *opposite* direction. If this turn is continued east or west, the compass card will begin to indicate a turn in the correct direction, but will *lag* behind the actual turn at a diminishing rate until within a few degrees of east or west. If the pilot makes a very gradual and shallow banked turn (less than five degrees) from a compass indication of north, it is possible to change the actual heading of the aircraft by 20° or more while still maintaining an indication of north on the compass.

When making a turn *from a heading of south*, the same set of forces will be in effect but the indications appear quite differently. The compass gives an indication of a turn in the *correct* direction, but at a much *faster* rate than is actually being experienced. As the turn is continued toward east or west, the compass indications will continue to *precede* the actual turn, but at a diminishing rate until within a few degrees of west or east.

ACCELERATION/DECELERATION ERRORS

Acceleration error can occur during airspeed changes and is most apparent on headings of east and west. Acceleration error is caused by a combination of inertia and the vertical component of the earth's magnetic field. Because of its pendulum-type mounting, the compass card tilts during speed changes. The momentary tilting of the card from the horizontal results in an error which is most apparent on headings of east and west. When *accelerating* on either of these headings, the error is in the form of an indication of a turn to the *north*, and when *decelerating*, the error is in the form of an indication of a turn to the *south*. A memory aid that helps pilots remember this relationship between airspeed change and the direction of error is the acronym "ANDS" — *Accelerate - North; Decelerate - South.*

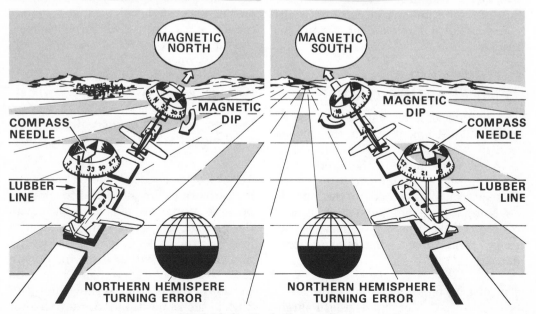

Fig. 2-12. Northerly Turning Error

TURNS USING THE MAGNETIC COMPASS

A maximum of 15° of bank should be used when turning to headings utilizing the magnetic compass. When a 15° bank is used in turning north or south, the amount of roll-out lead is proportional to the latitude. For example, in turning *from south to north* in the area of 30° north latitude, a pilot would start a roll-out 30° *prior* to reaching north, plus a 5° lead to allow time for roll-out from the banking attitude. For example, to roll the airplane to a wings-level attitude on a heading of north, the roll-out should be initiated at either 035° or 325° depending on the direction of the turn.

To roll out on a heading of south, it is necessary to fly past 180°, the number of degrees of latitude minus the 5° lead. For example, in making a right turn from north to south at 30° north latitude with 15° bank, a roll-out should begin on a heading of 205°.

If its errors and characteristics are thoroughly understood, the magnetic compass offers the pilot a most reliable means of determining the heading of his airplane. To accurately read the compass, however, the pilot must be certain the airplane is flown level and at a constant airspeed.

OUTSIDE AIR TEMPERATURE GAUGE

The outside air temperature gauge is an important instrument used in the calculation of true airspeed in flight. (See Fig. 2-13.) It is mounted in the aircraft in a position where the temperature sensing element can be exposed to the outside air. Normally, a bright metal shield is used to protect the sensitive element of the gauge from direct sunlight and from damage.

The dial on most gauges is calibrated in both degrees Celsius and degrees Fahrenheit. When making airspeed calculations

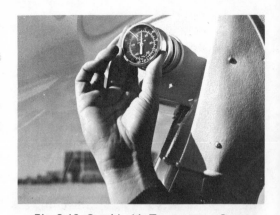

Fig. 2-13. Outside Air Temperature Gauge

in flight, it is important for the pilot to read the correct scale — normally Celsius.

PITOT-STATIC PRESSURE INSTRUMENTS

PITOT-STATIC SYSTEM

Instruments connected to the pitot-static system (see Fig. 2-14) include the airspeed indicator, altimeter, and vertical speed indicator. This system includes a *pitot tube* for measuring *impact* (ram) air pressure and one or more *static ports* for measuring barometric *static* pressure (air pressure at flight altitude).

The pitot tube is usually located on the leading edge of the wing where there is a minimum of airflow disturbance from the various parts of the aircraft. (See Fig. 2-14.) Its opening faces the line of flight of the airplane and supplies ram air pressure to the airspeed indicator. It should be remembered that the pitot tube is not involved with the operation of any instrument other than the airspeed indicator.

The altimeter, airspeed, and vertical speed indicator cases are all vented to allow air pressure inside the cases to equalize with the outside air as the aircraft gains or loses altitude. This venting is accomplished by connecting these instrument cases to static ports, normally located on the side of the fuselage, as shown in figure 2-14, in such a position that they will not pick up ram or turbulent air pressures.

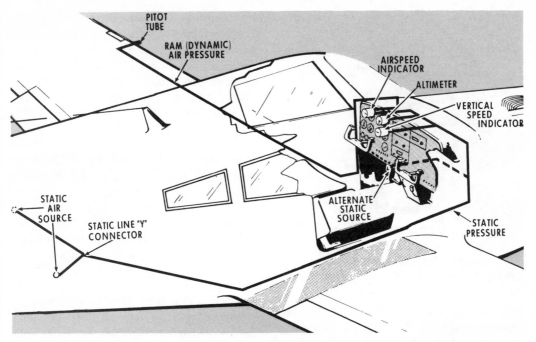

Fig. 2-14. Pitot-Static System

On some aircraft, the static vent holes are located in the pitot head, which also houses the pitot tube. In either location, the static vent is positioned to measure undisturbed atmospheric pressure only. Any change in pressures will be shown on the appropriate instrument after a lag of anywhere from one to nine seconds, depending on the design and construction of the instrument involved.

The pilot should check the pitot tube opening and the static vent openings during the preflight inspection to see that they are not clogged. Due to their location, the pitot tube and static vents are susceptible to outside elements such as dirt, water, and ice. Clogged or partially clogged openings may cause inaccurate instrument readings. The pilot should not blow into these openings since excess pressure can damage any of the three instruments.

AIRSPEED INDICATOR

The airspeed indicator measures the *difference* between impact and static pressure. When the airplane is parked on the ground, these two pressures are equal.

When the aircraft moves through the air, however, the pressure in the pitot line becomes greater than the pressure in the static line. (See Fig. 2-15.) This difference in pressure expands a small diaphragm into the area of lower pressure and this registers indicated airspeed on the instrument face in miles per hour, knots, or both. (See Fig. 2-16.) *The airspeed indicator displays the speed at which the airplane is moving through the air regardless of wind, and not the speed at which it is moving over the ground.*

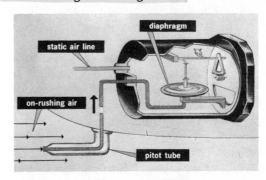

Fig. 2-15. Airspeed System

Airspeed Definitions

There are three airspeed definitions of particular importance to the aviator: indicated airspeed, calibrated airspeed, and true airspeed.

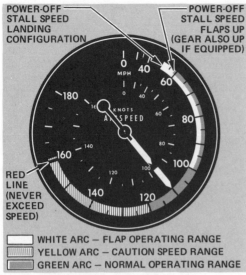

POWER-OFF STALL SPEED LANDING CONFIGURATION

POWER-OFF STALL SPEED FLAPS UP (GEAR ALSO UP IF EQUIPPED)

RED LINE (NEVER EXCEED SPEED)

☐ WHITE ARC — FLAP OPERATING RANGE
▨ YELLOW ARC — CAUTION SPEED RANGE
▧ GREEN ARC — NORMAL OPERATING RANGE

Fig. 2-16. Airspeed Markings

Indicated airspeed (IAS). Indicated airspeed is the direct instrument reading the pilot obtains from the airspeed indicator. It is uncorrected for variation in atmospheric density or any installation and instrument errors.

Calibrated airspeed (CAS). Calibrated airspeed is the indicated airspeed corrected for instrument and installation errors. Although aircraft and instrument manufacturers attempt to keep airspeed errors to a minimum, it is not possible to entirely eliminate these errors throughout the entire airspeed operating range.

As the aircraft flight attitude or configuration is changed, the airflow in the vicinity of the static inlet may introduce impact or negative pressure into the static source which results in erroneous airspeed indications. Airspeed errors may amount to several miles per hour and are generally greatest in the low airspeed range; however, in the cruising range, indicated airspeed and calibrated airspeed are quite close. To determine the calibrated airspeed, it is necessary for the pilot to consult the airspeed correction table found in the aircraft owner's manual.

Airspeed limitations, such as those found on the color-coded face of the airspeed indicator, on placards in the cabin, or in the airplane flight manual or owner's handbook, are calibrated airspeeds (sometimes referred to as TIAS — True Indicated Airspeed). Therefore, it is important for the pilot to refer to the airspeed calibration chart to allow for possible airspeed errors. The airspeed calibration chart is included in the airplane flight manual or owner's handbook.

True airspeed (TAS). True airspeed is the actual speed of the aircraft through the air mass. The airspeed indicator registers true airspeed only when there is no airspeed correction, and under standard sea level conditions (when the actual barometric pressure is 29.92 inches of mercury and the temperature is 15° Celsius).

Because air density decreases with an increase in altitude, the airplane must fly faster at higher altitudes for the same pressure difference between pitot impact pressure and static pressure to result. Thus, for a given true airspeed, indicated airspeed *decreases* as altitude increases. Conversely, true airspeed *increases* with a gain in altitude for a constant indicated airspeed.

Normally, the flight computer is used for true airspeed computations involving temperature and pressure altitude. Some aircraft, however, have an airspeed indicator which incorporates that portion of the computer which is necessary for determining true airspeed. This instrument is known as a *true airspeed indicator.*

Airspeed Indicator Markings

The airspeed indicator incorporates color-coded markings to indicate important reference speeds for the pilot. (See Fig. 2-16.) The following color codes are usually used:

1. A *red line* is placed at the *never-exceed speed.* This is the maximum speed at which the airplane can be operated in smooth air.

2. A *yellow arc* depicts the *caution speed range.* The aircraft should not be operated in this range except in smooth air.

3. A *green arc* is used to define the *normal operating speed range*. The low-speed end of the green arc is the power-off stalling speed with the landing gear and the wing flaps retracted. The upper airspeed limit of the green arc is the maximum structural cruising speed — the maximum speed for normal operations.

4. A *white arc* specifies the speed range where *fully-extended wing flaps* can be used safely. The low-speed end of the white arc represents the power-off stalling speed with landing gear and wing flaps extended. The upper limit of the white arc is the highest airspeed at which the pilot should extend flaps. Extension of full flaps at higher speeds could cause severe strain on the flap extension mechanism.

NOTE: The colored markings on airspeed indicators are based on calibrated airspeed, *not on indicated airspeed.*

Maneuvering Speed

An important airspeed limitation not marked on the face of the airspeed indicator is the *maneuvering speed* or the maximum speed at which full abrupt control travel at gross weight may be used without exceeding the load limits. This speed is generally found on placards or in the airplane flight manual or owner's handbook. If severe turbulence is encountered, the airspeed should be reduced to the maneuvering speed and the aircraft held in as constant an *attitude* as possible.

ALTIMETER

The altimeter (see Fig. 2-17) is a very important instrument in either VFR or IFR conditions. The altimeter is connected to the static system through an outlet in the back of the case. This outlet is *not* connected to the sensitive element at all, but only serves as a vent to allow the static atmospheric pressure to move into and out of the case as the aircraft climbs and descends.

Fig. 2-17. Altimeter

The altimeter utilizes a stack of aneroid capsules as its sensitive element. (An aneroid is a disc-shaped metallic capsule.) The aneroid barometer was designed for aircraft use because of its compact size and sensitivity to changing barometric pressures.

Changes in atmospheric pressure surrounding the aneroid cause it to expand or contract. As the aircraft climbs to altitude, the air moves out of the case due to the decrease in outside atmospheric pressure. This causes the sealed aneroid to expand, resulting in a pointer movement upscale. (See Fig. 2-18.) As the aircraft descends, air moves into the case due to the increase of outside atmospheric pressure. This causes the aneroid to contract, resulting in a pointer movement in the opposite direction, or downscale.

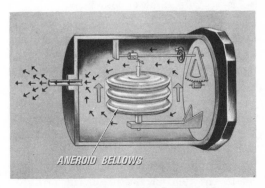

Fig. 2-18. Aneroid Expands

Setting the Altimeter

Most altimeters are equipped with a barometric pressure setting window (sometimes referred to as the Kollsman window) providing the pilot with a means to adjust his altimeter to compensate for the variation in atmospheric pressure. A knob is located at the bottom of the instrument for this adjustment. (See Fig. 2-17.)

FAA regulations specify the following concerning altimeter settings: "The cruising altitude of an aircraft below 18,000 feet MSL shall be maintained by reference to an altimeter that is set to the current reported altimeter setting of a station along the route of flight within 100 nautical miles. If there is no such station, the current reported altimeter setting of an appropriate available station shall be used. In an aircraft having no radio, the altimeter shall be set to the elevation of the departure airport or an appropriate altimeter setting available before departure." Above 18,000 feet, the pilot sets his altimeter to 29.92 in. Hg. regardless of station pressure.

If the altimeter does not read within 75 feet of the field elevation prior to takeoff, the pilot should request that an authorized mechanic make an adjustment on the altimeter so the altitude indication and the barometric scale are synchronized.

Types of Altitude

Pilots are usually concerned with five types of altitudes: indicated, pressure, density, true, and absolute.

Indicated altitude is the altitude displayed on the aircraft altimeter (uncorrected for temperature) above mean sea level, when the altimeter is set to the appropriate setting. When the aircraft is on the ground and the altimeter is set to the local altimeter setting, it should read field elevation (true altitude), assuming no instrument error. With the altimeter set to the field elevation on the ground, as soon as the aircraft becomes airborne,

it displays an indicated altitude because of non-standard temperature and pressure lapse rates within increasing altitude.

Pressure altitude is an altitude which, due to existing pressure, is equivalent to an elevation measured above a standard pressure level and obtained by applying the standard pressure to the altimeter. Pressure altitude is read from the altimeter when 29.92 is set in the barometric pressure window. This altitude is used for computer solutions for true airspeed, density altitude, true altitude, etc.

Density altitude is pressure altitude corrected for non-standard temperature variations. An aircraft will have the same performance characteristics as it would have in a standard atmosphere at this altitude. Density altitude is directly related to airplane performance and many aircraft performance charts are based on density altitude.

True altitude is the exact distance above mean sea level (MSL). The altitude of all fixed, non-changeable objects is given in true altitude. Field elevations and obstructions such as mountains, radio antennas, and towers are given as true altitude on the navigational charts. True altitude figures are not susceptible to change with variable atmospheric conditions.

Absolute altitude is the altitude of an aircraft above the surface or the terrain over which it is flying. This altitude may be abbreviated AGL, above ground level.

Altimeter Indications

Altimeters are calibrated on the basis of both a standard pressure (29.92 in. Hg.) and a standard temperature (59°F) at sea level with a standard lapse rate (reduction) as altitude is increased. If either of these factors is significantly different than standard for the altitude, an erroneous altitude will be indicated. When the pilot, while in flight, sets his altimeter to the station setting below, but does not compensate for a non-standard temperature lapse rate, his indicated altitude may be different than his true altitude. Therefore, if terrain or obstacle clearance is a factor in the selection of a cruising altitude, pilots must remember to anticipate that *colder* than standard temperatures will place the aircraft *lower* than the altitude displayed on the altimeter.

The same situation exists when a pilot flies from a high-pressure area to a lower pressure area without resetting the barometric pressure window. A pilot maintaining the same indicated altitude will descend, leading him to believe that he is flying higher than he actually is. (See Fig. 2-19.) A good rule of thumb used by many pilots is *from high to low, hot to cold, look out below.*

Since the actual barometric pressure decreases approximately one inch for every 1,000 feet of altitude, a change of one inch in the pressure setting will result in a change of 1,000 feet in the altimeter reading. Accordingly, a change of 0.1 inch of mercury in the pressure setting of an altimeter will result in a change of 100 feet in the altimeter reading.

The altimeter usually has three pointers to indicate altitude. (See Fig. 2-20.) The longest and largest pointer indicates hun-

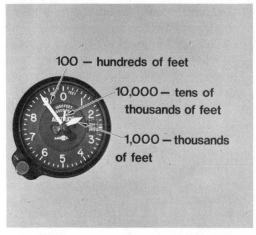

Fig. 2-20. Altimeter with Three Pointers

dreds of feet, the middle-sized pointer indicates thousands of feet, and the smallest pointer indicates tens of thousands of feet.

When the 100-foot pointer makes one complete revolution, the 1,000-foot pointer moves to one, or 1,000 feet. When the 1,000-foot pointer makes one revolution, the thin short hand will move to one, indicating 10,000 feet. This is why it is necessary to read the smallest pointer first, the middle pointer second, and the long pointer third. The middle-sized pointer and the thin pointer are disregarded in reading the altimeter until they reach "one" on the altimeter dial.

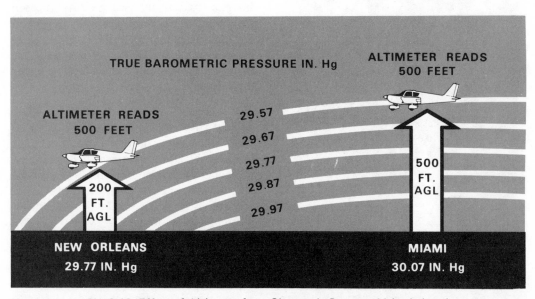

Fig. 2-19. Effect of Altimeter from Changes in Pressure Altitude Levels

In reading the altimeter in figure 2-20, the smallest pointer has not yet reached one, so the aircraft is less than 10,000 feet and the needle indication can be disregarded. The medium-sized pointer is between one and two which means the aircraft is above 1,000 feet but below 2,000 feet; therefore, the pointer indication is read as 1,000 feet plus whatever the long pointer indicates. In this case, the long pointer is on nine which signifies 900 feet. Putting together the information from the three pointers, the aircraft altitude is read as 1,900 feet.

VERTICAL SPEED INDICATOR

The vertical speed indicator (VSI) shows the rate of climb and rate of descent. (See Fig. 2-21.) This instrument is contained in a sealed case and also is connected to the static system. Inside the case, there is a diaphragm which is connected to the static pressure system. (See Fig. 2-22.) As the altitude changes, the static pressure changes to give pilots an indication of a *vertical* change. Just as the altimeter shows the amount of altitude change, the vertical speed indicator shows the *rate* of altitude change.

Fig. 2-21. Vertical Speed Indicator

This instrument actually measures the rate of change in the atmospheric pressure through the use of a *calibrated leak* built into the back of the instrument case. The rate of change is shown in feet per minute.

The VSI can be efficiently used as a trend instrument since it will *immediately* indicate any deviation from level flight. In turbulent air, the aircraft may be continually moving up and down, both because vertical gusts may change the pitch attitude and because the air itself is moving vertically. In turbulence, the VSI will make such rapid changes that it will be almost useless as a precision pitch instrument.

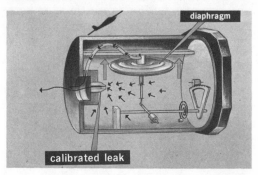

Fig. 2-22. Internal Components of Vertical Speed Indicator

The VSI will be very quick to show a climbing or descending *trend;* however, it has a characteristic lag of six to nine seconds before actually indicating the correct climb or descent rate. This lag is the result of the time necessary for pressure changes to take place inside the instrument. If pitch changes are small and are made slowly, the indication on the instrument will be a very close representation of the correct rate at any given instant. When a rapid or large pitch change or a combination of the two is made, the instrument may lag far behind the correct indication.

The vertical speed indicators used on most turbine-powered aircraft do not have the needle lag. The more precise and expensive VSI's give a correct instantaneous indication of the rate of climb or descent. These instruments use the principles of inertia and are known as instantaneous vertical speed indicators (IVSI).

GYRO INSTRUMENTS

Gyro flight instruments have played a major part in making instrument flight a practical part of modern-day flight. It is important for the pilot to understand their operation so he may use the gyro instruments correctly and effectively.

GYROSCOPIC PRINCIPLES

The primary trait of a rotating gyro rotor is its *rigidity in space* or gyroscopic inertia. Newton's First Law states in part: *A body in motion tends to move in a constant speed and direction unless disturbed by some external force.*

The rotor inside a gyro instrument maintains a constant attitude in space as long as no outside forces change its motion. This quality of stability is greater if the rotor has great mass and speed. Thus the gyros in aircraft instruments are of heavy construction and are designed to spin quite rapidly.

The three gyro instruments found in most general aviation airplanes are the attitude gyro, directional gyro, and turn-and-slip indicator or turn coordinator. (See Figs. 2-23, 2-24, and 2-25.)

The directional gyro and attitude indicator use gyros as an unchanging reference in space. Once the gyros are spinning, they stay in constant positions with respect to the actual horizon or direction. The aircraft heading and attitude can then be compared to these stable references. For example, the rotor of a universally-mounted gyro will remain in

Fig. 2-23. Attitude Gyro

Fig. 2-24. Directional Gyro

the same position while the surrounding *gimbals* or circular frames are moved.

Fig. 2-25. Turn Indicators

PRECESSION

Another characteristic of gyros is *precession* which is a tilting or turning of the gyro axis in reaction to applied forces. When a deflective force is applied to the rim of a gyro rotor that *is not turning*, the rotor naturally moves in the direction of the force. However, when the gyro rotor *is rotating*, the same applied force would cause the rotor to move in a direction as though the force had been applied at a point 90° around the rim in the direction of rotation. This turning movement, or precession, places the rotor in a new plane of rotation. Unavoidable precession is caused by maneuvers and internal friction in the attitude and directional gyros. This causes slow drifting and minor erroneous indications in these instruments.

SOURCES OF POWER FOR GYRO OPERATIONS

A gyroscopic instrument can be operated by either the vacuum system or the electrical system. Most modern training aircraft are equipped with a turn coordinator that is driven electrically, and attitude and directional gyros that are vacuum powered. The advantage of using two different sources of power is: if one power source fails, the other source still is available to provide directional reference.

VACUUM SYSTEM

The gyro instruments, operated by the vacuum system, are driven by air pressure differential. A vacuum pump creates a partial vacuum in the system. A central filtered air inlet to the system is located upstream from the gyros. As the result of the partial vacuum, air rushes into the system through the central filter and is routed to the gyro cases, causing the gyro wheels to spin. This air is exhausted into the atmosphere through the pump.

An engine-driven vacuum pump is the most common source of vacuum power for gyros installed in general aviation light aircraft. (See Fig. 2-26.) Pump capacity and size vary in different aircraft depending on the number of gyros to be operated. The system will usually have a relief valve to relieve excess vacuum.

A vacuum pressure gauge is mounted on the instrument panel to inform the pilot of the operating suction of the system. This gauge indicates the difference, in inches of mercury, between the pressure at the central filter and at the pressure relief valve.

Older vacuum gyros require a setting between 2.75 and 4.25 inches of mercury; however, the new gyro systems require *higher* vacuum settings. Recommended vacuum settings by various manufac-

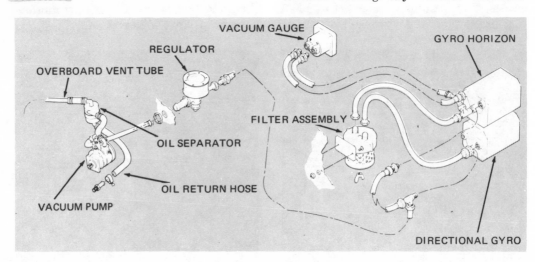

Fig. 2-26. Vacuum System

turers of new gyros may range from a low setting of 4.5 to a high setting of 5.8 inches of mercury. A reading below the recommended value indicates that the air flow is not spinning the gyros fast enough for reliable operation. On some aircraft, warning lights are used to indicate a low or high vacuum pressure.

VENTURI

A venturi can be used in place of a vacuum pump and has the advantage of relatively low cost due to its simplicity of installation and operation. One or more ventures are usually mounted on the side of the aircraft where the airflow from the propeller will pass through them. (See Fig. 2-27.) The venturi is shaped like a pair of airfoils so that the air velocity is increased as it flows through. The increase in air velocity inside the venturi causes a low pressure area in the venturi, result ing in a partial vacuum in the line leading from the ventures to the gyro cases.

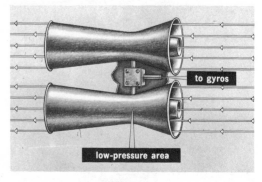

Fig. 2-27. Ventures

ELECTRICALLY-DRIVEN GYROS

Aircraft that normally fly at high altitudes do not have a vacuum system since its operation would be limited in the thin, cold air of high altitudes. Alternating current (AC) is used to drive the gyros in the directional and attitude instruments. The AC power is provided by inverters which convert direct current to alternating current or by engine-driven AC generators or alternators.

GYRO ROTORS

Gyro rotors are mounted two ways, depending upon how the gyroscopic proper-

ties are to be used in the operation of the instrument. A freely or universally-mounted gyro is set on three gimbals with the gyro free to rotate in any plane. (See Fig. 2-28.) Regardless of the position of the gyro base, the gyro tends to remain rigid in space. In the attitude indicator, the horizon bar is gyro-controlled to remain parallel to the natural horizon. Changes in the position of the aircraft are shown pictorially.

Fig. 2-28. Universally-Mounted Gyro

The semi-rigid or restricted mounting employs two gimbals limiting the rotor to two planes of rotation. In the turn indicator, the semi-rigid mounting is used to provide controlled precession of the rotor. The precession force, exerted upon the gyro by the turning aircraft, causes the needle to indicate a turn.

ATTITUDE GYRO

The *attitude gyro* is also referred to as the artificial horizon, the gyro horizon, or the attitude indicator. With its miniature airplane and horizon bar, it is the one instrument that portrays the picture of the actual attitude of the real airplane. Since it is connected to a gyroscope, the horizon bar remains *parallel* to the natural horizon. The relationship of the miniature aircraft to the horizon bar is the same as the relationship of the real airplane to the actual horizon. (See Fig. 2-29.) The horizon bar remains parallel to the actual horizon and the airplane and instrument case revolve around it.

An adjustment knob is provided to move the miniature aircraft upward or downward inside the case. Normally the miniature aircraft is adjusted so that the wings overlap the horizon bar when the real airplane is in straight-and-level cruising flight. The attitude gyro gives an instantaneous indication of even the smallest changes in attitude and has no lead or lag. It is very reliable if properly maintained. (See Fig. 2-29.)

When the airplane's nose is raised above level attitude, the pictorial representation on the attitude gyro shows the same attitude in miniature. Besides showing pitch motion around the lateral axis, the attitude gyro also shows movement around the longitudinal or roll axis. (See Fig. 2-30.)

The instrument has an index at the top of the case which indicates the degree of bank by pointing to bank marks on either side of the instrument centerline. Each large mark is equivalent to 30° bank, and the small marks between the first 30° index are each equal to 10° of bank.

When the airplane is banked to the left, the horizon bar in the instrument maintains its position level to the actual horizon so that the left wing of the little airplane on the instrument face moves below the horizon bar and the right wing moves above. (See Fig. 2-30, bottom right.) This is the same relationship that the full-sized aircraft has with the earth's horizon.

THE DIRECTIONAL GYRO

The *directional gyro* (DG), or heading indicator, is designed to operate without the magnetic dip errors inherent in the magnetic compass. The directional gyro has *no* direction-seeking properties and must be set to headings shown on the magnetic compass. The DG also operates on the gyroscopic principle of rigidity in space with a plane of rotation that is vertical. The compass rose card on the face of the directional gyro is geared to

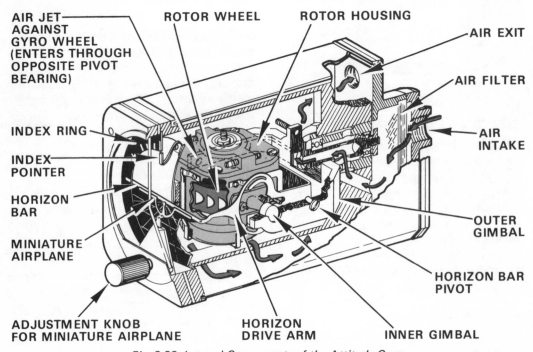

AIR JET AGAINST GYRO WHEEL (ENTERS THROUGH OPPOSITE PIVOT BEARING)
ROTOR WHEEL
ROTOR HOUSING
AIR EXIT
AIR FILTER
INDEX RING
AIR INTAKE
INDEX POINTER
HORIZON BAR
MINIATURE AIRPLANE
OUTER GIMBAL
HORIZON BAR PIVOT
ADJUSTMENT KNOB FOR MINIATURE AIRPLANE
HORIZON DRIVE ARM
INNER GIMBAL

Fig. 2-29. Internal Components of the Attitude Gyro

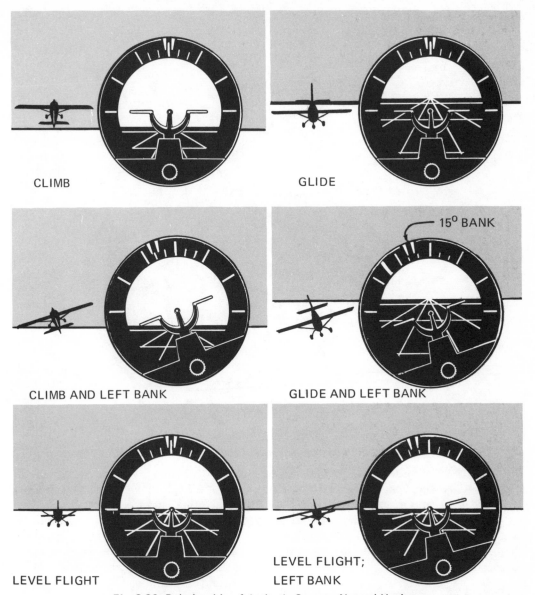

CLIMB

GLIDE

15° BANK

CLIMB AND LEFT BANK

GLIDE AND LEFT BANK

LEVEL FLIGHT

LEVEL FLIGHT;
LEFT BANK

Fig. 2-30. Relationship of Attitude Gyro to Natural Horizon

the gyro gimbal causing the card to turn as the airplane turns. (See Fig. 2-31.) To use the directional gyro properly, the pilot must be able to adjust the card. To do this, the pilot turns an adjustment knob which rotates the card to the correct magnetic heading.

The directional gyro instrument is subject to precession errors which cause the instrument to creep off the correct magnetic heading indication. While airborne, the directional gyro should be set to the magnetic compass only in straight-and-level, unaccelerated flight, or on the ground during run-up. Due to precession errors, the directional gyro should be checked at least every 15 minutes and reset when necessary. Even a DG in perfect condition can drift from its original setting.

TURN-AND-SLIP INDICATOR

The *turn-and-slip indicator* shown in figure 2-32 is one of the most important of the gyro instruments since it is considered essential to instrument flight.

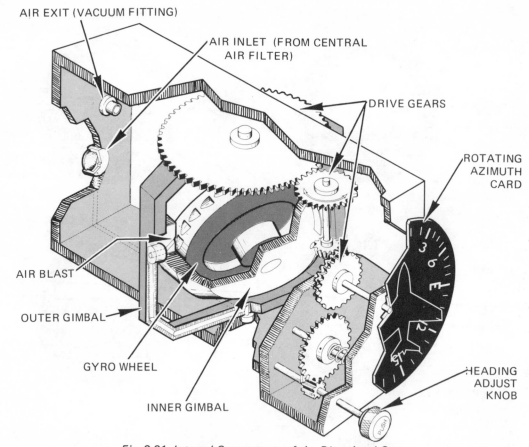

AIR EXIT (VACUUM FITTING)

AIR INLET (FROM CENTRAL AIR FILTER)

DRIVE GEARS

ROTATING AZIMUTH CARD

AIR BLAST

OUTER GIMBAL

GYRO WHEEL

INNER GIMBAL

HEADING ADJUST KNOB

Fig. 2-31. Internal Components of the Directional Gyro

The turn-and-slip indicator is sometimes referred to as the needle and ball, and can be a very important instrument to the pilot in bad weather. It can serve as a means of getting out of bad weather when it is the only gyro instrument in the aircraft.

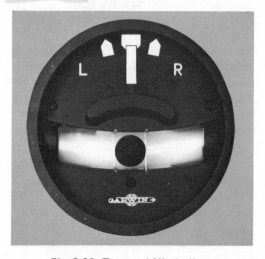

Fig. 2-32. Turn-and-Slip Indicator

The sensitive element of the turn-and-slip indicator is a spinning gyro that is mounted in a frame (more commonly called a gimbal ring). This frame is pivoted on each end and permits the gyro to rotate left and right. This freedom of movement gives the gyro the ability to indicate when the aircraft is turning.

A needle is located at the top to indicate left and right turns. An inclinometer at the bottom is used to indicate whether the aircraft is slipping to the inside or outside.

When the needle of the instrument is aligned with either the mark on the left or the mark on the right of the center-line, the aircraft will turn 180° in one minute, or 360° in two minutes. (See Fig. 2-33.) These reference marks to the left and right of the instrument centerline are commonly called *doghouses* because of their shape.

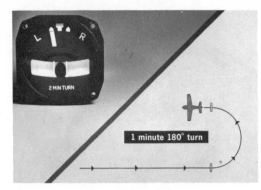

Fig. 2-33. Standard-Rate Turn

The inclinometer consists of a curved glass tube with a small black ball resting in the center. By observing this ball, the pilot can obtain the following information:

1. During a standard-rate turn or any type of turn where the aircraft's controls are properly coordinated, the ball will remain in the center position, indicating to the pilot that the proper forces are reacting on the aircraft.
2. Whenever the ball moves to the outside of the turn, it indicates that the aircraft is in a slip to the outside and that the aircraft controls are not properly coordinated.
3. When the ball moves to the inside of the turn, it indicates that the aircraft is in a slip to the inside and that the proper controls are not being applied to the aircraft.

TURN COORDINATOR

On many of the newer aircraft, the turn-and-slip indicator has been replaced with a *turn coordinator*. (See Fig. 2-34.) In this instrument, a small aircraft silhouette rotates to show that the aircraft is turning. It utilizes an electric gyro that is canted approximately 35°. Due to the canted gyro, the little airplane will bank whenever the aircraft is rotating about either the yaw or roll axis. (See Fig. 2-35.)

When the aircraft is turning left or right the small aircraft silhouette will bank in the direction of the turn. When the wing of the airplane silhouette is aligned with one of the lower index marks, the aircraft is in a standard-rate turn.

Roll rate is also sensed by this instrument because the gyro is tilted on its fore and aft axis, thus providing the pilot with an immediate indication of conditions that lead to a turn. After the bank angle for a turn has been established and the roll rate is zero, the airplane symbol indicates only the rate of turn of the aircraft.

Fig. 2-34. Turn Coordinator

The use of the turn-and-slip indicator or the turn coordinator for turns would be virtually impossible without the use of the aircraft clock. Instrument turns are made by noting the position of the turn indicator and the elapsed time that the aircraft remains in the turn. For example, by holding a standard-rate turn for one minute, the aircraft would turn through 180°.

The inclinometer in the turn coordinator gives the same indications as the inclinometer in the turn-and-slip indicator. (See Fig. 2-36.)

USING THE FLIGHT INSTRUMENTS IN COMBINATION

The flight instrument group can be divided into two sets of instruments: one for controlling the flight path profile (altitude), and the other for direction. The profile instruments, associated with the elevator and power controls, are the altimeter, the airspeed indicator, and the ver-

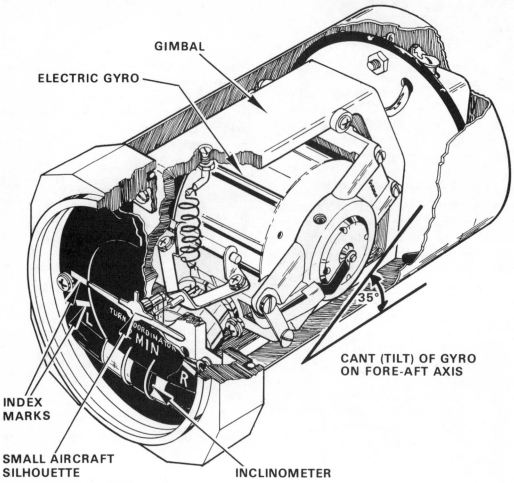

GIMBAL

ELECTRIC GYRO

35°

CANT (TILT) OF GYRO
ON FORE-AFT AXIS

INDEX
MARKS

SMALL AIRCRAFT
SILHOUETTE

INCLINOMETER

Fig. 2-35. Internal Components of Turn Coordinator

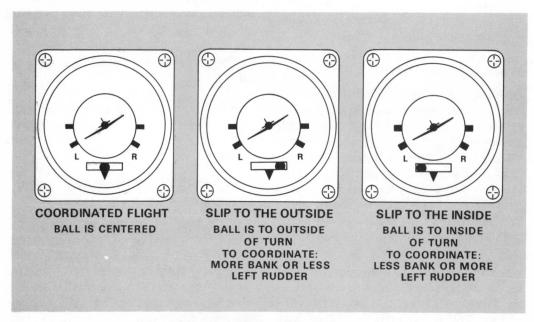

COORDINATED FLIGHT
BALL IS CENTERED

SLIP TO THE OUTSIDE
BALL IS TO OUTSIDE
OF TURN
TO COORDINATE:
MORE BANK OR LESS
LEFT RUDDER

SLIP TO THE INSIDE
BALL IS TO INSIDE
OF TURN
TO COORDINATE:
LESS BANK OR MORE
LEFT RUDDER

Fig. 2-36. Interpretations of the Turn Coordinator

tical speed indicator. The direction of the flight path, associated with aileron or bank control, is controlled by watching the indication of the directional gyro and the turn coordinator or the turn-and-slip indicator.

Each set of instruments is used in conjunction with the attitude indicator and/or outside visual reference. The techniques used for scanning the flight instruments and maneuvering the airplane are the same when flying by outside visual references as when flying by instrument references alone. The difference is that the attitude indicator, to the instrument pilot, replaces what the VFR pilot sees outside the airplane. (See Fig. 3-37.)

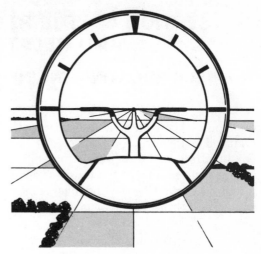

Fig. 2-37. Attitude Gyro Simulates Outside Visual References

SECTION C - THE RECIPROCATING ENGINE AND RELATED SYSTEMS

THE RECIPROCATING ENGINE

A reciprocating engine is the most widely used powerplant on general aviation aircraft today. (See Fig. 2-38 and 2-39.) The combination of the reciprocating engine and propeller is one of the most efficient means of converting the chemical energy of fuel into thrust for sustained flight.

In order to determine the power output of a reciprocating engine, a brake measuring device is attached to the drive shaft. Hence, the term *brake horsepower* (BHP) is used to denote the actual power developed by an engine. The output of any reciprocating engine is a direct function of the combination of engine torque, or twisting force, and rotation speed.

When the reciprocating engine is operated at higher density altitudes, efficiency of the engine diminishes as a result of a smaller volume of airflow being processed in the engine. Therefore, a turbocharger may be added to the engine to compress air before it is injected into the engine cylinders. The turbine on a turbocharged engine may be powered by exhaust gases from the engine. This compression increases air density so that an aircraft engine can operate at high altitudes and still produce the same amount of power as it would at sea level.

Engines designed for aircraft are similar in many ways to automotive engines; however, they are much lighter and a great deal more reliable. Airplane engines are built of relatively expensive materials and, because of their construction and the wide variety of environmental conditions in which they operate, must be operated more intelligently than the engine in a car. One noticeable difference between an aircraft engine and an automotive engine is that practically all aircraft engines are air-cooled rather than liquid-cooled.

COOLING THE ENGINE

The burning of fuel within the cylinder produces intense heat. The engines of a few older aircraft still use liquid coolant but the weight, complexity, and risk of losing the coolant in flight are high prices to pay for the advantages of this kind of engine. Much engine heat is expelled with the escaping gases, or exhaust, but to keep modern aircraft engines safely cooled, outside air must circulate around the cylinders.

The airplane engine is built with thin metal fins projecting from the cylinder

Fig. 2-38. Top View of Four-Cylinder Aircraft Reciprocating Engine

Fig. 2-39. Side View of Four-Cylinder Reciprocating Engine and Propeller Combination

wall allowing heat to be carried away as the air flows past. (See Fig. 2-40.) The cooling air enters the engine compartment through openings in the front of the cowling and is routed over the engine cylinders by tight fitting baffles and expelled through one or more openings in the bottom of the cowling.

Most lower-powered aircraft have a fixed opening in the bottom of the cowling. The size of this opening is determined by the amount of airflow needed to cool the engine during climb conditions when the forward speed is low and the engine power is high. However, during cruise, the aircraft speed is higher than during a climb, and the engine power is at a reduced setting. Thus, more cooling air flows through the cowling with a fixed opening than is needed to adequately cool the engine.

The cooling air flowing over the cylinders causes drag, and the greater the amount of air flowing through the cowling, the greater the drag. Thus, if the volume of air can be reduced during cruise, the drag can be reduced. The cruising speeds of lower-powered aircraft are in the performance range where the increased cooling drag during cruise is not excessive. Therefore, the additional cost of devices to restrict the cooling air is not warranted.

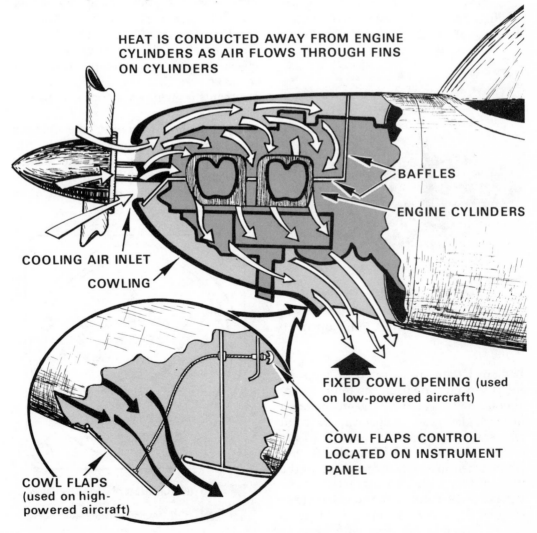

HEAT IS CONDUCTED AWAY FROM ENGINE CYLINDERS AS AIR FLOWS THROUGH FINS ON CYLINDERS

BAFFLES

ENGINE CYLINDERS

COOLING AIR INLET

COWLING

FIXED COWL OPENING (used on low-powered aircraft)

COWL FLAPS CONTROL LOCATED ON INSTRUMENT PANEL

COWL FLAPS (used on high-powered aircraft)

Fig. 2-40. Engine Cooling

Care should be exercised to prevent running the engine at a higher than designed maximum operating temperature, as this will cause loss of power, excessive oil consumption, and detonation. It will also lead to serious permanent damage such as scored cylinder walls, burned and scored pistons and rings, and burned and warped valves. Excessively high oil temperatures can be avoided in aircraft not equipped with cowl flaps by richening the mixture, increasing the airspeed and/or reducing power.

COWL FLAPS

In higher performance aircraft, the cooling drag becomes more critical, and *cowl flaps* (small doors that open out of the cowling) are usually installed to control the flow of cooling air. (See Fig. 2-41.)

Fig. 2-41. Cowl Flaps

The control is provided in the cabin, so the pilot can adjust the cowl flaps to any position desired from fully open to fully closed.

The cowl flaps are adjusted to keep the engine operation within the temperature limitations indicated by the green arc on the cylinder head temperature and oil temperature gauges. The cowl flaps are usually fully open for taxi, takeoff, and climb. When an engine is operating on the ground, very little air flows past the cylinders (particularly if the engine is closely cowled), and overheating is likely to occur, especially on days with high air temperatures. Overheating may also oc-

cur during prolonged climbs because the engine is usually developing high power at a relatively low airspeed.

For cruising flight, the cowl flaps are normally closed to reduce drag and to maintain normal engine operating temperatures. The cylinder head temperature will dictate the amount of opening for the proper operating temperature.

CYLINDER HEAD TEMPERATURE GAUGE

The cylinder head temperature gauge is installed in aircraft equipped with cowl flaps and is located in the engine instrument group. (See Fig. 2-42.) This instrument gives the pilot an accurate indication of engine cooling by measuring the temperature of one of the aft cylinder heads. This instrument can always be checked against the oil temperature gauge to make sure that *neither* of the two instruments is giving erroneous readings.

The pilot should monitor the cylinder head temperature gauge frequently during steep climbs and when the outside air temperature is high. High cylinder head temperatures decrease the life of an

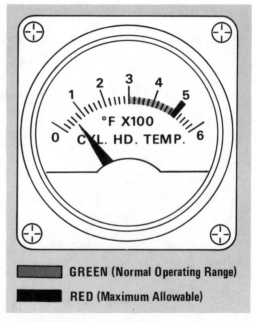

GREEN (Normal Operating Range)

RED (Maximum Allowable)

Fig. 2-42. Cylinder Head Temperature Gauge

engine and should be avoided by richening the mixture and opening the cowl flaps as needed.

CYLINDER BLOCK ARRANGEMENTS

The name for a type of internal combustion engine is derived from the arrangement of the cylinders around a central crankshaft. If the cylinders are arranged directly opposite to each other on the crankshaft, the engine is classified as an *opposed* type. (See Fig. 2-43.) If the

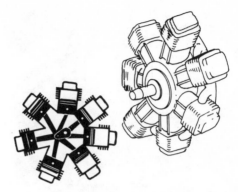

Fig. 2-44. Radial Engine

ENGINE POWER CYCLE

The reciprocating engine, like any other air-breathing engine, can be considered as a pump for air and fuel — the more air and fuel that is pumped through it per minute, the more energy is produced to turn the propeller to produce thrust. This pumping is carried on throughout the engine power cycle. The components of the engine necessary to complete a power cycle are shown in figure 2-45.

The engine power cycle consists of four strokes: *intake, compression, power,* and *exhaust.* (See Fig. 2-46.) The intake stroke occurs when the piston moves away from the cylinder head. During this stroke, the intake valve opens and the fuel/air mixture is drawn into the cylinder by the suction created by the movement of the piston.

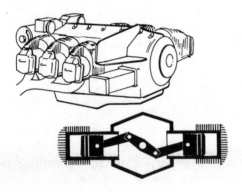

Fig. 2-43. Horizontally-Opposed Engine

cylinders, however, are mounted in a circle around the crankshaft, the engine is called a *radial* type. (See Fig. 2-44.) Most light aircraft engines are of the air-cooled, horizontally-opposed type, with four, six, or eight cylinders.

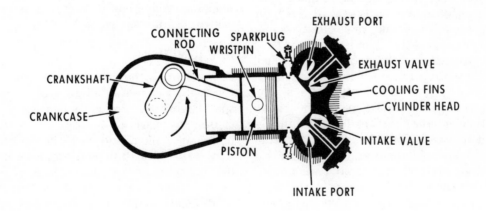

Fig. 2-45. Engine Components

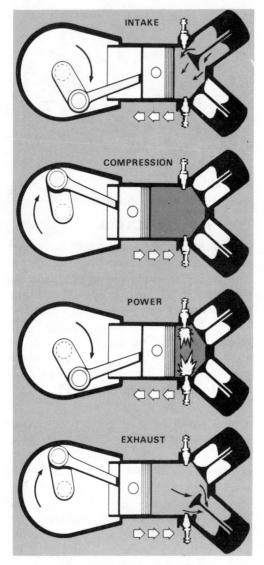

Fig. 2-46. Engine Power Cycle

craft is clockwise (when viewed from the pilot's seat).

The last stroke of the power cycle is called the exhaust stroke and consists of the piston moving toward the cylinder head for the second time in this cycle. The exhaust and burned gases are forced out as the piston returns to the top of the cylinder.

The power cycle just described takes place in each of the engine cylinders at different times thereby creating an *overlap* of power cycles during any single revolution of the propeller. (See Fig. 2-47) The manner in which the pistons are connected to the crankshaft in order to provide *smooth continuous* power to the propeller is also shown.

DETONATION

Detonation is defined as abnormally rapid combustion. (See Fig. 2-48.) If the burning rate of the fuel is too fast, the pressure in the cylinder will build up very rapidly and the remaining mixture will detonate (explode suddenly). Continued operation when detonation is present can result in engine damage.

Some of the principal causes of detonation are low grade fuel, fuel/air mixture too lean, overheated mixture temperature, high cylinder head temperatures, and opening the throttle abruptly when the engine is running at slow speeds. Proper engine servicing and correct engine operating procedures prevent detonation.

PRE-IGNITION

Pre-ignition is the uncontrolled firing of the fuel/air charge in advance of normal spark ignition. Pre-ignition usually is caused by carbon deposits on piston heads, spark plugs, and valves. Pre-ignition may start detonation, and paradoxically, detonation may start pre-ignition. Moreover, pre-ignition can be as destructive as detonation. Proper maintenance and correct operation procedures are necessary for the prevention of pre-ignition.

The compression stroke takes place when the piston moves back toward the cylinder head. The intake valve closes, sealing off the top of the cylinder, and the fuel/air mixture is compressed into a small space to prepare for the next stroke.

The power stroke begins as the piston approaches the closed end of the cylinder and the spark plugs fire. The spark from the spark plug ignites the compressed fuel/air mixture, and a rapid, even burning of the mixture takes place in the combustion chamber, pushing the piston and revolving the crankshaft. The direction of the crankshaft's revolution on most air-

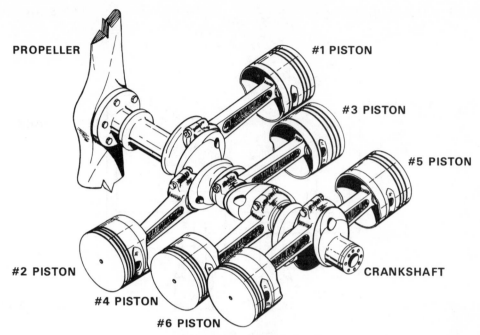

PROPELLER

#1 PISTON

#3 PISTON

#5 PISTON

#2 PISTON

#4 PISTON

#6 PISTON

CRANKSHAFT

Fig. 2-47. Crankshaft and Propeller

IGNITION SYSTEM

The function of the ignition system is to supply a spark to ignite the fuel/air mixture in the cylinders. To do this, the engine is equipped with an ignition system consisting of *two* magnetos, *two* spark plugs for *each* cylinder, ignition leads, and a magneto switch.

The basic source of current in an aircraft ignition system is a magneto wherein a permanent magnet is connected to a shaft rotated by the crankshaft of the engine. The current generated by the magnet is then strengthened by going through a set of breaker points and a secondary coiled circuit. At this point the voltage is approximately 20,000 volts, which is sufficient to jump a normal spark plug gap and ignite the fuel/air mixture in the

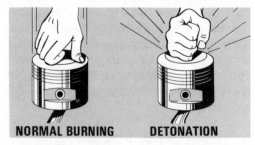

NORMAL BURNING   DETONATION

Fig. 2-48. Normal Burning and Detonation

cylinders. The high voltage produced in the secondary winding is then directed by the distributor to the proper ignition lead where it is conducted to the spark plug in the cylinder in sequence with the firing order of the engine.

The magneto has proved to be a very dependable type of ignition source since it operates completely independent of all other power sources. For aircraft engine applications, however, two magnetos are used, providing a dual ignition system that results in more efficient combustion for increased power and better engine performance. (See Fig. 2-49.) An aircraft engine will run on a single magneto, although a slight loss of engine r.p.m. and power will result. After the engine starts, the master switch can be turned off and the engine will continue to run using the magnetos for ignition.

Each cylinder has two spark plugs — one on top of the cylinder and the other on the bottom of the cylinder. Two plugs are provided for two reasons:

1. SAFETY FACTOR — one magneto fires one plug in each cylinder and the other magneto fires the other set of plugs making it possible for

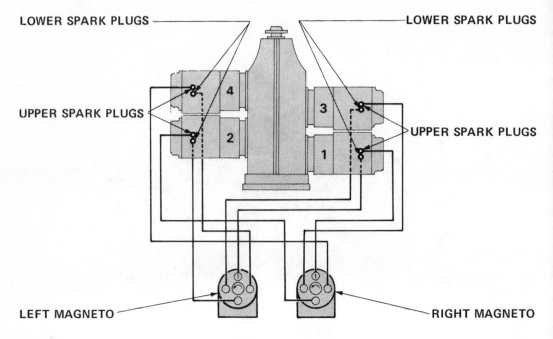

LOWER SPARK PLUGS

LOWER SPARK PLUGS

UPPER SPARK PLUGS

UPPER SPARK PLUGS

LEFT MAGNETO

RIGHT MAGNETO

Fig. 2-49. Dual Ignition

the engine to run on either set. Thus, if one magneto or a spark plug fails, the engine will continue to run safely. This is called dual ignition.

2. MORE EFFICIENCY — two spark plugs cause the fuel/air mixture in the combustion chambers to burn more evenly and produce more power.

The magneto switch is located on the instrument panel and provides the pilot with a means of switching from one of the ignition systems to the other, and back to both systems. (See Fig. 2-50.) It has four positions: OFF, RIGHT, LEFT,

and BOTH. The RIGHT position checks the right magneto and its set of plugs; the LEFT position checks the left magneto and its set of plugs. With the switch in the BOTH position, the engine operates on dual ignition.

A malfunction in the ignition system can be determined by watching the tachometer and noticing the r.p.m. decrease when the system is switched from both magnetos to the right magneto, and then from BOTH to LEFT. The permissible decrease in engine r.p.m. for operation on one set of plugs varies for different engines and the airplane operation manual should be consulted for this value.

If the engine quits completely when switched to one magneto or if it loses more than the specified amount of power, the airplane must not be flown until the problem is corrected. Some of the spark plugs could be fouled by combustion deposits, wiring may have broken or come loose, or the magnetos may not be timed to fire the spark plugs at the proper time.

It is imperative that the pilot remember to turn the ignition switch to BOTH for

Fig. 2-50. Magnetos and Magneto Switch

flight and to turn it completely OFF after shutting down the engine. Even with the master electrical switch turned off, the engine of a parked airplane could fire if the ignition switch were left on and someone were to move the propeller.

Ignition system wiring is shielded to prevent interference with the operation of the aircraft radio equipment. Broken shielding or a loose contact in the ignition system could be heard in the radio receivers as static.

ENGINE LUBRICATION SYSTEM

Oil, used primarily to lubricate the moving parts of the engine, also helps reduce engine temperature by reducing friction and removing some of the heat from the cylinders. The oil is supplied from a sump, or tank, by an oil pump and by splashing caused by the rotating crankshaft. (See Fig. 2-51.) The oil filler cap is readily accessible through an access door in the cowling.

There is also a dipstick for measuring oil quantity. (See Fig. 2-51.) It is important that pilots check oil quantity during the preflight inspection.

Many different types of aviation oil are available. The recommended type and weight are normally placarded on the cowling access door or listed in the owner's manual.

OIL PRESSURE GAUGE

The oil pressure gauge is one of the primary engine instruments and on most panels is found in a grouping with the oil temperature and fuel gauges. (See Fig. 2-52.) It indicates the pressure in pounds per square inch under which the oil of the engine lubricating system is being supplied to the moving parts of the engine. The oil pressure gauge also incorporates color markings to indicate when the oil pressure is in the proper range for safe engine operation. The red line at the lower end of the scale indicates minimum

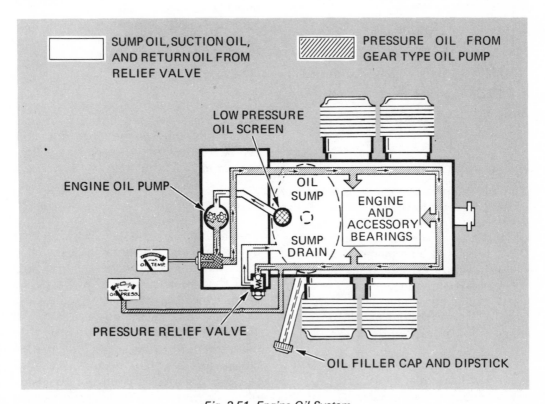

SUMP OIL, SUCTION OIL, AND RETURN OIL FROM RELIEF VALVE

PRESSURE OIL FROM GEAR TYPE OIL PUMP

LOW PRESSURE OIL SCREEN

ENGINE OIL PUMP

OIL SUMP

SUMP DRAIN

ENGINE AND ACCESSORY BEARINGS

OIL TEMP.

OIL PRESS.

PRESSURE RELIEF VALVE

OIL FILLER CAP AND DIPSTICK

Fig. 2-51. Engine Oil System

idling pressure and the red line at the upper end of the scale indicates the maximum allowable oil pressure. The green arc designates the normal operating range.

The oil pressure gauge is one of the first instruments that the pilot should observe after the engine is started. Most aircraft manufacturers recommend that if the oil pressure does not start to rise within 30 seconds in the summer and within one minute in the winter, the engine should be shut down and checked. In cold weather, the oil in the lines congeals and may become very heavy; therefore, a longer period of time is allowed during winter to permit the thick, cold oil to move up into the instrument.

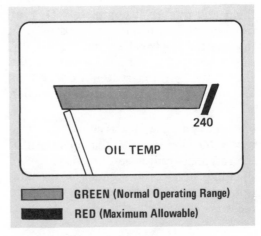

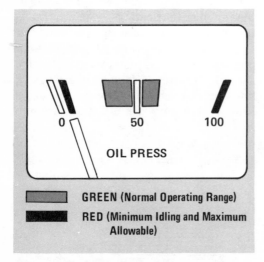

Fig. 2-52. Oil Pressure Gauge

OIL TEMPERATURE GAUGE

The oil temperature gauge is normally located adjacent to the oil pressure gauge on the instrument panel so that they can be observed at the same time. (See Fig. 2-53.) It should be remembered that the oil temperature will not react as quickly as the oil pressure gauge. In cold weather, the oil temperature on some aircraft need not be within the green arc until the takeoff roll is started.

The bulb for measuring the temperature is located in the oil circulation system of the engine on the pressure side of the oil pump, usually at the place where the

Fig. 2-53. Oil Temperature Gauge

oil leaves the oil cooler and enters the engine to be routed to the bearings. The temperature of the oil is measured as it enters the engine, not as it leaves the engine.

The pilot should monitor the oil temperature gauge closely during maneuvers that may cause high engine temperatures. High oil temperatures are usually noted along with high cylinder head temperatures, and any discrepancy in their indications probably indicates that one or the other of the instruments is malfunctioning.

The oil temperature gauge is also color-coded. The upper red line shows maximum allowable temperature for continuous operation. A green arc defines the normal operating range.

THROTTLE

The amount of fuel/air mixture that moves into the engine cylinders is regulated by the throttle valve. (See Fig. 2-54.) A throttle control in the pilot's compartment is connected to this valve to regulate the power output of the engine. Forward movement of the throttle opens the throttle valve and increases the engine speed, while aft movement slows the engine down. (A complete discussion of the carburetor is included in Section D of this chapter.)

Fig. 2-54. Throttle and Throttle Valve

After the fuel and air become mixed in the carburetor, the mixture travels past the throttle valve and through the intake manifold to the engine cylinders where it enters the combustion chambers (engine cylinders) through the intake valves. After the combustion takes place, the exhaust gases leave the combustion chamber on the exhaust stroke. These gases pass through the exhaust valve into the exhaust manifold to the engine muffler and then into the atmosphere through the exhaust pipe.

TACHOMETER

The tachometer measures the rate at which the engine crankshaft is revolving in revolutions per minute (r.p.m.) and is very important in making power adjustments for takeoffs, climbouts, descents, and cruise control. (See Fig. 2-55.) It is calibrated in hundreds of r.p.m.'s and usually incorporates a recording mechan-

Fig. 2-55. Tachometer

ism to keep an accurate record of the engine hours. This recording mechanism is very similar to the mileage indicator on the automobile speedometer.

On an aircraft using a fixed-pitch propeller, the tachometer is the *only* instrument available to the pilot for making power settings. The readings on this instrument are controlled by the throttle when a fixed-pitch propeller is used, or by the propeller control in the case of a constant-speed propeller.

The tachometer is also used when making magneto checks before takeoff or for maintenance checks on the engine. In addition, it will indicate the formation of carburetor ice for an aircraft equipped with a fixed-pitch propeller since the restriction of air flow to the engine cylinders due to carburetor ice causes a drop in engine r.p.m.

The tachometer is color-coded for easy interpretation. The green arc indicates the normal range, the yellow arc shows the caution range, and the red line is the maximum allowable r.p.m. Pilots should operate the aircraft engine at a continuous setting within the green arc.

MECHANICAL TYPE TACHOMETER

The mechanical type tachometer found on most single-engine aircraft uses a flexible driveshaft to transmit the crankshaft rotation to the instrument. The driveshaft is connected directly from the engine to the tachometer and is supported along its routings to prevent it from whipping as the shaft rotates.

MANIFOLD PRESSURE GAUGE

An engine with a fixed-pitch propeller turns at a speed determined by the throttle setting and atmospheric air pressure. With this kind of engine, the pilot has no way to control propeller pitch so the tachometer indications of engine speed are direct indications of power output. When a constant-speed propeller is used, however, the r.p.m. of the engine in cruising flight will remain constant, since the propeller automatically changes pitch to maintain the desired engine speed. In this case, a manifold pressure gauge is a required instrument in order to determine proper throttle settings. (See Fig. 2-56.)

The manifold pressure gauge indicates the engine's power and is controlled by the throttle. The tachometer shows the engine r.p.m. set with the propeller control by the pilot. The two instruments are located in close proximity so that reference can be made to both gauges when making power settings.

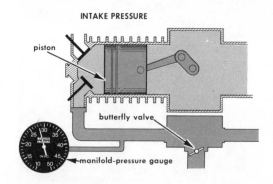

Fig. 2-57. Manifold Pressure Gauge Hook-Up

gauge. When carburetor ice forms, a decrease in manifold pressure results due to the fact that the air flow to the engine is restricted by the formation of ice. (The formation of carburetor ice is discussed in Section D.)

To help the pilot make power settings commensurate with the various flight conditions, the manifold pressure gauge has colored range markings on its dial to indicate the proper range of operation and danger areas.

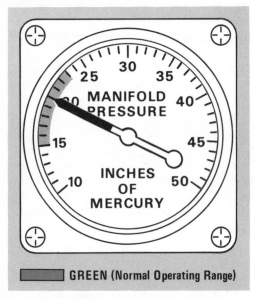

GREEN (Normal Operating Range)

Fig. 2-56. Manifold Pressure Gauge

The manifold pressure gauge indicates, in inches of mercury, the pressure inside the engine intake manifold that carries the fuel/air mixture to the cylinders. (See Fig. 2-57.) When the engine is not running, the manifold pressure gauge will indicate the existing atmospheric pressure. When the engine is running, the pistons create a partial vacuum, causing the manifold pressure to indicate below atmospheric pressure. A super-charger-equipped engine is capable of producing manifold pressure higher than atmospheric pressure.

Carburetor ice can be detected by a drop in the indication of the manifold pressure

AIRCRAFT PROPELLERS

The aircraft propeller functions to convert powerplant brake horsepower into thrust. The propeller achieves high thrust efficiency by processing a relatively large mass flow of air and imparting a relatively small velocity change. The action of the propeller can be visualized by assuming that the rotating propeller is a rotating airfoil. Figure 2-58 shows various cross sections of a typical propeller. It can be noted that the propeller cross sections have a well-defined airfoil shape. As the propeller blade rotates through the air, a low-pressure area is created on the front, or curved, side. Thus, thrust is produced as the airplane is actually pulled into this low-pressure area in front of the propeller.

Each small section of the propeller blade is set at a different angle to the relative wind, resulting in an infinite number of angles of attack. The sections near the

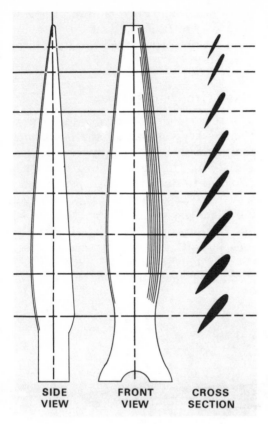

Fig. 2-58. Typical Propeller Blade

SIDE VIEW   FRONT VIEW   CROSS SECTION

tip rotate in a larger arc and travel at a greater speed so the angle of attack is smallest. This gradual decrease in blade angle gives the propeller blade its twisted appearance and maintains a uniform thrust throughout most of the diameter of the propeller.

A propeller with low blade angles, known as a *climb prop*, provides the best performance for takeoff and climb, while a propeller with high blade angles, known as a *cruise prop*, is more adapted to high speed cruise and high altitude flight. Aircraft manufacturers choose propellers that are best for the primary function of the aircraft; therefore, when using a fixed-pitch propeller, some compromises must be made.

The propeller installed in light aircraft may be of either the fixed-pitch variety (see Fig. 2-59) or the constant-speed type (see Fig. 2-60). The term *pitch*, when applied to propellers, should not be confused with pitch as it relates to an airplane's attitude. A propeller's pitch is the number of inches that it would move for-

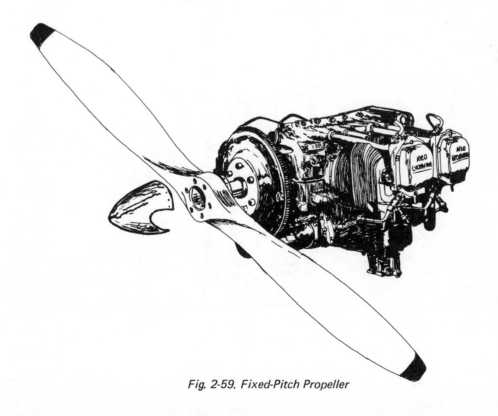

Fig. 2-59. Fixed-Pitch Propeller

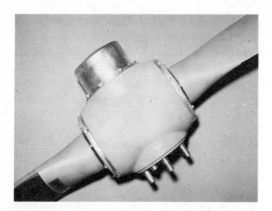

Fig. 2-60. Constant-Speed Propeller

ward in one revolution if it were moving through a solid medium and did not undergo any slippage or loss of efficiency

as it does in air. Pitch is proportional to blade angle (the angle between the chord line of the blade and the propeller's plane of rotation).

FIXED-PITCH PROPELLERS

The fixed-pitch propeller used on most training type airplanes has a blade angle that cannot be changed by the pilot (see Fig. 2-59). The blade angles are normally designed to obtain the best performance while the airplane is *cruising*. However, low blade angles are desirable on propellers for airplanes operating at high altitude airports and on airplanes equipped with floats where takeoff and climb performance is more critical. If the propeller

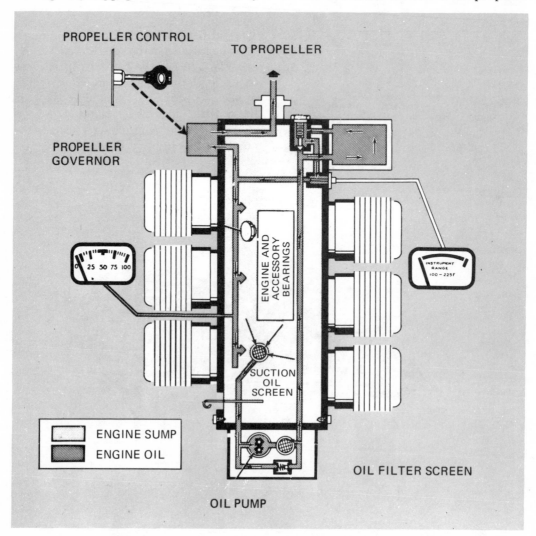

Fig. 2-61. Engine Oil System Schematic with Propeller Governor

is mounted directly to the crankshaft, it will turn at the same speed as the engine, so the throttle is the only power control and the tachometer is the only power indicator required in the cockpit.

CONSTANT-SPEED PROPELLERS

An airplane equipped with a constant-speed propeller has two power controls, a throttle and a propeller control. The throttle controls the power output of the engine, which is registered on the manifold pressure gauge. The propeller control regulates the r.p.m. of the propeller (and also the r.p.m. of the engine, which is registered on the tachometer). The desired r.p.m. is set by the pilot and a propeller governor automatically changes the blade angle to counteract any tendency of the engine to vary from the setting. As manifold pressure is increased as a result of applying more throttle, the propeller governor automatically increases the pitch of the blades to maintain the same r.p.m. The same is true conversely when the throttle setting is decreased.

The propeller control permits a low blade angle for takeoff and climb, and a higher blade angle for cruise. Therefore, a constant-speed propeller is more efficient than a fixed-pitch propeller in the various realms of flight.

Oil is supplied to the propeller governor from the engine under pressure supplied by the engine oil pump. (See Fig. 2-61.) The governor senses engine speed variations and directs oil into and out of the propeller dome as needed to move the propeller blades to maintain the desired engine speed.

Oil enters the dome and moves the piston aft, which rotates the blades to increase pitch whenever the engine tends to rotate at an r.p.m. higher than that set by the propeller control. (See Fig. 2-62.) This causes the engine to slow down since the propeller is taking a bigger "bite." When the engine slows down, the oil flows out of the dome and the piston moves forward. This action rotates the blades to a lower pitch and the engine speeds up.

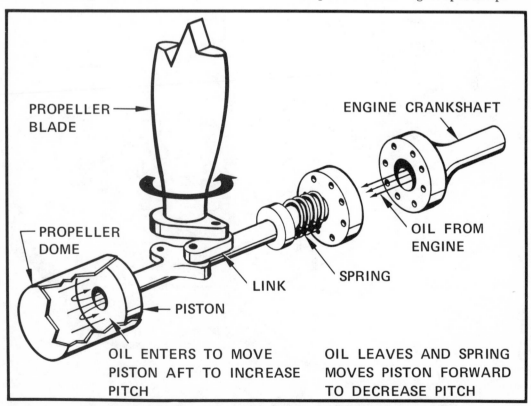

PROPELLER BLADE

ENGINE CRANKSHAFT

PROPELLER DOME

OIL FROM ENGINE

SPRING

LINK

PISTON

OIL ENTERS TO MOVE PISTON AFT TO INCREASE PITCH

OIL LEAVES AND SPRING MOVES PISTON FORWARD TO DECREASE PITCH

Fig. 2-62. Schematic of Typical Constant-Speed Propeller Components

SECTION D - AIRPLANE FUEL SYSTEMS

The primary function of the fuel system is to supply fuel to the engine. The two general types of fuel systems used in to-day's light aircraft are the gravity and fuel-pump types. (See Fig. 2-63 and 2-64.) Most low-horsepower, high-wing aircraft utilize the gravity-feed type fuel system, while most low-wing aircraft and high-horsepower, high-wing aircraft utilize the fuel-pump type system.

TYPES OF FUEL SYSTEMS

GRAVITY-TYPE FUEL SYSTEM

In the gravity-type fuel system illustrated in figure 2-63, the fuel tanks are usually mounted in the wings of high-wing air-craft, which places them *higher* than the engine carburetor. The resultant gravitational pressure from the height of the fuel above the engine is sufficient to supply all of the fuel needs of the engine. Fuel flows by means of gravity from the fuel tanks through a fuel selector valve and fuel strainer into the engine. (See Fig. 2-63, items 13, 12.)

In the carburetor (see Fig. 2-63, item 9), the fuel is mixed with air and the resultant fuel/air mixture is ducted to the engine cylinders through an intake manifold. Fuel quantity gauges, calibrated in gallons

1. Fuel Tank Vent	6. Throttle	11. Strainer (With Sediment Bowl)
2. Fuel Tank Filler Cap	7. Mixture Control	12. Fuel Strainer Drain Valve
3. Fuel Tank Interconnect Vent	8. Fuel Primer Nozzle	13. Fuel Selector Valve
4. Fuel Quantity Transmitter	(At Engine Cylinder)	14. Fuel Line Strainer
5. Fuel Primer Pump	9. Carburetor	15. Fuel Tank Drain Plug
	10. Fuel Quantity Gauges	

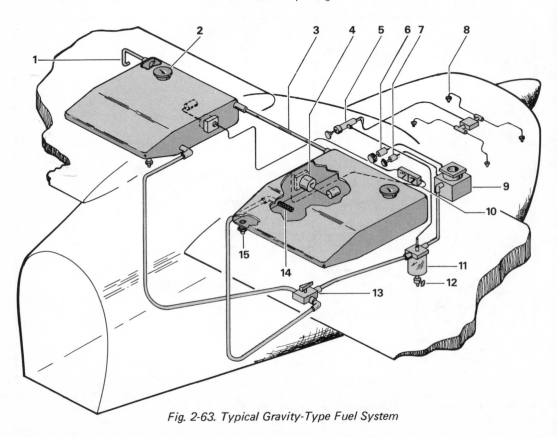

Fig. 2-63. Typical Gravity-Type Fuel System

or pounds, show the amount of fuel in each tank.

FUEL-PUMP TYPE FUEL SYSTEM

The fuel-pump system, shown in figure 2-64, incorporates an engine-driven pump (item 6) instead of gravity to supply fuel pressure. Since the fuel pressure derived from a fuel pump is not dependent upon the height of the fuel tanks above the engine, this system is especially beneficial to low-wing aircraft where the fuel tanks are located in the wings below the engine.

To provide a back-up system for the engine-driven pump and to start fuel flowing under pressure before the engine is running, an electrical fuel-boost pump is installed in the system with other components or submerged in the tanks themselves. The electrical boost pump is operated with a switch in the cockpit and is normally not required except for starting. As is true of components of any aircraft system, however, the operating procedures in the airplane flight manual for the fuel boost pumps should be followed.

FUEL SUPPLY SYSTEM

This system incorporates all of the components which are used to store and transport fuel to the engine. Included are the fuel tank lines, valves, and strainers.

1. Fuel Filler Cap
2. Fuel Tank
3. Fuel Primer
4. Throttle
5. Mixture Control
6. Electric Auxiliary Fuel Pump

7. Engine-Driven Fuel Pump
8. Primer Nozzle (At Cylinder)
9. Carburetor
10. Fuel Quantity Gauges
11. Fuel Strainer
12. Fuel Strainer Drain Valve

13. Fuel Quantity Transmitter
14. Fuel Tank Vent
15. Fuel Line Strainer
16. Fuel Tank Drain
17. Fuel Selector Valve

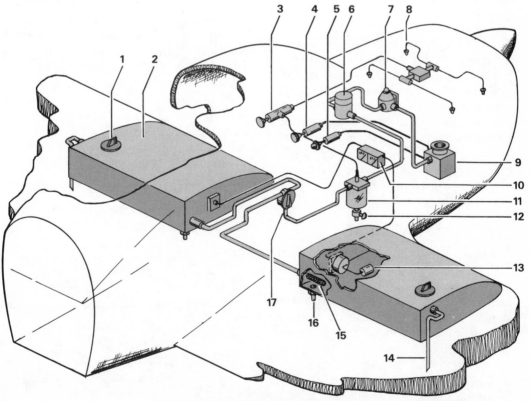

Fig. 2-64. Fuel Pump-Type Fuel System

FUEL TANKS

The fuel tanks are usually mounted in the wings and are serviced through filler caps located on top of the wings. The tanks should be filled each evening before storing the aircraft to prevent condensation of moisture in the tanks. Water and sediment are heavier than fuel and will settle in the lowest areas of the fuel system; therefore, drain plugs or valves are located on the bottom of the fuel tanks and are used to drain water or sediment at 100-hour intervals or when water is found in the fuel strainer.

FUEL TANK VENT AND OVERFLOW DRAINS

In order for fuel to flow from the tanks to the engine, the tanks must be vented to the outside atmosphere. Fuel tank vents may be found in the filler caps, or a tube extending from the wings may function as a vent. (See Fig. 2-63, item 1.) If only one opening to the outside is available for two tanks, then a cross-over vent between the two fuel tanks is supplied.

When the aircraft is parked and exposed to direct heating from the sun, fuel expands in the tanks. An overflow drain must be present to release fuel and prevent the tanks from bursting. This drain may be through the fuel tank vent or a separate opening. Many times, especially after refueling and on hot days when fuel expansion takes place, fuel may be seen dripping from these drains. This is quite normal.

FUEL QUANTITY GAUGE

Another important instrument is the fuel quantity gauge. (See Fig. 2-65.) It should be checked before every flight. In fact, prior to long flights, the pilot should not take for granted that this instrument is correct. A visual check of the fuel should be made.

A fuel-quantity transmitter is mounted in the fuel tank and senses fuel level. In most training aircraft, it is an electrical

Fig. 2-65. Fuel Quantity Gauge

unit and consists of a float, a mechanical linkage, and a transmitter, which contains an electrical resistance strip. The float moves up and down with the fuel level and rotates an arm in the transmitter. This causes the electrical resistance of the circuit to change. Electric wires transmit this electrical value to the fuel gauge which deflects the pointer to indicate the fuel quantity remaining in the tanks. (See Fig. 2-66.)

FUEL SELECTOR VALVES

A fuel selector valve is installed in a fuel system to provide a means for the pilot to select fuel from various tanks for efficient fuel management. There are many kinds of fuel valves, from the simple on-off valves to the complex multiple tank selector valves. A common type of valve used in general aviation aircraft has four positions: left tank, right tank, both tanks, and off. (See Fig. 2-67.) This valve is placed in the cabin of the airplane within easy reach of the pilot.

FUEL STRAINER

After passing from the fuel tank through the fuel selector valve, fuel flows through

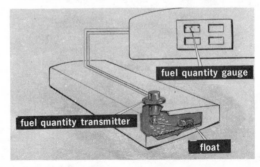

Fig. 2-66. Fuel Gauge System

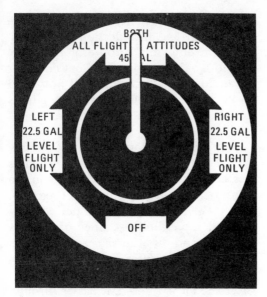

Fig. 2-67. Fuel Selector Valve

a fuel strainer before it reaches the carburetor. The *fuel strainer* is located at the lowest point in the fuel system and is used to trap any foreign particles or moisture that might accumulate in the system.

Since water is *heavier* than gasoline, it will settle out at the lowest point in the system. The location of the strainer at this point, plus the restrictive effect of the filter element, permits the water to be separated from the fuel and collected in a sump bowl at the base of the strainer assembly.

Before flight, the fuel strainer should be drained and checked for the presence of water. In cold weather, the water that accumulates could freeze and stop the flow of fuel. In warm weather, an excessive accumulation of water, if not drained, could flow into the carburetor and stop the engine. If water is found in the fuel strainer, there is probably water in one or more of the wing fuel tank sumps. Although the strainer has the capacity of collecting any water from the tanks before it reaches the engine, it is a good practice to drain enough fuel from the tanks to remove any water that may be present. The pilot should use a *clean*, transparent container to verify the content of drained matter. If an excessive amount of water or solid

matter is found, further checking is called for.

PRIMER

A small hand pump, called the primer, is located on the instrument panel. (See Fig. 2-68.) This device is used to pump fuel *directly* into the engine cylinders prior to starting and is especially useful for cold weather starts. Its sole purpose is to prime the engine so it will be more easily started.

Fig. 2-68. Fuel Primer

FUEL PRESSURE GAUGE

For those fuel systems which use a fuel pump, a fuel pressure gauge is installed to permit the pilot to monitor the fuel pressure. (See Fig. 2-69.) Should the fuel pressure drop, indicating that the engine-driven pump is malfunctioning, the pilot

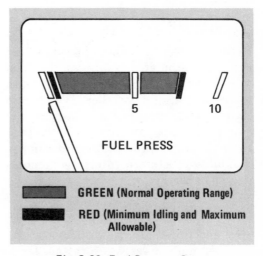

Fig. 2-69. Fuel Pressure Gauge

would switch to the electric fuel boost pump to keep the engine supplied with fuel.

The location of the fuel pressure gauge is shown in figure 2-70. Fuel flows from the fuel tanks to the pump where it is routed under pressure to the carburetor. The fuel pressure gauge is usually connected to the system at the carburetor. From here, a line carries fuel directly to the gauge, where its pressure is displayed on the instrument dial.

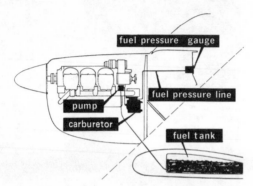

Fig. 2-70. Fuel Pressure Gauge Installation

ENGINE FUEL SYSTEM

This system serves the function of distributing fuel to the engine at the correct time and in the correct amounts. It is made up principally of the carburetor and associated parts, or the fuel injector and associated components.

CARBURETOR SYSTEM

It can be seen in figures 2-63 and 2-64 that the carburetor is one of the final steps in the flow of fuel from the tanks to the combustion chamber. The aircraft carburetor is designed to measure the correct quantity of fuel, mix it with air in the proper proportion, and atomize the mixture before it enters the combustion chamber.

Fuel enters the carburetor at the float valve and flows into the carburetor float chamber until it reaches a certain level. (See Fig. 2-71.) When the float rises to a

predetermined height, it shuts off the float valve, and no additional fuel is allowed to enter the carburetor until fuel is used by the engine. The float chamber is vented so that pressure will not build up in the chamber as the aircraft climbs to altitude.

The outside air, on its way to the engine, first passes through an air filter, which is located at the carburetor air intake in the front of the engine cowling. (See Fig. 2-72.) This filter should be periodically checked for cleanliness since operations on gravel and dirt runways and taxiways eventually cause filter clogging.

After the air is filtered, it passes through a venturi in the carburetor. (See Fig. 2-71.) This venturi creates a low-pressure area in the throat of a carburetor in a manner similar to a vacuum venturi. With the float chamber vented to the atmosphere and a low pressure in the venturi, the fuel is forced to flow by atmospheric pressure from the float chamber to the carburetor throat. The fuel makes its exit in the carburetor throat through the main jet and becomes mixed with the air flowing in from the outside.

MIXTURE CONTROL

A mixture control knob, located on the instrument panel, meters the amount of fuel that passes through the main jet in the carburetor and also regulates fuel

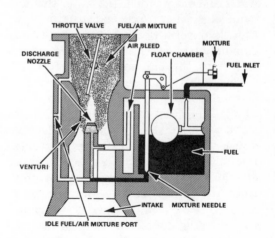

Fig. 2-71. Carburetor Components

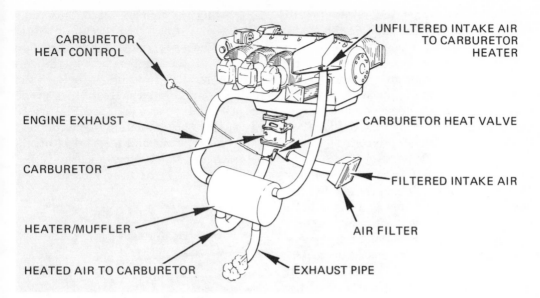

CARBURETOR
HEAT CONTROL

UNFILTERED INTAKE AIR
TO CARBURETOR
HEATER

ENGINE EXHAUST

CARBURETOR HEAT VALVE

CARBURETOR

FILTERED INTAKE AIR

HEATER/MUFFLER

AIR FILTER

HEATED AIR TO CARBURETOR

EXHAUST PIPE

Fig. 2-72. Air Intake System

consumption. (See Fig. 2-73.) The mixture control knob, usually found near the engine throttle on the aircraft panel, will normally be colored red (an indication to use caution). This control enables the pilot to adjust the *ratio* of the fuel/air mixture that goes into the cylinders.

If the fuel/air mixture is too lean (too little fuel for the amount of air in terms of weight), rough engine operation, sudden cutting out, backfiring, detonation, overheating, and appreciable loss of engine power may occur. Lean mixtures must be especially avoided when an engine is operating near its maximum output such as during takeoff, climbs, and go-arounds. At altitudes of less than 5,000 feet, an excessively lean mixture may cause serious overheating and loss of power.

If the fuel/air mixture is too rich (too much fuel for the amount of air in terms of weight), rough engine operation and an appreciable loss of engine power may also occur. Excessive fuel will cause below normal temperatures in the combustion chamber. Since spark plugs need sufficient heat to burn any excess carbon and lead, the plugs may become fouled if the mixture is too rich.

The most common mixture control system type is the needle type. The needle type is illustrated in figure 2-71. By moving the mixture control knob forward, the

Fig. 2-73. Mixture Control Knob

mixture needle retracts and allows fuel to flow freely to the discharge nozzle in the carburetor throat (richer mixture). By pulling the mixture control knob aft, the mixture needle moves into the fuel line to restrict the amount of fuel. With the same amount of air passing through the carburetor throat, the restricted fuel flow causes a leaner fuel/air mixture.

Carburetors are normally calibrated for sea level operation, which means that the correct mixture of fuel and air will be obtained at sea level with the mixture control in the *full rich* position. As altitude increases, the air density decreases, which means that a cubic foot of air will not weigh as much as at lower altitudes. This means that as the flight altitude increases, the weight of air entering the carburetor decreases, although the volume remains the same.

The amount of fuel entering the carburetor depends on the volume and *not* the weight of air. Therefore, as the flight altitude increases, the amount of fuel entering the carburetor remains approximately the same for any given throttle setting if the position of the mixture control remains unchanged. Since the same amount (weight) of fuel is entering the carburetor, but a lesser amount of air, the fuel/air mixture becomes richer as altitude increases.

To maintain the correct fuel/air ratio, the pilot must be able to adjust the amount of fuel mixed with the incoming air as his altitude increases. He accomplishes this by leaning the mixture. Reference should be made to the aircraft operating manual for proper fuel leaning technique.

CARBURETOR HEAT SYSTEM

Carburetor icing can cause loss of power or even engine stoppage in flight. The vaporization of fuel combined with the expansion of air as it passes through the carburetor causes a sudden cooling of the fuel/air mixture. The temperature of the

air passing through the carburetor may drop as much as 70° Fahrenheit. Water vapor in the air is condensed by this cooling, and if the temperature in the carburetor reaches freezing or below, the moisture is deposited as frost or ice inside the carburetor passages. (See Fig. 2-74.) Even a slight accumulation of this deposit will reduce power and may lead to complete engine failure, particularly if the throttle is partly or fully closed.

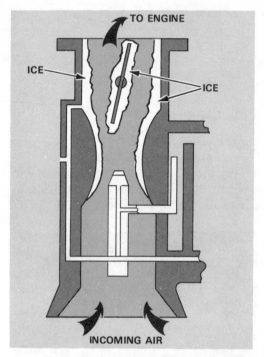

Fig. 2-74. Carburetor Ice

Dry days, or days when the temperature is well below freezing, are conditions *not* conducive to carburetor icing. If the temperature is between 20° and 70° Fahrenheit with visible moisture or high humidity, however, carburetor icing can occur. Carburetor icing is most severe when the temperature and dewpoint approach 68° Fahrenheit. It should be remembered that during low or closed throttle settings an engine is particularly susceptible to carburetor icing because the ice can more easily form on the throttle valve.

The pilot first becomes aware that carburetor ice may be forming by one of two indications. On aircraft equipped with

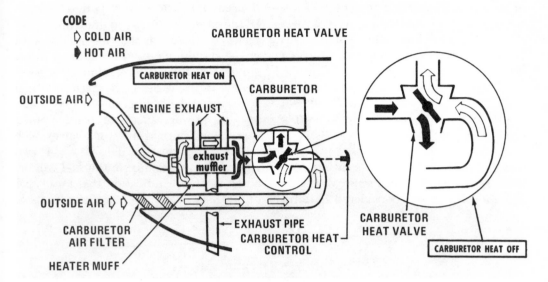

Fig. 2-75. Typical Carburetor Heat System

a fixed-pitch propeller, a loss of engine r.p.m. appears on the tachometer. On aircraft equipped with a constant-speed propeller, a loss of manifold pressure is noted. There is no reduction in r.p.m. in an engine with a constant-speed propeller since propeller pitch is automatically adjusted to compensate for the loss of power, thus maintaining constant r.p.m. In either of the two cases, a roughness in engine operation may develop later.

To eliminate this ice, a carburetor heat system is installed in the airplane. Outside air is ducted through a heater muff where the air is heated to a temperature that melts ice. After the air has been heated, it is routed to a valve that is controlled by a carburetor heat control knob located on the instrument panel. Pulling out the carburetor heat knob opens the carburetor heat valve, allowing hot air to pass through the carburetor where it mixes with the fuel, melts the ice, and continues on to the engine cylinders. (See Fig. 2-75.)

REFUELING

Static electricity formed by the flow of fuel through the hose and nozzle during refueling creates a fire hazard. To prevent a spark from igniting the fuel fumes, a *ground wire* should be attached to the aircraft before the cap is removed from the tank. The refueling nozzle should be grounded to the aircraft before refueling is begun and throughout the refueling process. The fuel truck should also be grounded to the aircraft and the ground. If the pilot must refuel from cans or drums, the fuel should be strained through a chamois skin to prevent contamination from entering the tanks.

USING THE PROPER FUEL

There are several kinds of aviation gasoline, and it can be dangerous to use the wrong kind. Although personnel who service airplanes are usually very careful about this, the *pilot* is the individual responsible for obtaining the proper grade of fuel. The flight manual always specifies the grade of fuel required for the aircraft engine, and this information is marked on placards next to the filler cap.

Each grade of gasoline is manufactured according to certain specifications so that engine designers can tailor their engines to the characteristics of a particular fuel grade. Each grade has an octane rating, or performance number, and is dyed a standard color so pilots and servicing personnel can recognize each of the grades of fuel. A reciprocating engine will not run on turbine fuel which is colorless and smells like kerosene. The various fuels and associated colors are listed below:

80/87	red
100/130	green
115/145	purple
Turbine fuel	colorless

Using aviation fuel which is a rating higher than specified does not improve engine operation but is not considered harmful if used only for a short period of time. If the grade of fuel required for an airplane is not available, the next highest grade should be used — *never a lower grade.* Using aviation gasoline of a lower octane rating is definitely harmful under any circumstances because it may cause loss of power, excessive heat, burned spark plugs, burned and stuck valves, high oil consumption, and detonation. If two grades of fuel are mixed, the fuel will become clear, an indication that two types of fuel have been placed in the same tank.

SECTION E - ELECTRICAL SYSTEMS

Electrical energy is supplied by a 12 or 24 volt direct-current system (12-volt system on most light aircraft) which is powered by an engine-driven generator or alternator. An electrical system schematic, shown in figure 2-76, shows the various circuits and sources of electrical power and can be found in most aircraft owner's manuals. An electrical storage battery serves as a standby power source, supplying current to the system when the generator or alternator is inoperative. The battery is used for actuating the starter and for operation of equipment when the engine is not running.

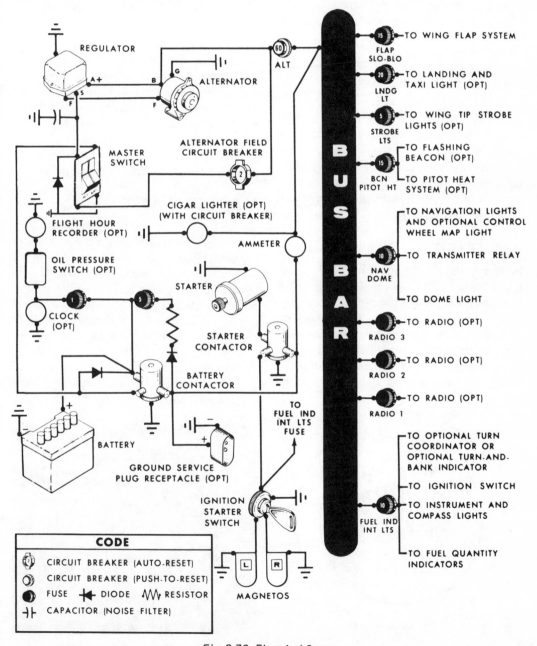

Fig. 2-76. Electrical System

ALTERNATORS/ GENERATORS

Alternators, as opposed to generators, have the advantages of lighter weight, no danger of overload, relatively constant power output (even at engine speeds approaching idle) and lower maintenance. Alternators produce alternating current (AC), which is changed into direct current (DC) for use in the electrical system. This energy is then transferred to a bus bar, which breaks down and distributes the current to the various electrical components of the aircraft. Control of the charging current and voltage is accomplished by a voltage regulator.

An alternator may be damaged by improper technique when attempting a "jumper start" with a low battery; therefore, the flight manual should be checked before any *non-routine* electrical procedure is attempted with an alternator-equipped aircraft.

AMMETER

Probably the most important of the electrical instruments, when it is utilized in the aircraft, is the ammeter. (See Fig. 2-77.) It indicates when the generator or alternator is charging the storage battery. When the pointer is deflected to the right side of the instrument, electrical energy is flowing into the battery from the generator or alternator. When the pointer is deflected to the left side, more electrical energy is being used than is being replaced by the generating system.

Normally, the generator starts charging at 1,000 or 1,200 r.p.m. Alternator systems, however, produce electrical power at a lower engine r.p.m. The aircraft operating manual indicates which system is used in that particular aircraft.

On many single-engine aircraft, a warning light is used in lieu of an ammeter. This light remains off at all times when the electrical-power generating system is functioning properly. It illuminates when the battery or external power is turned on prior to starting the engine and when there is insufficient engine r.p.m. to produce current. It also illuminates if the generator or alternator become defective.

MASTER SWITCH

A master switch controls the entire airplane electrical system with the exception of the ignition system which obtains its electrical energy from magnetos. Newer type aircraft have a split-type battery/alternator switch so that each electrical supply unit can be separately checked. (See Fig. 2-78.) Turning the master switch or battery/alternator switch to "on" supplies electricity to all electrical equipment circuits in the airplane. Some equipment which may use the electrical

Fig. 2-77. Ammeter.

Fig. 2-78. Split Type Battery/Alternator Switch

system as their source of power are landing lights, taxi lights, navigation lights, flashing beacon, instrument lights, dome lights, starter, pitot heater, stall warning indicator, cigarette lighter, radios, turn coordinator, fuel gauges, and fuel boost pumps.

FUSES/CIRCUIT BREAKERS

Electrical fuses or circuit breakers are provided in the various electrical circuits to protect them from electrical overloads. (See Fig. 2-79.) Spare fuses are carried in the pilot's compartment in the event

a fuse must be replaced in flight. In the case of circuit breakers, resetting the breaker *usually* will reactivate the circuit unless an overload or short exists. If an overload or short occurs, the circuit breaker will continue to pop, indicating an electrical problem.

A pilot is expected to know the location of the circuit protection device for each electrical component. Normally, the name and amperage for the circuit are written near the appropriate circuit breaker or fuse.

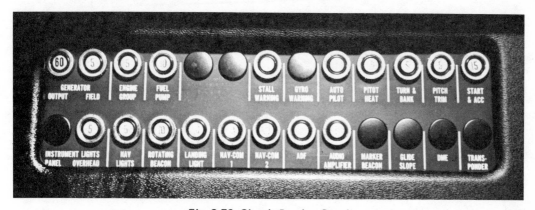

Fig. 2-79. Circuit Breaker Panel

chapter 3

AIRPORTS, COMMUNICATIONS and AIR TRAFFIC CONTROL

INTRODUCTION

In most cases, the airplane, unlike many other vehicles, has a highly predictable base of operations — the airport. Generally, flights originate and terminate at the airport and a portion of the fledgling pilot's new environment is the physical setting and characteristics of the particular airport at which he is taking his flight training. The knowledge gained from a study of airports will enable the pilot to quickly feel at home while operating in this new environment.

Furthermore, in order for the pilot to gain the greatest utility from his aircraft, it is important for him to understand radio communications. Today, the major portion of air traffic control is based on voice communications, making the mastery of this subject an important endeavor for the pilot.

SECTION A - AIRPORTS

There are more than 10,000 airports within the borders of the United States that serve all types of aviation activities. These range from large metropolitan airfields with several runways serving a variety of aircraft operations, to small grass strips suitable only for light aircraft. (See Fig. 3-1.) Learning about airports is an important aspect of flight training.

RUNWAY NUMBERING

Each runway is assigned a number which is determined from the magnetic direction of the runway. The runway's magnetic direction is rounded off to the nearest ten degrees and the last zero is omitted. For instance, the approximate magnetic direction of runway 21 is 210°, while runway 3 has a magnetic direction of approximately 030°. The actual magnetic direction for runway 3 could be any number from 025° to 034°, all of which, when rounded off to the nearest ten degree heading, would be 030°.

The runway number is usually displayed on the approach end of each runway in large numerals that are easily read from the air. The runway number, of course, is different at each end of a runway strip because the magnetic directions are 180° apart, or reciprocals of each other. For example, a runway aligned with magnetic north and south is called runway 36 on the end where landings are to the north, and runway 18 on the end where landings are made to the south.

The one or two number system is used for two reasons. Using one or two numbers instead of three makes the runway name shorter and easier to say. Also, this practice permits larger numerals on the end of the runway, which makes it easier to identify the runway from the air. (See Fig. 3-2.)

Fig. 3-1. Simple and Complex Airports

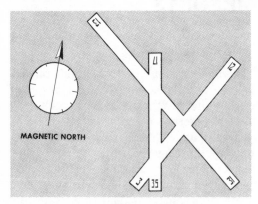

Fig. 3-2. Airport Runway Markings

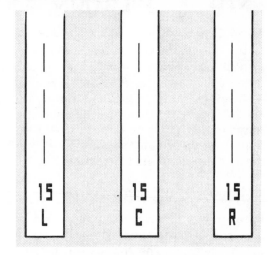

Fig. 3-3. Parallel Runways

If parallel runways exist, they must still have the *same* number since their directions are the same. For this reason, a suffix letter is added to differentiate between runways. For two parallel runways, the letter "L" designates left, and the letter "R" indicates right. In the case where three parallel runways exist, the letters "L," "C," and "R" are used, with the letter "C" standing for center. Reading from left to right, the runways shown in figure 3-3 are called "one five left," "one five center," and "one five right."

It is important to remember that runways are numbered in relation to the *magnetic* North Pole and not true north. Thus, the magnetic compass in the aircraft should agree with the runway designation (within ±5 degrees) when the aircraft is aligned with the runway. Also, navigation aids such as VOR stations, radio beacons, and instrument landing systems are laid out according to magnetic direction. This permits the various courses, emanating from these facilities, to be shown on charts in alignment with the runway used for an instrument approach. In addition, wind directions reported by control towers are given with reference to magnetic north so that a comparison between wind and the runway direction can be made.

ACTIVE RUNWAYS

The term *active runway* refers to the runway in use at a particular time. This is normally determined by wind direction. At some locations, however, a calm wind runway is designated when the wind is below a specific velocity, such as five miles per hour. In this case, the calm wind runway is used for all takeoffs and landings regardless of wind direction. Winds of higher velocity, however, will normally dictate the use of a different runway.

RUNWAY MARKINGS

BASIC RUNWAY MARKINGS

The *basic markings* on runways used for VFR (visual flight rules) operations consist of runway direction numbers and centerline marking. In addition to runway numbers located on the approach ends, centerlines are added to help the pilot align his aircraft with the runway during landings and takeoffs. (See Fig. 3-4.) At airports utilizing sod runways, the ends are often marked with L-shaped corner threshold markers to indicate when the aircraft is over the end of

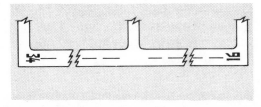

Fig. 3-4. Basic Runway Markings

the runway. (See Fig. 3-5.) In addition, runway numbers may be absent from the ends of grass, dirt, or sod runways, or indicated by inlaid white stones, concrete, or other material that will contrast with the surface.

Fig. 3-5. Sod Runway Markings

NON-PRECISION INSTRUMENT RUNWAY MARKINGS

Specific markings are added to the basic runway markings to designate additional information. For example, a *threshold marker*, consisting of two sets of four parallel marks, indicates the runway is served by a non-precision approach facility (such as a VOR station) and is intended for landings in instrument conditions, as well as for VFR landings. (See Fig. 3-6.)

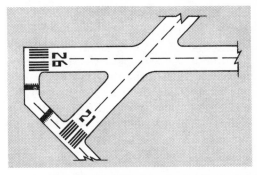

Fig. 3-6. Non-Precision Instrument Runway Markings

PRECISION INSTRUMENT RUNWAY MARKINGS

Touchdown zone marking, fixed distance marking and side stripes are normally added

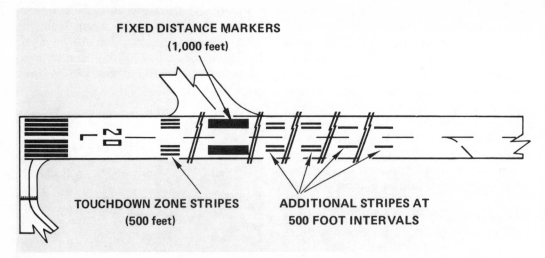

**FIXED DISTANCE MARKERS
(1,000 feet)**

**TOUCHDOWN ZONE STRIPES
(500 feet)**

**ADDITIONAL STRIPES AT
500 FOOT INTERVALS**

Fig. 3-7. Precision Runway Marking

to the runway markings to designate *precision instrument runways* which are served by a precision approach facility, such as an instrument landing system (ILS). (See Fig. 3-7.) The 500-foot spacing of these stripes can be used by the pilot to determine how much runway has been used on landing or takeoff. For example: touchdown zone stripes mean 500 feet have passed; fixed-distance markers—1,000 feet; first set of two stripes—1,500 feet; second set of two stripes—2,000 feet; first set of single stripes—2,500 feet; and the last set of stripes—3,000 feet.

DISPLACED THRESHOLDS

The *threshold* of a runway is sometimes displaced from what would appear to be the beginning of the runway. The threshold line is used to mark the start of the usable portion of the runway, and a pilot should not land until he has crossed this mark. The purpose of a displaced threshold is to insure that the initial landing touchdown is made farther down the runway.

The obvious question that comes to mind is, "If it is not to be used, why was the runway extended that far?" It must be remembered, however, that runways are used in both directions. As an overrun or a landing rollout area from the op-

posite direction, displaced thresholds are perfectly satisfactory.

Most thresholds are displaced because of obstacles at the end of the runway that necessitate additional clearances when approaching to land. The obstacle may be a road, trees, or towers. An obstacle may have been added, or a hazard may have developed after the runway was built. In most cases, however, the obstacle was there when the runway was built, and the runway was extended up to the obstacle so that the additional length could be used in a landing rollout from the other direction.

Displaced thresholds for basic VFR and instrument runways are marked as shown in figure 3-8. Note that the no-landing portion of each runway has arrows painted on the centerline to point to the threshold line. The top illustration in figure 3-8 shows two examples of displaced thresholds on VFR runways. The centerline arrows direct the pilot's attention to the displaced threshold. In addition, arrowheads are painted adjacent to the threshold lines to emphasize the beginning of the usable portion of the runway. These arrowheads are used only on VFR runways.

An example of a displaced threshold on an instrument runway is shown in the

Fig. 3-8. Displaced Threshold Marking

lization to prevent erosion. This stabilization may have the appearance of full strength pavement, but is not intended to support aircraft. Usually, the taxiway edge marking will define this area, but confusion may exist as to which side of the side stripe the full strength pavement is located. Where such a condition exists, the stabilized area is marked as shown in figure 3-9.

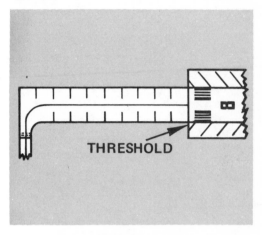

Fig. 3-9. Stabilized Area

lower picture in figure 3-8. Centerline arrows are also utilized on the no-landing portion to help the pilot locate the runway threshold. However, arrowheads are not used adjacent to the threshold, because the parallel marks designating an instrument runway provide adequate visual emphasis.

STABILIZED OR DECEPTIVE AREAS

Holding bays, aprons, and taxiways are sometimes provided with shoulder stabi-

OVERRUN, STOPWAY, AND BLASTPAD AREAS

Overrun, stopway, and blastpad areas of runways are marked with a chevron pattern as shown in figure 3-10. These areas are not structurally capable of withstanding loads created by landing, taking off, or taxiing aircraft. The shoulders or edges of some runways are also marked with chevrons. This means they are strictly stabilized shoulders or understrength areas of the runway, and are not meant to withstand landing loads.

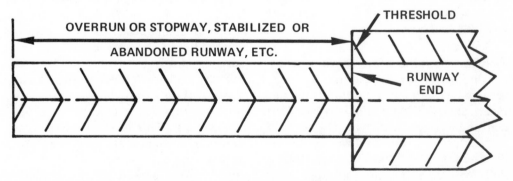

Fig. 3-10. Overrun and Stopway Areas

CLOSED RUNWAYS

A runway or taxiway that is *closed* to all operations is marked with a large "X" at each end. If the entire airport is closed, each runway could be "Xed" out, or the operator could place a large "X" near the wind tetrahedron, or in the center of the segmented circle. (See Fig. 3-11.)

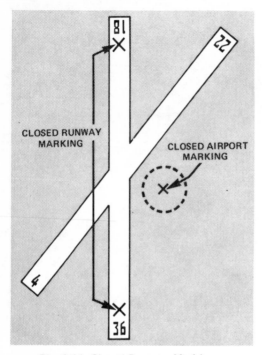

Fig. 3-11. Closed Runway Markings

TAXIWAYS

Taxiways are very important to airport operations, enabling traffic to move to and from runways without interfering with aircraft taking off or landing. These taxiways lead from the runways to aircraft storage, tiedown, maintenance, and fueling areas on the airport. The *open white* portion of figure 3-12 shows typical taxi routes.

Taxiway centerlines are marked with a continuous yellow line, while the edges are marked with two continuous lines six inches apart. Holding lines consist of two continuous lines and two dashed lines, perpendicular to the taxiway centerline. Standard holding lines appear 100 to 200 feet from the runway edge. However, at runways equipped for Category

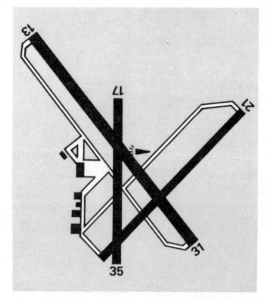

Fig. 3-12. Taxiways (White Areas)

II operations, the CAT II holding lines are two solid lines perpendicular to the taxiway centerline. These are connected by solid lines parallel with the taxiway, as shown in figure 3-13.

The placement of Category II holding lines is determined by the localizer, glide slope, and obstacle critical areas. The objective of these lines is to keep taxiing aircraft far enough away from the navigation facilities and the runways to eliminate any interference with the landing aircraft. When Category II operations are being conducted, taxiing aircraft will be advised by ground control.

PARKING AREAS

Parking areas are provided on airports and are used by transient and local unhangared aircraft. Servicing of aircraft for fuel and oil is also performed in this area. Ropes or chains attached to tiedown rings embedded in concrete are used to tie the airplane down and to keep it from overturning in high wind conditions. An alternate method of providing tiedowns is to string a long cable and attach ropes or chains to it. This method accomodates a large variety of aircraft sizes. A secure tiedown should be accomplished whenever an aircraft is parked outside and left unattended.

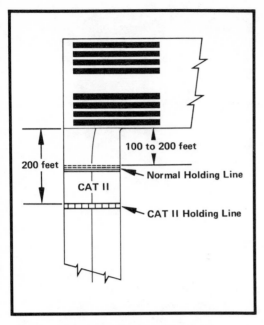

Fig. 3-13. Taxiway Markings

specified on the various navigation charts in feet above mean sea level (MSL). Before takeoff, the pilot should set the current altimeter setting in the Kollsman window. By doing this, the altimeter should read very close to field elevation. (See Fig. 3-14.) In addition, pilots should also check the field elevation prior to landing to determine the altitude used in the airport traffic pattern.

WIND DIRECTION INDICATORS

Since the wind direction determines the runway to be used for landings and takeoffs, there must be a means for the pilot to determine the direction from which the wind is blowing. Various devices are used for this purpose. During takeoffs and landings, airplanes fly as directly into the wind as possible.

FIELD ELEVATION

Airport elevation information is very important to the pilot. Airport (field) elevation is measured at the highest usable landing surface on the airport and is

WIND SOCK

The oldest and most common wind indicating device is the wind sock. The wind sock is nothing more than a cone-shaped

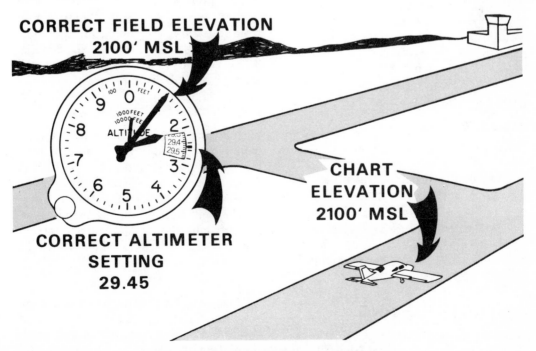

Fig. 3-14. Altimeter Setting Prior to Takeoff

Fig. 3-15. Wind Sock

Fig. 3-16. Tetrahedron

device built of durable, flexible material. When the wind blows through the large end of the cone, it causes the *small end* to stand out and point *downwind.* The amount of extension can also give an indication of the velocity of the wind. In addition, a fluttering wind sock indicates gusty conditions. Normally, lights are placed above the wind sock for night illumination. (See Fig. 3-15.)

TETRAHEDRON

The tetrahedron is a more elaborate type of wind indicator. This mechanism streamlines itself with the wind flow causing the *pointed end* to always turn *into* the wind. This is sometimes confusing to pilots because the pointed end of the wind sock points downwind, while the pointed end of the tetrahedron points upwind. It is important for pilots to exercise care when interpreting each of these indicators. (See Fig. 3-16.)

WIND TEE

Some locations use a wind tee to indicate wind direction. (See Fig. 3-17.) The tail of the airplane streamlines with the wind, and the wing-like structures face the wind. Like the tetrahedron, this device also points into the wind, or upwind. At most airports, there is at least one wind sock, plus either a tetrahedron or a wind tee.

AIRPORT LIGHTING

Airport *runway lights* are *white* and will illuminate the runway sufficiently for

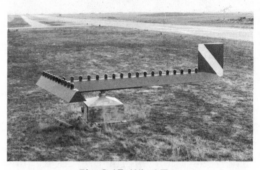

Fig. 3-17. Wind Tee

runway alignment. *Green* lights are placed on the *threshold* of the runway to show a pilot the beginning of the usable landing surface at night. When viewed from the opposite direction, the same threshold lights are seen as *red.* (See Fig. 3-18.) *Taxiway lights,* on the other hand, are *blue* which distinguishes them from runway lights.

With all the other lights on the ground, it would be difficult to find an airport at night without the *rotating beacon* used to designate the airport. The airport beacon rotates in a clockwise direction at a constant speed, thus producing the visual effect of light flashes at regular intervals. The rotation speed provides 12 to 15 flashes per minute.

Airport beacons incorporate color combinations which indicate various types of airports. For example, white and green alternating flashes denote a lighted land

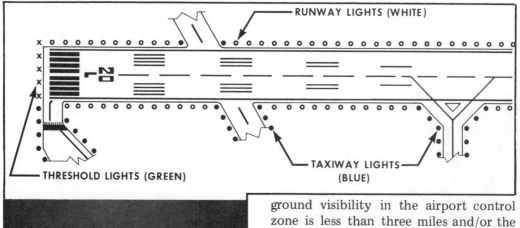

Fig. 3-18. Airport Runway Lights

airport (see Fig. 3-19), while white and yellow flashes inform the pilot that the beacon serves a lighted water airport. Military airport beacons flash *alternating* white and green, but are differentiated from civilian airport beacons by two quick white flashes between the green flashes. (See Fig. 3-19.)

Operation of the airport rotating beacon during daylight hours indicates that the airport is below VFR (visual flight rules) landing minimums. Therefore, the

ground visibility in the airport control zone is less than three miles and/or the ceiling is less than 1,000 feet. (This concept is explained in Chapter 5.) This means an ATC (air traffic control) clearance is required for landings and takeoffs at airports that are surrounded by controlled airspace. During the hours of darkness, instrument or IFR (instrument flight rules) conditions are indicated by flashing lights outlining the tetrahedron or wind tee.

A *flashing amber light*, near the center of the segmented circle or on top of the wind sock, control tower, or adjoining buildings, means that a right traffic pattern is in effect for the active runway.

Red beacons, with a flash rate of 12 to 40 flashes per minute, are used to mark flight hazards such as towers or tall buildings.

APPROACH LIGHTS

Various types of *approach lights* are installed on the extended centerline of run-

Fig. 3-19. Rotating Beacons

ways equipped with instrument landing systems to guide the pilot during IFR operations. These are especially helpful for runway alignment after the pilot descends through the clouds during an instrument approach. (See Fig. 3-20.)

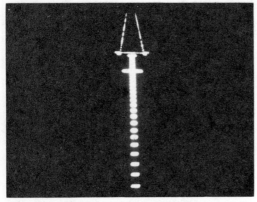

Fig. 3-20. Approach Lights

VASI SYSTEM (Visual Approach Slope Indicator)

A *visual approach slope indicator* system (VASI) is located on some runways to help the pilot maintain a normal glide path to a safe touchdown point on the runway. This system is *only* usable in VFR conditions or after a pilot executing an instrument approach is clear of clouds.

The VASI system consists of a set of upwind and downwind lights that indicate to the pilot when he is on the proper glide path. When approaching the runway at the proper angle, the downwind set of lights (those closest to the pilot) on both sides of the runway will appear white and the upwind lights (those farthest from the pilot) will be red. (See Fig. 3-21.) By maintaining a rate of descent to the runway corresponding to the VASI glide slope (normally $2\frac{1}{2}^{\mathrm{O}}$ to 4°), the airplane will touch down midway between the two sets of lights.

If the aircraft strays above the glide slope, the upwind lights will change to pink and then to white. When *well above* the approach path, both sets of lights will be illuminated white.

If the aircraft descends below the glide path, the downwind lights will change to pink and then to red. When operating *well below* the glide path, the upwind and downwind lights will appear red. In this situation, the pilot can expect to land short of the normal touchdown point. (See Fig. 3-21.)

AIRPORT TRAFFIC

The airport traffic pattern is very important as a means of separating aircraft using the same airport. A flight pattern is set up around each runway for use by approaching and departing traffic. An airport would be a disorganized, chaotic snarl if there were no means for approaching the field and flying a standard pattern. Traffic patterns are normally left; i.e., turns are made to the left. (See Fig. 3-22.) Nonstandard patterns require turns to the right. (See Fig. 3-23.)

TAKEOFF LEG

The airport traffic pattern can be broken down into various segments. After take-

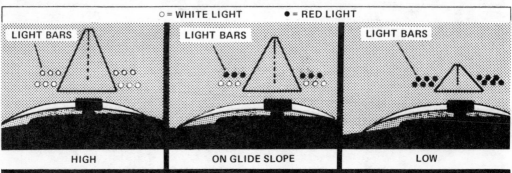

Fig. 3-21. The VASI System

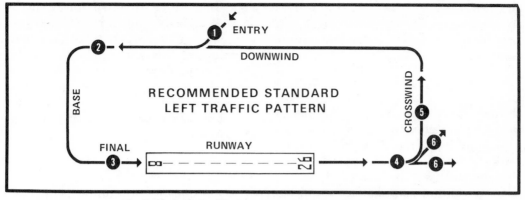

Fig. 3-22. Left Traffic Pattern at Uncontrolled Airport

1 Enter pattern in level flight, abeam the midpoint of the runway, at pattern altitude.

2 Maintain pattern altitude until abeam approach end of the landing runway, on downwind leg.

3 Complete turn to final at least 1/4 mile from runway.

4 Continue straight ahead until beyond departure end of runway.

5 If remaining in the traffic pattern, commence turn to crosswind leg beyond the departure end of the runway, within 300 feet of pattern altitude.

6 If departing the traffic pattern, continue straight out, or exit with a 45° left turn beyond the departure end of the runway, after reaching pattern altitude.

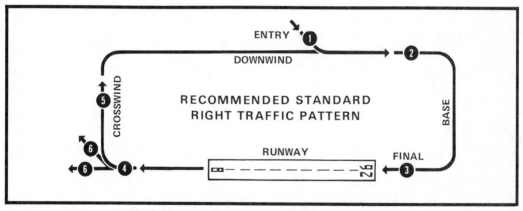

Fig. 3-23. Right Traffic Pattern at Uncontrolled Airport

off, it is recommended that the pilot climb straight ahead until within 300 feet of pattern altitude. This is called the *upwind or takeoff leg.*

CROSSWIND LEG

The *crosswind leg* is flown perpendicular to the takeoff leg and is flown after completing a 90° turn. When a pilot has established an appropriate interval behind any aircraft, he turns 90° and continues the climb to the traffic pattern altitude. The pilot should plan to be at pattern altitude before turning to the downwind leg.

DOWNWIND LEG

If the pilot desires to remain in the traffic pattern, he will turn another 90° to place the aircraft on the *downwind leg.* This leg is parallel to the runway, but flown in the opposite direction of the intended landing. Normally, the downwind leg is flown at 1,000 feet above the ground, but this height can vary to accommodate individual airport requirements.

The 1,000 foot AGL (above ground level) altitude is called the traffic pattern altitude. To determine the MSL (mean sea level) altitude, the 1,000 foot AGL pattern altitude should be added to the field elevation. This MSL altitude is the altitude that will be indicated on the altimeter while flying the downwind leg.

When the aircraft is opposite the touchdown point, the pilot reduces power, slows the aircraft to the speed used for the approach, and begins to descend for the landing.

BASE LEG

The 90° turn from downwind to a flight path *perpendicular* to the runway places the pilot on the *base leg*. The descent to the runway is continued on this leg and most final landing checks are performed.

FINAL APPROACH

After the pilot makes his descending turn from the base leg, his flight path should be aligned with the runway centerline This portion of the traffic pattern is known as the *final approach.*

TRAFFIC PATTERN ENTRY

Traffic pattern entry at airports with an operating control tower is specified by the tower operator. At uncontrolled airports, traffic pattern altitudes and entry procedures vary according to established local procedures. Pilots should familiarize themselves with these procedures and follow local flight instructor recommendations. At unfamiliar airports, airport advisory service or UNICOM, when available, should be utilized for receipt of traffic pattern and landing information.

TRAFFIC PATTERN DEPARTURE

When a control tower is in operation, the pilot can request and receive approach for a straight-out, downwind, or right-hand departure. This departure request should be made while asking for takeoff clearance. At airports without an operating control tower, the pilot must comply with the departure procedures established for that airport. These procedures usually are posted by airport operators so pilots can become familiar with the local rules.

SEGMENTED CIRCLE

A segmented circle is normally located around the wind sock, tetrahedron, or wind tee at *uncontrolled* airports. The use of L-shaped markers on the periphery of the circle indicates the direction of traffic patterns for the various runways on the airport. Figure 3-24 shows the segmented circle corresponding to two airport runways illustrating how L-shaped markers distinguish left and right-hand traffic patterns.

The leg of the "L" pointing inward toward the circle can be thought of as the final approach for the runway of corresponding direction. The leg perpendicular to the "final approach" leg of the "L" can be likened to the base leg of the appropriate runway. By noting whether the "L" marker indicates a left or right turn from "base" to "final," the pilot can determine the direction of traffic flow for a particular runway. For example, aircraft taking off to the east (runway 9) would use a left traffic pattern and aircraft landing to the north (runway 36) should fly a right pattern (See Fig. 3-24.)

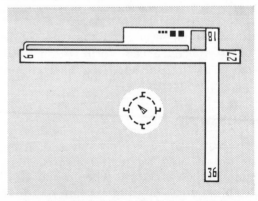

Fig. 3-24. Segmented Circle and L-Markers

SECTION B - RADIO COMMUNICATIONS

COMMUNICATIONS SYSTEMS

Through the use of diagrams, it is possible to study the flow of two-way communications. Figure 3-25 shows that voice signals travel from the microphone to the transmitter, where they are changed to radio signals. The radio signals are then carried by a wire to the antenna, where they are broadcast in all directions and received by radio receivers in control towers or other radio stations.

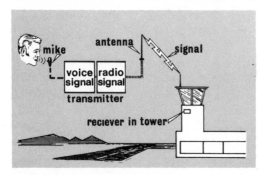

Fig. 3-25. Transmitting

Figure 3-26 shows that aircraft radio receivers are very similar to the radios utilized in a person's home and automobile. Radio signals are broadcast or transmitted from the transmitting station, such as a control tower, and received by the antenna on the airplane. From here, the signals are carried to the radio receiver where they are converted to voice signals which are heard by the pilot, either through earphones or through a speaker system.

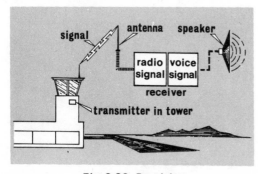

Fig. 3-26. Receiving

VHF COMMUNICATIONS EQUIPMENT

Airborne radios are different from automobile and home radios in that airborne communication requires two-way conversation. Therefore, aircraft radios generally consist of a *transmitter, receiver, antenna, microphone,* and *speaker* or *headset;* home and auto radios only provide a receiver with associated components such as an antenna and speakers.

When the transmitter and receiver are housed in one unit and have only one or a common frequency selector as shown in figure 3-27, the radio is known as a transceiver.

Fig. 3-27. Combined Transmitter and Receiver Unit

There are two communication terms with which pilot should become familiar. Single channel simplex communications involve using the same frequency for both transmitting and receiving. Dual channel or double channel communication consists of using a different frequency for transmitting than for receiving. (See Fig. 3-28.)

TUNING

Most modern VHF radios incorporate crystal-controlled tuning. With this system, the pilot automatically receives the desired frequency once he selects the proper setting on the tuning dial. (See Fig. 3-29.)

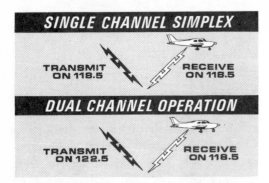

Fig. 3-28. Communication Terms

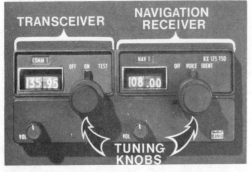

Fig. 3-29. Crystal-Controlled Tuning

Modern aircraft radios have what is commonly known as a 1½ system. The tuning knob on the left side of the radio is normally marked "COM," indicating that both a transmitter and receiver are provided. This is single channel simplex communication capability, or in other words, the "one" part of the 1½ system.

The frequency range of the "COM" side of the radio is from 118.0 to 135.95 MHz in the case of a 360 channel radio transceiver, or 118.0 to 126.9 MHz in the case of a 90 channel transceiver.

The right side of the radio has another tuning knob which controls frequencies from 108.0 to 117.9 MHz. This knob tunes only a receiver (the "one-half" part of the 1½ system) to be used in the reception of VOR and ILS navigation signals or VOR voice reception. The pilot *cannot* transmit on a frequency selected on the right side of the radio.

USING THE MICROPHONE

The first and most basic factor in good communication between pilots and con-

trollers is the proper use of the microphone.

The hand microphone is the most common type found in general aviation training aircraft. (See Fig. 3-30.) In order to obtain clear, readable transmission, most hand microphones should be held in a position so that the lips are lightly touching the face of the mike as shown in figure 3-31. Using the mike in this position allows the pilot transmitting to use a normal conversational volume level, and also prevents the transmission of most background noise such as that from the engine.

Also, the person transmitting should speak *directly* into the mike. Even though the microphone is held close to the lips, the transmission will sound fuzzy and indistinct if the words are spoken across the face of the mike.

The radio microphone has a key or button which must be depressed during trans-

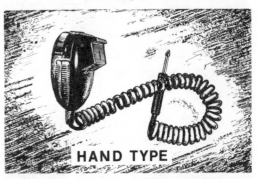

Fig. 3-30. Transmitting Microphone

Fig. 3-31. Correct Microphone Handling Technique

(ICAO) INTERNATIONAL PHONETIC ALPHABET

Letter	Code	Word	Pronunciation	Letter	Code	Word	Pronunciation
A	.—	Alfa	(Al-fah)	N	—.	November	(No-vem—ber)
B	—...	Bravo	(Brah-voh)	O	———	Oscar	(Oss-cah)
C	—.—.	Charlie	(Char-lee)	P	.——.	Papa	(Pah-pah)
D	—..	Delta	(Dell-tah)	Q	——.—	Quebec	(Keh-beck)
E	.	Echo	(Eck-oh)	R	.—.	Romeo	(Row-me-oh)
F	..—.	Foxtrot	(Foks-trot)	S	...	Sierra	(See-airrah)
G	——.	Golf	(Golf)	T	—	Tango	(Tang-go)
H	....	Hotel	(Hoh-tell)	U	..—	Uniform	(You-nee-form)
I	..	India	(In-dee-aH)	V	...—	Victor	(Vik-tah)
J	.———	Juliett	(Jew-lee-ett)	W	.——	Whiskey	(Wiss-key)
K	—.—	Kilo	(Key-loh)	X	—..—	X ray	(Ecks-ray)
L	.—..	Lima	(Lee-mah)	Y	—.——	Yankee	(Yang-key)
M	——	Mike	(Mike)	Z	——..	Zulu	(Zoo-loo)

Fig. 3-32. (ICAO) International Phonetic Alphabet

missions and released once the message is complete. This key triggers a switch in the radio transmitter which enables the radio to transmit voice signals on the selected frequency as long as the mike key is held in the depressed position. The pilot should be careful *never* to depress the transmitter key when another pilot or controller is transmitting on the same frequency. Doing this is known as *cutting out* another transmission. All that will be heard on the frequency when two stations are transmitting at the same time is a loud *squeal.*

RADIO PHRASEOLOGY

A pilot's world is a small one. Within a few hours, an airplane can travel hundreds of miles, and in certain parts of the world, international boundaries can be crossed many times in a single day. To overcome the language barriers, the various countries of the world organized the International Civil Aviation Organization (ICAO) and adopted English as the international aviation communications language. This international organization also adopted a phonetic alphabet to be used when making radio transmissions.

The phonetic alphabet is shown in figure 3-32 and consists of words in lieu of letters. This substitution is necessary because many letters of the English language sound alike. For example: B, C, D, E, T, V, and Z could be confused with one another during a radio transmission.

SPECIAL RADIO PHRASES

Due to the large number of radio-equipped aircraft, some radio frequencies have become very crowded, making it highly desirable to reduce the length of transmissions. For this reason, the following radio phraseology has been developed to aid in decreasing transmission time.

Radio Phrase	Meaning
ACKNOWLEDGE	Let me know that you have received and understood this message.
AFFIRMATIVE	Yes.
CORRECTION	An error has been made in this transmission. The correct version is . . .
GO AHEAD	Proceed with your message.
HOW DO YOU HEAR ME	Self-explanatory.

I SAY AGAIN .... Self-explanatory.

NEGATIVE .... That is not correct.

OUT......................... This conversation is ended and no response is expected. The word *out* is infrequently used since most often it is obvious when the conversation has terminated.

OVER .... My transmission is ended and I expect a response from you. The word *over* is omitted if the message obviously needs a reply.

READ BACK .... Repeat all of this message back to me.

ROGER .... I have received all of your last transmission. (This word is used to acknowledge receipt and should not be used for other purposes.)

SAY AGAIN .... Repeat what you have said.

SPEAK SLOWER................ Self-explanatory.

STAND-BY .... If used by itself means "I must pause for a few seconds." If the pause lasts longer than a few seconds, or if the phrase is used to prevent another station from transmitting, it must be followed by the ending *OUT*.

THAT IS CORRECT .... Self-explanatory.

VERIFY .... Check with originator.

WORDS TWICE .... a. As a request: "Communications are difficult. Please say every phrase twice."
b. As information: "Since communication is difficult, every phrase in this message will be spoken twice."

NUMBERS

Sometimes numbers also are difficult to understand in radio transmissions; therefore, they always should be enunciated clearly. Single-digit numbers are pronounced in the same way as they sound except for the number "nine" which is spoken as *niner* in radio conversations.

The ICAO adopted this substitute pronunciation because the sound of "nine" in German means "no."

Two or more digit numbers are generally spoken as single-digit numbers in series. Such as:

37 — *three seven.*

485 — *four eight five.*
0956 — *zero niner five six.*

VHF and UHF radio frequencies are spoken as follows:

122.6 — *one two two point six.*
118.7 — *one one eight point seven.*

For a frequency in the low/medium band, the word kiloHertz is added after the numbers. For example:

337 — *three three seven kiloHertz.*
242 — *two four two kiloHertz.*

An exception to this practice is the identification of airways. For example: *Victor Twelve* or *Victor Thirteen East* are both VHF airways.

Since altimeter settings always consist of two digits followed by a decimal point and two more digits, the decimal point is dropped. For example:

29.95 would be stated as *two niner niner five* and 30.15 would be reported as *three zero one five.*

Round numbers up to 9,000, such as for ceiling heights, flight altitudes and upper wind levels are spoken in accordance with the following examples:

800 — *eight hundred*
1,300 — *one thousand three hundred*
4,500 — *four thousand five hundred*
9,000 — *niner thousand*

Numbers above 9,000 are spoken by separating the digits preceding the word thousand. Some examples are:

10,000 — *one zero thousand*
18,500 — *one eight thousand five hundred*

The pronunciation of numbers also is involved when expressing time. To simplify this type of communication when there is no danger of confusion, time is stated as minutes past the hour such as *one zero* meaning ten minutes past the hour; or *four three* meaning 43 minutes past the hour.

TWENTY-FOUR HOUR CLOCK SYSTEM

The 24-hour clock system is used to eliminate the confusion resulting from the 12-hour system duplicated in both the a.m. and p.m. An expression of time using the 24-hour system consists of a four-digit number. The first two digits indicate the hour and the last two figures represent minutes past the hour. The hours are numbered consecutively from midnight to midnight, or from zero to 24. Some examples of time involving the a.m. are as follows:

0000 — *zero zero zero zero* (midnight)
0715 — *zero seven one five* (7:15 a.m.)
1200 — *one two zero zero* (noon)

Time in the p.m. is expressed according to the following examples:

1409 — *one four zero niner* (2:09 p.m.)
2259 — *two two five niner* (10:59 p.m.)

Since a pilot in an aircraft is capable of passing through several time zones during the course of a flight, a universal method of time was adopted to simplify the problems resulting from time changes. The time used for flights, flight plans, communications, etc., is the time at Greenwich, England, a city located on the zero degree line of longitude. This time is referred to as *Greenwich Mean Time* (GMT) and *Zulu Time.*

Because the sun rises over Greenwich, England, when it is still night time in the United States, the time is always later when expressed in Zulu than when expressed in local time in any time zone in the United States. To convert from the local time, then, it is necessary to add the appropriate number of hours to the local time to determine GMT or Zulu time. The location of the time zones in the United States is illustrated in figure 3-33.

To convert from one time to another, a pilot may use the following tables:

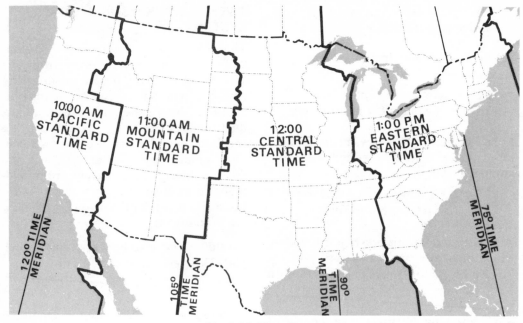

Fig. 3-33. Time Zones

To Convert From	To GMT
Eastern Stand. Time....	add 5 hours
Central Stand. Time....	add 6 hours
Mountain Stand. Time..	add 7 hours
Pacific Stand. Time	add 8 hours

The 24-hour clock system is used for local and Greenwich times. When a pilot requests local and Greenwich time from an FAA facility, he will hear Greenwich time stated first and then the local time. For example: time 0532 CST (1132 GMT) would be spoken as: *time one one three two Greenwich; zero five three two Central.* Current time checks, when requested, are stated to the nearest quarter minute.

FREQUENCY UTILIZATION

Radio frequencies are identified by a numerical value, depending upon the number of cycles per second (Hertz) made by the specific frequency. Due to the very high number of cycles involved, it is convenient to use the term *kiloHertz (kHz)* and *MegaHertz (MHz)* when referring to frequencies. One thousand cycles per second equals one kiloHertz and 1,000,000 cycles per second denotes one MegaHertz.

Radio frequency ranges refer to a group of frequencies with an upper and lower cycle limit. There are four frequency ranges which are utilized in aircraft communications. They are: L/MF (low/medium frequency range); HF (high frequency range); VHF (very high frequency range); and UHF (ultra high frequency range).

L/MF BAND

Figure 3-34 shows the frequency allocation chart as determined by the Federal Communications Commission. Notice that the low frequency range is from 30 to 300 kHz, and the medium frequency range extends from 300 to 3,000 kHz. Lower frequencies allocated by the FCC for aircraft communications extend from 190 to 415 kHz, including portions of both the low and medium frequency

Fig. 3-34. Low/Medium Frequency Band

TELEVISION AND FM RADIO

TELEVISION
54-87 MHz.

FM RADIO
88-107 MHz

GENERAL AVIATION

INSTRUMENT
LANDING SYSTEM (ILS)
108.0-111.9 MHz

VHF NAVIGATION
STATIONS (VOR)
108.0-117.9 MHz

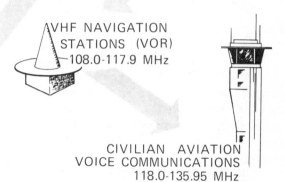

CIVILIAN AVIATION
VOICE COMMUNICATIONS
118.0-135.95 MHz

Fig. 3-35. Radio Frequency Bands in General Use in Aviation

range. This aircraft band is referred to as the L/MF band.

HF BAND

The high frequency (HF) band includes frequencies from 3,000 to 30,000 kHz. Although many frequencies in this range have been allocated to aviation, the use of high frequencies has gradually decreased to the point that the high frequency band is no longer considered an important system of communication for *domestic* operations.

VHF BAND

The very high frequency (VHF) band is the *most* important portion of the frequency spectrum to the *private pilot* and includes frequencies from 30 MHz to 300 MHz. For comparison, the relationship of the popular FM radio band and certain TV channels (two through six) to the aviation frequencies is shown. (See Fig. 3-35.)

The frequency range of 108.0 MHz through 117.9 MHz is used for air navigation aids, such as the instrument landing

system (ILS), VHF navigation stations (VOR), and provides simultaneous voice reception over these frequencies. The frequency range allotted for civilian aviation voice communication is from 118.0 MHz to 135.95 MHz. It should be emphasized that *the VHF band is by far the most commonly used in general aviation.*

UHF BAND

The ultra high frequency (UHF) band includes frequencies from 300 MHz to 3,000 MHz. This band of frequencies has been reserved for use by various government agencies including the military. In addition, distance measuring equipment (DME), and the glide slope (vertical guidance) portion of the ILS system operate with signals in the UHF band.

CHARACTERISTICS AND LIMITATIONS OF RADIO SIGNALS

Knowledge of the characteristics and limitations of the various frequency bands used by pilots is important in order to obtain the greatest efficiency from radio equipment. The low/medium and high frequencies can be bounced back from

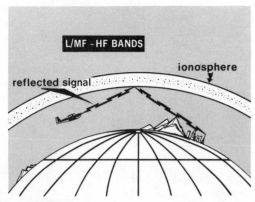

Fig. 3-36. Signal Bounce from Ionosphere

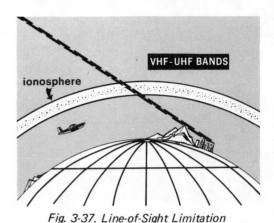

Fig. 3-37. Line-of-Sight Limitation

the ionosphere (a shell of electrically charged particles surrounding the earth), and are not affected by intermediate obstruction or line-of-sight restrictions. (See Fig. 3-36.) This frequency band provides its best advantage to the pilot for low altitude, long range reception of navigational signals and weather information. The greatest disadvantage is static, especially in areas of precipitation.

The greatest disadvantage of the VHF and UHF frequencies is the line-of-sight limitation. Any obstruction, including the curvature of the earth, will interfere with VHF communications. The line-of-sight feature restricts the reception distance of VHF and UHF signals for low altitude

communications and navigation. (See Fig. 3-37.)

The normal reception distances for VHF frequencies, listed below, are based upon the radio line-of-sight signal and assumes there are no intervening obstructions between the transmitter and receiver. Examples of various reception distances are tabulated below:

Feet Above Ground Station	Reception Distance (Nautical Miles)
1,000	39
3,000	69
5,000	87
10,000	122
15,000	152
20,000	174

SECTION C - AIR TRAFFIC CONTROL

Knowing how to use the various facilities provided by the FAA for air traffic control is one of the most important areas of aeronautical knowledge. The air traffic control (ATC) system is based upon voice communications between ATC controllers and pilots, necessitating an expansive network of communication facilities. These facilities can be divided into terminal and en route services, both of which are important for providing traffic separation and weather information.

AIRPORT TERMINAL FACILITIES

AUTOMATIC TERMINAL INFORMATION SERVICE

Automatic Terminal Information Service (ATIS) is recorded information provided at many major air terminals to give the arriving and departing pilot advance information on active runways, weather conditions, communication frequencies, and any Notices to Airmen affecting the airport at a particular time. ATIS has lessened ATC controller workloads by making it unnecessary for controllers to continually relay routine information. Instead, this information is recorded on a tape and transmitted continuously over an ILS, VOR, or other commonly used frequency in the airport areas.

As airport conditions, such as wind direction or velocity, altimeter setting or active runway change, a new tape is made to reflect the new conditions. Each ATIS recording is coded with a phonetic letter of the alphabet (example: *information Alpha* or *information Delta),* and the pilot, on initial call to the tower, ground control, or approach control should state that he has received the current information. The following is an example of a typical ATIS broadcast:

"This is Stapleton International Airport Information Quebec. Denver weather, measured ceiling one five thousand overcast, visibility five miles in smoke and haze. Wind two five zero degrees at six, altimeter two niner eight seven. ILS runway two six left approaches in use. Landing runway two six left for large aircraft; runway two six right for light aircraft. Tower frequency one one eight point three. Departures are on runway three five, tower frequency one one niner point five. Advise Denver Tower, Approach Control, or Ground Control on initial contact that you have received Information Quebec."

ATIS frequencies for airports incorporating the system can be found in the Airport/Facility Directory in Part 3 of the *Airman's Information Manual.* (See Fig. 3-38.)

On sectional charts the ATIS frequency is grouped with other elements of airport data. (See Fig. 3-39.)

```
§ STAPLETON INTL IFR  5E                    FSS:  DENVER on Fld
  5330  H115/17-35(3) (S-125, T-175, TT-300)
  BL5,6,8A,9,10,11,14,15  S5  F12,18,22,30,34  Ox1,2,3,4  U2
  VASI:  Rnwys 8R, 17   REIL:  Rnwy 17.   RVR:  Rnwy 17, 26L, 35
  Remarks:  U.S. Customs indg rgts arpt.  Rgt tfc rnwy 35, 26R,
    26L.  Rnwy 8R-26L GWT (S-200, T-200, TT-360).  Rwy 17-35
    clsd Mon & Thur 0730-0830 lcl.  Rnwy 17-35 lmtd to 285,000
    lbs.
  Denver Tower  118.3¹ 119.5⁵ 122.4R 126.9     Gnd Con  121.9
    ‡Clrnc Del:  125.3
  ATIS:  110.3
  Radar Services: (BCN)
    Denver App Con  119.3¹ 110.3 120.5² 116.3T 108.1T
    Denver Dep Con  124.8 126.9 122.4R 110.3T
    Stage I  Ctc App Con 20NM out
  ASR
  ILS  110.3  I-DEN  Apch Brg 257°    LOM:  362/DE
    108.1  I-SPO  Apch Brg 350°
  Denver (H) VORTAC  116.3 DEN  214°  8.1NM to fld.
    VHF/DF  Ctc twr
  Remarks:  ¹270-089°  ²090-269°.  ¹For lndg.  ⁵For tkof.  VOT:
    111.0
```

Fig. 3-38. ATIS Frequency in AIM

Fig. 3-39. ATIS Frequency Shown on Sectional Chart

Fig. 3-40. Tower Controller

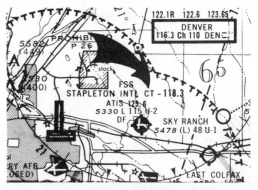

Fig. 3-41. Control Tower Radio Frequency Listing

CONTROL TOWER

After listening to ATIS, the next facility with which the arriving pilot would make contact is the *control tower* or local controller. (See Fig. 3-40.) After completing the pre-landing checklist, the radio transmitter and receiver would be switched to the appropriate tower frequency.

Here is an example of how an arriving aircraft should contact the tower after receiving ATIS service:

"Stapleton Tower, this is Cessna 1375 Golf. One zero miles northeast, landing with Information Quebec."

Towers are normally responsible for the control of all airborne traffic within a specified area, both landing and taking off.

During VFR conditions, the tower is in control of all the VFR traffic within the airport traffic area (five statute miles radius including airspace up to, but not including 3,000 feet above the airport). The IFR traffic may be controlled by *approach control* and *departure control* in this area. In addition, prior to taxiing onto an active runway, a pilot is *required* to receive takeoff clearance, or clearance to cross the active runway from tower personnel.

Tower frequencies can be found in several different sources. They are located in the Airport/Facility Directory (Part 3) of the *Airman's Information Manual* and can also be found in the radio frequency listing on aeronautical charts, as shown in figure 3-41.

With a transceiver-type radio, the pilot should select the appropriate tower frequency on the transceiver tuner. When using radios that have separate transmitters and receivers, and without the capability of transmitting on the assigned tower frequency, the pilot may transmit to most control towers on 122.5 MHz and listen on the published control tower frequency. In addition, 122.4 or 122.7 MHz can be used when listed for the particular airport.

In the event of two-way radio communication failure, every control tower has one or more powerful light guns which can transmit an intense, narrow beam of red, green or white light. This light gun has a sight somewhat similar to that on a rifle or shotgun and, providing the pilot is looking in the direction of the control tower, the controller can transmit a limited amount of instructions. (See Fig. 3-42.) Even though it is possible to fly for years without ever having to depend on light gun signals, the pilot should commit the signals and their meanings to memory. (See Fig. 3-43.)

GROUND CONTROL

After landing, the next frequency that would be utilized by an arriving pilot is the *ground control* frequency. The

Fig. 3-42. Controller Operating Light Gun

ground controller is located in the tower cab and is responsible for the separation of traffic moving on the *airport surface.*

The transmission to ground control would be similar to this:

"Stapleton Ground Control, Waxwing Seven Niner Eight Zero Whiskey, off two six right, taxi instructions to general aviation tie-down area."

Not every airport has a ground control frequency; however, due to the increase in traffic movement in recent years, it has become much more common at tower-equipped airports to utilize a specific ground control frequency. The frequency most commonly used for ground control is 121.9 MHz. Occasionally, 121.7, 121.8, and 121.6 MHz are used where there is a possibility of interference from another ground control station nearby.

Knowing that specific frequencies are assigned for controlling the movements of aircraft on the ground, the pilot must determine which frequency is used at each airport. Ground control frequencies are not published on aeronautical charts; however, these frequencies may be located in Part 3 of the *Airman's Information Manual.*

Ground control communications can be very useful in providing the pilot with valuable information. For example:

1. Ground control can provide the pilot with precise taxi instructions to the runway of intended takeoff.
2. It can give him information concerning any hazards that might exist along his route across the airport.
3. Ground control can supply transient pilots who are unfamiliar with the facilities, tiedown, and ramp service or maintenance. In addition, ground control can also give directions to various other places on the airport.

APPROACH AND DEPARTURE CONTROL

Some airports have approach and departure controls located at the tower facili-

Color and Type of Signal	On the Ground	In Flight
Steady Green	Clear for takeoff	Cleared to land
Flashing Green	Cleared to taxi	Return for landing (to be followed by steady green at proper time)
Steady Red	Stop	Give way to other aircraft and continue circling
Flashing Red	Taxi clear of landing area (runway) in use	Airport unsafe - do not land
Flashing White	Return to starting point on airport	(No Assigned Meaning)
Alternating Red and Green	General warning signal	Exercise extreme caution

Fig. 3-43. Light Gun Signals

ty. (See Fig. 3-44.) At other airports, however, approach and departure control are located separately from the tower. These facilities provide traffic separation through the use of radar and are most commonly used for IFR operations.

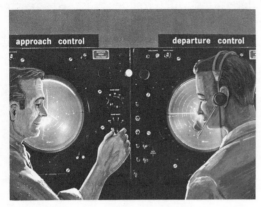

Fig. 3-44. Approach and Departure Control

At some busy terminals where it is desirable to exercise control of traffic until they leave the airport traffic area, the control tower will sometimes request departing VFR airplanes to contact departure control and arriving aircraft to contact approach control. The departure and approach control frequency will be given at the time the pilot is told to change frequencies.

Arriving and departing VFR aircraft may also *request* Stage 1 Service (radar advisory service for VFR aircraft) and it will be provided if IFR traffic load allows the controller to provide the service. Departure and approach control frequencies are listed in the Airport/Facility Directory (Part 3) of the *Airman's Information Manual.*

EN ROUTE VFR COMMUNICATIONS

FLIGHT SERVICE STATION

A major communications facility which the VFR pilot uses quite extensively is the *flight service station* (FSS). Flight service station attendants provide pilots with all types of flight information such as current altimeter settings, advisories on en route weather conditions which might be encountered, and any other information which would be of help during a cross-country flight. (See Fig. 3-45.) In addition, flight service stations can supply complete preflight weather briefings.

Fig. 3-45. FSS Communicating with Aircraft

Several radio navigation facilities may be *remotely controlled* by one flight service station. The controlling flight service station and its communication and navigation frequencies at each location are shown on the sectional charts and may be found in the *Airman's Information Manual.* The standard FSS communication frequencies and the codes used to indicate the available frequencies are listed in the "Radio Aids to Navigation" legend on the front panel of each sectional chart.

It should be noted that on the actual sectional chart, VHF radio frequencies and facilities are shown in blue, and information pertaining to low and medium frequency facilities is shown in red.

The flight service station communication frequency codes can be interpreted in the following manner: (See Fig. 3-46.)

1. An information box with a heavy border line (item 1) indicates a Flight Service Station. The standard FSS communication simplex frequency of 122.2 and the emergency frequency of 121.5 are available, unless otherwise noted.
2. Communication frequencies (item 2), in addition to the standard FSS

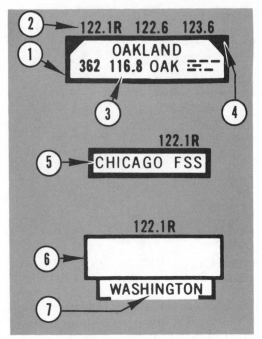

Fig. 3-46. FSS Frequency Codes

heavy line box containing only the name of the FSS (item 5).

6. An information box with a thin line border (item 6) denotes a navigation facility without a Flight Service Station or a navigation facility controlled by a remote FSS. A box without frequency and remote FSS notations indicates that no FSS communication is available.

7. The FSS controlling a remote navigation facility is shown in a bracket (item 7) under the thin line information box.

One common method used in contacting flight service stations is to transmit on frequency 122.1 MHz and receive the FSS on the VOR frequency. On the initial call to a flight service station, it is important for the pilot to indicate the frequency that he is monitoring.

A common frequency which often may be used as a simplex frequency is 122.6 MHz. When using this frequency, it is possible to transmit and receive the flight service station on the same frequency. Another simplex frequency common to many flight service stations is 123.6 MHz.

When arriving at a non-tower-equipped airport where a flight service station is located, it is possible for the pilot to receive airport advisory service on the frequency of 123.6 MHz. Flight service stations exercise no control over arriving or departing airport traffic; however, they can provide information on windspeed, direction, weather conditions, and any known traffic within the airport traffic pattern. It is wise for a pilot, arriving at an airport with a flight service station, to contact the FSS and request an airport advisory.

AIR ROUTE TRAFFIC CONTROL CENTERS

The safety of IFR traffic is the *prime* responsibility of *air route traffic control*

frequencies, are shown above the box. A frequency followed by the letter "R" means that the navigation facility receives this frequency only; likewise, the letter "T" indicates that the FSS facility can only transmit on this frequency. If no letter follows the frequency, it is a simplex frequency.

3. The navigation frequency (and TACAN channel if station is a VORTAC) is shown inside the box along with the station identification letters and Morse code (item 3). An underlined frequency indicates that there is no voice on the frequency; only the Morse code is audibly transmitted.

4. If the FSS has enroute flight advisory service (formerly enroute weather advisory service), this is noted by filling in the top corners of the information box (item 4). When this service is indicated, the pilot may contact the weather briefing service directly on the frequency of 122.0.

5. A Flight Service Station without a navigation facility is shown in a

centers (ARTCC). (See Fig. 3-47.) The United States is divided into 21 of these centers that jointly coordinate the flow of IFR traffic between their assigned areas. The centers throughout the country each have direct communications with adjacent centers. Each center is geographically divided into sectors with individual controllers directing IFR traffic in those sectors. As a flight leaves one sector for another, the controller's responsibility is passed to controllers working in the adjacent sectors.

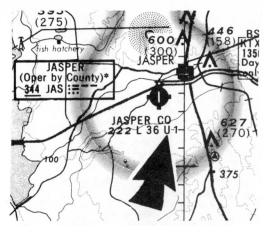

Fig. 3-48. "U-1" Means UNICOM Available

Fig. 3-47. Air Route Traffic Control Center

Today, almost all control by ARTCC of aircraft flying at sufficiently high altitude is exercised through the use of radar. VFR pilots, in some cases, may take advantage of radar service from air traffic control centers. Of course, this again is dependent upon the amount of IFR traffic a particular controller is handling.

UNICOM

There is one VHF frequency that seems to belong solely to general aviation. This simplex frequency of 122.8 is used by privately-operated radio stations and is known as *UNICOM*. The frequency of 122.8 MHz is used at many airports, but is assigned only to those airports not equipped with control towers. On the sectional aeronautical charts, UNICOM is designated by the letter "U-1" or "U-2" following the length of the runway in the airport data notation. (See Fig. 3-48.) On airports served by control towers, the frequency of 123.0 MHz is used in

place of 122.8 MHz for UNICOM air-to-ground communication.

The frequency 122.8 MHz is often used to provide the pilot with an airport advisory of wind directions and velocities, runway in use, and any known traffic in the airport traffic pattern. In addition, requests for transportation, information on aircraft service, personal messages, and availability of eating facilities can be transmitted on this frequency.

Frequency 123.0 MHz, since utilized only at airports equipped with control towers, is not used for airport advisory service. It should be remembered that *no* UNICOM frequency can be used for any type of traffic control; 122.8 is simply used to *advise* pilots about the active runway, winds, and weather conditions at the airport.

MULTICOM

When operating at an airport that has no facilities for radio communication, the pilot should transmit on 122.9 MHz which is the frequency assigned to Aeronautical Multicom Service. *Do not expect a reply.* When the aircraft approaches within 15 miles of the airport, the pilot should tune to 122.9 MHz and listen for other traffic transmission, and at five miles transmit his position, altitude, and intentions. This should be followed up by an announcement of his position on downwind, base, and final approach.

When departing the airport, he should tune to 122.9 MHz and listen before taxiing. The pilot should then broadcast his position on the airport and his intentions. After completing his pre-takeoff checklist, he should make an announcement before taxiing onto the active runway for takeoff.

EMERGENCY PROCEDURES

One of the frequencies that all pilots should memorize is 121.5 MHz. This is the international frequency for emergency use. Flight service stations, control towers, and air route traffic control centers maintain a listening watch on this single-channel communications frequency. The pilot's initial call-up should be *"Mayday, Mayday, Mayday"* if he has an emergency or is in distress. If he is not in immediate distress but is uncertain or desires to alert ground personnel, then he should preface his message with *"Pan, Pan, Pan."*

For aircraft equipped with a radar beacon transponder, the emergency transponder code is 7700. (See Fig. 3-49.)

If an aircraft so equipped experiences an emergency, the pilot should select code 7700. Doing this will automatically activate an alarm at the appropriate radar facility. In addition, the radar blip of the aircraft selecting the code will light up to a higher intensity on the radar screen.

D/F, or direction finding equipment, can be an invaluable aid to a lost or disoriented pilot. The equipment consists of a directional antenna system and a VHF and UHF radio receiver. The D/F display indicates the magnetic direction of the aircraft from the station each time the aircraft transmits. Where D/F equipment is tied into radar, a strobe of light is flashed from the center of the radar scope in the direction of the transmitting aircraft. In order to utilize D/F equipment, the pilot should contact the nearest tower or flight service station.

If an emergency or alert situation develops, the pilot should remember the four C's. They are:

1. *Confess* your predicament to any ground radio station. Don't wait too long.

2. *Communicate* with your ground radio link and pass on as much of the distress or alert message on the first transmission as possible.

3. *Climb*, if possible, for better radar and D/F detection. If flying at low altitude, the pilot's chance for establishing radio contact is improved by climbing.

4. *Comply* — especially comply, with the advice and instructions received.

When a pilot is in doubt of his position or feels apprehensive for his safety, he should not hesitate to request assistance.

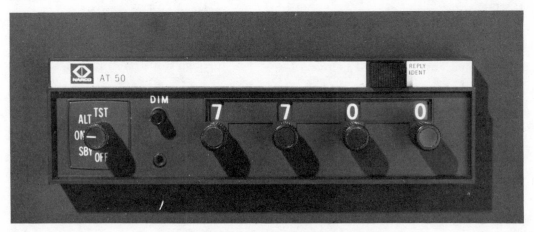

Fig. 3-49. Transponder Emergency Code

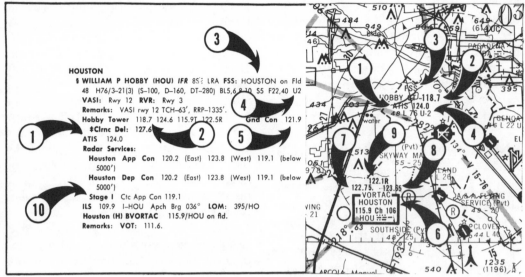

Fig. 3-50. Communication Codes and Frequencies

Search and rescue facilities including radar, radio, and D/F stations are ready and willing to help, and best of all, there is no *penalty* for using them.

CRASH LOCATOR BEACONS

Crash locator beacons are required by law for most light airplanes and are important equipment for pilots flying over mountainous or sparsely populated terrain. A certain "G" load is normally required to activate these beacons, or they may be activated from the cockpit. Once activated, they send out signals which, when picked up by search aircraft, will lead searchers directly to the downed aircraft.

REVIEW

A short review of radio frequencies as they are listed in Part 3 of the AIM and on sectional charts is listed below: (See Fig. 3-50.)

1. ATIS may be received on 124.0 MHz at Houston (item 1).

2. The primary Houston Tower frequency (item 2) is 118.7 MHz.

3. The Houston Flight Service Station is located on the field.

4. U-2 designates the frequency for UNICOM at a tower-controlled airport, 123.0 MHz.

5. Communication with Houston Ground Control is on 121.9 MHz.

6. The heavy line box indicates standard flight service station frequencies at Houston.

7. The non-standard frequency of 122.75 MHz is available.

8. The non-standard frequency of 123.65 MHz is also available.

9. The frequency of 122.1 only can be received by Houston.

10. Stage 1 radar service (radar advisory service for VFR aircraft) is available through Houston Approach Control.

Careful review of these two important sources of communication information will enable the pilot to quickly locate radio frequencies and interpret the coding.

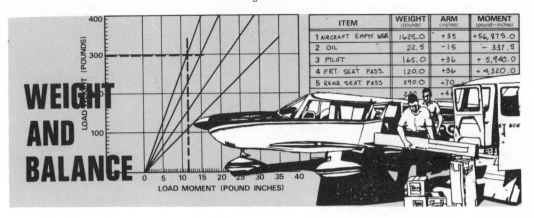

ITEM	WEIGHT (pounds)	ARM (inches)	MOMENT (pound-inches)
1 AIRCRAFT EMPTY WGT.	1625.0	+35	+56,875.0
2 OIL	22.5	-15	- 337.5
3 PILOT	165.0	+36	+ 5,940.0
4 FRT. SEAT PASS.	120.0	+36	+ 4,320.0
5 REAR SEAT PASS.	290.0	+70	
	240.0	+4	

INTRODUCTION

It is true that modern, light aircraft are designed so that they can be kept within their balance limits with a minimum of effort by pilots. However, as aircraft cabins become longer to provide greater payloads, better seating arrangements, and more comfort, weight and balance limits can be more easily exceeded by improper loading.

This chapter has been prepared to acquaint the student with the importance of keeping within authorized weight and balance limits and to show him how to check and control aircraft loading.

The first portion of this chapter explains weight and balance theory and how an aircraft will react to various loading conditions. A following section explains the various methods of checking weight and balance.

A pilot's responsibility for the safe operation of his aircraft begins before the aircraft leaves the ground. In addition to checking his aircraft for safe mechanical condition, the pilot should make sure the aircraft is loaded safely. Two factors must be checked to insure that the aircraft is loaded within safe limits. These are:
1. weight of the fully loaded aircraft; and
2. balance condition of the aircraft.

SECTION A - WEIGHT AND BALANCE THEORY

THE IMPORTANCE OF WEIGHT

The manufacturer of each aircraft has certificated the maximum weight above which the aircraft should not be flown. This weight is called *maximum gross weight*, and is the maximum certificated weight authorized by the FAA for operation of the aircraft involved.

The maximum gross weight is used by the manufacturer in conducting stress analysis, static tests, and flight tests to insure that the aircraft is strong enough

for safe operation at all flying weights up to and including this weight.

The aircraft manufacturers use a large safety factor in tests, and modern aircraft are actually capable of carrying loads that are well above those expected during normal operations. To exceed the aircraft structural limitations is usually quite difficult. However, the pilot should be aware that exceeding the maximum gross weight increases the chances that the structural limitations of the aircraft *could* be exceeded.

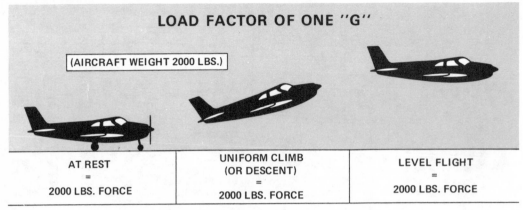

LOAD FACTOR OF ONE "G"

(AIRCRAFT WEIGHT 2000 LBS.)

AT REST = 2000 LBS. FORCE	UNIFORM CLIMB (OR DESCENT) = 2000 LBS. FORCE	LEVEL FLIGHT = 2000 LBS. FORCE

Fig. 4-1. Aircraft Experiencing One "G" Force

In order to understand how the weight of the aircraft determines the loads imposed on it during takeoff, flight, and landing, it is important that the student recall the term *load factor*. Load factor, as applied to aircraft stresses, is a ratio of actual load to aircraft weight. It is determined by dividing the actual load being applied to the structure of the aircraft by the total weight of the aircraft. Load factors are usually expressed in "G's."

A "G" is equal to the force of gravity or the actual weight of an object at rest. Therefore, if an aircraft was subject to a load factor of two "G's," the force would increase its effective weight to two times its normal weight.

For example, an aircraft has a load factor of one "G" when at rest on the ground, when in a uniform climb, or when flying straight and level in smooth air. (See Fig. 4-1.) In other words, the aircraft landing gear, or wings in flight, under these conditions, carries a load equal to one times the loaded weight of the aircraft.

When an aircraft in coordinated, level flight is turning in a 60° bank, it is subject to a load factor of two. (See Fig. 4-2.) This means the wings are lifting two times the weight experienced in straight-and-level flight. If the airplane weighed 2,000 pounds, the effective weight of the aircraft in a 60° bank is 4,000 pounds.

An airplane flying in gusty air is subject to sudden jolts that impose additional loads on the aircraft structure. The loads imposed by gusts are dependent upon the speed at which the aircraft penetrates them. It is somewhat like driving a car over a bumpy railroad track; if the person drives over it slowly, the car will bounce mildly; but if he drives over it fast, the car will bounce violently with the shock absorbers and springs hitting the frame. (See Fig. 4-3.)

Gusts are actually turbulent columns or layers of air that are rising, descending, or moving at different velocities. Flying through turbulent air at high speeds causes the wings to be subjected to severe bumps which result in high load factors. (See Fig. 4-4.)

It is for this reason that a pilot should slow his aircraft when he encounters severe gusty conditions. The pilot should adjust his airspeed when in turbulent con-

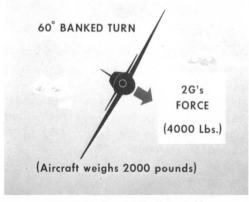

60° BANKED TURN

2G's
FORCE

(4000 Lbs.)

(Aircraft weighs 2000 pounds)

Fig. 4-2. Aircraft Experiencing Two "G's"

Ave. person : 170 lbs.

Fig. 4-3. Speed Affects Load Factor

ditions so that he doesn't exceed the *maneuvering speed* specified in the owner's manual.

CHECKING GROSS WEIGHT

To make it easy for the pilot to determine the maximum gross weight, the manufacturer generally publishes this information in two references usually kept in the aircraft:

1. Aircraft Weight and Balance Control Forms.
2. Aircraft Operation Manual or Owner's Handbook.

The pilot can easily check the operating gross weight of his aircraft by adding together the following:

1. Empty weight of the aircraft.
2. Weight of usable fuel on board.
3. Weight of oil on board.
4. Weight of removable equipment on board, such as oxygen bottles, towbars, etc.
5. Weight of pilot and passengers.
6. Weight of baggage and/or cargo.

The empty weight of the aircraft is shown on the Aircraft Weight and Balance Control Form in the aircraft data file.

Fuel weighs six pounds per gallon. Therefore, the total fuel weight can be computed by multiplying the number of gallons of usable fuel on board times six.

Oil weighs 7.5 pounds per gallon. Thus, the oil weight can be determined by multiplying 7.5 times the number of gallons of drainable oil on board. To determine the weight of the oil, if an aircraft has 12 quarts of oil on board, the pilot must first convert the quarts into gallons. Since 12 quarts equals three gallons, the oil weight is equal to three gallons x 7.5 pounds, or 22.5 pounds.

The weight of removable equipment in the aircraft must be determined by actually weighing it, or by referring to weight data supplied with the equipment.

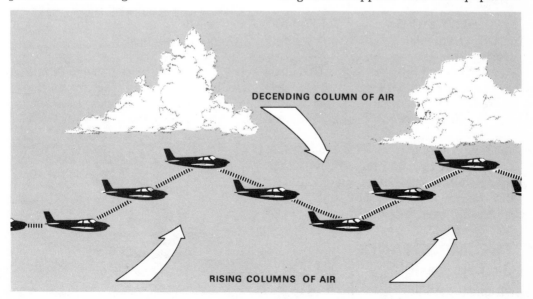

DECENDING COLUMN OF AIR

RISING COLUMNS OF AIR

Fig. 4-4. Gusts

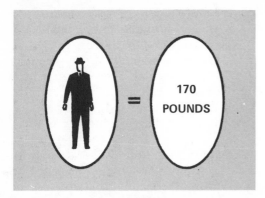

Fig. 4-5. Average Size Person

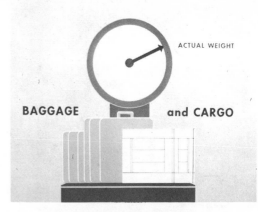

Fig. 4-6. Actual Weights of Baggage and Cargo

If the pilot and passengers are of average stature, a value of 170 pounds per person can be used to simplify weight and balance calculations. (See Fig. 4-5.) Children whose ages range from two to 12 years can be considered to have an average weight of 80 pounds. If extra large people are to be carried or if the aircraft is being loaded near its maximum gross weight, actual passenger weights should be used to get an accurate weight check.

Since baggage and cargo vary in weight, no standard weights can be used. Actual weights of the bags and cargo is the only accurate way to compute this portion of the aircraft's load. (See Fig. 4-6.)

An example of a gross weight check is shown here:

EMPTY WEIGHT	1,707.0 lbs.
OIL (12 quarts)	22.5 lbs.
PILOT	170.0 lbs.
PASSENGERS (3)	510.0 lbs.
FUEL (30 gallons)	180.0 lbs.
BAGGAGE	50.0 lbs.
TOTAL WEIGHT	2,639.5 lbs.

Assuming that an aircraft has a maximum gross weight of 2,650 pounds, its actual flying weight of 2,639.5 pounds is under the maximum gross weight, and the aircraft is *not* overloaded.

THE IMPORTANCE OF BALANCE

Although the pilot checks his load to see that the aircraft is under gross weight, he still doesn't know whether or not his aircraft is safe for flight. In addition to checking gross weight, it is necessary for the pilot to know that his aircraft is balanced within approved limits.

The balance condition of an aircraft can be determined by locating the center of gravity (CG) which is the imaginary point where all of the aircraft's weight is considered concentrated. (See Fig. 4-7.)

In the air, the aircraft is supported by its wings. The wings provide a safe support zone within which the aircraft is balanced in flight. This safe support zone is called the *center of gravity range* or *CG range*. The extremities of the CG range are called *forward CG limit* and *aft CG limit*. (See Fig. 4-8.)

The center of gravity range limits are usually specified in inches from a *datum*. A datum is an imaginary location in a vertical plane from which all pertinent

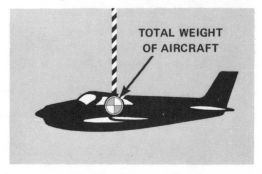

Fig. 4-7. CG = Center of Gravity

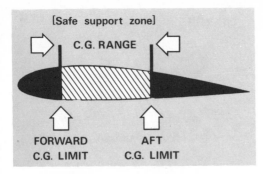

Fig. 4-8. Center of Gravity Range

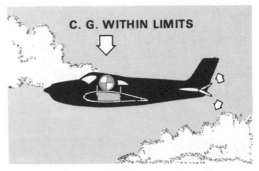

Fig. 4-10. Safe Elevator Control

horizontal measurements are made when the aircraft is in a level flight attitude.

There is no fixed rule as to the location of the datum. It is often located on the nose of the aircraft or on some point on the aircraft itself such as the firewall. In some cases, it is located a certain distance forward of the nose of the aircraft. The manufacturer has the choice of datum location. (See Fig. 4-9.)

As long as the center of gravity is maintained within the CG range specified by the manufacturer, the elevator control surfaces are capable of controlling the aircraft through all flight speed ranges — from stalling speeds to highest approved speeds. (See Fig. 4-10.)

FORWARD CENTER OF GRAVITY

The *longitudinal stability* of the aircraft is directly affected by the CG position.

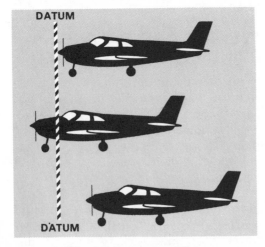

Fig. 4-9. Datum Location

Longitudinal stability is the resistance of an aircraft against pitching motions such as nose-up and nose-down. The longitudinal stability of an aircraft increases as the CG moves forward of the forward limit, eventually resulting in control forces beyond which the pilot cannot readily control the aircraft during landings. The more the CG moves forward, the more nose-heavy the airplane becomes. (See Fig. 4-11.)

If the CG is located forward of the forward CG limit, the following conditions exist, mildly at first, but becoming more severe as the CG moves farther and farther forward away from the CG range:

1. The aircraft becomes too stable. It has very little pitching tendency, but extra heavy elevator control forces are required to pull the nose up.
2. A CG that is forward of the forward CG limit is dangerous in that the aircraft will not have adequate elevator force to hold the tail down in the tail-low attitude during landing. (See Fig. 4-12.)
3. Also, a CG that is forward of the forward CG limit is dangerous for take-

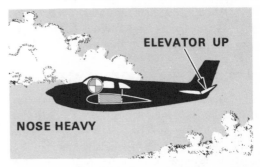

Fig. 4-11. Forward CG

Fig. 4-12. Unsafe for Landing

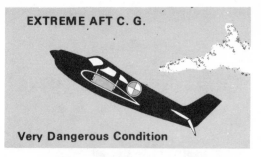

Fig. 4-13. Difficult Stall Recoveries

off. A nose-heavy aircraft will require longer ground runs and faster speeds before the nose can be raised for takeoff. If the runway is short, an extremely forward CG airplane could fail to become airborne in time to clear obstacles at the end of the runway.

4. Furthermore, if the weight distribution of the aircraft is located farther forward, additional down force on the tail must be developed through elevator back pressure in order to produce a critical angle of attack. If the elevators are already in the full-up position, the additional aerodynamic force must be generated by increased airflow over the horizontal tail surfaces. In order to do this, the airspeed must be greater; therefore, the *stall speed increases.*

This increase in stall speed could become a critical factor during takeoff, landing, or when performing a go-around from a missed approach to landing.

5. The nose-over tendency of conventional geared aircraft also increases as the CG moves farther forward of the approved limits.

AFT CENTER OF GRAVITY

If the aircraft should be loaded so that the CG is located behind the aft CG limits, the following conditions become noticeable, becoming more severe as the CG moves farther aft away from the CG range:

1. The aircraft becomes tail-heavy. This is a dangerous condition because it may be difficult to recover after maneuvering. (See Fig. 4-13.)

2. As the CG moves farther aft, the aircraft becomes more unstable and difficult to control.

3. Exceeding the aft CG limit seriously affects longitudinal stability and may reduce the chances of recovery from stalls and spins.

The aft CG limit is more apt to be exceeded with full gross weight conditions because loads in the back seats and baggage compartment have the greatest effect on moving the CG aft.

In order to keep the CG from moving too far back, a good rule of thumb to follow when loading the aircraft is *place the heaviest passengers and cargo forward and locate the lighter loads in the aft cabin or cargo area.*

WEIGHT AND BALANCE PRINCIPLES ILLUSTRATED

Children playing on a teeter-totter illustrate the basic principles involved in balance theory. The teeter-totter is in a balanced condition when the children adjust their distances from the *fulcrum,* or pivot point, so that their respective weights will balance.

For example, suppose that each child weighs 50 pounds. To balance the teeter-totter, they must each sit an equal distance from the fulcrum. (See Fig. 4-14.)

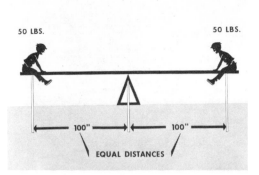

Fig. 4-14. Equal Weights and Distance

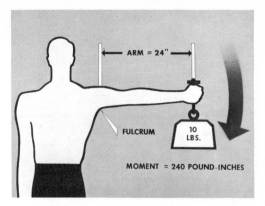

Fig. 4-16. Distance = Arm

However, if another child of 50 pounds sits at one end, the teeter-totter is unbalanced. In order to balance the teeter-totter, it is necessary that the fulcrum be moved toward the two children until the balanced condition is regained. (See Fig. 4-15.)

These examples show that, in addition to weight, the distance the weight is located from the fulcrum also affects the balance condition. This distance is usually called the *arm* in weight and balance calculations. (See Fig. 4-16.)

The tendency for an object to rotate is equal to its weight multiplied by its distance (or arm) from the fulcrum. This product is called *moment* and is usually stated in pound-inches.

For example, a 10-pound weight supported from a 24-inch arm would have a mo-

ment of 240 pound-inches (see Fig. 4-17). The teeter-totter is used to illustrate this. A 50-pound child located at an arm of 100 inches from the fulcrum has a moment of 5,000 pound-inches. (See Fig. 4-18.)

In order to balance this child, it is necessary to create a moment on the opposite side of the fulcrum of 5,000 pound-inches also. If a child weighing 100 pounds is to sit on the opposite side, where does he sit to balance the teeter-totter? (See Fig. 4-19.)

To determine the arm for the 100-pound child, the moment of 5,000 pound-inches is divided by the child's weight (5,000 ÷ 100 = 50 inches). Thus, the heavy child must sit at an arm of 50 inches from the fulcrum to balance the lighter child. This example illustrates how a weight can be located to balance an object when the fulcrum is fixed. (See Fig. 4-19.)

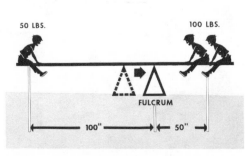

Fig. 4-15. Different Weights,
Move Fulcrum

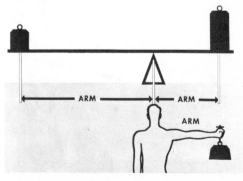

Fig. 4-17. Weight x Arm = Moment

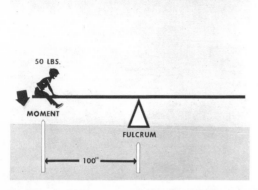

Fig. 4-18. Moment = 5,000 Pound-Inches

A different problem exists if the seats on the teeter-totter are fixed and only the fulcrum can be moved. For example, a 50-pound child is sitting on the left end and a 100-pound child is sitting on the right end. The distance between them is 200 inches and the teeter-totter itself weighs 40 pounds. When the children got on, the fulcrum was located 20 inches to the left of the center of the teeter-totter. Where is the new balance point? (See Fig. 4-20.)

In order to differentiate between the moments on one side of the teeter-totter from those on the opposite side, the arms (or distances) on the left side of the fulcrum will be labeled "minus" and those on the right side "plus." (See Fig. 4-21.)

The moment on the left side is equal to its arm, which is -80 inches (50 x [-80] = -4,000 pound-inches). Note that the moment is minus because it is located on the left side of the fulcrum.

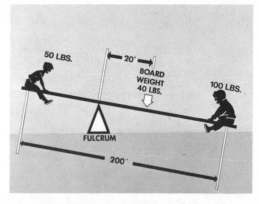

Fig. 4-20. Where is the Balance Point?

The child's moment on the right side is 100 pounds x (+120 inches) = +12,000 pound-inches.

Adding the moments of the child and the board, it is found that a total moment of +12,800 pound-inches exists on the right side of the fulcrum.

The weight of the teeter-totter board is 40 pounds and its center is located +20 inches from the fulcrum. Therefore, the board's moment is 40 pounds x (+20 inches) = +800 pound-inches.

Now, the moments on both sides of the fulcrum can be compared. On the left side is -4,000 pound-inches versus +12,800 pound-inches on the right. The teeter-totter is certainly not balanced. Where should it balance? (See Fig. 4-22.)

The arm of the balance point can be located as follows:

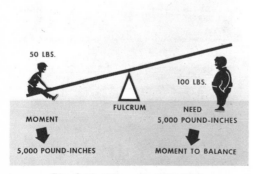

Fig. 4-19. Where Does He Sit?

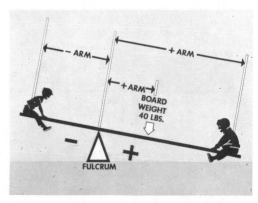

Fig. 4-21. Plus and Minus Arms

Fig. 4-22. Unbalanced

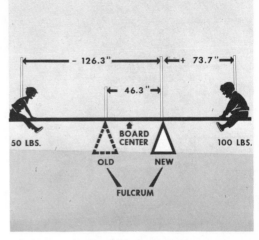

Fig. 4-23. Balanced

1. Add all the moments together, +12,800 - 4,000 = +8,800 pound-inches.
2. Next, add the weights of the children and board together. (50 + 100 + 40 = 190 pounds.)
3. Divide the total moment of +8,800 pound-inches by the total weight of 190 pounds. (+8,800 ÷ 190 = +46.3 inches.)

Since the arm is plus, it means that the balance point is 46.3 inches to the right of the old fulcrum. Relocating the fulcrum to this new position balances the teeter-totter. This results in the child on the left being located 126.3 inches from the fulcrum and the child on the right being 73.7 inches from the fulcrum. (See Fig. 4-23.)

A comparison can be made between the airplane and the teeter-totter.

1. The datum of the airplane can be likened to the fulcrum on the teeter-totter. Since the fulcrum is arbitrarily placed on the teeter-totter to provide a reference point for balance calculations, it could be located anywhere. In the case of the airplane, the manufacturer determines the datum location. (See Fig. 4-24.)
2. The center of gravity of the airplane is equivalent to the balance point of the teeter-totter. (See Fig. 4-25.)
3. The empty weight of the airplane can be compared to the weight of

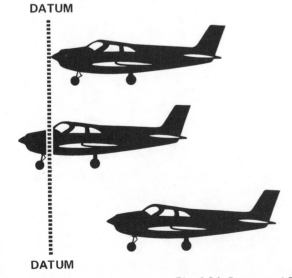

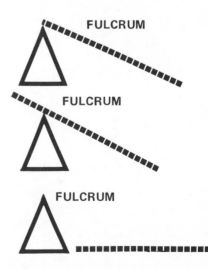

Fig. 4-24. Datum and Fulcrum Compared

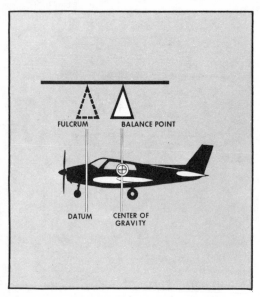

Fig. 4-25. Comparison: CG — Balance Point

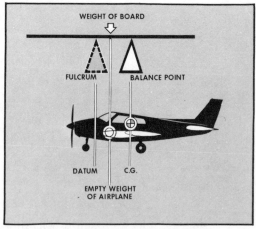

Fig. 4-26. Comparison: Empty Weight —
Weight of Board

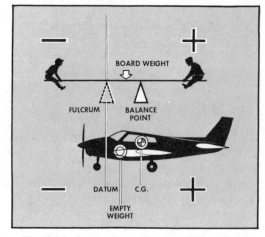

Fig. 4-27. Comparison: Plus and Minus Arms

the teeter-totter board. (See Fig. 4-26.)

4. The various loads the airplane carries, such as oil, pilot, passengers, cargo, baggage, and fuel can be compared to the children's weights.

5. All arms located to the left of the datum are labeled *minus* and all arms located to the right of the datum are labeled *plus*. (See Fig. 4-27.)

SECTION B
WEIGHT AND BALANCE CALCULATION

Since aircraft balance is so critical to the safe operation of an aircraft, it is important that the pilot know how to check this condition for each loading arrangement. There are three basic methods, provided by aircraft manufacturers, which can be used to check weight and balance of aircraft. These are:

1. computation method.
2. graph method.
3. table method.

The computation method can be used to check the weight and balance condition of any aircraft, regardless of size. Since this method is rather time-consuming, manufacturers usually supply the pilot with the graph or table method to save time in determining CG location. The computation method, however, is the basis for all weight and balance control. It is important that the pilot learn this method because it provides the basic steps involved in locating the aircraft center of gravity.

COMPUTATION METHOD

Understanding the theory used in weight and balance calculations provides a basis for understanding the computation method of checking weight and balance.

The steps involved in weight and balance control by the computation method are as follows:

1. List aircraft empty weight, fuel, oil, pilot, passengers, baggage, cargo, and loose equipment along with their respective weights and arms.
2. Multiply the weights and arms together to find the moment of each item.
3. Add the weights to get total loaded weight of the airplane.
4. Add the moments to find the total moment.
5. Then divide the total moment by the total weight to obtain the arm of the CG of the airplane fully loaded.

6. Compare the total weight to the approved gross weight of the airplane and CG arm with the approved CG range. Adjust the load as necessary to keep the aircraft loaded within approved limits.

SAMPLE PROBLEM USING COMPUTATION METHOD

The following information is obtained from the passengers and from the Aircraft Weight and Balance Control Form:

1. The datum is the firewall of the airplane.
2. Aircraft empty weight: 1,625 pounds; arm +35.0 inches.
3. Oil capacity: 12 quarts; arm, -15 inches. (Note that this is a minus arm because oil is located forward of the datum.)
4. Pilot and front seat passenger are located at +36.0 inches. The weight of the pilot is 165 pounds, and the weight of the front seat passenger is 120 pounds.
5. Rear seat passengers are located at +70.0 inches. The combined weight of the rear passengers is 290 pounds.
6. Fuel capacity (in excess of unusable): 40 gallons; arm, +48.0 inches.
7. Baggage area is located at +95.0 inches. The total baggage weight is 90 pounds.
8. Approved CG range is from +33.5 inches to +45.8 inches.
9. Maximum certificated gross weight: 2,700 pounds.

Given these facts, is the aircraft loaded safely? The problem is worked as follows:

1. Record the known facts on a work sheet which contains four columns: item, weight, arm, and moment. (See Fig. 4-28.)
2. After recording the description, weight, and arm of each item, the weights and arms are multiplied to-

WEIGHT and BALANCE FORM

ITEM	WEIGHT (pounds)	ARM (inches)	MOMENT (pound-inches)
1 AIRCRAFT EMPTY WEIGHT	1,625.0	+35	+ 56,875.0
2 OIL	22.5	-15	- 337.5
3 PILOT	165.0	+36	+ 5,940.0
4 FRONT SEAT PASSENGER	120.0	+36	+ 4,320.0
5 REAR SEAT PASSENGERS	290.0	+70	+ 20,300.0
6 FUEL	240.0	+48	+ 11,520.0
7 BAGGAGE	90.0	+95	+ 8,550.0
8			
9			
10			
TOTAL	2,552.5		+107,167.5
C.G. =	41.99	INCHES	

Fig. 4-28. Weight and Balance Worksheet

gether to obtain their moments. (See Fig. 4-28, item 1.)

3. Add all the values in the column. This yields a total weight of 2,552.5 pounds. Enter this at the bottom of the weight column. (See Fig. 4-28, item 2.)

4. Add all of the positive moments in the moment column. This value is +107,505 pound-inches. Then subtract the minus moment of -337.5 pound-inches from this value. (+107,505 - 337.5 = +107,167.5.) Enter the answer of +107,167.5 pound-inches at the bottom of the moment column. (See Fig. 4-28, item 3.)

5. Next, divide the total moment by the total weight. +107,167.5 ÷ 2,552.5 = +41.99). This gives +41.99 inches, which is the distance that the airplane's loaded CG is located aft of the datum. Enter the CG arm of +41.99 inches at the bottom of the work sheet. (See Fig. 4-28, item 4.)

6. Compare CG and total weight with approved CG range and maximum gross weight. The approved CG range extends from +33.5 inches to +45.8 inches. Since the value of +41.99 inches falls within these limits, the pilot knows that his aircraft is properly balanced. Also, since the aircraft weight of 2,552.5 pounds is below the 2,700 pound maximum certificated gross weight specified by the manufacturer, his aircraft is not overloaded.

From the solution to this problem, the pilot has determined that the aircraft is within the maximum gross weight and approved CG range. (See Fig. 4-29.)

GRAPH METHOD

Now that weight and balance theory and the computation method are known, the student can proceed with the other methods of checking weight and balance.

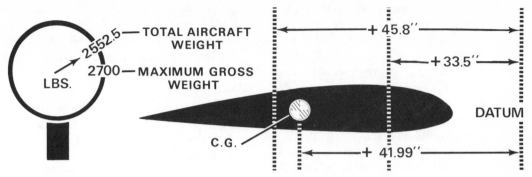

Fig. 4-29. CG and Gross Weight within Limits

The graph method is an easy way to perform weight and balance analyses. It eliminates the multiplication and division steps involved in the computation method because the moments can be read directly from a loading graph.

The loading graph contains a separate load line for each item. For example, the graph shown in figure 4-30 has the loading lines for oil, pilot, and front passenger, fuel, cargo area "B" for baggage, and cargo area "C."

To find the moment of an item using the graph method, proceed as follows:

1. First, locate its load line on the graph, such as the fuel load line shown in figure 4-30, item 1.
2. Then, find the weight of the item on the left side of the graph. (See Fig. 4-30, item 2.)

LOADING GRAPH

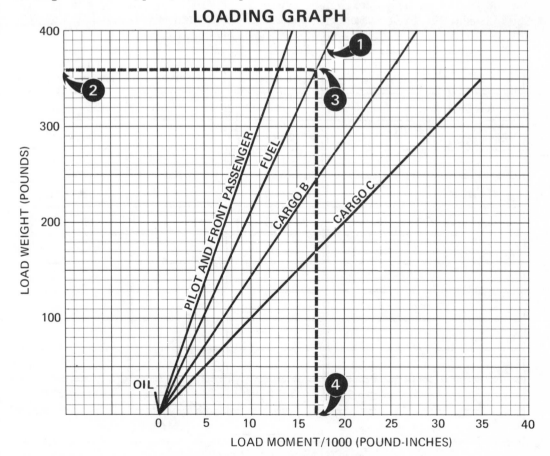

Fig. 4-30. Sample Loading Graph

3. Proceed horizontally along this weight line until the load line is reached. (See Fig. 4-30, item 3.)
4. At this point, drop vertically down and read the moment on the scale at the bottom of the chart. (See Fig. 4-30, item 4.) The value is two graduations to the right of 15. This graduation represents the number 17 since each graduation is equal to one unit. The moment is actually 17,000 pound-inches because the figures on this scale have been divided by 1,000 to make the numbers on the chart smaller, and therefore, more easily handled.

SAMPLE PROBLEM USING GRAPH METHOD

A sample problem should be worked to become better acquainted with the graph method. In this problem, the aircraft is loaded with two persons in the front seat. The pilot weighs 180 pounds and the front passenger weighs 120 pounds. The aircraft will carry a load of cargo with 540 pounds located in cargo area "B" and 180 pounds in area "C." Is this within approved gross weight and balance limits?

Using the graph method, the problem is worked as follows:
1. Enter the aircraft load items on a work sheet in the same manner as the computation method. (See Fig. 4-31, item 1.)
   a. The airplane empty weight and moment are 1,580 pounds and 55,500 pound-inches respectively. This data is found in the aircraft weight and balance forms.
   b. Pilot and front passenger combined weights are 300 pounds.
   c. The fuel load is 60 gallons, 360 pounds.
   d. Cargo in area "B" weighs 540 pounds.
   e. Cargo in area "C" weighs 180 pounds.

WEIGHT and BALANCE FORM

	ITEM	WEIGHT (pounds)	ARM (inches)	MOMENT (pound-inches)
1	AIRCRAFT EMPTY WEIGHT	1580.0		+55,500
2	PILOT & FRONT PASSENGER	300.0		+11,000
3	FUEL	360.0		+ 17,000
4	CARGO "B"	540.0		+38,000
5	CARGO "C"	180.0		+ 17,500
6	OIL	22.5		− 400
7				
8				
9				
10				
	TOTAL	2982.5		+138,600
	C.G. = _____ INCHES			

Fig. 4-31. Weight and Balance Worksheet

f. Oil tank capacity is 12 quarts or 22.5 pounds.

2. Now, find the moment for the pilot and front passenger as follows: (See Fig. 4-32.)

a. First, locate the load line labeled "pilot and front passenger" on the loading graph (item A).

b. On the left side of the graph, locate 300 pounds, which is the total weight of the pilot and front passenger (item B).

c. Proceed horizontally along this line to the "pilot and front passenger" load line (item C).

d. From the intersection point on the load line, proceed vertically downward to the horizontal scale at the bottom of the graph (item D). At this point the value is 11, which is actually 11,000 pound-inches.

e. Record 11,000 pound-inches in the moment column of the work sheet.

3. Proceed in a like manner to obtain the moments for the remaining load items and record them on the work sheet. (See Fig. 4-31, item 2.)

4. When all of the weights and moments are determined, add each column to get the total weight and total moment. This step is the same as that performed in the computation method. The total weight is 2,982.5 pounds, and the total moment is +138,600 pound-inches. (See Fig. 4-31, item 3.)

5. In order to check whether this is within the approved CG limits, locate the total weight and total moment on the *center of gravity moment envelope* provided by the manufacturer. No division is necessary.

LOADING GRAPH

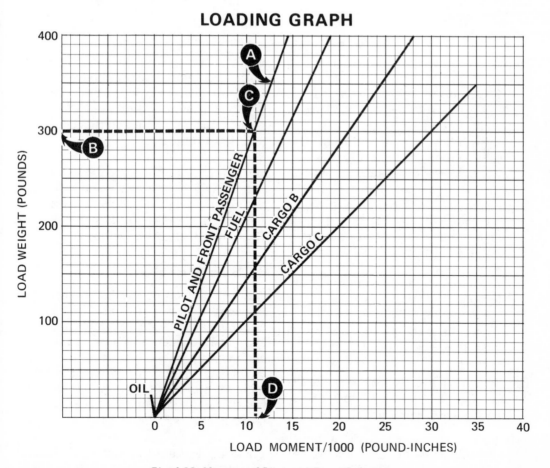

Fig. 4-32. Moment of Pilot and Front Passenger

CENTER OF GRAVITY MOMENT ENVELOPE

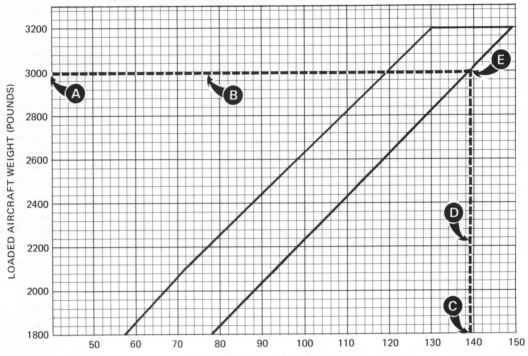

Fig. 4-33. Center of Gravity Location

The CG is located as follows: (See Fig. 4-33.)

1. First, locate the total weight of 2,982.5 pounds on the scale located on the left side of the graph (item A).
2. Draw a horizontal line through this value across the graph (item B).
3. Next, locate the total moment on the scale at the bottom of the graph. In order to do this, divide the moment of 138,600 pound-inches by 1,000. The answer is 138.6. Then locate 138.6 on the scale (item C).
4. A line is then drawn vertically through this point upward across the graph (item D).
5. Where the vertical and horizontal lines intersect is the CG position. (item E.)

The aircraft is correctly loaded as the CG location falls on the "aft" line of the center of gravity envelope.

If the student were to encounter a similar loading problem with the total weight within the maximum certificated gross weight, but the center of gravity aft of the maximum limit, he should attempt to reload the aircraft so the heavy objects are located more toward the front of the cabin or baggage area and the lighter objects toward the rear.

He can check his final aircraft loading by determining his new weight and moments, entering the values on his work sheet, and rechecking the CG location relative to the center of gravity envelope.

TABLE METHOD

Another method of checking weight and balance is the table method. This method also eliminates both the multiplication and division operation involved in the computation method. Instead of graphs, as used in the graph method, the table method provides tables which list the weights and moments for each type of useful load. (See Fig. 4-34.)

For example, item 1 of figure 4-34 shows a table for the baggage compartment.

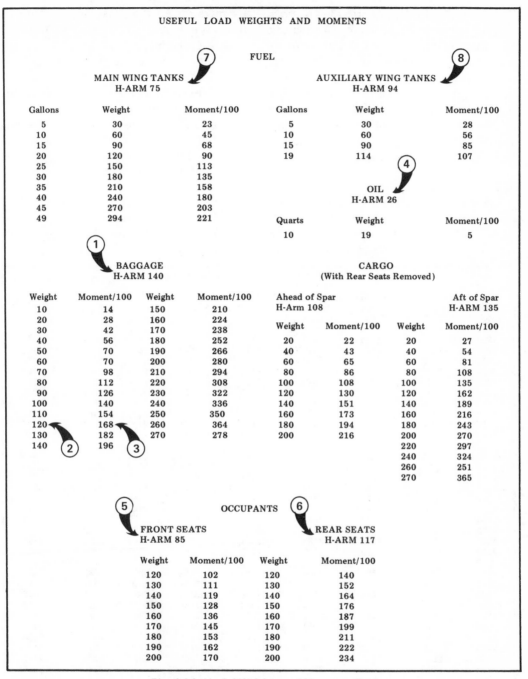

USEFUL LOAD WEIGHTS AND MOMENTS

FUEL

MAIN WING TANKS
H-ARM 75

Gallons	Weight	Moment/100
5	30	23
10	60	45
15	90	68
20	120	90
25	150	113
30	180	135
35	210	158
40	240	180
45	270	203
49	294	221

AUXILIARY WING TANKS
H-ARM 94

Gallons	Weight	Moment/100
5	30	28
10	60	56
15	90	85
19	114	107

OIL
H-ARM 26

Quarts	Weight	Moment/100
10	19	5

BAGGAGE
H-ARM 140

Weight	Moment/100	Weight	Moment/100
10	14	150	210
20	28	160	224
30	42	170	238
40	56	180	252
50	70	190	266
60	70	200	280
70	98	210	294
80	112	220	308
90	126	230	322
100	140	240	336
110	154	250	350
120	168	260	364
130	182	270	278
140	196		

CARGO
(With Rear Seats Removed)

Ahead of Spar H-Arm 108		Aft of Spar H-ARM 135	
Weight	Moment/100	Weight	Moment/100
20	22	20	27
40	43	40	54
60	65	60	81
80	86	80	108
100	108	100	135
120	130	120	162
140	151	140	189
160	173	160	216
180	194	180	243
200	216	200	270
		220	297
		240	324
		260	251
		270	365

OCCUPANTS

FRONT SEATS
H-ARM 85

REAR SEATS
H-ARM 117

Weight	Moment/100	Weight	Moment/100
120	102	120	140
130	111	130	152
140	119	140	164
150	128	150	176
160	136	160	187
170	145	170	199
180	153	180	211
190	162	190	222
200	170	200	234

Fig. 4-34. Useful Weights and Moments Table

Note that weights are provided in ten pound increments along with their respective moments. In order to simplify computations, the moments listed on this chart are divided by 100. This keeps the numbers small and easy to handle.

Assume that the aircraft is going to carry 120 pounds of baggage. What is its moment? Locate the weight of 120 pounds (see Fig. 4-34, item 2) and read the moment in the column to the right of the weight (see Fig. 4-34, item 3). It is found that the moment value for 120 pounds of baggage is 168. Since this value has been divided by 100, the moment is actually 16,800 pound-inches.

The table method also provides a *gross weight moment limits table* for insuring that the total aircraft weight and moment are within approved limits. (See Fig. 4-35.) This table provides minimum and maximum moments for each gross weight. If the aircraft total moment falls within the minimum and maximum moment values for the gross weight involved, the aircraft is safely loaded.

For example, assume that the aircraft has a loaded gross weight of 2,950 pounds and a total moment of 254,600 pound-inches. The pilot can check whether or not he is loaded within approved limits by referring to the gross weight moment limits table and proceeding as follows: (See Fig. 4-35.)

1. First, locate the gross weight of 2,950 pounds (item 1).
2. Observe that the minimum moment is 240,700 pound-inches (item 2) and the maximum moment permitted is 255,800 pound-inches (item 3). Note that the value in the table is multiplied by 100 to get the moments.
3. Since the aircraft total moment of 254,600 pound-inches is within these limits, the aircraft is safely loaded.

GROSS WEIGHT MOMENT LIMITS

Gross Weight	Minimum Moment 100	Maximum Moment 100	Gross Weight	Minimum Moment 100	Maximum Moment 100	Gross Weight	Minimum Moment 100	Maximum Moment 100
2000	1540	1734	2300	1771	1994	2600	2029	2254
2010	1548	1743	2310	1779	2003	2610	2039	2263
2020	1555	1751	2320	1786	2011	2620	2049	2272
2030	1563	1760	2330	1794	2020	2630	2060	2280
2040	1571	1769	2340	1802	2029	2640	2070	2289
2050	1579	1777	2350	1810	2037	2650	2081	2298
2060	1586	1786	2360	1817	2046	2660	2092	2306
2070	1594	1795	2370	1825	2055	2670	2102	2315
2080	1602	1803	2380	1833	2063	2680	2113	2324
2090	1609	1812	2390	1840	2072	2690	2123	2332
2100	1617	1821	2400	1848	2081	2700	2134	2341
2110	1625	1829	2410	1856	2089	2710	2145	2350
2120	1632	1838	2420	1863	2098	2720	2155	2358
2130	1640	1847	2430	1871	2107	2730	2166	2367
2140	1648	1855	2440	1879	2115	2740	2177	2376
2150	1656	1864	2450	1887	2124	2750	2188	2384
2160	1663	1873	2460	1894	2133	2760	2198	2393
2170	1671	1881	2470	1902	2141	2770	2209	2402
2180	1679	1890	2480	1910	2150	2780	2220	2410
2190	1686	1899	2490	1917	2159	2790	2231	2419
2200	1694	1907	2500	1925	2168	2800	2242	2428
2210	1702	1916	2510	1935	2176	2810	2253	2436
2220	1709	1925	2520	1946	2185	2820	2263	2445
2230	1717	1933	2530	1956	2194	2830	2274	2454
2240	1725	1942	2540	1966	2202	2840	2285	2462
2250	1733	1951	2550	1977	2211	2850	2296	2471
2260	1740	1959	2560	1987	2220	2860	2307	2480
2270	1748	1968	2570	1997	2228	2870	2318	2488
2280	1756	1977	2580	2008	2237	2880	2329	2497
2290	1763	1985	2590	2018	2246	2890	2340	2506
						2900	2351	2514
						2910	2362	2523
						2920	2373	2532
						2930	2385	2540
						2940	2396	2549
						2950	2407	2558
						2960	2418	2566
						2970	2429	2575
						2980	2441	2584
						2990	2452	2592
						3000	2463	2601

The above moment limits are based on the following weight and center of gravity limit data (landing gear down).

WEIGHT CONDITION	FORWARD CG LIMIT	AFT CG LIMIT
3000 lb. (max. take-off or landing)	82.1	86.7
2500 lb. or less	77.0	86.7

Fig. 4-35. Gross Weight Moment Limits Table

SAMPLE PROBLEM USING TABLE METHOD

With the aircraft loaded as listed below, what is the weight and balance condition of the aircraft?

1. Aircraft empty weight and moment: 1,889 pounds and 148,200 pound-inches respectively. This data was obtained from the weight and balance forms in the aircraft.
2. Oil: 10 quarts
3. Pilot: 170 pounds
4. Front passenger: 150 pounds
5. Rear passenger: 200 pounds
6. Rear passenger: 190 pounds
7. Baggage: 80 pounds
8. Fuel (main tanks): 49 gallons
9. Fuel (auxiliary tanks): 19 gallons

The problem is worked in two phases:

1. First, find the gross weight and total moment as follows: (See Fig. 4-36.)
   a. Enter the aircraft empty weight and moment on a work sheet.
   b. Next, enter the oil weight and moment on the work sheet. To get this information, refer to the Oil Section in the Useful Load Weights and Moments Table. (See Fig. 4-34, item 4.) Ten quarts of oil have a weight of 19 pounds and a moment of +500 pound-inches.
   c. Referring to the Front Seats Table (item 5), the student finds that the pilot's weight of 170 pounds has a moment of +14,500 pound-inches, and the front seat passenger's weight of 150 pounds

has a moment of +12,800 pound-inches. Add these values to the work sheet.

   d. In a like manner, the pilot finds the moment for the rear seat passengers and baggage by referring to the Rear Seats Table and Baggage Table respectively (see Fig. 4-34). The moments for the rear seat passengers are +23,400 and +22,200 pound-inches. The baggage moment is +11,200 pound-inches.
   e. The weight and moments for the main (item 7) and auxiliary fuel tanks (item 8) are obtained from their respective tables. (See Fig. 4-34.) Main fuel tank moment is +22,100 pound-inches; auxiliary fuel moment is +10,700 pound-inches.
   f. After all weights and moments are obtained, add them together to get the aircraft's gross weight and total moment for takeoff. The airplane has a gross weight of 3,106 pounds and a moment of +265,600 pound-inches.
2. Next, compare the gross weight and total moment with the approved limits shown on the Gross Weight and Moment Limits Table. First, look for the gross weight of 3,106 pounds on the table. Something is wrong! The highest weight shown is 3,000 pounds, which is the maximum gross weight approved for this airplane. (See Fig. 4-35.)

The aircraft is 106 pounds over maximum gross weight. This means that the gross weight must be reduced by 106 pounds before takeoff.

3. Referring to the work sheet, it is noted that the auxiliary fuel weighs 114 pounds. (See Fig. 4-36.) So, the effect of draining the auxiliary fuel tanks on the aircraft loading can be seen. Subtracting the auxiliary fuel weight and moment from the gross weight and total moment respectively, yields a new gross weight of 2,992 pounds, which is under the

WEIGHT and BALANCE FORM

ITEM	WEIGHT (pounds)	ARM (inches)	MOMENT (pound-inches)
1 AIRCRAFT EMPTY WEIGHT	1889		+ 148,200
2 OIL	19		+ 500
3 PILOT	170		+ 14,500
4 FRONT SEAT PASSENGER	150		+ 12,800
5 REAR SEAT PASSENGER	200		+ 23,400
6 REAR SEAT PASSENGER	190		+ 22,200
7 BAGGAGE	80		+ 11,200
8 MAIN FUEL	294		+ 22,100
9 AUX FUEL	114		+ 10,700
10			
TOTAL	3106		+265,600
C.G. =		INCHES	

Fig. 4-36. Weight and Balance Work Sheet

maximum gross weight, and a new moment of +254,900 pound-inches.

4. The new gross weight and moment are checked for being within approved limits as follows:

a. Referring to the gross weight moment limits chart, locate the weight on the chart which is the nearest to the gross weight of 2,992 pounds. The nearest chart value is 2,990 pounds. (See Fig. 4-37.)

b. Proceeding to the right of the 2,990 pound weight, find that the minimum approved moment is +245,200 pound-inches and the maximum approved moment is +259,200 pound-inches. Note that the chart values are multiplied by 100 to get the moments.

c. Since the aircraft total moment of +254,900 pound-inches falls within the approved minimum and maximum limits, the aircraft

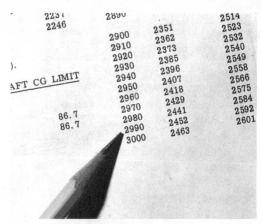

Fig. 4-37. Maximum Gross Weight

is loaded safely. The pilot can now proceed on his trip.

The preceding sample problem has shown how to use the table method for weight and balance control, and has illustrated how a pilot must lighten an overloaded plane to keep it within the maximum gross weight.

chapter 5

INTRODUCTION

An understanding of *meteorology* is one of the most important areas of knowledge for the aviator. Throughout the history of aviation, adverse weather has hampered flight operations, resulted in schedule delays for airlines, and has been the cause of accidents for "un-weather-conscious" pilots. Much research has been devoted to the development of sophisticated instruments and radio gear in an attempt to conquer the limitations to flight imposed by weather. Today, modern aircraft are capable of flying in most kinds of meteorological conditions and can perform instrument landings in extremely marginal conditions of ceiling and visibility.

In spite of man's technological achievements, certain *severe weather conditions* occasionally occur, making flight impossible. The violence of thunderstorms, the blindness caused by fog, and the turbulence resulting from high velocity winds can outmatch the capabilities of even the most sophisticated aircraft.

The study of the earth's atmosphere, weather, and basic meteorology is not only an exciting and challenging endeavor, but a prerequisite for even the novice pilot.

SECTION A - WEATHER THEORY

BASIC ATMOSPHERIC CIRCULATION

The weather on this planet is mainly produced by two factors: the *sun* and the *rotation of the earth.* The majority of atmospheric heating is the result of reradiation of the sun's rays from the surface of the earth. Approximately 85% of the heat absorbed by the atmosphere is caused by heating from the ground below, while about 15% is produced by the direct rays of the sun as they pass through the atmosphere on their path to the earth. (See Fig. 5-1.)

HEATING OF THE EARTH

Heating of the air surrounding the earth is not equal over the entire surface. Various types of surfaces over the face of the earth such as foliage, sand, mountains, and shadows tend to vary the amount of reradiation of the sun's heat energy. But

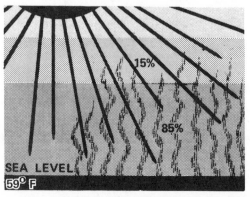

Fig. 5-1. Solar Reradiation

the major reason for uneven heating is the difference in the *angle of incidence* of the sun's rays as they strike various parts of the globe. The angle of incidence is the angle between the sun's rays and the surface they are striking. During the beginning of spring and fall, the sun is centered directly over the equator, and the angle of incidence between the sun's rays and the surface of the earth at the equator is 90°. In other words, the rays are striking the surface directly at the equator. At this point, a square foot of sunrays centers its heat over one square foot of the earth's surface. The greatest concentration of heat and the largest possible amount of reradiation result from this 90° angle of incidence. (See Fig. 5-2.)

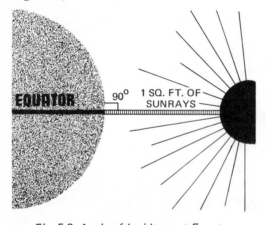

Fig. 5-2. Angle of Incidence at Equator

Due to the curvature of the earth, the angle of incidence of the sun's rays decreases toward both the North and South Poles. North and south of the equator, a square foot of sunrays is not concentrated over one square foot of the surface, but over a *larger area*. This lower concentration produces less reradiation of heat over a given surface area and less atmospheric heating results. At the North and South Poles, the angle of incidence is so small that very little heating of the atmosphere occurs, resulting in extremely cold air temperatures. Basically, then, overall air temperatures decrease moving south or north from the equator. (See Fig. 5-3.)

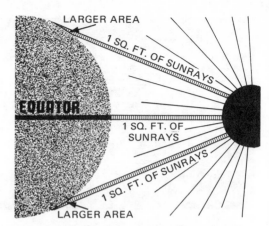

Fig. 5-3. Concentration Change of Solar Radiation

The unequal heating of the earth's atmosphere causes a *large air-cell circulation pattern*. The density of air determines whether the air will rise or descend. If air is more dense than the surrounding atmosphere, it will sink, while air that is less dense than its surroundings will begin to rise. Heated air at the equator becomes less dense and begins to rise. Conversely, cold air at the poles is very dense and sinks toward the surface of the earth as shown in figure 5-4.

The large cell circulation pattern around the earth can be likened to a pan of water over a burning flame. As the water directly over the flame at the bottom center of the pan is heated, it becomes less dense compared to the cooler water around the rim of the pan. The water over the flame rises to the top of the pan where it cools and flows outward toward the edge. As the water moves toward the edge, it cools further, becomes more dense, and begins to sink to the bottom. Because of the circulation pattern that is generated, this water works its way back toward the center to be heated again to continue the circulation. (See Fig. 5-4.)

THE ROTATING EARTH

If it were not for the rotation of the earth and other complicating factors, the same simple circulation pattern would take place over the globe. Air would move directly from the high density area (high pressure area) over the poles toward the

Fig. 5-4. Large Cell Circulation Pattern

low density area (low pressure area) at the equator to be reheated and circulated back toward the poles. In reality, however, a *three-cell circulation pattern* over each hemisphere develops.

As air rises over the equator, it begins to move northward and southward toward the poles. Due to the rotation of the earth, the air moving from the equator northward is deflected to the east in the Northern Hemisphere. At approximately 30° N. latitude, much of the air is moving east causing a pile-up of air which sinks and forms a relatively constant high pressure area at this latitude. The remainder of the air continues northward to the North Pole. The sinking air at 30° N. latitude reaches the surface and spreads outward, both north and south. The air that moves south at 30° N. latitude completes one cell of the circulation pattern (see Fig. 5-5, item 1). The air that moves north becomes part of another cell between 30° N. latitude and 60° N. latitude (see Fig. 5-5, item 2).

To complete the three-cell circulation pattern, the air descending from the North Pole southward collides with the air moving north from the 30° N. latitude high pressure area. As these two air masses meet and rise, a *semi-permanent low pressure area* is produced near 60° N. latitude. Therefore, a third circulation pattern between 60° N. latitude and the North Pole develops. (See Fig. 5-5, item 3.)

Due to the uneven heating of the earth's surface and the resultant three-cell circulation pattern of the air, four relatively permanent pressure systems exist — a low pressure area over the equator, a high pressure system at 30° N. latitude, a low at 60° N. latitude, and a high pressure area at the North Pole. If it were not for the rotation of the earth, air would constantly move directly from high pressure areas to low pressure areas. However, due to *Coriolis effect*, a circular flow around pressure centers develops.

CORIOLIS EFFECT

The natural rotation of the earth causes an apparent force to be exerted on the atmosphere which deflects the air toward the right in the Northern Hemisphere, and toward the left in the Southern Hemisphere. In fact, any free-moving body has an apparent deflection as it moves above the rotating earth. When viewed from some point out in space, a free-moving body would appear to follow a straight line. However, to a viewer on the earth, the path of a free body would appear to curve. Therefore, this "force" is only apparent, but the deflection certainly is in effect when using the earth as a reference point.

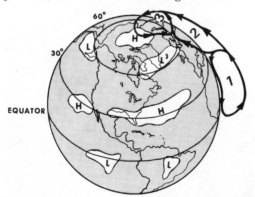

Fig. 5-5. Three-Cell Circulation Pattern

Due to Coriolis effect, the air moving around the high pressure system at the North Pole circulates clockwise, and due to pressure differential moves outward from the center of the high. This produces the prevailing wind system known as the *polar easterlies*. (It is important to remember that wind directions are labeled according to the direction *from* which they are blowing.) At 60° N. latitude, the circulation about the semi-permanent low is counterclockwise and inward toward the center of the low. At 30°N. latitude, air circulates clockwise around the high and produces the *prevailing westerlies* and *northeast tradewinds*. (See Fig. 5-6.)

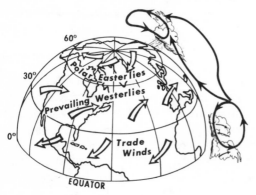

Fig. 5-6. Prevailing Wind Systems

AIR MASSES

An air mass is an extensive body of air whose temperature and moisture characteristics are approximately the same at a given altitude throughout the body of air. Air masses are formed over large areas of water or land. An air mass may range in size from hundreds to thousands of square miles in area, with a depth of approximately three to eight miles. (See Fig. 5-7.)

If a body of air having uniform properties of temperature and moisture comes to rest or moves very slowly over a large area, it begins to take on the corresponding characteristics of the surface. In general, the weather conditions prevailing at a certain location at a particular time are dependent upon the characteristics of the prevailing air mass, or the *interaction* between two air masses. The interaction of two air masses is known as a *front* and will be discussed later in this chapter.

Except where fronts exist, the weather conditions on the surface covered by an air mass will be similar throughout the entire area. There will be, however, variations produced by local geographic features such as large bodies of water, mountains, or valleys.

AIR MASS SOURCE REGIONS

A source region is the locale where an air mass acquires its identifying properties of temperature and moisture. The most common air mass source regions are large snow or ice covered polar regions, tropical oceans, and large desert areas. (See Fig. 5-8.) The mid latitudes, which have

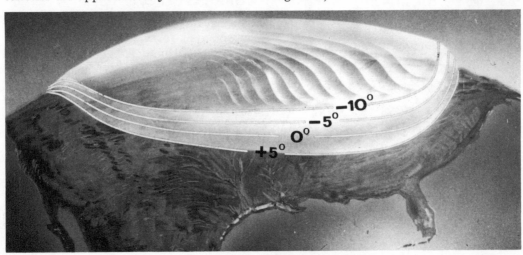

Fig. 5-7. Air Mass

irregular surface features, are less common source regions since air has little opportunity to stagnate in the mid latitudes because weather systems are constantly on the move in these areas.

Air masses formed in the area of 60° north latitude and northward will be cold and are called *polar* air masses. Also, air masses originating over land will be relatively dry and are termed *continental*. For example, one source of *continental polar* air masses is the area between the Hudson Bay and the Rocky Mountains. (See Fig. 5-8.)

If an air mass is formed over water it will acquire moisture and be known as *maritime*. As illustrated in figure 5-8, one source region for *maritime polar* air masses is over the waters south of the Aleutian Islands. *Tropical* air masses are formed over land and water in the region from about 20° to 50° north latitude and naturally these air masses

are quite warm. (See Fig. 5-8.) Air masses may be further designated by source, such as Atlantic, Pacific, or Gulf.

AIR MASS MODIFICATION

Air masses that form over a given source region vary in their properties of temperature and moisture from season to season, as the characteristics of the source region vary.

The extent of air mass modification is dependent upon the length of time a mass remains over a source region and the difference between the original temperature of the air and that of the underlying surface.

Once an air mass has formed in a source region it may move out from this source region on a trajectory of course across the country and carry with it the temperature and moisture characteristics of its *origin*. Much of this movement is due to the basic atmospheric circulation pattern

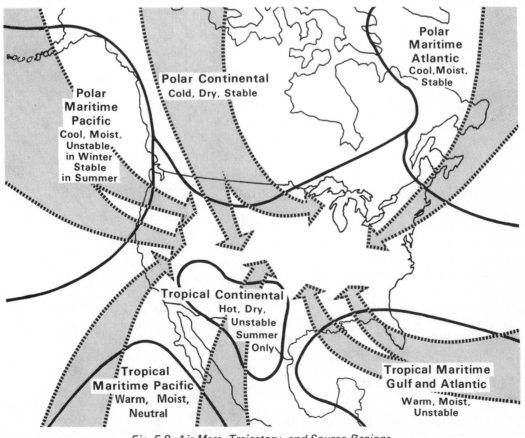

Fig. 5-8. Air Mass, Trajectory, and Source Regions

Fig. 5-9. Cumulus Clouds

Fig. 5-10. Stratus Clouds

discussed previously. The air mass is then further modified by the *surface over which it moves*, although its original characteristics tend to persist.

As air masses build up and start to move they are *also* termed *cold* or *warm*, depending on whether they are warmer or colder than the *surface over which they move*. The cold type air mass will be warmed from below and convection currents will form. Some typical characteristics of the cold type air mass are: (1) instability [tendency to overturn as convection currents form within]; (2) turbulent or rough air up to about 10,000 feet; (3) good visibility, except in showers and duststorms; (4) cumiliform clouds; and (5) precipitation in the form of showers, thunderstorms, hail, sleet, and snow flurries, when enough moisture is available within the air mass.

The warm type of air mass, cooled from below by the surface over which it travels, exhibits different characteristics. This type of air mass tends to maintain its original properties and is modified only in the lower few thousand feet as it moves. Some typical characteristics are: (1) stability [resists vertical movement]; (2) generally smooth air; (3) poor visibility as smoke and dust are held in the lower few thousand feet; (4) stratiform clouds and fog; and (5) drizzle.

CLOUD FORMATION AND TYPES

To an experienced pilot, clouds are *signposts* that indicate the forces at work in the atmosphere. Their location and appearance (size, shape, and color) can show a weather-wise aviator where turbulence might be found, whether a frontal system is approaching, or where hail might be encountered. It is often possible through experience to read the existing weather situation and how it may change by watching the clouds.

CLASSIFICATIONS OF CLOUDS

One way of classifying clouds is according to how they are formed. Two basic types of clouds are *cumulus* and *stratus*.

Cumulus clouds are formed by rising air currents and are prevalent in unstable air that favors vertical development. The currents of air create cumuliform clouds and give them a piled-up or bunched appearance. (See Fig. 5-9.)

Stratus clouds are formed when a layer of moist air is cooled below its saturation point. Stratiform clouds lie mostly in horizontal layers or sheets, resisting vertical development. The word stratus is derived from a form of the Latin word for layer. (See Fig. 5-10.)

The prefixes and suffixes which are combined to make up the names of clouds tell something about the character of a cloud. Clouds which produce rain or snow contain the term *nimbus*. Broken or fragmented clouds are prefixed by *fracto*. *Lenticular* clouds are formed over mountains during high wind conditions and have a lens-shaped appearance.

Fig. 5-11. Cirrus Clouds

Fig. 5-12. Altostratus Clouds

Fig. 5-13. Stratus Clouds

Clouds are also named according to their characteristic elevation. *Cirrus* clouds are the high clouds normally found above 20,000 feet. Because these clouds are found in the cold air at high elevations, they are composed of ice crystals and take on a thin, wispy appearance. (See Fig. 5-11.) *Alto*-prefixed clouds are the middle clouds found between 6,500 feet and 20,000 feet. Clouds without either of these two prefixes are low clouds normally found below 6,500 feet.

HIGH CLOUDS

The high clouds are composed almost entirely of ice crystals due to the extremely cold temperatures at higher elevations. These clouds include *cirrus, cirrocumulus,* and *cirrostratus.* The bases of the high clouds average about 25,000 feet, but may go as low as 16,500 feet in the polar latitudes and as high as 45,000 feet in the equatorial regions. Because they are composed of ice crystals, cirrus clouds do not present an icing hazard to aircraft. Generally, these clouds are so thin that the outline of the sun or moon may be seen through them, producing a halo effect. (See Fig. 5-11.)

MIDDLE CLOUDS

The middle cloud group consists of *altocumulus* and *altostratus.* Middle cloud bases generally range from 6,500 to 16,500 feet. Altocumulus clouds can form in several different ways and take on various appearances. Generally, however, they are patchy layers of puffy, roll-like clouds of a whitish or grayish color. Often they resemble cirrocumulus, but the puffs or rolls are much larger. Instead of being made of ice crystals, middle clouds are made of very small water droplets. Altocumulus clouds are generally associated with unfavorable flying weather and precipitation. Altostratus clouds are simply stratus clouds that form in the middle altitudes. (See Fig. 5-12.)

LOW CLOUDS

The low cloud group consists of *stratus, stratocumulus,* and *nimbostratus* clouds. Bases may range from near the surface to about 6,500 feet. Stratus, or low clouds, are of the greatest importance to the aviator since they create low ceilings and visibility restrictions near the surface of the earth. The height of these cloud bases has been known to change rapidly. If any of the low clouds form below *50 feet,* they are reclassified as *fog.*

The stratus cloud is quite uniform and resembles fog. (See Fig. 5-13.) It has a fairly uniform base and a dull gray appearance. Stratus clouds make the sky appear heavy and will occasionally produce fine drizzle or very light snow with fog. Because there is little or no vertical movement in the stratus clouds, however, they usually do not produce precipitation in the form of heavy rain or snow.

Nimbostratus clouds are the true rain or snow clouds. These appear wet and have a darker appearance than ordinary stratus clouds. Precipitation usually reaches the ground in the form of continuous rain, snow, or sleet. They are often accompanied by fractostratus, commonly known as *low scud*, when the wind is blowing strongly.

Stratocumulus clouds are irregular masses of clouds spread out in a rolling or puffy layer. These do not produce rain by themselves, but sometimes change to nimbostratus types which do. (See Fig. 5-14.) They are very whitish in color, normally with darker spots, and are usually composed of rounded masses which fuse together and become indistinct when rain appears.

The size of the droplets in stratus clouds is significantly smaller than in cumulus clouds.

CLOUDS WITH VERTICAL DEVELOPMENT

Cumulus and cumulonimbus clouds are characterized by vertical development. Cloud bases generally range between 1,000 and 10,000 feet; however, the tops of cumulonimbus clouds may rise to an elevation of 75,000 feet. Their vertical development is caused by some type of lifting action such as convective currents found on hot summer afternoons, or when air is forced to rise up the slope of a mountain, or possibly the lifting action that may be present in a frontal system.

The cumulus cloud has a puffy and cauliflower-shaped appearance whose structure is constantly changing. Over land, cumulus clouds are often formed by the rising air currents and, therefore, disappear at night when the lifting action recedes. These are referred to as fair-weather cumulus unless they begin to pile up to form cumulonimbus.

Cumulonimbus clouds form the familiar *thunderhead* and produce thunderstorm activity. (See Fig. 5-15.) These clouds are characterized by violent updrafts which carry the tops of the clouds to extreme elevations. Tornadoes, hail, and severe rainstorms are all products of this type of cloud. At the top of the cloud, a flat *anvil-like* form caused by the winds aloft appears as the thunderhead begins to dissipate.

Usually, isolated cumulonimbus clouds present no flight problem since they can be circumnavigated without difficulty. However, should these clouds develop in groups or lines known as *squall lines*, they may be impossible to fly around. Sometimes cumulonimbus clouds are em-

Fig. 5-14. Stratocumulus Clouds

Fig. 5-15. Cumulonimbus Clouds

bedded and hidden in stratiform clouds, resulting in very hazardous conditions for flight in instrument conditions.

ELEMENTS OF WEATHER

MOISTURE

The moisture content of the atmosphere is an important weather determinant. Water enters the atmosphere in a gaseous state called *water vapor*. This vapor is produced from evaporation of water surfaces, mainly the oceans. In addition, some water vapor is derived from plants and from the moist earth. The amount of water vapor present in the atmosphere may vary from a trace to approximately five percent by volume.

A measure of the amount of water vapor present in the atmosphere is called *relative humidity*. The amount of water vapor that an air mass is capable of holding is dependent upon temperature. An air mass with a high temperature has a greater ability to contain water vapor than does an air mass of lower temperature.

An air mass with a relative humidity of 90% means the air is 90% *saturated*, or is holding 90% of the water vapor it is capable of holding. If more water vapor was added to the air or if the temperature was reduced, the relative humidity would increase until condensation would take place and visible moisture would appear. The *dewpoint* is the temperature to which the air mass must be cooled in order for condensation to take place. The dewpoint is dependent upon the amount of water vapor present in the atmosphere.

TEMPERATURE

The temperature of the air is primarily regulated by the surface of the earth. Only 15% of the incoming solar radiation (sun's heat) is absorbed by the atmosphere. Eighty-five percent of the atmospheric heating is caused by heating from the earth's surface below. This accounts for the fact that the temperature of still air decreases approximately 3.5° Fahrenheit or 2° Celsius for each 1,000 foot rise in elevation. This is known as the average or *standard lapse rate*.

STABILITY

Stability refers to the ability of an air mass to *resist vertical development*. A stable air mass will resist vertical lifting; however, an unstable air mass will tend to promote vertical development. When air is forced to rise due to *orographic* (air forced to rise over a land surface such as mountains), or *convectional* (air rising due to heating from below) lifting, it may become warmer than the surrounding air. If this should occur, the rising air has reached the level of free convection and will become unstable and continue to rise even though the initial lifting force is no longer present. This situation will produce unstable air with resultant cumulus-type clouds if moisture is present.

TEMPERATURE INVERSIONS

In cases where the temperature increases with an *increase* in altitude, an inversion is present. (See Fig. 5-16.) Since air is a poor conductor of heat, an *inversion* may persist for days or possibly even weeks before the air is cooled above. This,

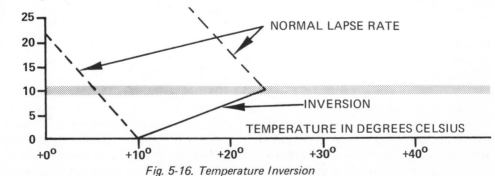

Fig. 5-16. Temperature Inversion

of course, is assuming no horizontal movement of the air mass. An inversion may cause smog or fog to lie in an area for days.

Inversions may occur at the surface or above the surface when warm air overlies a colder layer of air. Inversions result in an extremely stable condition since the warmer air on top will tend to rise and the colder air on the bottom will tend to descend, thus allowing very little mixing of air in the vicinity of the inversion layer.

Inversions are often produced on cold, clear nights over snow-covered ground. Because the ground is much colder than the air, the air just above the surface becomes cooler than the air at altitude.

PRESSURES AND WINDS

Wind is simply air in motion caused by differences in pressure from one place to another. In the United States, two types of pressure systems are present, both of which are caused by the *unequal* heating of the earth's surface, including the ocean. These are the semi-permanent high and low systems and the variable day-to-day high and low pressure systems.

Atmospheric pressure is the force exerted by the weight of the atmosphere and is due to the gravitational force of the earth. The weight of the atmosphere over a given point is continually fluctuating. These fluctuations result from air mass motion and changes in the temperature and moisture content of the air. A column of air one square inch across at sea level weighs approximately 14.7 pounds.

A device used to measure atmospheric pressure is a *barometer*, which means weight meter.

The simplest kind of barometer — the mercury type — consists of a dish filled with mercury into which is placed the open end of a glass tube with the air removed. (See Fig. 5-17.) The weight of the

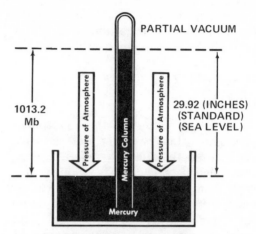

Fig. 5-17. Simple Barometer

atmosphere presses down on the mercury in the dish and supports the column of mercury inside the tube. At sea level, the pressure of the atmosphere on a standard day will cause the mercury in the column to rise to 29.92 inches. This value is accepted as the standard value of atmospheric pressure at sea level. As altitude increases, the weight of the atmosphere decreases. The height of the mercury column will decrease by nearly one inch for each 1,000 foot ascent. This is known as the *pressure lapse rate.*

In scientific work, a convenient unit of measurement of atmospheric pressure is the *millibar.* This unit is used internationally for pressure measurement. The standard atmospheric pressure at sea level is 1013.2 millibars (mb), corresponding to 29.92 inches of mercury or 14.7 pounds per square inch. These pressures are all equal, but are expressed in different units. One inch of mercury is equivalent to approximately 34 millibars. Weather charts normally show barometric pressure in millibars, while teletype weather reports usually express barometric pressure in millibars *and* inches of mercury.

STATION PRESSURE

Station pressure is the actual atmospheric pressure taken at a given place. The pressure observed depends upon the height of the station above sea level, the effect of gravity, and the amount of air above the station. Since air becomes thin-

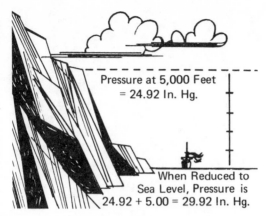

Pressure at 5,000 Feet
= 24.92 In. Hg.

When Reduced to
Sea Level, Pressure is
24.92 + 5.00 = 29.92 In. Hg.

*Fig. 5-18. Altimeter Setting
Converted to Sea Level*

ner with altitude, the pressure at higher elevations will be less than at lower elevations if standard atmospheric conditions are present at both places.

Station pressure is of little value to the meteorologist, however, since the pressure at higher elevations is normally less than at lower elevations even if standard atmospheric conditions persist at both places. For instance, if the barometric pressure was 29.92 inches of mercury at sea level, the pressure at 5,000 feet would be approximately five inches less, or 24.92 if standard atmospheric conditions existed at both elevations. (See Fig. 5-18.)

If all weather observation stations were located at sea level, then the individual barometric readings would give a correct picture of the amount of pressure exerted on a common level. However, to be meaningful, atmospheric pressure measurements are converted to a standard reference, or the *mean sea level altitude.* This eliminates differences in pressure caused by differences in station elevation.

TYPES OF PRESSURE SYSTEMS

The amount of atmospheric pressure at any given time varies for three main reasons. The first is the passage of a well-developed pressure system. In this situation, it is not uncommon to find a change of atmospheric pressure of one inch of mercury in a day's time. The second is an actual change in the intensity of the pressure system. An example would be the deepening of a low pressure area. And last, a daily variation can occur where the pressure is lowest during the late evening and early morning hours, and highest from about mid-morning to around 10 p.m. In the United States, a daily variation may amount to as much as 0.04 of an inch of mercury. It is important to realize that the daily variation is normal and does not necessarily indicate the approach of a storm or a change in the pressure system.

Isobars, or lines connecting points of equal pressure, are drawn on weather maps. "Iso" means equal and "bar" means barometric pressure. These lines of equal pressure are drawn at selected intervals, usually four millibars apart, and indicate the configuration of the pressure system over a given area. The isobar pattern is never the same on any two weather maps; however, their patterns are similar and have definite meanings for the aviator. The five types of pressure systems are defined as follows: (See Fig. 5-19.)

1. LOW — a center of low pressure surrounded on all sides by higher pressure. An area of low pressure is like a depression or a valley in the atmosphere.

2. HIGH — a center of high pressure surrounded on all sides by lower pressure. It can be visualized as a mountain surrounded by valleys or areas of low pressure on all sides.

3. TROUGH — an elongated area of low atmospheric pressure, usually extending from the center of a low pressure system.

4. RIDGE — an elongated area of relatively high pressure, extending from the center of a high-pressure system.

5. COL — a neutral area that exists between two high-pressure areas or low-pressure areas.

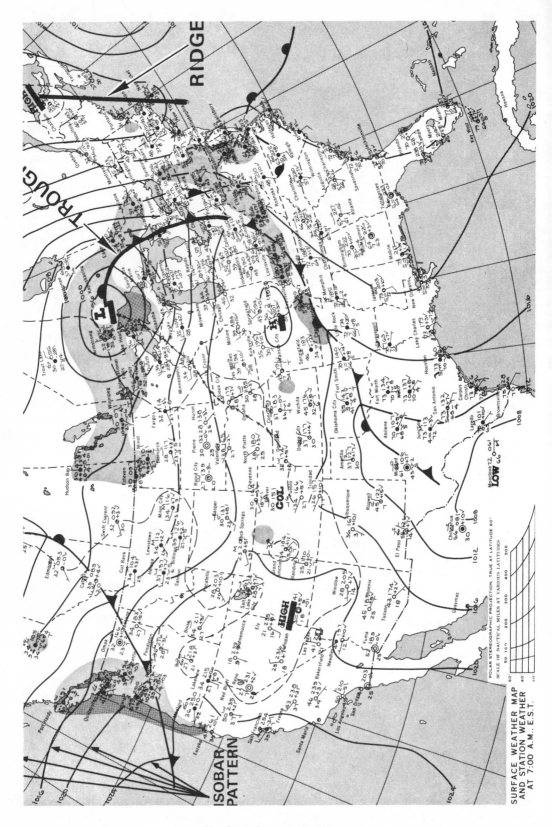

Fig. 5-19. Pressure Systems

PRESSURE SYSTEM CHARACTERISTICS

Pilots should know the general character-istics associated with different pressure systems in the Northern Hemisphere. In the Northern Hemisphere, the wind flow is counterclockwise around the low pressure system. In addition, unfavorable flying conditions may be present in the form of low clouds, poor visibility, rain or snow, fog, strong gusty winds, and turbulence.

Conversely, the winds around a high-pressure system in the Northern Hemi-sphere flow clockwise. Flying conditions are generally much more favorable than in low-pressure areas. This is because there are normally fewer clouds, better visibility, and only calm or light winds. and the areas of turbulence are much less severe.

Where a trough is present, the weather is frequently very violent; whereas along a high-pressure ridge, favorable flying weather will generally prevail.

PRESSURE GRADIENT

If it were not for the earth's rotation, air from high-pressure areas would tend to flow directly toward areas of low pres-sure. The speed at which this wind would blow would be dependent upon the steep-ness of the *pressure gradient*, or the amount of change of pressure over a given distance. (See Fig. 5-20.)

The spacing of isobars on a surface weath-er map gives a good impression of the steepness of the pressure gradient. Where the isobars are spread widely apart, the pressure gradient is very *shallow;* how-ever, where isobars are close together, the pressure gradient is very *steep.* The steep-ness of the pressure gradient gives an indication of wind velocity. A steep pres-sure gradient will normally produce strong wind, and a shallow pressure grad-ient will result in a light wind.

In general, upper winds tend to flow parallel to the isobars. This is due to the resultant effects of two opposing forces — centrifugal force and pressure gradient. As air spins counterclockwise around a low, centrifugal force is produced out-ward from the center of the low. The pres-sure gradient, on the other hand, direct-ly opposes centrifugal force and the re-sulting winds flow *parallel* to the isobars. (See Fig. 5-21.)

On weather maps, the upper wind direc-tion can be determined by noting the pattern of the isobars surrounding the high- and low-pressure areas. Remember, winds flow *parallel* to the isobars and *clockwise* around a high and *counter-clockwise* around a low-pressure area.

FRICTION EFFECT

Friction between the air masses and the surface of the earth retards the movement of the air. This causes the low-level winds to blow at an angle across the isobars from a high-pressure area to a low-pres-sure area instead of parallel to the isobars as in the upper atmosphere. The amount which the wind veers due to the friction effect varies with the type of ground sur-face and velocity of the wind.

Over water, where the friction is slight, the wind will veer only about ten degrees. Over land, however, depending on the roughness of the terrain, the amount of veer may be as much as 30 to 50 degrees from parallel to the isobars. Naturally, the friction effect is greatest near the sur-face of the earth, but exists up to 1,500 to 2,500 feet above the ground. Above this level, the friction effect decreases

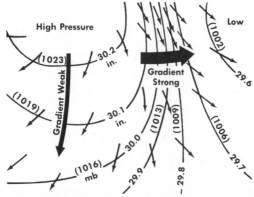

Fig. 5-20. Pressure Gradient

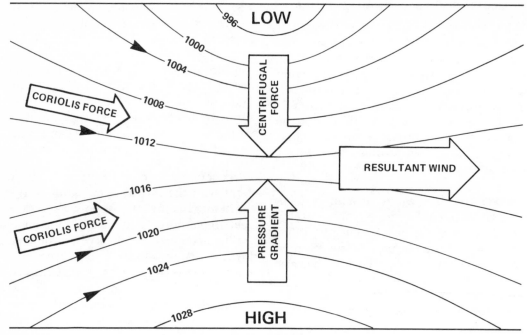

Fig. 5-21. Counterbalancing Forces

very rapidly, and, for all practical purposes, can be considered negligible. Therefore, air about 2,500 feet or more above the surface of the earth will usually tend to flow parallel to the isobars.

LOCAL WIND SYSTEMS

Local wind systems are created by the immediate geographic features in a given area, such as mountains, valleys, or large bodies of water. These systems account for significant weather changes in some localities.

LAND AND SEA BREEZES

Because land and water surfaces absorb heat at different rates, *convection* over the two kinds of surfaces varies and causes local winds. During the daytime, the temperature of the land rises faster than the bordering water and cools faster during the night. This difference between the temperature of the land and of the water produces a corresponding difference in the local pressure gradient. As the earth is heated during the day and convection begins to take place, the pressure over the land becomes lower than over the colder water.

This sets up an air circulation pattern during the day as shown on the left side of figure 5-22. Air moves from the high-pressure area over the cooler sea toward the low-pressure area over the warmer land surface. This creates the characteristic sea breeze that is normally encountered during the daytime hours on the coast. As this air rises over the warmer land surface, it cools and descends again over the water to continue the circulation.

The land breeze, as shown on the right half of figure 5-22, is produced at night from a circulation pattern opposite the circulation pattern which occurs during the day. During the evening hours, the land cools faster than the sea creating a high pressure over the land surface and a lower pressure area over the ocean. Air moving from the "land high" to the "sea low" creates a breeze outward from the land toward the ocean. This is known as a land breeze.

EFFECTS OF ELEMENTS OF WEATHER

After gaining an understanding of the elements producing the earth's weather, the pilot is concerned with the effects these

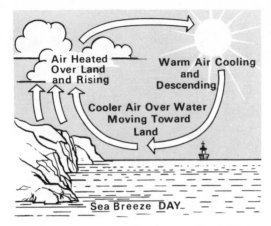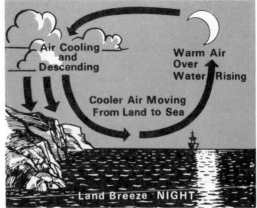

Fig. 5-22. Land and Sea Breezes

elements have on the atmosphere in which he must fly. Pilots are mainly concerned with two major factors: the ceiling (the height above the earth's surface of the *lowest* layer of clouds or obscuring phenomena that covers more than one-half of the sky) and visibility. One of the major visibility restrictions that has been a hazard for pilots throughout the history of aviation is *fog*.

FOG

Fog is a stratiform cloud within 50 feet of the surface. Fog consists of small water droplets or ice crystals suspended in the atmosphere. Although they may each be too small to see with the naked eye, they are so numerous that visibility is reduced. The existence of fog is a serious problem in terminal areas because of the restricted visibility during the takeoff or landing phase of flight. A knowledge of the different types of fog, how they are formed, where they are most likely to occur, and the dissipation properties are extremely important in the planning of a safe flight.

For fog to form, several conditions must be present. First, a relatively high humidity must exist in order for condensation to occur. Therefore, it is necessary for a small *temperature/dewpoint spread* to be present. Anytime the temperature and dewpoint are within 4° Fahrenheit or less of each other, pilots should expect the possibility of fog. Fog is most frequently encountered, however, when the spread is

less than 2°F. The temperature/dewpoint spread can be ascertained from aviation weather reports which are described later in this chapter.

Generally, a light surface wind is necessary for fog formation. Wind provides a mixing action which spreads and increases the thickness of the fog. If no wind is present, the fog will most likely be shallow and close to the ground. Strong surface winds, however, are not conducive to fog formation. High winds break up the fog layer and move the stratus clouds away from the surrounding areas.

The third necessary ingredient for the formation of fog is an abundance of *condensation nuclei* such as particles of salt or smoke which are suspended in the atmosphere and provide the nuclei around which moisture can condense.

The high relative humidity necessary for fog formation can occur in two distinct ways: (1) when the air is cooled to its dewpoint; (2) when the moisture is added to the air near the surface.

The different types of fog are named according to the processes by which they are formed.

Radiation fog forms as the earth rapidly loses its heat on clear nights. If the sky is overcast, the radiation escaping into the

Fig. 5-23. Radiation Fog

Fig. 5-24. Advection Fog

atmosphere is limited and most of it is reflected back to the earth. When the earth cools on a clear night, however, the temperature near the surface may drop more than 20° Fahrenheit. Although the dewpoint remains constant during the night, the temperature/dewpoint spread decreases as the air is cooled by direct contact with the earth's surface and radiation fog develops.

Radiation fog is sometimes referred to as *ground fog*. Often this fog formation occurs near sunrise when the temperature is at its lowest reading. Radiation fog normally lifts and dissipates as the temperature begins to rise within a few hours after sunrise. (See Fig. 5-23.)

Advection fog is formed when warm, moist air flows over a cold surface. This type of fog is most often found along coastal regions where the temperature of the land and the temperature of the water differ widely. As warm, moist air flows inward over the continent, it is cooled by direct contact with the cold ground. If the temperature of the air reaches the dewpoint temperature, fog forms.

Advection fog can be very persistent during both day and night. Like radiation fog, it deepens as the wind speed is increased up to about 15 knots. If the velocity of the wind is about 15 knots or greater, however, advection fog will usually be lifted to form a layer of low stratus clouds. If advection fog forms

over water and then blows ashore, it is sometimes referred to as *sea fog*. (See Fig. 5-24.)

Upslope fog occurs when moist air flows uphill resulting in a temperature drop through adiabatic cooling or cooling by expansion. In order for upslope fog to form, the pressure gradient must be great to maintain an uphill airflow. This upslope wind is not only necessary for the formation of upslope fog, but also for its continued existence. (See Fig. 5-25.)

Steam fog occurs when cold air moves over water that is much warmer than air. This causes intense evaporation of the moisture into the cold air until it becomes saturated and the excess water vapor condenses as fog. Because steam fog rises upward from the water and looks much like rising smoke, it is sometimes called *sea smoke*.

Steam fog is commonly found over rivers and lakes during the fall and over the ocean during the winter when the wind is blowing from offshore. Since its formation is brought about by the movement of a cold air mass over a warm surface, the air is unstable. Turbulence and icing conditions are frequently encountered in this type of fog.

Rain-induced fog results from the addition of moisture to the air or by the evaporation of cold rain falling from above. This type of fog usually becomes more

Fig. 5-25. Upslope Fog

Fig. 5-26. Rain-Induced Fog

dense as night approaches and the air near the ground cools. (See Fig. 5-26.)

HAZE AND SMOKE

Haze consists of very fine dust or salt particles suspended in the atmosphere. Stable air is necessary in order for haze to become a very serious restriction to visibility since the particles can only be concentrated in stable air. Haze is often a characteristic of an inversion condition. Although it is usually only found at low altitudes, the top of a haze layer can go as high as 15,000 feet above the surface.

Flight visibility on hazy days depends upon several things: the height of the haze layer, its intensity, the position of the sun, and the direction in which the pilot is looking. Visibility is often much less when the sun is low and unobscured by clouds above the haze.

Smoke is usually found near large industrial areas and restricts the forward visibility in much the same way as haze. When smoke is produced by forest fires, it is normally in concentrated layers with generally good visibility above and below these layers.

Smog is technically a mixture of smoke and fog and can produce very poor visibility. Today, however, the term "smog" has been applied to any heavy concentration of atmospheric pollution. Normally present in large industrial areas, flight within a smog area is similar to that in haze or smoke. The amount of visibility depends upon its density and whether an individual is facing toward or away from the sun. When the top layer of smog is well defined, visibility above is usually very good.

VISIBILITY RESTRICTIONS DUE TO WIND

Wind, blowing snow, dust, or sand can produce very low visibility. Blowing snow makes it difficult for the pilot to distinguish objects and sometimes produces optical illusions. Blowing dust occurs in dry areas where the air is unstable and wind speed is high. An optical illusion can result when the pilot tries to top a layer of blowing dust. The pilot may sense that climbing a few hundred feet more will carry him to the top of the dust layer; however, as the climb is continued, he may find that it will be necessary to climb several thousand feet before actually topping the layer.

OBSCURED SKY CONDITIONS

The word *obscured* as used in aviation meteorology means that the sky or clouds are partially or totally hidden from an observer on the ground. An obscuration may extend upward from the surface of the earth and is usually caused by various forms of visibility restrictions such as drizzle, snow, smog, fog, etc. The phrase *total obscuration* means the sky is

Fig. 5-27. Obscured Sky

totally covered to the observer on the ground.

The reported ceiling in this situation is the vertical visibility looking up from the ground. If an observer can see part of the sky or clouds through the obscuring phenomenon, the sky condition is reported as a *partial obscuration*. In this case, the height of clouds above is reported rather than the vertical visibility through the obscuring phenomena. (See Fig. 5-27.)

WEATHER NEAR LARGE BODIES OF WATER

Large bodies of water sometimes have pronounced effects on local weather conditions. For example, during the winter season, the Great Lakes are warmer than the surrounding land areas and the cold air masses that move southward from Canada. Heat and moisture are added to these cold air masses as they pass over the lakes. The resulting convection causes cumulus-type clouds and rain or snow showers and squalls. The areas on the downwind side of large bodies of water are usually affected by this type of weather.

FRONTS

Air masses which have different properties of temperature and/or moisture content do not mix to any appreciable degree. When different air masses collide they form a *front*. A front or frontal zone may be hundreds of miles long and many miles wide.

If a cold air mass is advancing toward a warm air mass, the cold, heavy air attempts to push under and lift up the warm, relatively lighter air. This is known as a *cold front*. Conversely, a warm air mass advancing toward cold, retreating air pushes up over the cold air, thus creating a *warm front*. If the boundary between the two air masses does not move, then a *stationary front* has developed.

In the case of warm and cold fronts, a clash between the two air masses may exist, breeding unsettled and stormy weather. The different types of air masses which may clash were discussed in a previous paragraph. (See Fig. 5-28.)

Fig. 5-28. Front

Fronts are of great importance to pilots since many weather hazards may be present. Usually frontal weather produces clouds and precipitation and in some cases helps to breed thunderstorms.

DISCONTINUITIES ACROSS FRONTS

Fronts are located and classified by the difference in properties between the air masses creating the front. These factors include temperature, moisture, pressure, wind, and type of clouds present. When a front passes, a meteorological change is experienced from the properties of one mass to the properties of another air mass.

This can be very rapid and may occur in a distance of less than a mile. If the change is gradual, however, the transition zone or area of discontinuity may be over 200 miles wide.

The most easily recognized discontinuity across a frontal zone is *temperature.* On the surface of the earth, a notable change in temperature occurs with the passage of a front. With a rapidly-moving front, the rate of temperature change is usually quite rapid. In the case of a slow-moving front, the temperature change is gradual.

Pressure changes will also occur with the passage of a front. Because a front is normally located along a low-pressure trough, the pressure on each side of the front will be higher. When a front is approaching, the pressure will usually be decreasing and, at the point of frontal passage, the pressure will be at its lowest reading. As the front passes, the pressure will begin to rise and the direction of the wind will change. In the Northern Hemisphere, the wind shift is toward the right, or in a clockwise direction as the front passes.

FACTORS WHICH INFLUENCE FRONTAL WEATHER

The factors which influence frontal weather are dependent upon moisture availability, the stability of the air which is being lifted, the frontal slope, the speed of the front, and the amount of temperature and moisture difference between the two air masses.

It should be understood that not all frontal weather is severe or contains hazardous flying conditions. Fronts may be invisible with little or no clouds or precipitation present. They can be extremely hazardous to aviation activities, however, when thunderstorms, low clouds, and poor visibility exist. Where there is sufficient moisture available, the front will have an abundance of clouds and precipitation resulting in poor visibility.

The stability of the air being lifted by the frontal system will determine the type of clouds which will form, either stratiform or cumuliform type. With stratiform clouds, there is little or no turbulence and if precipitation is present, it is steady. Cumuliform clouds possess the shower type of precipitation and are an indication that turbulence may be present.

The slope of the front determines the extent of clouds and precipitation areas. In a shallow frontal surface, there will be large areas of cloudiness and precipitation. With a steep frontal surface moving rapidly, however, there will be narrow bands of cloudiness and showery precipitation. Air masses which are vastly different in their properties will have a steep frontal slope, as illustrated in figure 5-29.

COLD FRONTS

A cold front is a wedge of advancing dense, cold air replacing warm air at the surface. It is called a cold front because the cold air is forcing its way under the warm air. The cold air is capable of doing this because it is more dense. (See Fig. 5-29.) In the Northern Hemisphere, intense cold fronts are normally oriented in a line from the northeast to the southwest and their movement is toward the east or southeast.

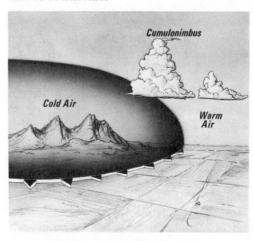

Fig. 5-29. Cold Front Profile

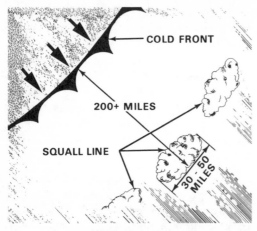

Fig. 5-30. Squall Line

Cold fronts are classified as either *fast moving* or *slow moving*. Fast-moving cold fronts may contain winds as high as 60 m.p.h. and can move across the surface at speeds above 30 m.p.h. Generally, their speed is faster in the winter than in the summer, causing the slope to steepen resulting in a narrow band of weather. A thunderstorm may develop ahead of a fast-moving cold front. A line of thunderstorms or a squall line will have cloud tops which may average 40,000 feet or higher. Some of these cloud tops have been observed as high as 75,000 feet.

A squall line may develop between 50 and 300 miles ahead of the front and will lie roughly parallel to the front itself. (See Fig. 5-30.) With a fast-moving cold front, the weather behind will clear rapidly because the air is so much colder than the surface over which it is lying. Gusty and turbulent winds will be present upon passage of a fast-moving cold front.

With a slow-moving cold front, the frontal slope is less steep and the warm air is not lifted as quickly or violently as with the fast-moving cold front. This results in a broad cloud cover extending well over the front. If the warm air is stable as it is being lifted, stratiform clouds will develop. If, however, the warm air is moist and unstable, cumulus-type clouds will normally be present.

WARM FRONTS

A warm front is created when a warm air mass is overtaking and replacing a cold air mass. The leading edge of the advancing warm air is called the warm front. The speed of a warm front is approximately half that of a cold front or about 15 m.p.h., or even slower. The less dense, warm air gradually moves up the slope and does not produce the typical bulge of cold air which is associated with a cold front. This occurs because ground friction drags the bottom edge of the retreating cold air. This, in turn, creates a broad cloud system which typically extends from the surface portion of the front to about 500 to 700 miles in advance of the surface front.

The type of clouds that form are dependent upon the moisture content and the stability of the air as it rises up the slope. (See Fig. 5-31.) If the air is warm, moist, and unstable, however, cumuliform clouds will be mixed with stratiform clouds. When the air is stable, the sequence of clouds encountered with the approach of a warm front is as follows: cirrus, cirrostratus, altostratus, and nimbostratus. Precipitation will increase gradually with the approach of a warm front and will usually continue until after frontal passage.

If the air is unstable, the sequence of the clouds will be as follows: cirrus, cirrocumulus, altocumulus, and cumulonim-

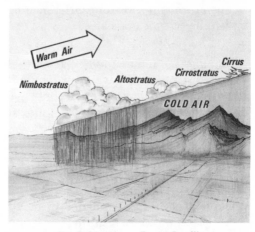

Fig. 5-31. Warm Front Profile

Fig. 5-32. Occluded Front Profile and Plan View

bus. Cumulonimbus clouds are frequently embedded in the cloud masses as the warm front approaches. This is the major hazard for pilots flying in unstable warm front areas.

Showery precipitation will usually occur with the advance of an unstable warm front. Because of the large area of precipitation ahead of the warm front surface, low clouds in the form of stratus and fog will sometimes form creating areas of poor visibility and low ceilings over a great area. If the retreating cold air has temperatures below freezing, the precipitation may be in the form of freezing rain or sleet, another hazard to aircraft flying in the vicinity.

STATIONARY FRONTS

When a front has little or no movement, it is then referred to as a stationary front. This condition exists when opposing air masses are of *equal* pressure. In the case of a stationary front, the surface wind tends to blow parallel to the front instead of against or away from it. The weather phenomena associated with stationary fronts are comparable to those found in a warm front but are usually less intense in nature and cover a much wider area.

OCCLUDED FRONTS

Occluded fronts develop from a complex storm system which consists briefly of a warm front followed by a cold front

both radiating outward from the center of the same low-pressure system. (See Fig. 5-32.) Many times the rain remains in the center of a low-pressure system. They are of two types, cold and warm, designated by the front that develops at the surface.

The occlusion starts at the center of the low and progresses outward as the faster moving cold front overtakes the warm front. This produces the occluded front profile as shown in figure 5-32A. The weather associated with occluded fronts will usually consist of widespread rains and cloud cover with poor visibility and possible icing.

THUNDERSTORMS

Thunderstorms produce the most severe type of weather known to mankind. Tornadoes with winds reaching 350 miles per hour, hailstones the size of baseballs, and extreme turbulence are all conditions which can result from thunderstorm formation.

Thunderstorms are a hazard for all types of flight operation since tops of the cumulonimbus clouds may reach elevations in excess of 75,000 feet. Even jet airliners capable of topping all other types of weather phenomena are subject to thunderstorm hazards.

The hazards that exist with thunderstorm activity are not confined just to the

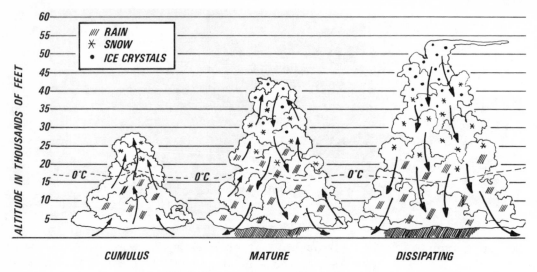

Fig. 5-33. Thunderstorm Development

storm area itself. Pilots flying outside the thunderstorm cell, but near the storm area, can encounter hail and/or extreme turbulence.

Several factors must exist for the formation of a thunderstorm. First, for the formation of the cumulus cloud and for its continuing build-up, some sort of *lifting action* must be present. This lifting action may be orographic, convectional, or frontal.

Orographic lifting consists of air forced to rise over the surface of the earth, such as air rising over a mountain. *Convectional lifting* occurs when air being heated from below is forced to rise. *Frontal lifting* takes place when moist air is forced aloft by the wedge of an advancing cold front. In addition, the formation of a thunderstorm requires the availability of moisture and unstable air.

STAGES OF THE THUNDERSTORM

There are three stages to the development of a thunderstorm. The first stage is known as the *cumulus stage*. The main feature of this stage is the cumulus cloud and the updraft which may extend from near the earth's surface to several thousand feet above visible cloud tops. Water droplets are very small, but grow into

raindrops as the clouds build upward. Many times the raindrops remain in the liquid state even above the freezing level. These raindroplets are suspended by the currents within the clouds. (See Fig. 5-33.)

The second or *mature* stage begins as rain begins to fall at the earth's surface. Raindrops and ice particles by this point have grown to such a size that they can no longer be supported by the updrafts. The mature stage occurs approximately 10 to 15 minutes after the cloud has built beyond the freezing level in the atmosphere. Occasionally, during the mature stage, a cloud may build as high as 50,000 to 60,000 feet, but 25,000 to 30,000 feet is the norm.

Severe up and downdrafts occur in the mature stage. As the raindrops fall, they pull air with them and create downdrafts that may exceed 2,500 feet per minute. This causes gusty winds at the surface as the downdrafts strike the earth and spread outward.

The *dissipating* stage is characterized by the collapse of the cumulonimbus cloud. Downdrafts continue to develop and spread vertically and horizontally while updrafts weaken and finally dissipate completely. Soon the entire thunderstorm becomes an area of downdrafts. Rain decreases, then ceases, and the thunderstorm begins to dissipate. The top of the thunderstorm, at this point, begins to develop the characteristic anvil appearance with the point of the anvil in the direction of the prevailing winds. (See Fig. 5-33.)

WEATHER IN MOUNTAINOUS AREAS

Terrain, as well as fronts, can cause air to be lifted and cooled until cloudiness results. When the wind blows against a mountain barrier, the moisture content of the air and its stability determine the amount and height of clouds, their thickness, and their type. As the air descends on the leeward side of the terrain, it is warmed and clouds disappear.

Fronts and low-pressure systems, moving through mountainous regions, create more cloudiness and precipitation, turbulence, and icing conditions on the windward slopes than occur as the systems move over flat terrain. This is because the terrain lifting effect is added to the frontal lifting forces at work.

Because of the rapid changes of elevation in mountainous areas, weather conditions frequently change rapidly in short distances. Cloud heights above ground level, in particular, vary significantly from place to place even though the mean sea level elevation of clouds may be uniform. Pilots should understand the effects of terrain on weather especially in mountainous areas.

MOUNTAIN WAVE

The movement of air over the mountain ridges also causes turbulence, and may generate a mountain wave. The air, being cooled and forced up over a mountain, has a tendency to roll, or tumble, down the leeward side of the mountain causing tremendous downdrafts and turbulence. It also causes

a wave or ripple in the atmosphere which can extend for miles downwind from the mountain range.

The only visual means of knowing that this tremendous turbulence exists is by noticing the various clouds that form in this area. (See Fig. 5-34.) A lenticular cloud, which is a lens-shaped cloud, is usually found at the top in the crest of one of the standing mountain waves. At about the same level as the mountain peak, or the ridge of the mountain range, roll or rotor clouds are formed. A pilot should be on the lookout for this type of warning in mountainous areas. These downdrafts and turbulence are very severe and could cause an aircraft to go out of control and crash into the mountains.

A mountain wave should be anticipated whenever winds of 40 knots or greater are blowing perpendicular to the mountain range. Aircraft should fly at a level at least 50% greater than the height of the range to avoid mountain turbulence. For example, a pilot should fly at 15,000 feet for a 10,000-foot range of mountains. (See Fig. 5-34.) Small aircraft, whose performance ceiling is close to the suggested height, should not attempt to cross the mountain ranges. Also, an oxygen system should be available to the occupants of the aircraft for extended flight above 10,000 feet.

Fig. 5-34. Mountain Waves

The effects of the mountain wave are dangerous since they can produce downdrafts in excess of 5,000 feet per minute and extreme turbulence. Therefore, pilots should avoid flight in mountainous areas when winds at mountain top level are at high velocities. If moderate or severe turbulence is encountered, however, it is important for the pilot to immediately reduce airspeed to the maneuvering speed and maintain a constant attitude to prevent excessive load factors.

ICING

During flight in areas of visible moisture, when the outside air temperature is below +4° Celsius, structural icing can occur. Ice can build up on any exposed surface of an airplane causing a loss of lift, an increase in weight, and a control problem. Any aircraft operating in icing conditions should be equipped with anti-icing and de-icing equipment.

Two types of ice can form on the airplane wings, tail surfaces, propellers, and other components: *rime ice* and *clear ice*. Rime ice is normally encountered in stratus-type clouds and results from instantaneous freezing of tiny water droplets striking the airplane surface. Rime ice has a characteristic opaque color caused by air being trapped in the water droplet as it instantly freezes. The major hazard of rime ice is its ability to change the shape of an airfoil, destroying the airfoil's lift. Since rime ice instantly freezes, it builds up on the leading edge of airfoils but does not flow back following the basic curvature.

Clear ice is normally found in cumulus-type clouds and is formed by the relatively slow freezing of large water droplets. Clear ice can glaze airplane surfaces when the large water droplets slowly freeze as they flow over the airplane structure.

Clear ice is the most serious of the various forms of ice because it adheres tenaciously to the aircraft and is more difficult to remove than rime ice.

SECTION B - WEATHER REPORTS AND FORECASTS

WEATHER SERVICES

One of the important steps in learning about weather is becoming acquainted with facilities and services available to the pilot. The National Weather Service maintains a comprehensive weather observing program and a nationwide aviation weather forecasting service. However, at some locations, the actual weather observations are made by FAA flight service specialists or by the FAA traffic controllers associated with the airport traffic control towers.

The FAA flight service stations are pilot communication centers that provide pre-flight weather briefing for pilots and scheduled aviation weather broadcasts over en route navigation aids. These stations are strategically located throughout the nation. If an airport has no flight service station or traffic control tower but does have a Weather Service Office, pilot briefing services are provided by the National Weather Service. The National Weather Service and flight service stations are well equipped with a vast communications system, measuring instruments, maps, charts, and radar information to perform many important services for the pilot.

The National Weather Service has stations throughout the United States and coordinates the use of weather information from Canada, Alaska, and many ships at sea. These stations are the primary means of determining current weather and predicting future weather. At most of these stations, meteorologists are on duty 24 hours a day making observations and sending hourly reports to central locations. The National Weather Service is responsible for operating the national weather teletypewriter systems.

TELETYPE REPORTS

Along the airways and important points off the airways, weather observations are taken and transmitted at periodic intervals. A series of teletype circuits is used to provide weather stations throughout the country with up-to-date weather information. Each station is part of a circuit which consists of all stations in a geographical region. The United States is divided into regions. Each station in a local circuit sends its own report in assigned order. Monitoring stations on the circuit then relay selected reports from other stations, beginning with nearby stations and following with stations at greater distances. The result is that stations on the local circuit receive reports from all stations on their own circuit and selected reports from other circuits.

HOURLY AVIATION WEATHER REPORTS

Hourly aviation weather reports are used to compile information concerning existing surface weather at various observation stations. These reports are given every hour. To help explain this type report, figure 5-35 shows a sample from the teletype machine. A plain-language interpretation of the report is given in the following paragraph. Then, on following pages, aviation weather reports are described in detail.

"Wichita . . . special report . . . 300 feet scattered; measured 800 feet broken; 2,000 feet overcast . . . One and one-half miles visibility with thunderstorms and moderate rain showers . . . Barometer reading, in millibars, converted to sea level, 1013.2 . . . Temperature 72° F . . . Dewpoint 64° F . . . Wind is from 020° at 14 knots with gusts to 18 knots . . . The altimeter setting is 29.92 in. Hg. (con-

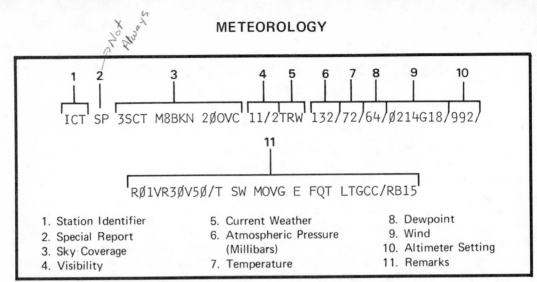

1. Station Identifier
2. Special Report
3. Sky Coverage
4. Visibility
5. Current Weather
6. Atmospheric Pressure (Millibars)
7. Temperature
8. Dewpoint
9. Wind
10. Altimeter Setting
11. Remarks

Fig. 5-35. Sample Aviation Weather Report

verted to sea level) . . .Runway 01 visual range is variable from 3,000 to 5,000 feet. A thunderstorm is southwest of the station, moving eastward and has frequent lightning from cloud to cloud. The rain began at 15 minutes after the previous hour."

STATION IDENTIFICATION

The first part of the aviation weather report is the three letter *station identification code*. (See Fig. 5-35, item 1.) The one shown in the sample sequence report is "ICT" which stands for Wichita, Kansas. A list of station codes is usually placed in the flight service station or National Weather Service along the edge of clipboards or bulletin boards where the reports are mounted to help pilots identify the stations to which the hourly weather reports are referring.

An "SP" in the second segment of an hourly weather report indicates it is a special report. (See Fig. 5-35, item 2.) This would mean that there was a significant change in the weather condition at the field since the last report was issued. Sometimes these specials are transmitted on the teletype separately; however, they are usually embedded in a series of hourly weather reports.

SKY COVERAGE

The sky coverage segment of the aviation weather report indicates the amount and height of clouds in the sky. The symbols indicating the extent of sky coverage are shown in figure 5-36. The portion of the sample report (see Fig. 5-35, item 3) indicates scattered clouds at 300 feet, broken clouds at 800 feet, and an overcast layer at 2,000 feet. The heights of the clouds are given in hundreds of feet *above ground level* (AGL). Two zeros must be added to the number given in the report; i.e., 3 designates 300 feet.

○	CLR	Clear: Less than 0.1 sky cover.
◔	SCT	Scattered: 0.1 to 0.5 sky cover.
◑	BKN	Broken: 0.6 to 0.9 sky cover.
⊕	OVC	Overcast: More than 0.9 sky cover.
−		Thin (When prefixed to the above symbols.)
−X		Partial obscuration: 0.1 to less than 1.0 sky hidden by precipitation or obstruction to vision (bases at surface).
X		Obscuration: 1.0 sky hidden by precipitation or obstruction to vision (bases at surface).

Fig. 5-36. Sky Cover Symbols

CEILING

For all practical purposes, the pilot can think of a *ceiling* as being the lowest height at which more than half of the sky is covered by clouds. A further definition is: ceiling is the height above the ground of the base of the lowest layer of clouds or obscuring phenomena aloft that is reported as broken or overcast and not classified as scattered, thin, or partial. The term *thin* means that the cloud cover

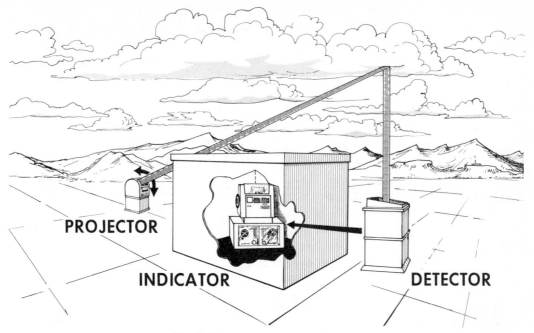

Fig. 5-37. Rotating Beam Ceilometer

is predominantly transparent and does not constitute a ceiling. This condition is indicated by a minus sign (-) preceding the sky cover symbol.

The ceiling figures in the aviation weather reports are always preceded by a code letter which indicates how the ceiling was determined. A code letter will not precede a non-ceiling cloud cover. These code letters are:

M - measured
E - estimated
W - indefinite

MEASURED CEILINGS

A *measured ceiling* (M) means that the ceiling was determined through the use of a rotating beam *ceilometer* which is an electronic device used to measure cloud bases. (See Fig. 5-37.) The resultant readings may be remotely indicated in the control tower, the flight service station, and the National Weather Service.

ESTIMATED CEILINGS

Aircraft-estimated ceilings are reported to a local approach controller or FAA radio station and then forwarded to the weather observer. The MSL altitude given by a pilot is converted to an above

ground level (AGL) altitude for reporting cloud *bases*.

A *balloon-estimated ceiling* is determined by releasing a weather balloon and noting the time it requires to reach the cloud base. The accuracy of this method, however, is rather limited because vertical air currents often cause the balloon to rise at speeds other than normal rates. (See Fig. 5-38.)

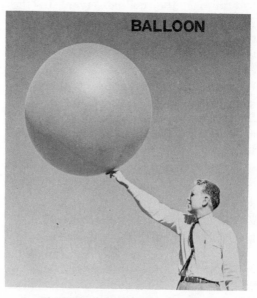

Fig. 5-38. Ballon-Estimated Ceiling

In the absence of the above methods of determining cloud heights, the meteorologist can estimate the ceiling height by relying on his experience and knowledge of cloud forms. Balloon, aircraft and observer-determined ceilings are collectively termed *estimated ceilings* (E).

INDEFINITE CEILINGS

An *indefinite ceiling* (W) indicates only the maximum vertical visibility into a surface-based obstruction to vision. These obstructions are normally indicated by an "X" in the sky coverage portion of aviation weather reports which stands for an *obscured* sky. This symbol can be prefixed with a minus (-) sign meaning *partial obscuration*. Obscured means the overlying sky conditions cannot be observed due to fog, dust, blowing snow, or other restrictions.

If a ceiling is *variable*, the letter "V" will follow the ceiling height. For example, the hourly weather report " ICT E15VOVC " indicates that there is an estimated ceiling of 1,500 feet overcast with varying ceilings in the area.

The ceiling over an airport is actually determined by the first totally obscured, broken, or overcast layer of clouds prefixed on the weather report by one of the letters indicating the method of determining the ceiling layer. In the Wichita example shown in figure 5-35, the 800 foot broken layer is prefixed by the letter "M" indicating measurement by ceilometer. Therefore, the ceiling at Wichita is 800 feet broken. If there had been a minus sign between the 800 and the broken sky cover symbol, this would mean light or thin and would *not* constitute a ceiling. In addition, if the broken layer was classified as thin, the "M" would not have preceded the number "8."

VISIBILITY

The next segment in the sample aviation weather report is the visibility. (See Fig. 5-35, item 4.) This is measured in *statute miles*. In the Wichita example, the visibility was one and one-half statute miles.

PRESENT WEATHER

The next segment of the aviation weather report could be one letter or a combination of letters to indicate the weather at the station at the time of observation. This could also include an obscuration or visibility restriction. The "TRW" shown in figure 5-35, item 5, stands for thunderstorms and moderate rain showers. If a minus sign immediately follows the "TRW," it indicates that showers are light. Weather conditions in the aviation weather report are shown by one or more of the following letter codes:

R — rain   C — crystals
S — snow   Y — spray
H — haze
A — hail
B — blowing (this could be combined with other letter identifiers)
F — fog
D — dust
L — drizzle
G — ground (this could be combined with F to indicate ground fog)
W — showers
K — smoke
I — ice
IP — ice pellets
Z — freezing (this could be combined with R to indicate freezing rain)
T — thunderstorm
+ — heavy intensity
- — light intensity
-- — very light intensity
no sign — moderate intensity

ATMOSPHERIC PRESSURE

The next entry in the sample aviation weather report (see Fig. 5-35, item 6) is the atmospheric pressure in millibars (an alternate method of measuring barometric pressure). The numbers given in the report (132) are the last three digits of the pressure reading, 1013.2 millibars. The method of determining the barometric pressure from the weather report is as follows:

1. If the first two numbers are less than 56, prefix the number with 10. For example, 132 would become 1013.2.

2. If the first two numbers are 56 or greater, prefix the number with 9. For example, 584 would become 958.4. Note that the last digit of the barometric pressure reading is always in tenths.

∅∅∅∅	— Calm
3416	— From 340° at 16 knots
2727	— From 270° at 27 knots
3414G28	— From 340° at 14 knots, gusts to 28 knots.

Fig. 5-40. Wind Reports

TEMPERATURE

Next on the report is the surface temperature in *degrees Fahrenheit.* In the Wichita example, the temperature at the station is 72° F. (See Fig. 5-35, item 7.)

DEWPOINT

Following the temperature on the sample report is "64" which is the dewpoint in *degrees Fahrenheit.* (See Fig. 5-35, item 8.) The dewpoint is the temperature to which the air must be cooled before the moisture in the air will condense and become visible. Pilots observe the temperature and dewpoint to note the difference between them. If the difference is small, some restriction to visibility such as drizzle or fog can be anticipated.

WIND

The next two digits indicate wind direction reported to the nearest ten degrees. In the "ICT" example (see Fig. 5-35, item 9, the wind is blowing from 020°. This direction is measured with reference to the true North Pole. A zero must be added to the two digit number in the report to obtain the correct wind direction. (See Fig. 5-39.)

NORTH	(360°)	36
EAST	(090°)	09
SOUTH	(180°)	18
WEST	(270°)	27
NNW	(340°)	34

Fig. 5-39. Wind Direction Symbols

Following the wind direction is the wind velocity in *knots.* At the Wichita station, the wind velocity is 14 knots with gusts to 18 knots. (See Fig. 5-35, item 9.) The "G" denotes gusts with the speed of the highest gust in the past 15 minutes shown immediately following the "G." Wind velocities are more commonly listed without gusts; however, gusts are shown when this information is of value to pilots.

To denote the passing of a squall (a comparatively sustained wind), a "Q" is used immediately following the wind velocity. The speed of the strongest gust occurring during the squall is given after the "Q".

In the bottom example shown in figure 5-40, the wind velocity is 14 knots with the strongest gust at 28 knots. A calm wind is shown by ∅∅∅∅. (See Fig. 5-40.) An "E" precedes a wind group in which some portion has been estimated.

ALTIMETER SETTINGS

The next item in the aviation weather report is the altimeter setting in *inches of mercury.* (See Fig. 5-35, item 10.) The information in this segment consists of the last three digits of the altimeter setting, "992" indicating that the altimeter setting is "29.92."

To read the altimeter value in the aviation weather report, if the first number is eight or nine, prefix the number with a two. For example, 896 in the report would indicate an altimeter setting of 28.96. If the first number is zero, or one, prefix the number with a three. For example, 102 means the altimeter setting is 31.02 in. Hg. Notice that the last two digits of the altimeter reading are in hundreds. All altimeter settings on aviation weather reports are converted to sea level. (See Fig. 5-41.)

If first number is 8 or 9 — prefix with Number 2

     Example: 896   28.96

       Prefix with No. 2

If first number is 0 or 1 — prefix with Number 3

     Example: 102   31.02

       Prefix with No. 3

Examples: 2 $\frac{8.96}{9.54}$ 3 $\frac{0.15}{1.04}$

Fig. 5-41. Reading Altimeter Setting

REMARKS

The next segment of the aviation weather report includes remarks regarding weather conditions existing at the airport at the time of observation. (See Fig. 5-42.) In the Wichita example, the remarks include information concerning runway visual range or RVR. Runway visual range is a visibility value expressed in feet that is instrumentally measured at some landing runways. In the example shown in figure 5-42, at the approach end of runway one (indicated by R01) the RVR is 2,000 feet variable (indicated by the V) to 5,000 feet. These values are averages recorded during a ten-minute period.

The remarks section may also include information on cloud types, thunderstorms, types of precipitation, and lightning. In the Wichita example, T SW MOVG E FQT LTGCC means that thunderstorms southwest of the airport are moving east, and there is frequent lightning from cloud to cloud.

The final remark on the aviation weather report shows that rain began at 15 minutes past the previous hour.

Here are other typical remarks found on the aviation weather reports:

BINOVC — breaks in overcast
CB — cumulonimbus clouds
CU FRMG — cumulus forming
OCNL SPKL — occasional sprinkle
RB 48 — rain began 48 minutes past the hour
RE 10 — rain ended 10 minutes past the hour
CIG RGD — ceiling ragged

NOTAMs

NOTAMs (Notices to Airmen) are included at the end of some aviation report sequences to indicate information of interest to the pilot regarding the airport. Closed or shut down radio or instrument approach facilities, runway construction, and closed runways are all types of information that may be deciphered from NOTAMs. NOTAM information is usually written in code to save space on the teletype page. An explanation of this code can normally be found at the flight service station next to the display boards where aviation weather reports are posted.

Figure 5-43 shows a NOTAM added to the Wichita example aviation weather report. The horizontal arrow pointing to the right indicates the beginning of the NOTAM. The letters "ICT" following the arrow denote the station reporting the notice (Wichita). Following the station letters is the NOTAM status code arrow indicating whether the NOTAM is current, new, or cancelled. An arrow pointing to the right and down indicates a current notice. If the NOTAM is being cancelled, the letter "C" will be used in place of the arrows. The 9/4 is simply a reference number for the NOTAM.

. . . . RØ1VR2ØV5Ø/T SW MOVG E FQT LTGCC/RB15

Fig. 5-42. Remarks

To find out just what the Notice to Airmen says, the pilot must consult the NOTAM listed on the aviation weather report and printed and transmitted on the same teletypes which print the weather reports. The NOTAM, referred to in figure 5-43, is decoded as: "Wichita NOTAM No. 9/4, runways 01 and 19 closed."

Figure 5-44 shows examples of actual aviation weather reports as they come off the teletype machine first on a local circuit and then by state. The system is programmed to prevent duplications between local and state groupings.

TERMINAL FORECASTS

Another type of weather information provided in the teletype format is the *terminal forecast*. A sample teletype page of terminal forecasts is shown in figure 5-45. The terminal forecast is a forecast of future weather for the various large air terminals throughout the country and is

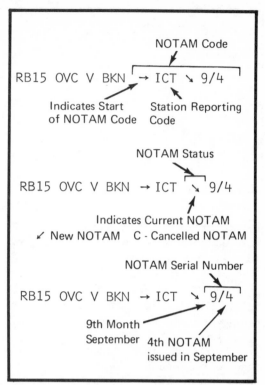

Fig. 5-43. NOTAM Explanation

SA27 010900
DEN M140BKN 20 144/64/52/3509/019
EGE E70BKN 20 177/56/50/0307/034→EGE↘7/7
BFF SP 50SCT E75OVC 25T 154/65/58/0705/014/TB50 SE
 MOVG E OCNL LTGICCG SE AND N/→BFF↘7/11
AKO 110SCT 15 157/60/58/0000/015→AKO↘7/4

→NOSUM 010902
→EGE 7/7 HDN ARPT CLSD
→BFF 7/11 BFF S 700 17-35 CLSD
→ AKO 7/4 STK ARPT CLSD THRU 8/9

AZ 010910
FLG 40SCT 15 161/61/42/2908/033
PHX CLR 40 088/89/55/3203/986 →PHX↘ 7/8 7/10
→ PHX 7/8 DVT NON-STD RWY LGTS SW 3000 7R-25L
→ PHX 7/10 DVT ATCT 0600-SS TIL 10/1
 PRC CLR 15 147/63/47/0000/019
 TUS CLR 30 099/76/48/1605/997→TUS↘6/10 7/4
→ TUS 6/10 TUS NW 1200 11R-29L CLSD
→ TUS 7/4 TUS 3-21 CLSD

Fig. 5-44. Sample Hourly Aviation Weather Reports as Transmitted

```
FT
OR Ø11Ø4Ø

BKE Ø11111 3ØSCT 5ØSCT 3315 5ØSCT OCNL BKN CHC C2ØOVC 5SW-- TIL ØØZ.   Ø5Z VFR..
MFR Ø11111 4ØSCT 25ØSCT.  23Z 5Ø-BKN.  Ø5Z VFR..
OTH Ø11111 CLR 151Ø.  21Z 5ØSCT 25Ø-BKN 3312.  Ø5Z VFR..
PDX Ø11111 C15BKN. .18Z 2ØSCT SCT OCNL BKN.  2ØZ 3ØSCT 25Ø-BKN.  Ø4Z
  5ØSCT C12ØBKN.  Ø5Z VFR..
RDM Ø11111 25Ø-SCT.  21Z 4ØSCT 25Ø-BKN.  Ø5Z VFR..
TTD Ø11111 15SCT.  2ØZ 3ØSCT 25Ø-BKN.  Ø4Z 5ØSCT C12ØBKN.  Ø5Z VFR..
```

Fig. 5-45. Terminal Forecast — VFR

issued three times daily. Airport weather is forecast for a 24-hour period. The last six hours is an outlook stated in terms of IFR, LIFR (low IFR), MVFR (marginal VFR), or VFR. The cause of LIFR, IFR or MVFR is also given by ceiling or visibility restrictions or both.

The following steps indicate how to analyze the heading for the VFR terminal forecast in figure 5-45:

1. This FT (terminal forecast) was issued for the state of OR (Oregon).

2. These forecasts were issued on the 1st day of the month.

3. Time of the release was 1040 GMT time, or 0440 Central Standard Time. GMT (Greenwich Mean Time) is used by pilots as a standard time reference to alleviate the confusion associated with flying through several time zones.

4. The time that this terminal forecast covers is from 1100 Zulu or 5 a.m. Central Standard Time through 1100 Zulu or 5 a.m. Central Standard Time on the following day.

Figure 5-46 shows a series of terminal forecasts which predict marginal VFR and IFR conditions in Colorado.

Figure 5-47 shows the decoding of a typical forecast body. It is read as follows:

1. The ceiling in a terminal forecast is denoted with the same sky coverage symbols and format as used in the hourly weather reports with one major exception. In terminal forecasts, the forecast ceiling will always be preceded with the letter "C."

2. The wind is forecast to be 150° true at 15 knots.

```
FT
CO 11Ø945
ALS DLAD TIL 112ØZ
COS 1110Ø1Ø C3X 1/2ZL--S-F OCNL C1ØX 11/2S--F.  15Z C5X
  1/2S-F OCNL C1ØX 11/2S--F.  Ø4Z IFR CIG VSBY SNW..
DEN 1110Ø1Ø 1ØSCT C2ØOVC 5F OCNL C5X 1S--F.  2ØZ -X
  C1ØOVC 21/2S--F OCNL C5X 3/4S- Ø412.  Ø4Z IFR CIG
  VSBY SNW..
GJT 1110Ø1Ø 12SCT 2ØSCT C4ØBKN SCT V BKN OCNL S--.
  15Z 2ØSCT C4ØBKN CHC C2ØOVC RW-.  22Z C2ØBKN
  OCNL S-- CHC C1ØBKN 11/2SW-.  Ø4Z MVFR CIG..
PUB 1110Ø1Ø C3X 1S-F OCNL C1ØX 3S--F.  Ø4Z IFR CIG VSBY SNW..
```

Fig. 5-46. Terminal Forecast — MVFR/IFR

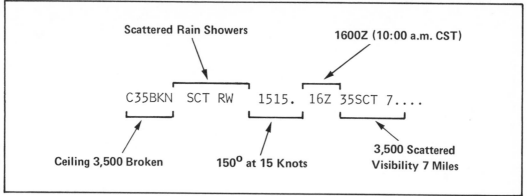

Fig. 5-47. Terminal Forecast Explanation

3. At 1600 Zulu or 10 a.m. Central Standard Time, the weather is forecast to become 3,500 feet scattered and the visibility is forecast to be seven statute miles.

The terminal forecast continues in this manner until a complete forecast of the weather is given for the *valid period* noted in the beginning of the forecast.

The symbols and notations used in terminal forecasts are the same as those used in hourly weather reports and appear in the same order. However, the wind forecast is only included when it is expected to be ten knots or greater. The prevailing visibility will appear in a terminal forecast only if it is forecast to be six statute miles or less. Terminal forecasts are originated in one or more forecast centers per state.

AREA FORECASTS

Certain National Weather Service Offices are designated as forecast centers to predict aviation weather within their *areas*. These forecasts, called *area forecasts* are issued every twelve hours covering a period of 18 hours, plus a general outlook for an additional 12 hours. Area forecasts are distributed via teletype and are presented in a standard format consisting of separate paragraphs for each type of information.

The normal order of presentation begins with the heading followed by a designation of the forecast area. Then, a synopsis of the weather-producing structures appears such as fronts, troughs, and air masses. Statements follow concerning the amount of cloud cover, bases and tops, weather, obstructions to visibility, and surface winds when significant. If icing is anticipated, the freezing level, types of ice, and intensities are specified. Turbulence is not predicted but is *always* implied when thunderstorms are forecast.

The heading identifies the type of forecast, the originating forecast center, the date and the month, the time the forecast was issued, and the period covered by the forecast. All times are in Greenwich Mean Time, as indicated by the letter "Z." For example, in the heading pointed out by item 1 in figure 5-48, "FA" indicates this is an area forecast. "MSY" stands for New Orleans, the forecast center, 13 is the thirteenth day of the month, and 0040 is the time the forecast was issued.

This forecast is valid from 0100Z until 1900Z on Tuesday. Note the forecast period is for 18 hours.

Item 2 of figure 5-48 shows the next three items in the forecast. They include the geographic area covered, a standard notation, and a synopsis of weather-producing conditions. They read as follows: "This forecast is for Tennessee, Arkansas, Louisiana, Mississippi, Alabama, Florida west of 85° west longitude, and the coastal waters.

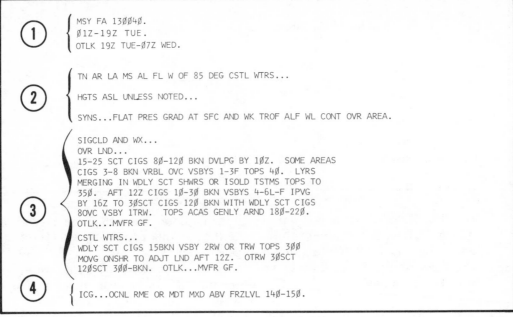

① {
MSY FA 130040.
01Z-19Z TUE.
OTLK 19Z TUE-07Z WED.
}

② {
TN AR LA MS AL FL W OF 85 DEG CSTL WTRS...

HGTS ASL UNLESS NOTED...

SYNS...FLAT PRES GRAD AT SFC AND WK TROF ALF WL CONT OVR AREA.
}

③ {
SIGCLD AND WX...
OVR LND...
15-25 SCT CIGS 80-120 BKN DVLPG BY 10Z. SOME AREAS
CIGS 3-8 BKN VRBL OVC VSBYS 1-3F TOPS 40. LYRS
MERGING IN WDLY SCT SHWRS OR ISOLD TSTMS TOPS TO
350. AFT 12Z CIGS 10-30 BKN VSBYS 4-6L-F IPVG
BY 16Z TO 30SCT CIGS 120 BKN WITH WDLY SCT CIGS
80VC VSBY 1TRW. TOPS ACAS GENLY ARND 180-220.
OTLK...MVFR GF.

CSTL WTRS...
WDLY SCT CIGS 15BKN VSBY 2RW OR TRW TOPS 300
MOVG ONSHR TO ADJT LND AFT 12Z. OTRW 30SCT
120SCT 300-BKN. OTLK...MVFR GF.
}

④ {
ICG...OCNL RME OR MDT MXD ABV FRZLVL 140-150.
}

Fig. 5-48. Area Forecast

All heights are above sea level unless noted otherwise." The synopsis states: "There is a flat pressure gradient at the surface and a weak trough aloft which will continue over the area."

The next section (see Fig. 5-48, item 3) describes the amount and height of sky cover. It gives cloud tops, surface visibility, state of weather, visibility obstructions, surface winds, and other information needed to describe flying conditions expected during the forecast period. It is read as follows:

"Clouds and weather: Over land, 1,500 to 3,000 feet scattered, ceiling 8,000 to 12,000 broken developing by 1000Z. Some areas, ceiling 300 to 800 feet broken, variable to overcast with visibilities from one to three miles with fog. Cloud tops will be 4,000 feet. Cloud layers will be merging in widely scattered showers or isolated thunderstorms with tops to 35,000 feet. After 1200Z, the ceilings will be 1,000 to 3,000 broken with visibilities from four to six miles, with light drizzle and fog, improving by 1600Z. After that time,

clouds will be 3,000 feet scattered, ceilings 12,000 broken with widely scattered areas having ceilings of 800 feet overcast, visibility one mile with moderate thunderstorms and rain showers. Tops of altocumulus and altostratus generally around 18,000 to 22,000 feet. Outlook marginal VFR conditions due to ground fog.

"Over coastal waters, widely scattered ceilings of 1,500 broken, two miles visibility with moderate rain showers or thunderstorms, cloud tops at 30,000 feet moving onshore to adjacent land after 1200Z. Otherwise 3,000 scattered, 12,000 scattered, 30,000 thin broken. Outlook marginal VFR conditions due to ground fog.

Area forecasts include a statement concerning expected icing conditions plus the height of the freezing level. Item 4 in figure 5-48 shows a very brief statement read as follows: "Icing: Occasional rime or moderate mixed icing above freezing level at 14,000 to 15,000 feet." Mixed refers to a combination of rime and clear ice.

At about 0500Z, or midnight local time, the ground will have radiated enough heat and be cooled sufficiently to chill the adjacent warm moist air to its dewpoint. Ground fog or radiation fog will bring marginal V.FR conditions.

WINDS ALOFT FORECASTS

Another teletype service is the winds aloft forecast. Figure 5-49 shows the data time, time of teletype transmission, valid time, and usable time for each forecast. The *data time* is the actual time the observation of wind velocity and speed is made and on which the forecast is based. The *teletype transmission time* is the time that the winds aloft information was transmitted over the teletype system. The valid time designates the hour at which the winds, as indicated in the forecast, are to be as forecast. This forecast will be usable, however, for winds aloft information during the *use period*.

The winds aloft forecasts begin with the heading shown in figure 5-50. The forecast information is printed on the teletype in vertical columns according to altitude and is given for 3,000, 6,000 9,000, 12,000, 18,000, 24,000, 30,000, 34,000, and 39,000 feet *above sea level* (ASL). The first level for which winds aloft are issued for a particular station is 1,500 feet or more *above the station elevation*. The temperature is forecast in

Data Time	Report Time	Valid Time	For Use (Period)	Heading
0000Z	0545Z	1200Z	0600-1500Z	FDUS1
0000Z	0545Z	1800Z	1500-2100Z	FDUS2*
0000Z	0945Z	0000Z	2100-0300Z	FDUS3*
1200Z	1745Z	0000Z	1800-0300Z	FDUS1
1200Z	1745Z	0600Z	0300-0900Z	FDUS2‡
1200Z	1945Z	1200Z	0900-1500Z	FDUS3‡

* At 1800Z, discontinue using FD forecasts based on 0000Z data.

‡ At 0600Z, discontinue using FD forecasts based on 1200Z data.

Fig. 5-49. Winds Aloft Forecast Times

Celsius for all wind levels that are 2,500 feet or more above the level of the station with the exception of the 3,000-foot level. Since temperatures are always below $0°C$ at 30,000 feet and above, the minus signs are deleted from the temperatures at the 30,000, 34,000, and 39,000 foot levels.

The winds aloft forecast is fairly simple to read. For example, a segment of a sample winds aloft forecast is shown in figure 5-51. The first two numbers in the winds aloft forecast indicate the true direction from which the wind is blowing. The number "14" indicates the wind direction of 140° (a zero must be added). This

Winds Aloft Forecast
    \   Date Transmission Time (Zulu)
     \   \   /
FDUS3   KWBC   161945
DATA BASED ON 161200Z
VALID 171200Z   FOR USE 0900-1500Z.   TEMPS NEG ABV 24000

FT	3000	6000	9000	12000	18000	24000	30000	34000	39000
BOS	2431	2638-03	2646-08	2654-13	2669-24	2783-35	279747	770355	279258
EMI	2619	2728-01	2635-05	2641-09	2654-21	2665-32	268046	269353	269658
JFK	2526	2736-02	2642-06	2649-11	2662-22	2674-33	269046	770154	279858
MLT	2327	2635-08	2644-12	2653-17	2767-27	2780-37	279250	279356	287957
OAL			3308+07	3311+00	2730-10	2752-21	277737	279246	278857

Fig. 5-50. Sample Winds Aloft Forecast

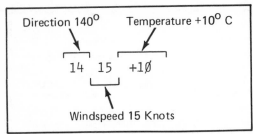

Fig. 5-51. Winds Aloft Segment

direction is in relation to the *true* North Pole.

The next item is the wind velocity, which is 15 knots and can be read as printed. If the wind is five knots or less at a particular altitude, the number 9900 is used to indicate the wind is light and variable. The last item in this segment is the temperature, +10°C. Forecast winds of 100 to 199 knots are deciphered by adding 100 to the speed given, and subtracting 50 from the two numbers in the direction space. 7545, for example, would be read 250 degrees, 145 knots.

IN-FLIGHT ADVISORIES

When it becomes apparent to weather forecasters and flight service station personnel that weather conditions in a certain area contain hazardous elements or are significantly worse than forecast, it becomes the responsibility of the National Weather Service to issue an inflight advisory and of FSS to disseminate this information to pilots flying or intending

to fly in the affected area. This is accomplished through the issuance of SIGMETs and AIRMETs. These advisories are broadcast over flight service station navigation frequencies and are an important part of preflight weather briefings.

SIGMETs

SIGMET stands for *significant meteorology*. This category of in-flight advisories includes weather phenomena of particular significance to *all* aircraft. SIGMETs cover tornadoes, embedded thunderstorms, squall lines, hail three-fourths of an inch in diameter or more, severe and extreme turbulence, severe icing, and widespread duststorms or sandstorms which lower visibility below three miles. An example of a SIGMET is shown in figure 5-52.

AIRMETs

AIRMETs are issued for meteorological conditions less severe than for SIGMETs. The conditions are potentially hazardous to aircraft having *limited capabilities* because of lack of equipment or instrumentation or pilot qualification and are at least of operational interest to *all* aircraft. AIRMETs cover moderate icing, moderate turbulence over an extensive area, large areas where visibility is less than three miles or ceilings less than 1,000 feet including mountain ridges or passes, and sustained winds of 30 knots

---

```
CHI WS 152220
152220-160200

SIGMET ALFA 1. FLT PRCTN BECAUSE OF LN TSTMS CNTRL AND SRN INDIANA.
BKN LN OF TSTMS ABT 20 MI WD WITH THE LEADING EDGE 20NW LAFAYETTE-
INDIANAPOLIS-30SSW EVANSVILLE AT 22Z WL MOV EWD ABT 30 KT ACRS THE
S TWO-THIRDS OF IND TO BE BYD IND BY 02Z. CB TOPS ABV 35 THSD FT
AND LCLLY ABV 45 THSD FT. CNL ADVY AT 02Z
```

It is read as follows: .
SIGMET Alpha One: Flight precaution because of a line of thunderstorms in central and southern Indiana. Broken line of thunderstorms about 20 miles wide with the leading edge 20 miles northwest of Lafayette and Indianapolis, and 30 miles south southwest of Evansville at 2200 ZULU. Will move eastward about 30 knots across the southern two-thirds of Indiana, to be beyond Indiana by 0200 ZULU. Cumulonimbus tops above 35,000 feet, and locally above 45,000 feet. Cancel advisory at 0200 ZULU.

Fig. 5-52. SIGMET

```
GSW WA 161210
161210-161400

AIRMET ALFA 1. FLT PRCTN OVR NRN TEX AND OKLA BECAUSE OF STG WNDS
AND LOW CIGS VSBYS. OVR NWRN N CNTRL NERN TEX AND OKLA WNDS 40 KTS
OR MORE WITHIN 2 THSD FT OF SFC. OVR NERN TEX CIGS BLO 1 THSD FT
VSBY OCNLY 2 MIS OR LESS IN FOG. IPVG BY 1400Z. CNL AT 1400Z.
```

It is read as follows:
AIRMET Alpha One: Flight precaution over northern Texas and Oklahoma because of strong winds and low ceilings and visibilities. Over northwestern, north central, and northeastern Texas, and all of Oklahoma; winds at four zero knots or more within two thousand feet of the surface. Over northeastern Texas; ceilings below one thousand feet, visibility occasionally two miles or less in fog. Improving by fourteen hundred ZULU. Cancel this AIRMET at fourteen hundred ZULU.

Fig. 5-53. AIRMET

at or within 2,000 feet of the surface. An example of an AIRMET is shown in figure 5-53.

FAA flight service stations broadcast SIGMETs at 15-minute intervals for the first hour after issuance. Then they are broadcast as alert notices or flight precautions at 15 and 45 minutes past the hour for the remainder of the period for which the SIGMETs are valid. When broadcasting SIGMETs as alert notices, the coded name of the advisory is given and is designated current; i.e., "SIGMET Bravo 2 is current."

AIRMETs are broadcast twice each hour, at 15 and 45 minutes past the hour during the first hour. Thereafter, they are carried as alert notices at 15 and 45 minutes past the hour for the valid period of the advisory.

PIREPs

PIREPs (pilot reports) contain significant weather information reported by pilots in flight. This information can be very valuable since pilots can supply firsthand information on cloud tops, turbulence, etc., that weather forecasters and FSS personnel would be unable to predict. These reports, in turn, are compiled by FSS personnel and made part of pilot weather briefings and broadcast aviation weather reports. Pilots are urged to cooperate and volunteer reports of cloud tops, upper cloud layers, thunderstorms, ice, turbulence, strong winds, and other significant flight condition information.

Pilots usually report height values as above mean sea level since they determine elevations by the altimeter. In the reports disseminated as PIREPs, elevation references are always given the same as received from pilots — above mean sea level. An example of a PIREP at the end of an aviation weather report is shown in figure 5-54. It is read as follows: "Pilot reports tops of overcast at 2,300 feet mean sea level." If the pilot were to report the bottom of an overcast, or other cloud layer, the sky coverage code would appear *after* the elevation figure. When the sky coverage code appears before the altitude figure, tops of clouds have been reported.

PIREP IN REMARKS SECTION

. . . /1807/993 OVC23

Fig. 5-54. PIREP

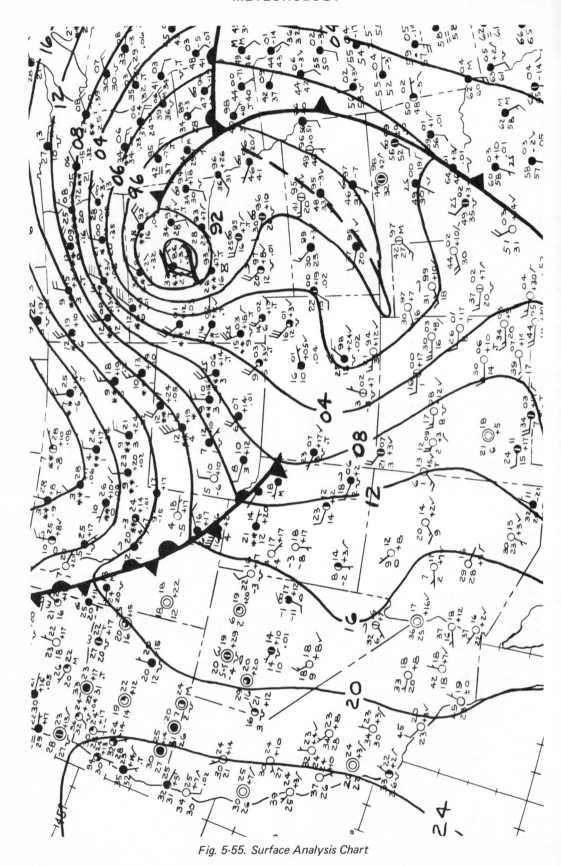

Fig. 5-55. Surface Analysis Chart

SECTION C - WEATHER CHARTS

In addition to teletype reports and forecasts, the National Weather Service also has a variety of weather charts that provide valuable meteorological information. By referring to these charts, the pilot is provided with a pictorial representation of the weather across the nation and specifically, along his route of flight. Weather charts are produced by National Weather Service personnel in the National Meteorological Center in Suitland, Maryland, and transmitted to weather stations across the country. These charts are received and reproduced in the various weather stations on facsimile machines like the one shown in figure 5-56, or increasingly, on computer screens similar to television sets. These charts include the surface analysis chart, the weather depiction chart, and the low-level significant weather prognosis chart.

Fig. 5-56. Facsimile Machine

SURFACE ANALYSIS CHART

The *surface analysis chart* is of fundamental importance to the pilot. The velocity and direction of the wind, the temperature and humidity, dewpoint, and other weather data are indicated on the surface weather chart. In addition, this chart provides a general picture of the atmospheric pressure pattern at the surface of the earth by showing the position of highs, lows, and fronts.

A sample chart is shown on the preceding page. (See Fig. 5-55.) The basic unit of information on the chart is the station model. City designators are included on the newest surface charts to assist pilots in briefing themselves. Figure 5-57 shows a close-up of the station model which is read as follows:

1. The station is reporting a wind from the west at 15 knots. The stem of the wind arrow always points in the direction from which the wind is blowing. The velocity of the wind can be determined by the feathers pointing away from the stem. A full-length feather represents 10 knots while a half-length feather is equivalent to five knots. A pennant attached to the wind arrow represents 50 knots of wind.

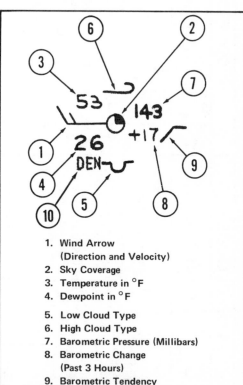

1. Wind Arrow
   (Direction and Velocity)
2. Sky Coverage
3. Temperature in °F
4. Dewpoint in °F
5. Low Cloud Type
6. High Cloud Type
7. Barometric Pressure (Millibars)
8. Barometric Change
   (Past 3 Hours)
9. Barometric Tendency
10. Location Identifier

Fig. 5-57. Station Model

HIGH CLOUDS

Code No.	C_H	DESCRIPTION
1	(symbol)	Filaments of Ci, or "mares tails," scattered and not increasing
2	(symbol)	Dense Ci in patches or twisted sheaves, usually not increasing, sometimes like remains of Cb; or towers or tufts
3	(symbol)	Dense Ci, often anvil-shaped, derived from or associated with Cb
4	(symbol)	Ci, often hook-shaped, gradually spreading over the sky and usually thickening as a whole
5	(symbol)	Ci and Cs, often in converging bands, or Cs alone; generally overspreading and growing denser; the continuous layer not reaching 45° altitude.
6	(symbol)	Ci and Cs, often in converging bands, or Cs alone; generally overspreading and growing denser; the continuous layer exceeding 45° altitude
7	(symbol)	Veil of Cs covering the entire sky
8	(symbol)	Cs not increasing and not covering entire sky
9	(symbol)	Cc alone or Cc with some Ci or Cs, but the Cc being the main cirriform cloud

MIDDLE CLOUDS

Code No.	C_M	DESCRIPTION
1	(symbol)	Thin As (most of cloud layer semitransparent)
2	(symbol)	Thick As, greater part sufficiently dense to hide sun (or moon), or Ns
3	(symbol)	Thin Ac, mostly semi-transparent, cloud elements not changing much and at a single level
4	(symbol)	Thin Ac in patches; cloud elements continually changing and/or occurring at more than one level
5	(symbol)	Thin Ac in bands or in a layer gradually spreading over sky and usually thickening as a whole
6	(symbol)	Ac formed by the spreading out of Cu or Cb
7	(symbol)	Double-layered Ac, or a thick layer of Ac, not increasing, or Ac with As and/or Ns
8	(symbol)	Ac in the form of Cu-shaped tufts or Ac with turrets
9	(symbol)	Ac of a chaotic sky, usually at different levels; patches of dense Ci are usually present also

LOW CLOUDS

Code No.	C_L	DESCRIPTION
1	(symbol)	Cu of fair weather, little vertical development and seemingly flattened
2	(symbol)	Cu of considerable development, generally towering, with or without other Cu or Sc bases all at same level
3	(symbol)	Cb with tops lacking clear-cut outlines, but distinctly not cirriform or anvil-shaped; with or without Cu, Sc, or St
4	(symbol)	Sc formed by spreading out of Cu, Cu often present also
5	(symbol)	Sc not formed by spreading out of Cu
6	(symbol)	St or Fs or both, but no Fs of bad weather
7	(symbol)	Fs and/or Fc of bad weather (scud)
8	(symbol)	Cu and Sc (not formed by spreading out of Cu) with bases at different levels
9	(symbol)	Cb having a clearly fibrous (cirriform) top, often anvil-shaped, with or without Cu, Sc, St, or scud

CLOUD ABBREVIATIONS

St—STRATUS
Fs—FRACTOSTRATUS
Sc—STRATOCUMULUS
Cu—CUMULUS
Fc—FRACTOCUMULUS
Cb—CUMULONIMBUS
Ac—ALTOCUMULUS
Ns—NIMBOSTRATUS
As—ALTOSTRATUS
Ci—CIRRUS
Cs—CIRROSTRATUS
Cc—CIRROCUMULUS

Fig. 5-58. Cloud Type Symbols

2. The sky coverage symbol represents the amount of sky covered by clouds in tenths. The symbol shown represents two to three-tenths sky coverage. Other symbols used are shown in figure 5-59.

SKY COVERAGE
(Total Amount)

	No clouds
	Less than one-tenth or one-tenth
	Two-tenths or three-tenths
	Four-tenths
	Five-tenths
	Six-tenths
	Seven-tenths or eight-tenths
	Nine-tenths or overcast with openings
	Completely overcast
	Sky obscured

Fig. 5-59. Sky Coverage Symbols

3. The temperature at the station is 53° F.

4. The dewpoint is 26° F.

5. The station is reporting stratocumulus low clouds. Other cloud type symbols are shown in figure 5-58 on the preceding page.

6. There are high, scattered cirrus clouds or mare's tails over the station.

7. The barometric pressure expressed in millibars is 1014.3. The prefix number for the millibar designation is determined in the same way as on aviation weather reports. Thus, barometric pressure is given to the nearest tenth of a millibar.

8. The barometric pressure has risen 1.7 millibars in the past three hours.

9. The barometric tendency in the past three hours is represented by two connected lines. In the example shown in figure 5-57, the barometric pressure has been rising but is now steady. This can be determined from the pattern of the two indicator lines. Notice that the first line rises and is followed by a second line which is level (rising pressure followed by steady pressure). On the other hand, if the pressure had been remaining steady and then dropped, the tendency symbol would be a horizontal line followed by a descending line. Through this pictorial representation, the pilot can easily determine barometric pressure tendency on the station model.

Isobars (lines connecting points of equal pressure) are drawn on the surface weather map at four millibar intervals and are labeled with the last two digits of the pressure reading. For example, figure 5-60 shows an isobar that is labeled "08." This represents 1008 millibars. High and low pressure areas are labeled as "H" or "L" on the surface analysis chart. Fronts are shown with conventional symbols as illustrated in figure 5-62. Before reading a surface weather chart, the time the chart was issued should be noted.

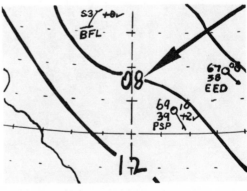

Fig. 5-60. 1008 Millibars

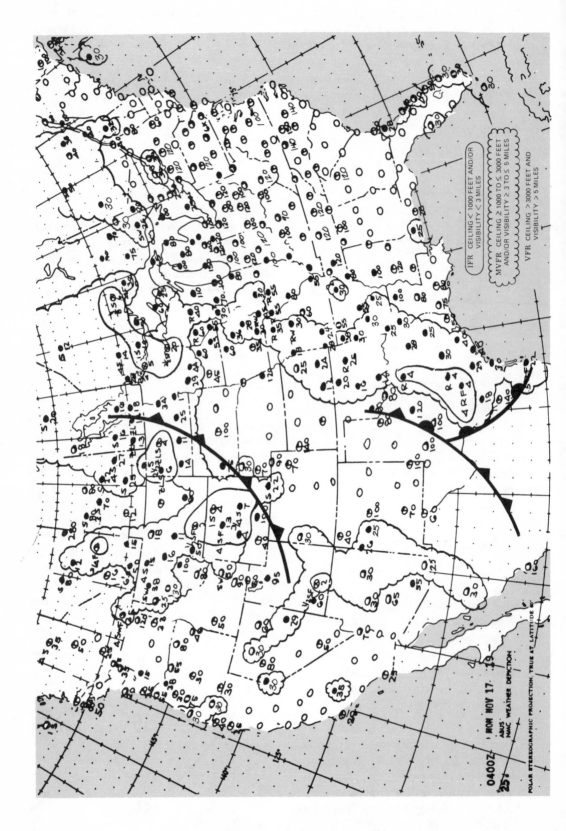

Fig. 5-61. Weather Depiction Chart

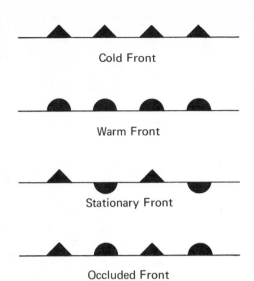

Fig. 5-62. Front Symbols

WEATHER DEPICTION CHART

Another very important weather chart for pilots is the weather depiction chart which is issued every three hours in the National Meteorological Center and reproduced on facsimile machines in various National Weather Service Offices. (See Fig. 5-61 on preceding page.) An *abbreviated* station model is used on a weather depiction chart to show visibility, type of weather, amount of sky cover, and the height of the cloud base. A typical station model is shown in figure 5-63. It is read as follows:

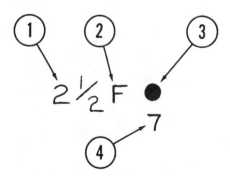

1. Visibility
2. Visibility Restriction
3. Sky Coverage Symbol
4. Ceiling Height

Fig. 5-63. Station Model for Weather Depiction Chart

1. The visibility is two and one-half miles. The visibility value is deleted when it is greater than six miles.
2. The present weather is fog. For other symbols, refer to Present Weather paragraph on page 5-28.
3. The sky coverage symbol as shown indicates the sky is completely overcast. Other sky coverage symbols used on weather depiction charts are shown in figure 5-64.

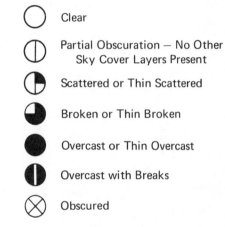

Fig. 5-64. Weather Depiction Chart Symbols

4. The height of the ceiling is given below the cloud cover symbol in hundreds of feet. If the cover does not constitute a ceiling, the height then refers to the altitude of the lowest layer of clouds. Item 4 indicates 700 feet.

A *scalloped* boundary line is drawn around areas in which the ceiling height is between 1,000 and 3,000 feet. (See Fig. 5-65.) A *smooth* boundary line is

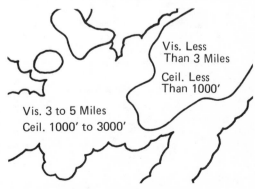

Fig. 5-65. Visibility and Ceiling Boundary Codes

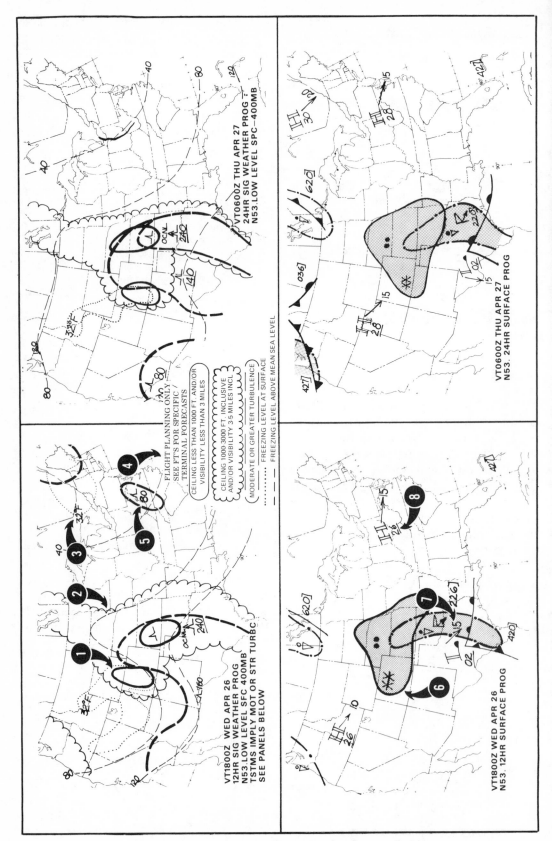

Fig. 5-66. Low Level Significant Weather Prognostic Chart

drawn around areas in which the visibility is below three miles *or* the ceiling is less than 1,000 feet. These conditions indicated by the smooth line would require a pilot to have an instrument clearance to fly in the area outlined. By observing the smooth and scalloped lines, the pilot can see at a glance marginal weather areas across the nation. The scalloped and solid line designations are always explained in the lower, right-hand corner of the weather depiction chart. (See Fig. 5-62.)

LOW-LEVEL SIGNIFICANT WEATHER PROGNOSIS CHARTS

The low level significant weather prognostic chart shown in figure 5-66 is divided into *two* forecast periods. The two panels on the left show the weather prognosis for a 12-hour period and those on the right for a 24-hour period. The valid times and titles for each panel are shown.

The panels on the top are the forecaster's best estimate of where low or middle clouds will exist. Also depicted are freezing levels, plus areas and altitudes of turbulence. The lower panels are the forecaster's best estimate of the location of frontal and pressure systems as well as the areas and types of precipitation. The two panels on the right (valid for 24 hours) will contain the same information as the two panels on the left (valid for 12 hours), but will indicate the movement *expected between* the two time frames.

Low level prognostic charts are issued four times a day or every six hours. The valid time of each panel appears on the lower left-hand corner of that panel. For example, the weather "picture" on the left-hand panel containing "1800 WED APR 26" represents a prediction of the weather situation that is expected to exist at 1800 Zulu time on Wednesday, April 26th.

On the two upper panels, as shown by item 1 of figure 5-66, the smooth lines enclose areas where the ceiling is expected to be below 1,000 feet AGL and/or visibility of less than three miles.

Those areas with expected ceilings from 1,000 to 3,000 feet AGL and/or visibilities of three to five miles are enclosed in scalloped lines drawn around the area, as shown by item 2.

A dotted line broken only by the notation "32°F," illustrated by item 3, indicates the predicted location of the freezing isotherm at the surface. An isotherm is a line of equal temperature (just as isobar is a line of equal pressure).

The broken lines represent heights in hundreds of feet of lowest forecast freezing level. As shown by item 4, a line broken by "40" shows the freezing level would be at 4,000 feet MSL. All temperatures will be above freezing at altitudes below 4,000 feet at all locations along the broken line. A pilot in the vicinity of 4,000 feet from that line northward should expect to encounter icing.

Various symbols are used to portray forecast fog or low stratus clouds and also to depict areas of turbulence. A spike, as shown by item 5 of figure 5-66, with the underlined number "80" means that moderate turbulence is predicted below 8,000 feet MSL in that region.

On the lower panels, those areas which are expected to have showers are enclosed by a broken line of dots and dashes. If the precipitation is expected to be more *persistent* (continuous or intermittent), the area will be shaded, as shown by item 6. Various symbols may be used within the precipitation areas to indicate rain showers, snow showers, or thunderstorms. These symbols may be found in figure 5-67.

Arrows may be used to show the predicted direction of movement of the pressure system, as shown by item 7 of figure 5-66. The speed of the low pressure area (15 knots in this example) is often shown near the arrows. The predicted atmospheric pressure in millibars is depicted near the major pressure systems. For example, as shown in item 8, 26 stands for 1026 millibars. The rule previously discussed for prefixing a 9 or 10 still applies.

While verification of these charts has improved greatly over the past decade, the pilot should remember that these charts are *still forecasts*. If the pilot compares an *actual* synoptic chart for a particular time period with the *prognostic*

chart for the same time period, they will probably look quite different. For example, if a weather system decelerates and deepens, it could bring more widespread weather. Conversely, if the weather system accelerates and diminishes in size, the bad weather area would be less extensive.

The individual position of the freezing levels on these two prognostic charts can be helpful to the pilot in planning his flight. For example, the freezing level ordinarily *lowers* behind a cold front. This chart will indicate how much it is expected to lower.

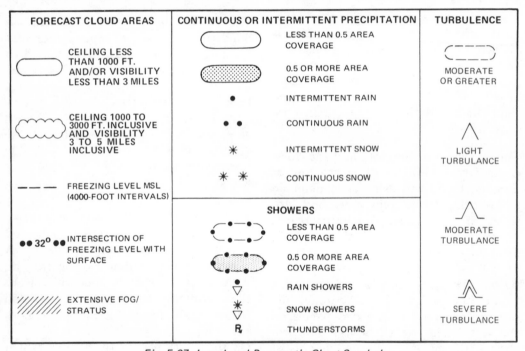

Fig. 5-67. Low Level Prognostic Chart Symbols

chapter 6

INTRODUCTION

Just as the driver must learn the laws concerning the operation of his automobile, the pilot must know the Federal Aviation Regulations that pertain to the operation of aircraft. In this chapter, excerpts from Parts 1, 61, and 91 have been included that are considered most pertinent to student and private pilots. Also included is the National Transportation Safety Board Part 830 which contains rules pertaining to aircraft accidents, incidents and overdue aircraft. A small airplane silhouette is used to denote those regulations of particular interest to the private pilot applicant.

SECTION A-TOPICAL LISTING OF FARs

Accident Notification, NTSB 830.5
Accident Reporting, NTSB 830.15
Acrobatic Flight, 91.71
Aeronautical Experience, 61.63, 61.109
Aeronautical Knowledge, 61.103, 61.105
Aeronautical Skill, 61.87, 61.107
Age—pilot minimum, 61.83, 61.103
Air Agencies—graduates of, 61.71
Air Traffic Clearances, 91.75
Air Traffic Control Instructions, 91.75
Aircraft and Engine Maintenance Records, 91.173
Aircraft Approaching Head-on, 91.67, 91.69
Aircraft Converging, 91.67
Aircraft in Distress, 91.67
Aircraft Landing, 91.67
Aircraft Lights, 91.33, 91.73
Aircraft Ratings, 61.5
Aircraft Requirements, 61.45, 91.33
Aircraft Speed, 91.70

Aircraft Used in Flight Tests, 61.45
Airplane Class Ratings, 61.5
Airplane—High Performance, 61.31
Airport—operation in vicinity of, 91.85, 91.87, 91.89, 91.90
Airworthiness Certificate, 91.27, 91.33
Airworthy Condition, 91.29
Altimeter Setting, 91.81
Altitudes
  airport traffic area, 91.87
  minimum safe, 91.79
  VFR cruising, 91.109
Anchor Lights, 91.73
Anti-collision Lights, 91.33
Arrival—notification of, 91.83
Approaching
  head-on, 91.67, 91.69
  to land, 91.87
Authority of Pilot, 91.3
Avoidance of Disaster Areas, 91.91

Basic VFR Minimum Weather Conditions, 91.105
Belts—fastening of safety, 91.14

SECTION B-FAR PART 1

Part 1 of the FAR(s) alphabetically lists the definitions of terms used in the subsequent regulations. In addition, any abbreviations or symbols used in the FARs are defined and explained in Part 1. Only those definitions of significant importance or of general interest to the student and private pilot have been included.

1.1 GENERAL DEFINITIONS

AIR CARRIER means a person who undertakes directly by lease, or other arrangement, to engage in air transportation.

AIRCRAFT means a device that is used or intended to be used for flight in the air.

AIRCRAFT ENGINE means an engine that is used or intended to be used in propelling aircraft. It includes engine appurtenances and accessories necessary for its functioning, but does not include propellers.

AIRFRAME means the fuselage, booms, nacelles, cowlings, fairings, airfoil surfaces (including rotors but excluding propellers and rotating airfoils of engines), and landing gear of an aircraft and their accessories or controls.

AIRPLANE means an engine-driven fixed-wing aircraft heavier than air that is supported in flight by the dynamic reaction of the air against its wings.

AIRPORT means an area of land or water that is used or intended to be used for the landing and takeoff of aircraft, and includes its buildings and facilities, if any.

AIRPORT TRAFFIC AREA means, unless otherwise specifically designated in Part 93, that airspace within a horizontal radius of 5 statute miles from the geographical center of any airport at which a control tower is operating, extending from the surface up to, but not including, 3,000 feet above the elevation of the airport.

AIRSHIP means an engine-driven, lighter-than-air aircraft that can be steered.

AIR TRAFFIC means aircraft operating in the air or on an airport surface, exclusive of loading ramps and parking areas.

AIR TRAFFIC CLEARANCE means an authorization by air traffic control for the purpose of preventing collision between known aircraft, for an aircraft to proceed under specified traffic conditions within controlled airspace.

AIR TRAFFIC CONTROL means a service operated by appropriate authority to promote the safe, orderly, and expeditious flow of air traffic.

AIR TRANSPORTATION means interstate, overseas, or foreign air transportation or the transportation of mail by aircraft.

ALTERNATE AIRPORT means an airport at which an aircraft may land if a landing at the intended airport becomes inadvisable.

AREA NAVIGATION (RNAV) means a method of navigation that permits aircraft operations on any desired course within the coverage of station-referenced navigation signals or within the limits of self-contained system capability.

BALLOON means a lighter-than-air aircraft that is not engine-driven.

BRAKE HORSEPOWER means the power delivered to the propeller shaft (main drive or main output) of an aircraft engine.

CALIBRATED AIRSPEED means indicated airspeed of an aircraft, corrected for position and instrument error. Cali-

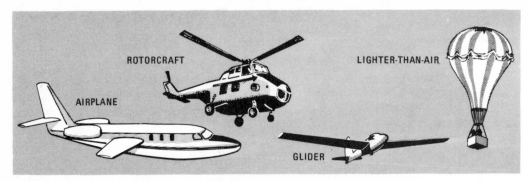

Fig. 6-1. Examples of Airman Certification Categories

brated airspeed is equal to true airspeed in standard atmosphere at sea level.

CATEGORY:

(1) As used with respect to the certification, ratings, privileges, and limitations of airmen, means a broad classification of aircraft. Examples include: airplane; rotorcraft; glider; and lighter-than-air; and

(2) As used with respect to the certification of aircraft means a grouping of aircraft based upon intended use or operating limitations. Examples include: transport; normal; utility; acrobatic; limited; restricted; and provisional.

It is important to distinguish between the term "category" as it pertains to pilot certificates (1) and as it refers to aircraft certification (2). When a pilot receives an airman's certificate, it is endorsed with a category limitation. If the pilot learns to fly a glider only, then his pilot certificate is endorsed and only valid for piloting gliders. The four category ratings for pilot certificates are specified in the regulation.

Aircraft category certification refers to the type of flying and structural strength requirements of the aircraft. For instance, an airplane certified in the acrobatic category must have certain extra structural strength requirements over an airplane certified in the normal category. Restricted category airplanes include aerial application aircraft (agricultural spray planes). Utility category airplanes, although not acrobatic, have greater structural integrity than normal category airplanes. Examples of aircraft categories are given in the regulation.

CLASS:

(1) As used with respect to the certification, ratings, privileges, and limitations of airmen, means a classification of aircraft within a category having similar operating characteristics. Examples include: single-engine; multi-engine; land, water; gyroplane; helicopter; airship; and free balloon; and

(2) As used with respect to the certification of aircraft, means a broad

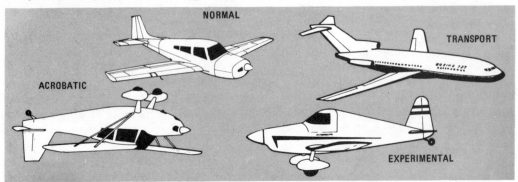

Fig. 6-2. Examples of Aircraft Certification Categories

grouping of aircraft having similar characteristics of propulsion, flight or landing. Examples include: airplane; rotorcraft; glider; balloon; landplane; and seaplane.

COMMERCIAL OPERATOR means a person who, for compensation or hire, engages in the carriage by aircraft in air commerce of persons or property, other than as an air carrier or foreign air carrier or under the authority of Part 375 of this Title. Where it is doubtful that an operation is for "compensation or hire," the test applied is whether the carriage by air is merely incidental to the person's other business or is, in itself, a major enterprise for profit.

CONTROLLED AIRSPACE means airspace, designated as continental control area, control area, control zone, terminal control area, or transition area, within which some or all aircraft may be subject to air traffic control.

CRITICAL ENGINE means the engine whose failure would most adversely affect the performance or handling qualities of an aircraft.

EXTENDED OVER-WATER OPERATION means an operation over water at a horizontal distance of more than 50 nautical miles from the nearest shore line.

FLAP EXTENDED SPEED means the highest speed permissible with wing flaps in a prescribed extended position.

FLIGHT CREWMEMBER means a pilot, flight engineer, or flight navigator assigned to duty in an aircraft during flight time.

FLIGHT LEVEL means a level of constant atmospheric pressure related to a reference datum of 29.92 inches of mercury. Each is stated in three digits that represent hundreds of feet. For example, flight level 250 represents a barometric altimeter indication of 25,000 feet; flight level 255, an indication of 25,500 feet.

FLIGHT PLAN means specified information, relating to the intended flight of an aircraft, that is filed orally or in writing with air traffic control.

FLIGHT TIME means the time from the moment the aircraft first moves under its own power for the purpose of flight until the moment it comes to rest at the next point of landing. ("Block-to-block" time.)

FLIGHT VISIBILITY means the average forward horizontal distance, from the cockpit of an aircraft in flight, at which prominent unlighted objects may be seen and identified by day and prominent lighted objects may be seen and identified by night.

GLIDER means a heavier-than-air aircraft, that is supported in flight by the dynamic reaction of the air against its lifting surfaces and whose free flight does not depend principally on an engine.

GROUND VISIBILITY means prevailing horizontal visibility near the earth's surface as reported by the United States Weather Bureau or an accredited observer.

GYROPLANE means a rotorcraft whose rotors are not engine-driven except for initial starting, but are made to rotate by action of the air when the rotorcraft is moving; and whose means of propulsion, consisting usually of conventional propellers, is independent of the rotor system.

HELICOPTER means a rotorcraft that, for its horizontal motion, depends principally on its engine-driven rotors.

HELIPORT means an area of land, water, or structure used or intended to be used for the landing and takeoff of helicopters.

IFR CONDITIONS means weather conditions below the minimum for flight under visual flight rules.

INDICATED AIRSPEED means the speed of an aircraft as shown on its pitot static airspeed indicator calibrated to reflect standard atmosphere adiabatic compressible flow at sea level uncorrected for airspeed system errors.

INSTRUMENT means a device using an internal mechanism to show visually or aurally the attitude, altitude, or operation of an aircraft or aircraft part. It includes electronic devices for automatically controlling an aircraft in flight.

KITE means a framework, covered with paper, cloth, metal, or other material, intended to be flown at the end of a rope or cable, and having as its only support the force of the wind moving past its surfaces.

LANDING GEAR EXTENDED SPEED means the maximum speed at which an aircraft can be safely flown with the landing gear extended.

LANDING GEAR OPERATING SPEED means the maximum speed at which the landing gear can be safely extended or retracted.

Both the landing gear extension speed and landing gear operating speeds are calibrated airspeeds and are found in the owner's handbook or airplane flight manual. The landing gear extension speed refers to the maximum speed at which the airplane can be flown with the gear "down and locked," while landing gear operating speed refers to the maximum speed at which the gear can be retracted or extended. Exceeding either of these speeds may cause damage to the landing gear mechanism.

LARGE AIRCRAFT means aircraft of more than 12,500 pounds, maximum certificated takeoff weight.

LIGHTER - THAN - AIR AIRCRAFT means aircraft that can rise and remain suspended by using contained gas weighing less than the air that is displaced by the gas.

LOAD FACTOR means the ratio of a specified load to the total weight of the aircraft. The specified load is expressed in terms of any of the following: aerodynamic forces, inertia forces, or ground or water reactions.

Load factor is usually expressed in terms of "G-Loading." One "G" refers to the weight of the airplane when sitting on the ground. If certain maneuvers increase the effective weight of an airplane to twice its weight, however, the load factor would be two "G's."

MACH NUMBER means the ratio of true airspeed to the speed of sound.

MAJOR ALTERATION means an alteration not listed in the aircraft, aircraft engine, or propeller specifications that might appreciably affect weight, balance, structural strength, performance, powerplant operation, flight characteristics, or other qualities affecting airworthiness.

MANIFOLD PRESSURE means absolute pressure as measured at the appropriate point in the induction system and usually expressed in inches of mercury.

MEDICAL CERTIFICATE means acceptable evidence of physical fitness on a form prescribed by the Administrator.

NIGHT means the time between the end of evening civil twilight and the beginning of morning civil twilight, as published in the American Air Almanac, converted to local time.

PARACHUTE means a device used or intended to be used to retard the fall of a body or object through the air.

PILOTAGE means navigation by visual reference to landmarks.

PILOT IN COMMAND means the pilot responsible for the operation and safety of an aircraft during flight time.

PITCH SETTING means the propeller blade setting as determined by the blade angle measured in a manner, and at a radius, specified by the instruction manual for the propeller.

POSITIVE CONTROL means control of all air traffic, within designated airspace, by air traffic control.

PROHIBITED AREA means designated airspace within which the flight of aircraft is prohibited.

PROPELLER means a device for propelling an aircraft that has blades on an engine-driven shaft and that, when rotated, produces by its action on the air, a thrust approximately perpendicular to its plane of rotation. It includes control components normally supplied by its manufacturer, but does not include main and auxiliary rotors or rotating airfoils of engines.

RATING means a statement that, as a part of a certificate, sets forth special conditions, privileges, or limitations.

REPORTING POINT means a geographical location in relation to which the position of an aircraft is reported.

RESTRICTED AREA means airspace designated under Part 91 (present Part 608) of this chapter within which the flight of aircraft, while not wholly prohibited, is subject to restriction.

ROCKET means an aircraft propelled by ejected expanding gases generated in the engine from self-contained propellants and not dependent on the intake of outside substances. It includes any part which becomes separated during the operation.

ROTORCRAFT means a heavier-than-air aircraft that depends principally for its support in flight on the lift generated by one or more rotors.

SMALL AIRCRAFT means aircraft of 12,500 pounds or less, maximum certificated takeoff weight.

STANDARD ATMOSPHERE means the atmosphere defined in U.S. Standard Atmosphere, 1962 (Geopotential altitude tables).

TAKEOFF POWER:
(1) With respect to reciprocating engines, means the brake horsepower that is developed under standard sea level conditions, and under the maximum conditions of crankshaft rotational speed and engine manifold pressure approved for the normal takeoff, and limited in continuous use to the period of time shown in the approved engine specifications; and

(2) With respect to turbine engines, means the brake horsepower that is developed under static conditions at a specified altitude and atmospheric temperature, and under the maximum conditions of rotorshaft rotational speed and gas temperature approved for the normal takeoff, and limited in continuous use to the period of time shown in the approved engine specifications.

TRUE AIRSPEED means the airspeed of an aircraft relative to undisturbed air.

TYPE:
(1) As used with respect to the certification, ratings, privileges, and limitations of airmen, means a specific make and basic model of aircraft, including modifications thereto that do not change its handling or flight characteristics. Examples include: DC-7; 1049; and F-27; and

(2) As used with respect to the certification of aircraft, means those aircraft which are similar in design. Examples include: DC-7 and DC-7C, 1049G and 1049H, and F-27 and F-27F.

VFR OVER-THE-TOP, with respect to the operation of aircraft, means the operation of an aircraft over-the-top under VFR when it is not being operated on an IFR flight plan.

1.2 ABBREVIATIONS AND SYMBOLS

ALS means approach light system.

ASR means airport surveillance radar.

ATC means air traffic control.

CAS means calibrated airspeed.

DME means distance measuring equipment compatible with TACAN.

FAA means Federal Aviation Administration.

FM means fan marker.

GS means glide slope.

HIRL means high-intensity runway light system.

ICAO means International Civil Aviation Organization.

IFR means instrument flight rules.

ILS means instrument landing system.

LMM means compass locator at middle marker.

M means mach number.

NDB(ADF) means nondirectional beacon (automatic direction finder).

OM means ILS outer marker

PAR means precision approach radar.

RAIL means runway alignment indicator light system.

RBN means radio beacon.

REIL means runway end identification lights.

TVOR means very high frequency terminal omnirange station.

The following designations are commonly called "V-speeds." V stands for velocity and the small letter designators next to the V refer to the specific type of speed. These "V-speeds" are simply shorthand methods of indicating the various speeds associated with today's modern aircraft.

V_A *means design maneuvering speed.*

V_B *means design speed for maximum gust intensity.*

V_C *means design cruising speed.*

V_F *means design flap speed*

V_{FE} *means maximum flap extended speed.*

V_{LE} *means maximum landing gear extended speed.*

V_{LO} *means maximum landing gear operating speed.*

V_{LOF} *means lift-off speed.*

V_{MC} *means minimum control speed with the critical engine inoperative.*

V_{NE} *means never-exceed speed.*

V_{S_0} *means the stalling speed or the minimum steady flight speed in the landing configuration.*

V_{S_1} *means the stalling speed or the minimum steady flight speed obtained in a specified configuration.*

V_X *means speed for best angle of climb.*

V_Y *means speed for best rate of climb.*

VFR means visual flight rules.

VHF means very high frequency.

VOR means very high frequency omnirange station.

VORTAC means co-located VOR and TACAN.

SECTION C - FAR PART 61

Part 61 is important because it prescribes the requirements for the issuance of the student, private, commercial, airline transport pilot, flight instructor certificates, instrument ratings, and aircraft type ratings. In addition, Part 61 lists the requirements for medical certificates and pilot logbooks. An understanding of Part 61 is especially important to the beginning pilot and any pilot wishing to upgrade his certificate or add a new rating to his license. The prospective private pilot should be especially interested in those sections dealing with student and private pilot qualifications, privileges, and limitations, and the medical certificate requirements.

SUBPART A — GENERAL

61.1 APPLICABILITY.

(a) This part prescribes the requirements for issuing pilot and flight instructor certificates and ratings, the conditions under which those certificates and ratings are necessary, and the privileges and limitations of those certificates and ratings.

(b) Except as provided in section 61.71 of this part, an applicant for a certificate or rating may, until November 1, 1974, meet either the requirements of this part, or the requirements in effect immediately before November 1, 1973. However, the applicant for a private pilot certificate with a free balloon class rating must meet the requirements of this part.

61.3 REQUIREMENT FOR CERTIFICATES, RATING, AND AUTHORIZATIONS. ✈

(a) *Pilot certificate.* No person may act as pilot in command or in any other capacity as a required pilot flight crewmember of a civil aircraft of United States registry unless he has in his personal possession a current pilot certificate issued to him under this part. However, when the aircraft is operated within a foreign country a current pilot license issued by the country in which the aircraft is operated may be used.

(b) *Pilot certificate: foreign aircraft.* No person may, within the United States, act as pilot in command or in any other capacity as a required pilot flight crewmember of a civil aircraft of foreign registry unless he has in his personal possession a current pilot certificate issued to him under this part, or a pilot license issued to him or validated for him by the country in which the aircraft is registered.

(c) *Medical certificate.* Except for free balloon pilots piloting balloons and glider pilots piloting gliders, no person may act as pilot in command or in any other capacity as a required pilot flight crewmember of an aircraft under a certificate issued to him under this part, unless he has in his personal possession an appropriate current medical certificate issued under Part 67. However, when the aircraft is operated within a foreign country with a current pilot license issued by that country, evidence of current medical qualification for that license, issued by that country, may be used. In the case of a pilot certificate issued on the basis of a foreign pilot license under 61.75, evidence of current medical qualification accepted for the issue of that license is used in place of a medical certificate.

(e) *Instrument rating.* No person may act as pilot in command of a civil aircraft under instrument flight rules, or in weather conditions less than the minimums prescribed for VFR flight unless—

    (1) In the case of an airplane, he holds an instrument rating or an airline transport pilot certificate with an airplane category rating on it;

(h) *Inspection of certificate.* Each person who holds a pilot certificate, flight instructor certificate, medical certificate, authorization or license required by this part shall present it for inspection upon the request of the Administrator, an authorized representative of the National Transportation Safety Board, or any Federal, State, or local law enforcement officer.

61.5 CERTIFICATES AND RATINGS ISSUED UNDER THIS PART.

(a) The following certificates are issued under this part:

    (1) Pilot certificates:
        (i)   Student pilot.
        (ii) Private pilot.
        (iii) Commercial pilot.
        (iv) Airline transport pilot.
    (2) Flight instructor certificates.

(b) The following ratings are placed on pilot certificates (other than student pilot) where applicable:

    (1) Aircraft category ratings:
        (i)   Airplane.
        (ii) Rotorcraft.
        (iii) Glider.
        (iv) Lighter-than-air.
    (2) Airplane class ratings:
        (i)   Single-engine land.
        (ii) Multiengine land.
        (iii) Single-engine sea.
        (iv) Multiengine sea.
    (3) Rotorcraft class ratings:
        (i)   Helicopter.
        (ii) Gyroplane.

    (4) Lighter-than-air class ratings:
        (i)   Airship.
        (ii) Free balloon.
    (5) Aircraft type ratings are listed in Advisory Circular 61-1 entitled "Aircraft Type Ratings." This list includes ratings for the following:
        (i)   Large aircraft, other than lighter-than-air.
        (ii) Small turbojet-powered airplanes.
        (iii) Small helicopters for operations requiring an airline transport pilot certificate.
        (iv) Other aircraft type ratings specified by the Administrator through aircraft type certificate procedures.
    (6) Instrument ratings (on private and commercial pilot certificates only):
        (i)   Instrument—airplanes.
        (ii) Instrument—helicopter.

(c) The following ratings are placed on flight instructor certificates where applicable:

    (1) Aircraft category ratings:
        (i)   Airplane.
        (ii) Rotorcraft.
        (iii) Glider.
    (2) Airplane class ratings:
        (i)   Single-engine (land and sea).
        (ii) Multiengine (land and sea).
    (3) Rotorcraft class ratings:
        (i)   Helicopter.
        (ii) Gyroplane.
    (4) Instrument ratings:
        (i)   Instrument—airplane.
        (ii) Instrument—helicopter.

61.11 EXPIRED PILOT CERTIFICATES AND REISSUANCE.

(a) No person who holds an expired pilot certificate or rating may exercise the privileges of that pilot certificate, or rating.

(c) A private or commercial pilot certificate or a special purpose pilot certificate, issued on the basis of a foreign pilot license, expires on the expiration date stated thereon. A certificate without an expiration date is issued to the holder of the expired certificate only if he meets the requirements of 61.75 for the issue of a pilot certificate based on a foreign pilot license.

61.13 APPLICATION AND QUALIFICATION.

(a) Application for a certificate and rating, or for an additional rating under this part is made on a form and in a manner prescribed by the Administrator.

(b) An applicant who meets the requirements of this part is entitled to an appropriate pilot certificate with aircraft ratings. Additional aircraft category, class, type and other ratings, for which the applicant is qualified, are added to his certificate. However, the Administrator may refuse to issue certificates to persons who are not citizens of the United States and who do not reside in the United States.

(c) An applicant who cannot comply with all of the flight proficiency requirements prescribed by this part because the aircraft used by him for his flight training or flight test is characteristically incapable of performing a required pilot operation, but who meets all other requirements for the certificate or rating sought, is issued the certificate or rating with appropriate limitations.

(d) An applicant for a pilot certificate who holds a medical certificate under 67.10 with special limitations on it, but who meets all other requirements for that pilot certificate, is issued a pilot certificate containing such operating limitations as the Administrator determines are necessary because of the applicant's medical deficiency.

(f) Unless authorized by the Administrator—
  (1) A person whose pilot certificate is suspended may not apply for any pilot or flight instructor certificate or rating during the period of suspension;

(g) Unless the order of revocation provides otherwise—
  (1) A person whose pilot certificate is revoked may not apply for any pilot or flight instructor certificate or rating for 1 year after the date of revocation;

61.15 OFFENSES INVOLVING NARCOTIC DRUGS, MARIHUANA, AND DEPRESSANT OR STIMULANT DRUGS OR SUBSTANCES.

(a) No person who is convicted of violating any Federal or State statute relating to the growing, processing, manufacture, sale, disposition, possession, transportation, or importation of narcotic drugs, marihuana, and depressant or stimulant drugs or substances, is eligible for any certificate or rating issued under this Part for a period of 1 year after the date of final conviction.

(b) No person who commits an act prohibited by 91.12(a) of this chapter is eligible for any certificate or rating issued under this part for a period of 1 year after the date of that act.

(c) Any conviction specified in paragraph (a) of this section or the commission of the act referenced in paragraph (b) of this section, is grounds for suspending or revoking any certificate or rating issued under this part.

61.17 TEMPORARY CERTIFICATE.

(a) A temporary pilot or flight instructor certificate, or a rating, effective for a period of not more than 90 days, is issued to a qualified applicant pending a review of his qualifications and the issuance of a permanent certificate or rating by the Administrator. The permanent certificate or rating is issued to an applicant found qualified and a denial thereof is issued to an applicant found not qualified.

(b) A temporary certificate issued under paragraph (a) of this section expires—

   (1) At the end of the expiration date stated thereon: or

   (2) Upon receipt by the applicant, of—

      (i) The certificate or rating sought; or

      (ii) Notice that the certificate or rating sought is denied.

61.19 DURATION OF PILOT AND FLIGHT INSTRUCTOR CERTIFICATES. ✈

(a) *General.* The holder of a certificate with an expiration date may not, after that date, exercise the privileges of that certificate.

(b) *Student pilot certificate.* A student pilot certificate expires at the end of the 24th month after the month in which it is issued.

(c) *Other pilot certificates.* Any pilot certificate (other than a student pilot certificate) issued under this part is issued without a specific expiration date. However, the holder of a pilot certificate issued on the basis of a foreign pilot license may exercise the privileges of that certificate only while the foreign pilot license on which that certificate is based is effective.

(e) *Surrender, suspension, or revocation.* Any pilot certificate or flight instructor certificate issued under this part ceases to be effective if it is surrendered, suspended, or revoked.

(f) *Return of certificate.* The holder of any certificate issued under this part that is suspended or revoked shall, upon the Administrator's request, return it to the Administrator.

✶ 61.23 DURATION OF MEDICAL CERTIFICATES. ✈

(a) A first-class medical certificate expires at the end of the last day of—

   (1) The sixth month after the month of the date of examination shown on the certificate, for operations requiring an airline transport pilot certificate;

   (2) The 12th month after the month of the date of examination shown on the certificate, for operations requiring only a commercial pilot certificate; and

   (3) The 24th month after the month of the date of examination shown on the certificate, for operations requiring only a private or student pilot certificate.

(b) A second-class medical certificate expires at the end of the last day of—

   (1) The 12th month after the month of the date of examination shown on the certificate, for operations requiring a commercial pilot certificate; and

   (2) The 24th month after the month of the date of examination shown on the certificate, for operations requiring only a private or student pilot certificate.

(c) A third-class medical certificate expires at the end of the last day of the 24th month after the month of the date of examination shown on the certificate, for operations requiring a private or student pilot certificate.

61.25 CHANGE OF NAME.

An application for the change of a name on a certificate issued under this part must be accompanied by the applicant's current certificate and a copy of the marriage license, court order, or other document verifying the change. The documents are returned to the applicant after inspection.

61.27 VOLUNTARY SURRENDER OR EXCHANGE OF CERTIFICATE.

The holder of a certificate issued under this part may voluntarily surrender it for cancellation, or for the issue of a certificate of lower grade, or another certificate with specific ratings deleted. If he so requests, he must include the following signed statement or its equivalent:

This request is made for my own reasons, with full knowledge that my (insert name of certificate or rating, as appropriate) may not be reissued to me unless I again pass the tests prescribed for its issue.

61.29 REPLACEMENT OF LOST OR DESTROYED CERTIFICATE.

(a) An application for the replacement of a lost or destroyed airman certificate issued under this part is made by letter to the Department of Transportation, Federal Aviation Administration, Airman Certification Branch, Post Office Box 25082, Oklahoma City, OK 73125. The letter must—
(1) State the name of the person to whom the certificate was issued, the permanent mailing address (including zip code), social security number (if any), date and place of birth of the certificate holder, and any available information regarding the grade, number,

and date of issue of the certificate, and the ratings on it; and
(2) Be accompanied by a check or money order for $2, payable to the Federal Aviation Administration.
(b) An application for the replacement of a lost or destroyed medical certificate is made by letter to the Department of Transportation, Federal Aviation Administration, Aeromedical Certification Branch, Post Office Box 25082, Oklahoma City, OK 73125, accompanied by a check or money order for $2.

(c) A person who has lost a certificate issued under this part, or a medical certificate issued under Part 67 or both, may obtain a telegram from the FAA confirming that it was issued. The telegram may be carried as a certificate for a period not to exceed 60 days pending his receipt of a duplicate certificate under paragraph (a) or (b) of this section, unless he has been notified that the certificate has been suspended or revoked. The request for such a telegram may be made by letter or prepaid telegram, including the date upon which a duplicate certificate was previously requested, if a request had been made, and a money order for the cost of the duplicate certificate. The request for a telegraphic certificate is sent to the office listed in paragraph (a) or (b) of this section, as appropriate. However, a request for both airman and medical certificates at the same time must be sent to the office prescribed in paragraph (a) of this section.

61.31 GENERAL LIMITATIONS.

(a) *Type ratings required.* A person may not act as pilot in command of any of the following aircraft unless he holds a type rating for that aircraft:

(1) A large aircraft (except lighter-than-air).

(2) A helicopter, for operations requiring an airline transport pilot certificate.

(3) A turbojet powered airplane.

(4) Other aircraft specified by the Administrator through aircraft type certificate procedures.

(b) *Authorization in lieu of a type rating.*

   (1) In lieu of a type rating required under paragraphs (a) (1), (3), and (4) of this section, an aircraft may be operated under an authorization issued by the Administrator, for a flight or series of flights within the United States, if—

   (i) The particular operation for which the authorization is requested involves a ferry flight, a practice or training flight, a flight test for a pilot type rating, or a test flight of an aircraft, for a period that does not exceed 60 days;

   (ii) The applicant shows that compliance with paragraph (a) of this section is impracticable for the particular operations; and

   (iii) The Administrator finds that an equivalent level of safety may be achieved through operating limitations on the authorization.

   (2) Aircraft operated under an authorization issued under this paragraph—

   (i) May not be operated for compensation or hire; and

   (ii) May carry only flight crewmembers necessary for the flight.

   (3) An authorization issued under this paragraph may be reissued for an additional 60-day period for the same operation if the applicant shows that he was prevented from carrying out the purpose of the particular operation before his authorization expired.

   The prohibition of paragraph (b)(2)(i) of this section does not prohibit compensation for the use of an aircraft by a pilot solely to prepare for or take a flight test for a type rating.

(c) *Category and class rating: Carrying another person or operating for compensation or hire.* Unless he holds a category and class rating for that aircraft, a person may not act as pilot in command of an aircraft that is carrying another person or is operated for compensation or hire. In addition, he may not act as pilot in command of that aircraft for compensation or hire.

(d) *Category and class rating: Other operations.* No person may act as pilot in command of an aircraft in solo flight in operations not subject to paragraph (c) of this section, unless he meets at least one of the following:

   (1) He holds a category and class rating appropriate to that aircraft.

   (2) He has received flight instruction in the pilot operations required by this part, appropriate to the category and class of aircraft for first solo, given to him by a certificated flight instructor who found him competent to solo that category and class of aircraft and has so endorsed his pilot logbook.

   (3) He has soloed and logged pilot-in-command time in that category and class of aircraft before November 1, 1973.

(e) *High performance airplanes.* A person holding a private or commercial pilot certificate may not act as pilot in command of an airplane that has more than 200 horsepower, or that has a retractable landing gear, flaps, and a controllable propeller, unless he has received flight instruction from an authorized flight instructor who has certified in his logbook that he is competent to pilot an airplane that has more than 200 horsepower, or that has a retractable landing gear, flaps, and a controllable propeller, as the case may be. However, this instruction is not required if he has logged flight time as pilot in command in high performance airplanes before November 1, 1973.

(f) *Exception.* This section does not require a class rating for gliders, or category and class ratings for aircraft that are not type certificated as airplanes, rotorcraft, or lighter-than-air aircraft. In addition, the rating limitations of this section do not apply to—

(1) The holder of a student pilot certificate;

(2) The holder of a pilot certificate when operating an aircraft under the authority of an experimental or provisional type certificate;

(3) An applicant when taking a flight test given by the Administrator; or

(4) The holder of a pilot certificate with a lighter-than-air category rating when operating a hot air balloon without an airborne heater.

61.33 TESTS: GENERAL PROCEDURE.

Tests prescribed by or under this part are given at times and places, and by persons, designated by the Administrator.

61.35 WRITTEN TEST: PREREQUISITES AND PASSING GRADES.

(a) An applicant for a written test must—

(1) Show that he has satisfactorily completed the ground instruction or home study course required by this part for the certificate or rating sought:

(2) Present as personal identification an airman certificate, driver's license, or other official document; and

(3) Present a birth certificate or other official document showing that he meets the age requirement prescribed in this part for the certificate sought not later than 2 years from the date of application for the test.

(b) The minimum passing grade is specified by the Administrator on each written test sheet or booklet furnished to the applicant.

This section does not apply to the written test for an airline transport pilot certificate or a rating associated with that certificate.

61.37 WRITTEN TESTS: CHEATING OR OTHER UNAUTHORIZED CONDUCT.

(a) Except as authorized by the Administrator, no person may—

(1) Copy, or intentionally remove, a written test under this part;

(2) Give to another, or receive from another, any part or copy of that test;

(3) Give help on that test to, or receive help on that test from, any person during the period that test is being given;

(4) Take any part of that test in behalf of another person;

(5) Use any material or aid during the period that test is being given; or

(6) Intentionally cause, assist, or participate in any act prohibited by this paragraph.

(b) No person whom the Administrator finds to have committed an act prohibited by paragraph (a) of this section is eligible for any airman or ground instructor certificate or rating, or to take any test therefor, under this chapter for a period of 1 year after the date of that act. In addition, the commission of that act is a basis for suspending or revoking any airman or ground instructor certificate or rating held by that person.

61.39 PREREQUISITES FOR FLIGHT TESTS.

(a) To be eligible for a flight test for a certificate, or an aircraft or instrument rating issued under this part, the applicant must—

(1) Have passed any required written test since the beginning of the 24th month before the month in which he takes the flight test;

(2) Have the applicable instruction and aeronautical experience prescribed in this part;

(3) Hold a current medical certificate appropriate to the certificate he seeks or, in the case of a rating to be added to his pilot certificate, at least a third-class medical certificate issued since the beginning of the 24th month before the month in which he takes the flight test;

(4) Except for a flight test for an airline transport pilot certificate, meet the age requirement for the issuance of the certificate or rating he seeks; and

(5) Have a written statement from an appropriately certificated flight instructor certifying that he has given the applicant flight instruction in preparation for the flight test within 60 days preceding the date of application, and finds him competent to pass the test and to have satisfactory knowledge of the subject areas in which he is shown to be deficient by his FAA airman written test report. However, an applicant need not have this written statement if he—

(i) Holds a foreign pilot license issued by a contracting State to the Convention on International Civil Aviation that authorizes at least the pilot privileges of the airman certificate sought by him;

(ii) Is applying for a type rating only, or a class rating with an associated type rating; or

(iii) Is applying for an airline transport pilot certificate or an additional aircraft rating on that certificate.

(b) Notwithstanding paragraph (a)(1) of this section, an applicant for an airline transport pilot certificate or an additional aircraft rating on that certificate who has been since passing the written examination, continuously employed as a pilot, or as a pilot assigned to flight engineer duties by, and is participating in an approved pilot training program of a U.S. air carrier or commercial operator, or who is rated as a pilot by, and is participating in a pilot training program of a U.S. scheduled military air transportation service, may take the flight test for that certificate or rating.

61.41 FLIGHT INSTRUCTION RECEIVED FROM FLIGHT INSTRUCTORS NOT CERTIFICATED BY FAA.

Flight instruction may be credited toward the requirements for a pilot certificate or rating issued under this part if it is received from—

(a) An Armed Force of either the United States or a foreign contracting State to the Convention on International Civil Aviation in a program for training military pilots; or

(b) A flight instructor who is authorized to give that flight instruction by the licensing authority of a foreign contracting State to the Convention on International Civil Aviation and the flight instruction is given outside the United States.

61.43 FLIGHT TESTS: GENERAL PROCEDURES.

(a) The ability of an applicant for a private or commercial pilot certificate, or for an aircraft or instrument rating on that certificate to perform the required pilot operations is based on the following:

   (1) Executing procedures and maneuvers within the aircraft's performance capabilities and limitations, including use of the aircraft's systems.

   (2) Executing emergency procedures and maneuvers appropriate to the aircraft.

   (3) Piloting the aircraft with smoothness and accuracy.

   (4) Exercising judgment.

   (5) Applying his aeronautical knowledge.

   (6) Showing that he is the master of the aircraft, with the successful outcome of a procedure or maneuver never seriously in doubt.

(b) If the applicant fails any of the required pilot operations in accordance with the applicable provisions of paragraph (a) of this section, the applicant fails the flight test. The applicant is not eligible for the certificate or rating sought until he passes any pilot operations he has failed.

(c) The examiner or the applicant may discontinue the test at any time when the failure of a required pilot operation makes the applicant ineligible for the certificate or rating sought. If the test is discontinued the applicant is entitled to credit for only those entire pilot operations that he has successfully performed.

61.45 FLIGHT TESTS: REQUIRED AIRCRAFT AND EQUIPMENT.

(a) *General.* An applicant for a certificate or rating under this part must furnish, for each flight test that he is required to take, an appropriate aircraft of United States registry that has a current standard or limited airworthiness certificate. However, the applicant may, at the discretion of the inspector or examiner conducting the test, furnish an aircraft of U.S. registry that has a current airworthiness certificate other than standard or limited, an aircraft of foreign registry that is properly certificated by the country or registry, or a military aircraft in an operational status if its use is allowed by an appropriate military authority.

(b) *Required equipment (other than controls).* Aircraft furnished for a flight test must have—

   (1) The equipment for each pilot operation required for the flight test;

   (2) No prescribed operating limitations that prohibit its use in any pilot operation required on the test;

   (3) Pilot seats with adequate visibility for each pilot to operate the aircraft safely, except

as provided in paragraph (d) of this section; and

(4) Cockpit and outside visibility adequate to evaluate the performance of the applicant, where an additional jump seat is provided for the examiner.

(c) *Required controls.* An aircraft (other than lighter-than-air) furnished under paragraph (a) of this section for any pilot flight test must have engine power controls and flight controls that are easily reached and operable in a normal manner by both pilots, unless after considering all the factors, the examiner determines that the flight test can be conducted safely without them. However, an aircraft having other controls such as nose-wheel steering, brakes, switches, fuel selectors, and engine air flow controls that are not easily reached and operable in a normal manner by both pilots may be used, if more than one pilot is required under its airworthiness certificate, or if the examiner determines that the flight can be conducted safely.

(d) *Simulated instrument flight equipment.* An applicant for any flight test involving flight maneuvers solely by reference to instruments must furnish equipment satisfactory to the examiner that excludes the visual reference of the applicant outside of the aircraft.

(e) *Aircraft with single controls.* At the discretion of the examiner, an aircraft furnished under paragraph (a) of this section for a flight test may, in the cases listed herein, have a single set of controls. In such case, the examiner determines the competence of the applicant by observation from the ground or from another aircraft.

(1) A flight test for addition of a class or type rating, not involving demonstration of instrument skills, to a private or commercial pilot certificate.

61.47 FLIGHT TESTS: STATUS OF FAA INSPECTORS AND OTHER AUTHORIZED FLIGHT EXAMINERS.

An FAA inspector or other authorized flight examiner conducts the flight test of an applicant for a pilot certificate or rating for the purpose of observing the applicant's ability to perform satisfactorily the procedures and maneuvers on the flight test. The inspector or other examiner is not pilot in command of the aircraft during the flight test unless he acts in that capacity for the flight, or portion of the flight, by prior arrangement with the applicant or other person who would otherwise act as pilot in command of the flight, or portion of the flight. Notwithstanding the type of aircraft used during a flight test, the applicant and the inspector or other examiner are not, with respect to each other (or other occupants authorized by the inspector or other examiner), subject to the requirements or limitations for the carriage of passengers specified in this chapter.

61.49 RETESTING AFTER FAILURE.

An applicant for a written or flight test who fails that test may not apply for retesting until after 30 days after the date he failed the test. However, in the case of his first failure he may apply for retesting before the 30 days have expired upon presenting a written statement from an authorized instructor certifying that he has given flight or ground instruction as appropriate to the applicant and finds him competent to pass the test.

61.51 PILOT LOGBOOKS.

(a) The aeronautical training and experience used to meet the requirements for a certificate or rating, or the recent flight experience requirements of this part must be shown by a reliable record. The logging of other flight time is not required.

(b) *Logbook entries.* Each pilot shall enter the following information for each flight or lesson logged:

(1) *General.*
  (i) Date.
  (ii) Total time of flight.
  (iii) Place, or points of departure and arrival.
  (iv) Type and identification of aircraft.

(2) *Type of pilot experience or training.*
  (i) Pilot in command or solo.
  (ii) Second in command.
  (iii) Flight instruction received from an authorized flight instructor.
  (iv) Instrument flight instruction from an authorized flight instructor.
  (v) Pilot ground trainer instruction.
  (vi) Participating crew (lighter-than-air).
  (vii) Other pilot time.

(3) *Conditions of flight.*
  (i) Day or night.
  (ii) Actual instrument.
  (iii) Simulated instrument conditions.

(c) *Logging of pilot time —*

(1) *Solo flight time.* A pilot may log as solo flight time only that flight time when he is the sole occupant of the aircraft. However, a student pilot may also log as solo flight time that time during which he acts as the pilot in command of an airship requiring more than one flight crew-member.

(2) *Pilot-in-command flight time.*
  (i) A private or commercial pilot may log as pilot in command time only that flight time during which he is the sole manipulator of the controls of an aircraft for which he is rated, or when he is the sole occupant of the aircraft, or when he acts as pilot in command of an aircraft on which more than one pilot is required under the type certification of the aircraft, or the regulations under which the flight is conducted.

(3) *Second-in-command flight time.* A pilot may log as second in command time all flight time during which he acts as second in command of an aircraft on which more than one pilot is required under the type certification of the aircraft, or the regulations under which the flight is conducted.

(4) *Instrument flight time.* A pilot may log as instrument flight time only that time during which he operates the aircraft solely by reference to instruments, under actual or simulated instrument flight conditions. Each entry must include the place and type of each instrument approach completed, and the name of the safety pilot for each simulated instrument flight. An instrument flight instructor may log as instrument time that time during which he acts as instrument flight instructor in actual instrument weather conditions.

(5) *Instruction time.* All time logged as flight instruction, instrument flight instruction, pilot ground trainer instruction, or ground instruction time must be certified by the appropriately rated and certificated instructor from whom it was received.

(d) *Presentation of logbook.*

(1) A pilot must present his logbook (or other record required by this section) for in-

spection upon reasonable request by the Administrator, an authorized representative of the National Transportation Safety Board, or any State or local law enforcement officer.

(2) A student pilot must carry his logbook (or other record required by this section) with him on all solo cross-country flights, as evidence of the required instructor clearances and endorsements.

61.53 OPERATIONS DURING MEDICAL DEFICIENCY. ✈

No person may act as pilot in command, or in any other capacity as a required pilot flight crewmember while he has a known medical deficiency, or increase of a known medical deficiency, that would make him unable to meet the requirements for his current medical certificate.

*61.57 RECENT FLIGHT EXPERIENCE: PILOT IN COMMAND. ✈

(a) *Flight review.* After November 1, 1974, no person may act as pilot in command of an aircraft unless, within the preceding 24 months, he has—

(1) Accomplished a flight review given to him, in an aircraft for which he is rated, by an appropriately certificated instructor or other person designated by the Administrator; and

(2) Had his logbook endorsed by the person who gave him the review certifying that he has satisfactorily accomplished the flight review.

However, a person who has, within the preceding 24 months, satisfactorily completed a pilot proficiency check conducted by the FAA, an approved pilot check airman or a U.S. armed force

for a pilot certificate, rating or operating privilege, need not accomplish the flight review required by this section.

(b) *Meaning of flight review.* As used in this section, a flight review consists of—

(1) A review of the current general operating and flight rules of Part 91; and

(2) A review of those maneuvers and procedures which in the discretion of the person giving the review are necessary for the pilot to demonstrate that he can safely exercise the privileges of his pilot certificate.

(c) *General experience.* No person may act as pilot in command of an aircraft carrying passengers, nor of an aircraft certificated for more than one required pilot flight crewmember, unless within the preceding 90 days, he has made three takeoffs and three landings as the sole manipulator of the flight controls in an aircraft of the same category and class and, if a type rating is required, of the same type. If the aircraft is a tailwheel airplane, the landings must have been made to a full stop in a tailwheel airplane. For the purpose of meeting the requirements of the paragraph a person may act as pilot in command of a flight under day VFR or day IFR if no persons or property other than as necessary for his compliance thereunder, are carried. This paragraph does not apply to operations requiring an airline transport pilot certificate, or to operations conducted under Part 135.

(d) *Night experience.* No person may act as pilot in command of an aircraft carrying passengers during the period beginning 1 hour after sunset and ending 1 hour before sunrise (as published in the American Air Almanac) unless, within the preceding 90 days, he has made at least three takeoffs and three landings to a full stop during that period in the category and class of aircraft to be used. This paragraph does not apply to operations requiring an airline transport pilot certificate.

61.59 FALSIFICATION, REPRODUCTION, OR ALTERATION OF APPLICATIONS, CERTIFICATES, LOGBOOKS, REPORTS, OR RECORDS.

(a) No person may make or cause to be made—

    (1) Any fraudulent or intentionally false statement on any application for a certificate, rating, or duplicate thereof, issued under this part;

    (2) Any fraudulent or intentionally false entry in any logbook, record, or report that is required to be kept, made, or used, to show compliance with any requirement for the issuance, or exercise of the privileges, or any certificate or rating under this part;

    (3) Any reproduction, for fraudulent purpose, of any certificate or rating under this part; or

    (4) Any alteration of any certificate or rating under this part.

(b) The commission by any person of an act prohibited under paragraph (a) of this section is a basis for suspending or revoking any airman or ground instructor certificate or rating held by that person.

61.60 CHANGE OF ADDRESS.

The holder of a pilot or flight instructor certificate who has made a change in his permanent mailing address may not after 30 days from the date he moved, exercise the privileges of his certificate unless he has notified in writing the Department of Transportation, Federal Aviation Administration, Airman Certification Branch, Box 25082, Oklahoma City, OK 73125, of his new address.

SUBPART B — AIRCRAFT RATINGS AND SPECIAL CERTIFICATES

61.61 APPLICABILITY.

This subpart prescribes the requirements for the issuance of additional aircraft ratings after a pilot or instructor certificate is issued, and the requirements and limitations for special pilot certificates and ratings issued by the Administrator.

61.63 ADDITIONAL AIRCRAFT RATINGS (OTHER THAN AIRLINE TRANSPORT PILOT).

(a) *General.* To be eligible for an aircraft rating after his certificate is issued to him an applicant must meet the requirements of paragraphs (b) through (d) of this section, as appropriate to the rating sought.

(b) *Category rating.* An applicant for a category rating to be added on his pilot certificate must meet the requirements of this Part for the issue of the pilot certificate appropriate to the privileges for which the category rating is sought. However, the holder of a category rating for powered aircraft is not required to take a written test for the addition of a category rating on his pilot certificate.

(c) *Class rating.* An applicant for an aircraft class rating to be added on his pilot certificate must—

    (1) Present a logbook record certified by an authorized flight instructor showing that the applicant has received flight instruction in the class of aircraft for which a rating is sought and has been found competent in the pilot operations appropriate to the pilot certificate to which his category rating applies; and

(2) Pass a flight test appropriate to his pilot certificate and applicable to the aircraft category and class rating sought.

A person who holds a lighter-than-air category rating with a free balloon class rating, who seeks an airship class rating, must meet the requirements of paragraph (b) of this section as though seeking a lighter-than-air category rating.

(d) *Type rating.* An applicant for a type rating to be added on his pilot certificate must meet the following requirements:

(1) He must hold, or concurrently obtain, an instrument rating appropriate to the aircraft for which a type rating is sought.

(2) He must pass a flight test showing competence in pilot operations appropriate to the pilot certificate he holds and to the type rating sought.

(3) He must pass a flight test showing competence in pilot operations under instrument flight rules in an aircraft of the type for which the type rating is sought or, in the case of a single pilot station airplane, meet the requirements of paragraph (d)(3)(i) or (ii) of this section, whichever is applicable.

(i) The applicant must have met the requirements of this subparagraph in a multiengine airplane for which the type rating is required.

(ii) If he does not meet the requirements of paragraph (d)(3)(i) of this section and he seeks a type rating for a single-engine airplane, he must meet the requirements of this subparagraph in either a single or multiengine air-

plane, and have the recent instrument experience set forth in 61.57 (e), when he applies for the flight test under paragraph (d)(2) of this section.

(4) An applicant who does not meet the requirements of paragraphs (d)(1) and (3) of this section may obtain a type rating limited to "VFR only." Upon meeting these instrument requirements or the requirements of 61.73 (e)(2), the "VFR only" limitation may be removed for the particular type of aircraft in which competence is shown.

(5) When an instrument rating is issued to the holder of one or more type ratings, the type ratings on the amended certificate bear the limitation described in paragraph (d)(4) of this section for each airplane type rating for which he has not shown his instrument competency under this paragraph.

61.69 GLIDER TOWING: EXPERIENCE AND INSTRUCTION REQUIREMENTS.

No person may act as pilot in command of an aircraft towing a glider unless he meets the following requirements:

(a) He holds a current pilot certificate (other than a student pilot certificate) issued under this part.

(b) He has an endorsement in his pilot logbook from a person authorized to give flight instruction in gliders, certifying that he has received ground and flight instruction in gliders and is familiar with the techniques and procedures essential to the safe towing of gliders, including airspeed limitations, emergency procedures, signals used, and maximum angles of bank.

(c) He has made and entered in his pilot logbook—

(1) At least three flights as sole manipulator of the controls of an aircraft towing a glider (while accompanied by a pilot who has met the requirements of this section), and at least 10 flights as pilot in command of an aircraft towing a glider; or

(2) At least three flights as sole manipulator of the controls of an aircraft simulating glider towing flight procedures (while accompanied by a pilot who meets the requirements of this section), and at least three flights as pilot or observer in a glider being towed by an aircraft.

However, any person who, before May 17, 1967, made, and entered in his pilot logbook, 10 or more flights as pilot in command of an aircraft towing a glider in accordance with a certificate or waiver need not comply with paragraphs (c)(1) and (2) of this section.

(d) If he holds only a private pilot certificate he must have had, and entered in his pilot logbook at least—

(1) 100 hours of pilot flight time in powered aircraft; or

(2) 200 total hours of pilot flight time in powered or other aircraft.

(e) Within the preceding 12 months he has—

(1) Made at least three actual or simulated glider tows while accompanied by a qualified pilot who meets the requirements of this section; or

(2) Made at least three flights as pilot in command of a glider towed by an aircraft.

61.71 GRADUATES OF CERTIFICATED FLYING SCHOOLS: SPECIAL RULES.

(a) A graduate of a flying school that is certificated under Part 141 of this chapter is considered to meet the applicable aeronautical experience requirements of this part if he presents an appropriate graduation certificate within 60 days after the date he is graduated. However, if he applies for a flight test for an instrument rating he must hold a commercial pilot certificate, or hold a private pilot certificate and meet the requirements of 61.65(e)(1) and 61.123 (except paragraphs (d) and (e) thereof). In addition, if he applies for a flight instructor certificate he must hold a commercial pilot certificate.

(b) An applicant for a certificate or rating under this part is considered to meet the aeronautical knowledge and skill requirements, or both, applicable to that certificate or rating, if he applies within 90 days after graduation from an appropriate course given by a flying school that is certificated under Part 141 of this chapter and is authorized to test applicants on aeronautical knowledge or skill or both.

However, until January 1, 1977, a graduate of a flying school certificated and operated under the provisions of section 141.29 of this chapter, is considered to meet the aeronautical experience requirements of this part, and may be tested under the requirements of Part 61 that were in effect prior to November 1, 1973.

61.73 MILITARY PILOTS OR FORMER MILITARY PILOTS: SPECIAL RULES.

(a) *General.* A rated military pilot or former rated military pilot who applies for a private or commercial pilot certificate, or an aircraft or instrument rating, is entitled to that certificate with appropriate ratings or to the addition of a rating on the pilot certificate he holds, if he meets the applicable requirements of this section. This section does not apply to a military pilot or former military pilot who has been removed from flying status for lack of proficiency or because of disciplinary action involving aircraft operations.

61.75 PILOT CERTIFICATE ISSUED ON BASIS OF A FOREIGN PILOT LICENSE.

(a) *Purpose.* The holder of a current private, commercial, senior commercial, or airline transport pilot license issued by a foreign contracting State to the Convention on International Civil Aviation may apply for a pilot certificate under this section authorizing him to act as a pilot of a civil aircraft of U.S. registry.

The remainder of 61.73 and 61.75 has been omitted from this publication. For additional information, refer to the complete regulation.

SUBPART C—STUDENT PILOTS

61.81. APPLICABILITY.

This subpart prescribes the requirements for the issuance of student pilot certificates, the conditions under which those certificates are necessary, and the general operating rules for the holders of those certificates.

61.83 ELIGIBILITY REQUIRE-MENTS: GENERAL. ✈

To be eligible for a student pilot certificate, a person must—

(a) Be at least 16 years of age, or at least 14 years of age for a student pilot certificate limited to the operation of a glider or free balloon;

(b) Be able to read, speak, and understand the English language, or have such operating limitations on his student pilot certificate as are necessary for the safe operation of aircraft, to be removed when he shows that he can read, speak, and understand the English language; and

(c) Hold at least a current third-class medical certificate issued under Part 67 of this chapter, or, in the case of glider or free balloon operations, certify that he has no known medical defect that makes him unable to pilot a glider or a free balloon.

61.85 APPLICATION.

An application for a student pilot certificate is made on a form and in a manner provided by the Administrator and is submitted to—

(a) A designated aviation medical examiner when applying for an FAA medical certificate; or

(b) An FAA operations inspector or designated pilot examiner, accompanied by a current FAA medical certificate, or in the case of an application for a glider or free balloon pilot certificate it may be accompanied by a certification by the applicant that he has no known medical defect that makes him unable to pilot a glider or free balloon.

61.87 REQUIREMENTS FOR SOLO FLIGHT. ✈

(a) *General.* A student pilot may not operate an aircraft in solo flight until he has complied with the requirements of this section. As used in this subpart the term solo flight means that flight time during which a student pilot is the sole occupant of the aircraft, or that flight time during which he acts as pilot in command of an airship requiring more than one flight crewmember.

(b) *Aeronautical knowledge.* He must have demonstrated to an authorized instructor that he is familiar with the flight rules of Part 91 of this chapter which are pertinent to student solo flights.

(c) *Flight proficiency training.* He must have received ground and flight instruction in at least the following procedures and operations:

(1) *In airplanes.*

    (i) Flight preparation procedures, including preflight inspection and powerplant operation;

    (ii) Ground maneuvering and runups;

    (iii) Straight and level flight, climbs, turns, and descents;

    (iv) Flight at minimum controllable airspeeds, and stall recognition and recovery;

    (v) Normal takeoffs and landings;

    (vi) Airport traffic patterns, including collision avoidance precautions and wake turbulence; and

    (vii) Emergencies, including elementary emergency landings.

Instruction must be given by a flight instructor who is authorized to give instruction in airplanes.

(d) *Flight instructor endorsements.* A student pilot may not operate an aircraft in solo flight unless his student pilot certificate and pilot logbook are endorsed, and unless within the preceding 90 days his pilot certificate has been endorsed by an authorized flight instructor who—

(1) Has given him instruction in the make and model of aircraft in which the solo flight is made;

(2) Finds that he has met the requirements of this section; and

(3) Finds that he is competent to make a safe solo flight in that aircraft.

61.89 GENERAL LIMITATIONS. ✈

(a) A student pilot may not act as pilot in command of an aircraft—

(1) That is carrying a passenger;

(2) That is carrying property for compensation or hire;

(3) For compensation or hire;

(4) In furtherance of a business; or

(5) On an international flight, except that a student pilot may make solo training flights from Haines, Gustavus, or Juneau, Alaska, to White Horse, Yukon, Canada, and return, over the province of British Columbia.

(b) A student pilot may not act as a required pilot flight crewmember on any aircraft for which more than one pilot is required, except when receiving flight instruction from an authorized flight instructor on board an airship and no person other than a required flight crewmember is carried on the aircraft.

61.91 AIRCRAFT LIMITATIONS: PILOT IN COMMAND.

A student pilot may not serve as pilot in command of any airship requiring more than one flight crewmember unless he has met the pertinent requirements prescribed in 61.87.

61.93 CROSS-COUNTRY FLIGHT REQUIREMENTS. ✈

(a) *General.* A student pilot may not operate an aircraft in a solo cross-country flight, nor may he, except in an emergency, make a solo flight landing at any point other than the airport of takeoff, until he meets the requirements prescribed in this section. However, an authorized flight instructor may allow a student pilot to practice solo landings and takeoffs at another airport within 25 nautical miles from the airport at which the student pilot receives instruction if he finds that the student pilot is competent to make those landings and takeoffs. As used in this section the term cross-country flight means a flight beyond a radius of 25 nautical miles from the point of takeoff.

(b) *Flight training.* A student pilot must receive instruction from an authorized instructor in at least the following pilot operations pertinent to the aircraft to be operated in a solo cross-country flight:

   (1) For solo cross-country in airplanes—

     (i) The use of aeronautical charts, pilotage, and elementary dead reckoning using the magnetic compass;

     (ii) The use of radio for VFR navigation, and for two-way communication;

     (iii) Control of an airplane by reference to flight instruments;

     (iv) Short field and soft field procedures, and crosswind takeoffs and landings;

     (v) Recognition of critical weather situations, estimating visibility while in flight, and the procurement and use of aeronautical weather reports and forecasts; and

     (vi) Cross-country emergency procedures.

(c) *Flight instructor endorsements.* A student pilot must have the following endorsements from an authorized flight instructor:

   (1) An endorsement on his student pilot certificate stating that he has received instruc-

tion in solo cross-country flying and the applicable training requirements of this section, and is competent to make cross-country solo flights in the category of aircraft involved.

   (2) An endorsement in his pilot logbook that the instructor has reviewed the preflight planning and preparation for each solo cross-country flight, and he is prepared to make the flight safely under the known circumstances and the conditions listed by the instructor in the logbook. The instructor may also endorse the logbook for repeated solo cross-country flights under stipulated conditions over a course not more than 50 nautical miles from the point of departure if he has given the student flight instruction in both directions over the route, including takeoffs and landings at the airports to be used.

SUBPART D — PRIVATE PILOTS

61.101 APPLICABILITY.

This subpart prescribes the requirements for the issuance of private pilot certificates and ratings, the conditions under which those certificates and ratings are necessary, and the general operating rules for the holders of those certificates and ratings.

61.103 ELIGIBILITY REQUIREMENTS: GENERAL.

To be eligible for a private pilot certificate, a person must—

(a) Be at least 17 years of age, except that a private pilot certificate with a free balloon or a glider rating only may be issued to a qualified applicant who is at least 16 years of age;

(b) Be able to read, speak, and understand the English language, or have such operating limitations placed on his pilot certificate as are necessary for the safe operation of aircraft, to be removed when he shows that he can read, speak, and understand the English language;

(c) Hold at least a current third-class medical certificate issued under Part 67 of this chapter, or, in the case of a glider or free balloon rating, certify that he has no known medical defect that makes him unable to pilot a glider or free balloon, as appropriate;

(d) Pass a written test on the subject areas on which instruction or home study is required by 61.105;

(e) Pass an oral and flight test on procedures and maneuvers selected by an FAA inspector or examiner to determine the applicant's competency in the flight operations on which instruction is required by the flight proficiency provisions of 61.107; and

(f) Comply with the sections of this part that apply to the rating he seeks.

61.105 AERONAUTICAL KNOWLEDGE. ✈

An applicant for a private pilot certificate must have logged ground instruction from an authorized instructor, or must present evidence showing that he has satisfactorily completed a course of instruction or home study in at least the following areas of aeronautical knowledge appropriate to the category of aircraft for which a rating is sought.

(a) *Airplanes.*
  (1) The Federal Aviation Regulations applicable to private pilot privileges, limitations, and flight operations, accident reporting requirements of the National Transportation Safety Board, and the use of the "Airman's Information Manu-

al" and the FAA Advisory Circulars;

  (2) VFR navigation, using pilotage, dead reckoning, and radio aids;

  (3) The recognition of critical weather situations from the ground and in flight and the procurement and use of aeronautical weather reports and forecasts; and

  (4) The safe and efficient operation of airplanes, including high density airport operations, collision avoidance precautions, and radio communication procedures.

61.107 FLIGHT PROFICIENCY. ✈

The applicant for a private pilot certificate must have logged instruction from an authorized flight instructor in at least the following pilot operations. In addition, his logbook must contain an endorsement by an authorized flight instructor who has found him competent to perform each of those operations safely as a private pilot.

(a) *In airplanes.*
  (1) Preflight operations, including weight and balance determination, line inspection, and airplane servicing;

  (2) Airport and traffic pattern operations, including operations at controlled airports, radio communications, and collision avoidance precautions;

  (3) Flight maneuvering by reference to ground objects;

  (4) Flight at critically slow airspeeds, and the recognition of and recovery from imminent and full stalls entered from straight flight and from turns;

  (5) Normal and crosswind take-offs and landings;

  (6) Control and maneuvering an airplane solely by reference to instruments, including descents

and climbs using radio aids or radar directives;

(7) Cross-country flying, using pilotage, dead reckoning, and radio aids, including one 2-hour flight:

(8) Maximum performance take-offs and landings;

(9) Night flying, including take-offs, landings, and VFR navigation; and

(10) Emergency operations, including simulated aircraft and equipment malfunctions.

61.109 AIRPLANE RATING: AERONAUTICAL EXPERIENCE. ✈

An applicant for a private pilot certificate with an airplane rating must have had at least a total of 40 hours of flight instruction and solo flight time which must include the following:

(a) Twenty hours of flight instruction from an authorized flight instructor, including at least—

(1) Three hours of cross country;

(2) Three hours at night, including 10 takeoffs and landings for applicants seeking night flying privileges, and

(3) Three hours in airplanes in preparation for the private pilot flight test within 60 days prior to that test.

An applicant who does not meet the night flying requirement in paragraph (a)(2) of this section is issued a private pilot certificate bearing the limitation "Night flying prohibited." This limitation may be removed if the holder of the certificate shows that he has met the requirements of paragraph (a)(2) of this section.

(b) Twenty hours of solo flight time, including at least—

(1) Ten hours in airplanes;

(2) Ten hours of cross-country flights, each flight with a landing more than 50 nautical

miles from the point of departure, and one with landings at three points, each of which is more than 100 nautical miles from each of the other two points; and

(3) Three solo takeoffs and landings to a full stop at an airport with an operating control tower.

61.111 CROSS-COUNTRY FLIGHTS: PILOTS BASED ON SMALL ISLANDS.

(a) An applicant who shows that he is located on an island from which the required flights cannot be accomplished without flying over water more than 10 nautical miles from the nearest shoreline need not comply with paragraph (b)(2) of 61.109. However, if other airports that permit civil operations are available to which a flight may be made without flying over water more than 10 nautical miles from the nearest shoreline, he must show that he has completed two round trip solo flights between those two airports that are farthest apart, including a landing at each airport on both flights.

(b) The pilot certificate issued to a person under paragraph (a) of this section contains an endorsement with the following limitation which may be subsequently amended to include another island if the applicant complies with paragraph (a) of this section with respect to that island:
Passenger carrying prohibited on flights more than 10 nautical miles from (appropriate island).

(c) If an applicant for a private pilot certificate under paragraph (a) of this section does not have at least 3 hours of solo cross-country flight time, including a round trip flight to an airport at least 50 nautical miles from the place of departure with at least two full stop landings at different points along the route, his pilot certificate is also endorsed as follows:

Holder does not meet the cross-country flight requirements of ICAO.

(d) The holder of a private pilot certificate with an endorsement described in paragraph (b) or (c) of this section, is entitled to a removal of the endorsement, if he presents satisfactory evidence to an FAA inspector or designated pilot examiner that he has complied with the applicable solo cross-country flight requirements and has passed a practical test on cross-country flying.

61.118 PRIVATE PILOT PRIVILEGES AND LIMITATIONS: PILOT IN COMMAND.

Except as provided in paragraphs (a) through (d) of this section, a private pilot may not act as pilot in command of an aircraft that is carrying passengers or property for compensation or hire; nor may he, for compensation or hire, act as pilot in command of an aircraft.

(a) A private pilot may, for compensation or hire, act as pilot in command of an aircraft in connection with any business or employment if the flight is only incidental to that business or employment and the aircraft does not carry passengers or property for compensation or hire.

(b) A private pilot may share the operating expenses of a flight with his passengers.

(c) A private pilot who is an aircraft salesman and who has at least 200 hours of logged flight time may demonstrate an aircraft in flight to a prospective buyer.

(d) A private pilot may act as pilot in command of an aircraft used in a passenger-carrying airlift sponsored by a charitable organization, and for which the passengers make a donation to the organization, if—

(1) The sponsor of the airlift notifies the FAA General Aviation District Office having jurisdiction over the area concerned, at least 7 days before the flight, and furnishes any essential information that the office requests;

(2) The flight is conducted from a public airport adequate for the aircraft used, or from another airport that has been approved for the operation by an FAA inspector;

(3) He has logged at least 200 hours of flight time;

(4) No acrobatic or formation flights are conducted;

(5) Each aircraft used is certificated in the standard category and complies with the 100-hour inspection requirement of 91.169 of this chapter; and

(6) The flight is made under VFR during the day.

For the purpose of paragraph (d) of this section, a "charitable organization" means an organization listed in Publication No. 78 of the Department of the Treasury called the "Cumulative List of Organizations described in section 170(c) of the Internal Revenue Code of 1954," as amended from time to time by published supplemental lists.

61.120 PRIVATE PILOT PRIVILEGES AND LIMITATIONS: SECOND IN COMMAND OF AIRCRAFT REQUIRING MORE THAN ONE REQUIRED PILOT.

Except as provided in paragraphs (a) through (d) of 61.118 a private pilot may not, for compensation or hire, act as second in command of an aircraft that is type certificated for more than one required pilot, nor may he act as second in command of such an aircraft that is carrying passengers or property for compensation or hire.

SECTION D - FAR PART 91

Part 91 contains the general operating and flight rules governing the operation of aircraft (other than moored balloons, kites, unmanned rockets, and unmanned balloons) within the United States. This part is especially important and is the subject of careful study by pilots. Since changes and amendments are often made to Part 91, it is important for aviators to remain abreast of such revisions.

SUBPART A—GENERAL

91.1 APPLICABILITY

(a) Except as provided in paragraph (b) of this section, this part describes rules governing the operation of aircraft (other than moored balloons, kites, unmanned rockets, and unmanned free balloons) within the United States.

(b) Each person operating a civil aircraft of U.S. Registry outside of the United States shall:

  (1) When over the high seas, comply with Annex 2 (Rules of the Air) to the Convention on International Civil Aviation and with 91.70(c) and 91.90 of Subpart B;

  (2) When within a foreign country, comply with the regulations relating to the flight and maneuvers of aircraft there in force; and

  (3) Except for 91.15(b), 91.17, 91.38, and 91.43, comply with Subparts A and C of this part so far as they are not inconsistent with applicable regulations of the foreign country where the aircraft is operated or Annex 2 to the Convention on International Civil Aviation.

91.3 RESPONSIBILITY AND AUTHORITY OF THE PILOT IN COMMAND

(a) The pilot in command of an aircraft is directly responsible for, and is the final authority as to, the operation of that aircraft.

(b) In an emergency requiring immediate action, the pilot in command may deviate from any rule of this Subpart or of Subpart B to the extent required to meet that emergency.

(c) Each pilot in command who deviates from a rule under paragraph (b) of this section shall, upon the request of the Administrator, send a written report of that deviation to the Administrator.

This is one of the more important regulations for all pilots to be aware of. This regulation gives the pilot in command of an airplane authority to deviate from any regulation in this part to the extent required to safely meet an emergency. This is known as the emergency authority of a pilot in command.

91.5 PREFLIGHT ACTION

Each pilot in command shall, before beginning a flight, familiarize himself with all available information concerning that flight. This information must include:

(a) For a flight under IFR or a flight not in the vicinity of an airport, weather reports and forecasts, fuel requirements, alternatives available if the planned flight cannot be completed, and any known traffic delays of which he has been advised by ATC.

(b) For any flight, runway lengths at airports of intended use, and the following takeoff and landing distance information:

  (1) For civil aircraft for which an approved airplane or rotorcraft flight manual containing takeoff and landing distance data is required, and takeoff

and landing distance data contained therein; and

(2) For civil aircraft other than those specified in subparagraph (1) of this paragraph, other reliable information appropriate to the aircraft, relating to aircraft performance under expected values of airport elevation and runway slope, aircraft gross weight, and wind and temperature.

91.8 PROHIBITION AGAINST INTERFERENCE WITH CREWMEMBERS

(a) No person may assault, threaten, intimidate, or interfere with a crewmember in the performance of his duties aboard an aircraft being operated in air commerce.

(b) No person may attempt to cause or cause the flight crew of an aircraft being operated in air commerce to divert its flight from its intended course or destination.

91.9 CARELESS OR RECKLESS OPERATION

No person may operate an aircraft in a careless or reckless manner so as to endanger the life or property of another.

91.10 CARELESS OR RECKLESS OPERATION OTHER THAN FOR THE PURPOSE OF AIR NAVIGATION

No person may operate an aircraft other than for the purpose of air navigation, on any part of the surface of an airport used by aircraft for air commerce (including areas used by those aircraft for receiving or discharging persons or cargo), in a careless or reckless manner so as to endanger the life or property of another.

91.11 LIQUOR AND DRUGS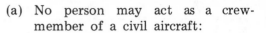

(a) No person may act as a crewmember of a civil aircraft:
   (1) Within 8 hours after the consumption of any alcoholic beverage;
   (2) While under the influence of alcohol; or
   (3) While using any drug that affects his faculties in any way contrary to safety.

(b) Except in an emergency, no pilot of a civil aircraft may allow a person who is obviously under the influence of intoxicating liquors or drugs (except a medical patient under proper care) to be carried in that aircraft.

91.12 CARRIAGE OF NARCOTIC DRUGS, MARIHUANA, AND DEPRESSANT OR STIMULANT DRUGS OR SUBSTANCES.

(a) Except as provided in paragraph (b) of this section, no person may operate a civil aircraft within the United States with knowledge that narcotic drugs, marihuana, and depressant or stimulant drugs or substances as defined in Federal or State statutes are carried in the aircraft.

(b) Paragraph (a) of this section does not apply to any carriage of narcotic drugs, marihuana, and depressant or stimulant drugs or substances authorized by or under any Federal or State agency.

91.13 DROPPING OBJECTS

No pilot in command of a civil aircraft may allow any object to be dropped from that aircraft in flight that creates a hazard to persons or property. However, this section does not prohibit the dropping of any object if reasonable precautions are taken to avoid injury or damage to persons or property.

91.14 FASTENING OF SAFETY BELTS ✈

(a) Unless otherwise authorized by the Administrator—

(1) No pilot may take off or land a U.S. registered civil aircraft (except an airship) unless the pilot in command of that aircraft ensures that each person on board has been notified to fasten his safety belt.

(2) During the takeoff and landing of U.S. registered civil aircraft (except airships) each person on board that aircraft must occupy a seat or berth with a safety belt properly secured about him. However, a person who has not reached his second birthday may be held by an adult who is occupying a seat or berth, and a person on board for the purpose of engaging in sport parachuting may use the floor of the aircraft as a seat.

91.15 PARACHUTES AND PARACHUTING ✈

(a) No pilot of a civil aircraft may allow a parachute that is available for emergency use to be carried in that aircraft unless it is an approved type and:

(1) If a chair type (canopy in back), it has been packed by an appropriately rated parachute rigger within the preceding 120 days.

(2) If any other type, it has been packed by an appropriately rated parachute rigger within the preceding 60 days.

(b) Except in an emergency, no pilot in command may allow, and no person may make, a parachute jump from an aircraft within the United States except in accordance with Part 105 (New) of this chapter.

(c) Unless each occupant of the aircraft is wearing an approved parachute, no pilot of a civil aircraft, carrying any person (other than a crewmember) may execute any intentional maneuver that exceeds:

(1) A bank of 60° relative to the horizon; or

(2) A nose-up or nose-down attitude of 30° relative to the horizon.

(d) Paragraph (c) of this section does not apply to:

(1) Flight tests for pilot certification or rating; or

(2) Spins and other flight maneuvers required by the regulations for any certificate or rating when given by:

(i) A certificated flight instructor; or

(ii) An airline transport pilot instruction in accordance with 61.169 of this chapter.

91.21 FLIGHT INSTRUCTION: SIMULATED INSTRUMENT FLIGHT AND CERTAIN FLIGHT TESTS

(a) No person may operate a civil aircraft that is being used for flight instruction unless that aircraft has fully functioning dual controls.

(b) No person may operate a civil aircraft in simulated instrument flight unless:

(1) An appropriately rated pilot occupies the other control seat as safety pilot;

(2) The safety pilot has adequate vision forward and to each side of the aircraft, or a competent observer in the aircraft adequately supplements the vision of the safety pilot; and

(3) Except in the case of a lighter-than-air aircraft, that aircraft is equipped with functioning dual controls.

91.24 ATC TRANSPONDER EQUIPMENT.

(a) *All airspace: U.S. registered civil aircraft.* For operations not conducted under Parts 121, 123, 127, or 133 of this chapter, ATC transponder equipment installed after January 1, 1974, in U.S. registered civil aircraft not previously equipped with an ATC transponder, and all ATC transponder equipment used in U.S. registered civil aircraft after July 1, 1975, must meet the performance and environmental requirements of any class of TSO-C74b or any class of TSO-C74c as appropriate, except that the Administrator may approve the use of TSO-C74 or TSO-C74a equipment after July 1, 1975, if the applicant submits data showing that such equipment meets the minimum performance standards of the appropriate class of TSO-C74c and environmental conditions of the TSO under which it was manufactured.

(b) *Controlled airspace: all aircraft.* Except for persons operating helicopters in terminal control areas at or below 1,000 feet AGL under the terms of a letter of agreement, and except for persons operating gliders above 12,500 feet MSL but below the floor of the positive control area, no person may operate an aircraft in controlled airspace, after the applicable dates prescribed in subparagraphs (b)(1) through (b)(4) of this paragraph, unless that aircraft is equipped with an operable coded radar beacon transponder having a Mode 3/A 4096 code capability, replying to Mode 3/A interrogation with the code specified by ATC, and is equipped with automatic pressure altitude reporting equipment having a Mode C capability that automatically replies to Mode C interrogations by transmitting pressure altitude information in 100-foot increments. This requirement applies—

(1) After January 1, 1975, in Group I Terminal Control Areas governed by 91.90(a);

(2) After July 1, 1975, in Group II Terminal Control Areas governed by 91.90(b);

(3) After July 1, 1975, in Group III Terminal Control Areas governed by 91.90(c), except as provided therein; and

(4) After July 1, 1975, in all controlled airspace of the 48 contiguous States and the District of Columbia, above 12,500 feet MSL, excluding the airspace at and below 2,500 feet AGL.

(c) *ATC authorized deviations.* ATC may authorize deviations from paragraph (b) of this section—

(1) Immediately, to allow an aircraft with an inoperative transponder to continue to the airport of ultimate destination, including any intermediate stops, or to proceed to a place where suitable repairs can be made, or both; and

(2) On a continuing basis, or for individual flights, for operations not involving an inoperative transponder, in which cases the request for a deviation must be submitted to the ATC facility having jurisdiction over the airspace concerned at least four hours before the proposed operation.

91.27 CIVIL AIRCRAFT: CERTIFICATIONS REQUIRED

(a) Except as provided in 91.28, no person may operate a civil aircraft unless it has within it:

(1) An appropriate and current airworthiness certificate.

(2) A registration certificate issued to its owner.

(b) No person may operate a civil aircraft unless the airworthiness certificate required by paragraph (a) of this section or a special flight authorization issued under 91.28 is displayed at the cabin or cockpit entrance so that it is legible to passengers or crew.

91.29 CIVIL AIRCRAFT AIRWORTHINESS ✈

(a) No person may operate a civil aircraft unless it is in an airworthy condition.
(b) The pilot in command of a civil aircraft is responsible for determining whether that aircraft is in condition for safe flight. He shall discontinue the flight when unairworthy mechanical or structural conditions occur.

91.31 CIVIL AIRCRAFT OPERATING LIMITATIONS AND MARKING REQUIREMENTS ✈

(a) Each person operating a civil aircraft shall comply with the operating limitations for that aircraft prescribed by the certificating authority of the country of registry.
(b) No person may operate a U.S. registered civil aircraft unless there is available in the aircraft a current FAA approved Aircraft Flight Manual for that aircraft, placards, listings, instrument markings, or any combination thereof, containing each operating limitation prescribed for that aircraft by the Administrator, including the following:
   (1) Powerplant (e.g., RPM, manifold pressure, gas temperature, etc.);
   (2) Airspeeds (e.g., normal operating speed, flaps extended speed, etc.);
   (3) Aircraft weight, center of gravity, and weight distribution, including the composition of the useful load in those combinations and ranges intended to insure that the weight and center of gravity position will remain within approved limits (e.g., combinations and ranges of crew, oil, fuel, and baggage);
   (4) Minimum flight crew;
   (5) Kinds of operation;
   (6) Maximum operating altitude;
   (7) Maneuvering flight load factors;
   (8) Rotor speed (for rotorcraft); and
   (9) Limiting height-speed envelope (for rotorcraft).

91.32 SUPPLEMENTAL OXYGEN ✈

(a) *General.* No person may operate a civil aircraft of U.S. registry:
   (1) At cabin pressure altitudes above 12,500 feet (MSL) up to and including 14,000 feet (MSL), unless the required minimum flight crew is provided with and uses supplemental oxygen for that part of the flight at those altitudes that is of more than 30 minutes duration;
   (2) At cabin pressure altitudes above 14,000 feet (MSL), unless the required minimum flight crew is provided with and uses supplemental oxygen during the entire flight time at those altitudes; and
   (3) At cabin pressure altitudes above 15,000 feet (MSL), unless each occupant of the aircraft is provided with supplemental oxygen. (See Fig. 6-3).
(b) *Pressurized cabin aircraft:*
   (1) No person may operate a civil aircraft of U.S. registry with a pressurized cabin:
      (i) At flight altitudes above flight level 250, unless at least a 10-minute supply of supplemental oxygen in addition to any

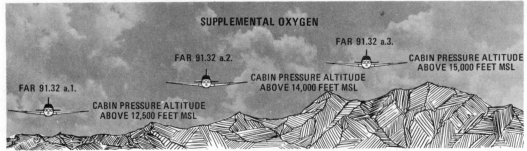

Fig. 6-3. Supplemental Oxygen

oxygen required to satisfy paragraph (a) of this section, is available for each occupant of the aircraft for use in the event that a descent is necessitated by loss of cabin pressurization.

91.33 POWERED CIVIL AIRCRAFT WITH STANDARD CATEGORY U.S. AIRWORTHINESS CERTIFICATES; INSTRUMENT AND EQUIPMENT REQUIREMENTS

(a) *General.* Except as provided in paragraphs (c)(3) and (e) of this section, no person may operate a powered civil aircraft with a standard category U.S. airworthiness certificate in any operation described in paragraphs (b) through (f) of this section unless that aircraft contains the instruments and equipment specified in those paragraphs (or FAA-approved equivalents) for that type of operation, and those instruments and items of equipment are in operable condition.

(b) *Visual flight rules (day).* For VFR flight during the day the following instruments and equipment are required:

(1) Airspeed indicator;

(2) Altimeter;

(3) Magnetic direction indicator;

(4) Tachometer for each engine;

(5) Oil pressure gauge for each engine using pressure system;

(6) Temperature gauge for each liquid-cooled engine;

(7) Oil temperature gauge for each air-cooled engine;

(8) Manifold pressure gauge for each altitude engine;

(9) Fuel gauge indicating the quantity of fuel in each tank;

(10) Landing gear position indicator, if the aircraft has a retractable landing gear;

(11) If the aircraft is operated for hire over water and beyond power-off gliding distance from shore, approved floatation gear readily available to each occupant, and at least one pyrotechnic signaling device; and

(12) Approved safety belts for all occupants who have reached their second birthday. The rated strength of each safety belt shall not be less than that corresponding with the ultimate load factors specified in the current applicable aircraft airworthiness requirements considering the dimensional characteristics of the safety belt installation for the specific seat or berth arrangement. The webbing of each safety belt shall be replaced as required by the Administrator.

(c) *Visual flight rules (night).* For VFR flight at night the following instruments and equipment are required:

(1) Instruments and equipment specified in paragraph (b) of this section;

(2) Approved position lights;

(3) An approved aviation red or aviation white anti-collision light system on all large aircraft, on all small aircraft when required by the aircraft's airworthiness certificate, and on all other small aircraft after August 11, 1972. Anti-collision light systems initially installed after August 11, 1971, on aircraft for which a type certificate was issued or applied for before August 11, 1971, must at least meet the anti-collision light standards of Parts 23, 25, 27, or 29, as applicable, that were in effect on August 10, 1971, except that the color may be either aviation red or aviation white. In the event of failure of any light of the anti-collision light system, operations with the aircraft may be continued to a stop where repairs or replacement can be made.

(4) If the aircraft is operated for hire; one electric landing light;

(5) An adequate source of electrical energy for all installed electric and radio equipment; and

(6) One spare set of fuses, or three spare fuses of each kind required.

(d) *Instrument flight rules.* For IFR flight the following instruments and equipment are required:

(1) Instruments and equipment specified in paragraph (b) of this section and for night flight, instruments and equipment specified in paragraph (c) of this section;

(2) Two-way radio communications system and navigation equipment appropriate to the ground facilities to be used;

(3) Gyroscopic rate-of-turn indica-tor, except on large aircraft with a third attitude instrument system usable through flight attitudes of 360° of pitch and roll and installed in accordance with 121.305(j) of this title;

(4) Slip-skid indicator;

(5) Sensitive altimeter adjustable for barometric pressure;

(6) Clock with sweep-second hand;

(7) Generator of adequate capacity;

(8) Gyroscopic bank and pitch indicator (artificial horizon); and

(9) Gyroscopic direction indicator (directional gyro or equivalent).

(e) *Flight at and above 24,000 feet MSL.* If VOR Navigational equipment is required under paragraph (d)(2) of this section, no person may operate a U.S. registered aircraft at and above 24,000 feet MSL, unless that aircraft is equipped with an approved distance measuring equipment (DME). When DME required by this paragraph fails at and above 24,000 feet MSL, each pilot shall notify ATC immediately, and may then continue operations at and above 24,000 feet MSL to the next airport of intended landing at which repairs or replacement of the equipment can be made.

Above 24,000 feet, DME is required on any aircraft that is using VOR as its method of navigation.

91.52 EMERGENCY LOCATOR TRANSMITTERS

(a) Except as provided in paragraphs (e), (f), and (g) of this section:

(1) After December 30, 1971, no person may operate a U.S. registered civil airplane manufactured or imported after that date unless it meets the

applicable requirements of paragraphs (b), (c), and (d) of this section.

(2) After December 30, 1973, no person may operate a U.S. registered civil airplane unless it meets the applicable requirements of paragraphs (b), (c), and (d) of this section.

(b) *(Not applicable to the private pilot.)*

(c) Each emergency locator transmitter required by paragraphs (a) and (b) of this section must be attached to the airplane in such a manner that the probability of damage to the transmitter, in the event of crash impact, is minimized. Fixed and deployable automatic type transmitters must be attached to the airplane as far aft as practicable.

(d) Batteries used in the emergency locator transmitters required by paragraphs (a) and (b) of this section must be replaced (or recharged, if the battery is rechargeable)—

(1) When the transmitter has been in use for more than one cumulative hour; or

(2) When 50 percent of their useful life (or, for rechargeable batteries, 50 percent of their useful life of charge), as established by the transmitter manufacturer under 37.200(g)(2) of this chapter, has expired.

The new expiration date for the replacement (or recharged) battery must legibly marked on the outside of the transmitter. Subparagraph (d)(2) of this paragraph does not apply to batteries (such as water-activated batteries) that are essentially unaffected during probable-storage intervals.

(e) Notwithstanding paragraphs (a) and (b) of this section, a person may—

(1) Ferry a newly acquired airplane from the place where

possession of it was taken to place where the emergency locator transmitter is to be installed; and

(2) Ferry an airplane with an inoperative emergency locator transmitter from a place where repairs or replacement cannot be made to a place where they can be made.

No person other than required crewmembers may be carried aboard an airplane being ferried pursuant to this paragraph (e).

(f) Paragraphs (a) and (b) of this section do not apply to—

(1) Turbojet engine-powered airplanes;

(2) Scheduled operations (other than charter flights) conducted by a domestic or flag air carrier certificated under Part 121 of this chapter;

(3) Training flights conducted within a 20-mile radius of the airport from which the flight began; or

(4) Agricultural aircraft operations conducted under Part 137 of this chapter.

(g) Until December 30, 1975, a U.S. registered civil airplane may be operated with an emergency locator transmitter that does not meet the requirements of paragraphs (b), (c), and (d) of this section, if—

(1) Its installation was approved before October 21, 1971;

(2) It was manufactured under a TSO Authorization issued against TSO-C61a of Part 37 of this chapter;

(3) It transmits simultaneously on 121.5 and 243.0 MHz; and

(4) After the effective dates prescribed in paragraph (a) of this section, as applicable, it is attached to the airplane and is in operable condition.

91.53 REPORT ON IDENTIFICATION AND ACTIVITY OF AIRCRAFT

(a) Except as provided in paragraph (b) of this section, the owner of each aircraft registered in the United States should (but is not required to) submit an Aircraft Registration Eligibility, Identification, and Activity Report, Part 2, AC Form 8050-73, to the FAA Aircraft Registry before July 1 of each of the years 1970 and 1971, and April 1 of each year thereafter stating—

    (1) The name and address of the principal operator of the aircraft if other than the owner;

    (2) The make and model of the engines installed in the aircraft;

    (3) The identification of the communications and navigational aids capability of equipment installed in the aircraft;

    (4) Airport where the aircraft is based; and

    (5) Activity of the aircraft as shown by hours flown and purpose of flight for the previous calendar year.

(b) The owner of an aircraft operated under Part 121 or 127 should include in his report under paragraph (a) of this section only the item listed in subparagraph (2) of that paragraph.

SUBPART B—FLIGHT RULES: GENERAL

91.61 APPLICABILITY

This subpart prescribes flight rules governing the operation of aircraft within the United States.

91.63 WAIVERS

(a) The Administrator may issue a certificate of waiver authorizing the operation of aircraft in deviation of any rule of this subpart if he finds that the proposed operation can be safely conducted under the terms of the certificate of waiver.

(b) An application for a certificate of waiver under this section is made on a form and in a manner prescribed by the Administrator and may be submitted to any FAA office.

(c) A certificate of waiver is effective as specified on that certificate.

91.65 OPERATING NEAR OTHER AIRCRAFT

(a) No person may operate an aircraft so close to another aircraft as to create a collision hazard.

(b) No person may operate an aircraft in formation flight except by arrangement with the pilot in command of each aircraft in the formation.

(c) No person may operate an aircraft, carrying passengers for hire, in formation flight.

(d) Unless otherwise authorized by ATC, no person operating an aircraft may operate his aircraft in accordance with any clearance or instruction that has been issued to the pilot of another aircraft for radar Air Traffic Control purposes.

91.67 RIGHT-OF-WAY: EXCEPT WATER OPERATIONS

(a) *General.* When weather conditions permit, regardless of whether an operation is conducted under Instrument Flight Rules, vigilance shall be maintained by each person operating an aircraft so as to see and avoid other aircraft in compliance with this section. When a rule of this section gives another aircraft right-of-way, he shall give way to that aircraft and may not pass over, under, or ahead of it, unless well clear.

(b) *In distress.* An aircraft in distress has the right-of-way over all other air traffic.

(c) *Converging.* When aircraft of the same category are converging at approximately the same altitude (except head-on, or nearly so), the aircraft to the other's right has the right-of-way. If the aircraft are of different categories:

    (1) A balloon has the right-of-way over any other category of aircraft;

    (2) A glider has the right-of-way over an airship, airplane or rotorcraft; and

    (3) An airship has the right-of-way over an airplane or rotorcraft.

    However, an aircraft towing or refueling other aircraft has the right-of-way over all other engine-driven aircraft.

(d) *Approaching head-on.* When aircraft are approaching each other head-on, or nearly so, each pilot of each aircraft shall alter course to the right to pass well clear.

(e) *Overtaking.* Each aircraft that is being overtaken has the right-of-way and each pilot of an overtaking aircraft shall alter course to the right to pass well clear.

(f) *Landing.* Aircraft, while on final approach to land, or while landing, have the right-of-way over other aircraft in flight or operating on the surface. When two or more aircraft are approaching an airport for the purpose of landing, the aircraft at the lower altitude has the right-of-way, but it shall not take advantage of this rule to cut in front of another which is on final approach to land, or to overtake that aircraft.

(g) *Inapplicability.* This section does not apply to the operation of an aircraft on water.

91.69 RIGHT-OF-WAY RULES: WATER OPERATIONS

(a) *General.* Each person operating an aircraft on the water shall, insofar as possible, keep well clear of all vessels and avoid impeding their navigation, and shall give way to any vessel or other aircraft that is given the right-of-way by any rule of this section.

(b) *Crossing.* When aircraft, or an aircraft and a vessel, are on crossing courses, the aircraft or vessel to the other's right has the right-of-way.

(c) *Approaching head-on.* When aircraft, or an aircraft and a vessel, are approaching head-on or nearly so, each shall alter its course to the right to keep well clear.

(d) *Overtaking.* Each aircraft or vessel that is being overtaken has the right-of-way, and the one overtaking shall alter course to keep well clear.

(e) *Special circumstances.* When aircraft, or an aircraft and a vessel, approach so as to involve risk of collision, each aircraft or vessel shall proceed with careful regard to existing circumstances, including the limitations of the respective craft.

91.70 AIRCRAFT SPEED

(a) Unless otherwise authorized by the Administrator, no person may operate an aircraft below 10,000 feet MSL at an indicated airspeed of more than 250 knots (288 MPH).

(b) Unless otherwise authorized or required by ATC, no person may operate an aircraft within an airport traffic area at an indicated airspeed of more than:

    (1) In the case of a reciprocating engine aircraft, 156 knots (180 MPH); or

    (2) In the case of a turbine-powered aircraft, 200 knots (230 MPH).

(c) No person may operate aircraft in the airspace beneath the lateral limits of any terminal control area at an indicated airspeed of more than 200 knots (230 MPH).

However, if the minimum safe airspeed for any particular operation is greater than the maximum speed prescribed in this section, the aircraft may be operated at that minimum speed.

91.71 ACROBATIC FLIGHT

No person may operate an aircraft in acrobatic flight:
(a) Over any congested area of a city, town or settlement;
(b) Over an open air assembly of persons;
(c) Within a control zone or Federal Airway;
(d) Below an altitude of 1,500 feet above the surface; or
(e) When flight visibility is less than three miles.

For the purposes of this section, acrobatic flight means an intentional maneuver involving an abrupt change in an aircraft's attitude, an abnormal attitude, or abnormal acceleration not necessary for normal flight.

91.73 AIRCRAFT LIGHTS

No person may, during the period from sunset to sunrise (or, in Alaska, during the period a prominent unlighted object cannot be seen from a distance of three statute miles or the sun is more than six degrees below the horizon):
(a) Operate an aircraft unless it has lighted position lights;
(b) Park or move an aircraft in, or in dangerous proximity to, a night flight operations area of an airport unless the aircraft:
(1) Is clearly illuminated;
(2) Has lighted position lights; or
(3) Is in an area which is marked by obstruction lights; or
(c) Anchor an aircraft unless the aircraft:
(1) Has lighted anchor lights; or
(2) Is in an area where anchor lights are not required on vessels.

91.75 COMPLIANCE WITH ATC CLEARANCES AND INSTRUCTIONS

(a) When an ATC clearance has been obtained, no pilot in command may deviate from that clearance, except in an emergency, unless he obtains an amended clearance. However, except in positive controlled airspace, this paragraph does not prohibit him from cancelling an IFR flight plan if he is operating in VFR weather conditions.
(b) Except in an emergency, no person may, in an area in which air traffic control is exercised, operate an aircraft contrary to an ATC instruction.
(c) Each pilot in command who deviates, in an emergency, from an ATC clearance or instruction shall notify ATC of that deviation as soon as possible.
(d) Each pilot in command who (though not deviating from a rule of this subpart) is given priority by ATC in an emergency, shall, if requested by ATC, submit a detailed report of that emergency with 48 hours to the chief of that ATC facility.

91.77 ATC LIGHT SIGNALS

ATC light signals have the meaning shown in figure 6-4.

91.79 MINIMUM SAFE ALTITUDES: GENERAL

Except when necessary for takeoff or landing, no person may operate an aircraft below the following altitudes:
(a) *Anywhere.* An altitude allowing, if a power unit fails, an emergency landing without undue hazard to persons or property on the surface.
(b) *Over congested areas.* Over any congested area of a city, town, or settlement, or over any open air assembly of persons, an altitude of 1,000 feet above the highest

Color and type of signal	Meaning with respect to aircraft on the surface	Meaning with respect to aircraft in flight
Steady green	Cleared for takeoff	Cleared to land
Flashing green	Cleared to taxi	Return for landing (to be followed by steady green at proper time)
Steady red	Stop	Give way to other aircraft and continue circling
Flashing red	Taxi clear of runway in use	Airport unsafe — do not land
Flashing white	Return to starting point on airport	Not applicable
Alternating red and green	Exercise extreme caution	Exercise extreme caution

Fig. 6-4. ATC Light Signals

obstacle within a horizontal radius of 2,000 feet of the aircraft.

(c) *Over other than congested areas.* An altitude of 500 feet above the surface, except over open water or sparsely populated areas. In that case, the aircraft may not be operated closer than 500 feet to any person, vessel, vehicle, or structure.

(d) *Helicopters.* Helicopters may be operated at less than the minimums prescribed in paragraph (b) or (c) of this section if the operation is conducted without hazard to persons or property on the surface. In addition, each person operating a helicopter shall comply with routes or altitudes specifically prescribed for helicopters by the Administrator.

91.81 ALTIMETER SETTINGS ✈

(a) Each person operating an aircraft shall maintain the cruising altitude or flight level of that aircraft, as the case may be, by reference to an altimeter that is set, when operating:

(1) Below 18,000 feet MSL, to:

  (i) The current reported altimeter setting of a sta-

tion along the route and within 100 nautical miles of the aircraft;

(ii) If there is no station within the area prescribed in subdivision (i) of this subparagraph, the current reported altimeter setting of an appropriate available station; or

(iii) In the case of an aircraft not equipped with a radio, the elevation of the departure airport or an appropriate altimeter setting available before departure; or

(2) at or above 18,000 feet MSL, to 29.92" Hg.

Above 18,000 feet, flight levels begin for all aircraft operating VFR or IFR. The pilot will set his altimeter to 29.92 in. regardless of the local station pressure. Altitudes are then referred to as flight levels. For instance, 23,000 feet would be flight level 230.

91.83 FLIGHT PLAN: INFORMATION REQUIRED ✈

(a) Unless otherwise authorized by ATC, each person filing an IFR or VFR flight plan shall include in it the following information:

(1) The aircraft identification number and, if necessary, its radio call sign.

(2) The type of the aircraft or, in the case of a formation flight, the type of each aircraft and the number of aircraft in the formation.

(3) The full name and address of the pilot in command or, in the case of a formation flight, the formation commander.

(4) The point and proposed time of departure.

(5) The proposed route, cruising altitude (or flight level), and true airspeed at that altitude.

(6) The point of first intended landing and the estimated elapsed time until over that point.

(7) The radio frequencies to be used.

(8) The amount of fuel on board (in hours).

(9) In the case of an IFR flight plan, an alternate airport, except as provided in paragraph (b) of this section.

(10) In the case of an international flight, the number of persons in the aircraft.

(11) Any other information the pilot in command or ATC believes is necessary for ATC purposes.

When a flight plan has been filed, the pilot in command, upon canceling or completing the flight under the flight plan, shall notify the nearest FAA Flight Service Station or ATC facility.

Even though the filing of a flight plan is not required by regulation, in the event of an emergency landing, it could save your life by making rescue easier.

91.84 FLIGHTS BETWEEN MEXICO OR CANADA AND THE UNITED STATES

Unless otherwise authorized by ATC, no person may operate a civil aircraft between Mexico or Canada and the United States without filing an IFR or VFR flight plan, as appropriate.

91.85 OPERATING ON OR IN THE VICINITY OF AN AIRPORT: GENERAL RULES

(a) Unless otherwise required by Part 93 of this chapter, each person operating an aircraft on or in the vicinity of an airport shall comply with the requirements of this section and of 91.87 and 91.89.

(b) Unless otherwise authorized or required by ATC, no person may operate an aircraft within an airport traffic area except for the purpose of landing at, or taking off from, an airport within that area. ATC authorizations may be given as individual approval of specific operations or may be contained in written agreements between airport users and the tower concerned.

91.87 OPERATION AT AIRPORTS WITH OPERATING CONTROL TOWERS

(a) *General.* Unless otherwise authorized or required by ATC, each person operating an aircraft to, from, or on an airport with an operating control tower shall comply with the applicable provisions of this section.

(b) *Communications with control towers operated by the United States.* No person may, within an airport traffic area, operate an aircraft to, from, or on an airport having a control tower operated by the United States unless two-way radio communications are maintained between that aircraft and the control tower. However, if the aircraft radio fails in flight, he may operate that aircraft and land if weather conditions are at or above basic VFR weather minimums, he maintains visual contact with the tower, and he receives clearance to land.

(c) *Communications with other control towers.* No person may, within an airport traffic area, operate an aircraft to, from, or on an airport having a control tower that is operated by any person other than the United States unless:

   (1) If that aircraft's radio equipment so allows, two-way radio communications are maintained between the aircraft and the tower; or

   (2) If that aircraft's radio equipment allows only reception from the tower, the pilot has the tower's frequency monitored.

(e) *Approaches.* When approaching to land at an airport with an operating control tower, each pilot of:

   (1) An airplane, shall circle the airport to the left; and

   (2) A helicopter, shall avoid the flow of fixed-wing aircraft.

(f) *Departures.* No person may operate an aircraft taking off from an airport with an operating control tower except in compliance with the following:

   (1) Each pilot shall comply with any departure procedures established for that airport by the FAA.

   (2) Unless otherwise required by the departure procedure or the applicable distance from clouds criteria, each pilot of a turbine-powered airplane and each pilot of a large airplane shall climb to an altitude of 1,500 feet above the surface as rapidly as practicable.

(h) *Clearances required.* No pilot may, at an airport with an operating control tower, taxi an aircraft on a runway, or take off or land an aircraft, unless he has received an appropriate clearance from ATC. A clearance to "taxi to" the runway is a clearance to cross all intersecting runways but is not a clearance to "taxi on" the assigned runway.

91.89 OPERATION AT AIRPORTS WITHOUT CONTROL TOWERS

(a) Each person operating an aircraft to or from an airport without an operating control tower shall:

   (1) In the case of an airplane approaching to land, make all turns of that airplane to the left unless the airport displays approved light signals or visual markings indicating that turns should be made to the right, in which case the pilot shall make all turns to the right.

A left-hand traffic pattern is considered "standard" at uncontrolled airports. However, right-hand patterns are designated for some runways due to conflict with other airport patterns, terrain clearance or noise abatement over a city. A right-hand pattern will be indicated by an appropriate L-marker, or at night, by a rotating amber light on the tetrahedron or near the segmented circle.

---

   (2) In the case of a helicopter approaching to land, avoid the flow of fixed-wing aircraft; and

   (3) In the case of an aircraft departing the airport, comply with any FAA traffic pattern for that airport.

91.90 TERMINAL CONTROL AREAS.

(a) *Group I terminal control areas.*

    (1) *Operating rules.* No person may operate an aircraft within a Group I terminal control area designated in Part 71 of this chapter except in compliance with the following rules:

        (i) No person may operate an aircraft within a Group I terminal control area unless he has received an appropriate authorization from ATC prior to the operation of that aircraft in that area.

        (ii) Unless otherwise authorized by ATC, each person operating a large turbine engine powered airplane to or from a primary airport shall operate at or above the designated floors while within the lateral limits of the terminal control area.

    (2) *Pilot requirements.* The pilot in command of a civil aircraft may not land or take off that aircraft from an airport within a Group I terminal control area unless he holds at least a private pilot certificate.

    (3) *Equipment requirements.* Unless otherwise authorized by ATC in the case of in-flight VOR, TACAN, or two-way radio failure; or unless otherwise authorized by ATC in the case of a transponder failure occurring at any time, no person may operate an aircraft within a Group I terminal control area unless that aircraft is equipped with—

        (i) An operable VOR or TACAN receiver (except in the case of helicopters);

        (ii) An operable two-way radio capable of communicating with ATC on appropriate frequencies for that terminal control area; and

        (iii) On and before the applicable dates specified in paragraphs (a) and (b)(1) of 91.24, an operable coded radar beacon transponder having at least a Mode 3/A 64-code capability, replying to Mode 3/A interrogation with the code specified by ATC. On and before those dates, this requirement is not applicable to helicopters operating within the terminal control area, or to IFR flights operating to or from an airport outside of the terminal control area but which is in close proximity to the terminal control area, when the commonly used transition, approach, or departure procedures to such airport require flight within the terminal control area. After the applicable dates specified in paragraphs (a) and (b)(1) of 91.24, the applicable provisions of that section shall be complied with, notwithstanding the exceptions in this section.

(b) *Group II terminal control areas.*

    (1) *Operating rules.* No person may operate an aircraft within a Group II terminal control area designated in Part 71 of this chapter except in compliance with the following rules:

        (i) No person may operate an aircraft within a Group II Terminal Control Area unless he has received an appropriate authorization from ATC prior to operation of that aircraft in that area, except that, after the applicable dates in 91.24 (b)(2), authorization is not required if the aircraft is VFR, is equipped as required by 91.24(b), and does not land or take off within the terminal control area.

    (ii) Unless otherwise authorized by ATC, each person operating a large turbine engine powered airplane to or from a primary airport shall operate at or above the designated floors while within the lateral limits of the terminal control area.

(2) *Equipment requirements.* Unless otherwise authorized by ATC in the case of in-flight VOR, TACAN, or two-way radio failure; or unless otherwise authorized by ATC in the case of a transponder failure occurring at any time, no person may operate an aircraft within a Group II terminal control area unless that aircraft is equipped with—

    (i) An operable VOR or TACAN receiver (except in the case of helicopters);

    (ii) An operable two-way radio capable of communicating with ATC on the appropriate frequencies for that terminal control area; and

    (iii) On and before the applicable dates specified in paragraphs (a) and (b)(2) of 91.24, an operable coded radar beacon transponder having at least a Mode 3/A 64-code capability, replying to Mode 3/A interrogation with the code specified by ATC. On and before those dates, this requirement is not applicable to helicopters operating within the terminal control area, or to VFR aircraft operating within the terminal control area, or to IFR flights operating to or from a secondary airport located within the terminal

control area, or to IFR flights operating to or from an airport outside of the terminal control area but which is close proximity to the terminal control area, when the commonly used transition, approach, or departure procedures to such airport require flight within the terminal control area. After the applicable dates in paragraphs (a) and (b)(2) of 91.24, that section shall be complied with, notwithstanding the exceptions in this section.

(c) *Group III terminal control areas.* After the date specified in 91.24(b)(3), no person may operate an aircraft within a Group III Terminal Control Area designated in Part 71 unless the applicable provisions of 91.24(b) are complied with, except that such compliance is not required if two-way radio communications are maintained, within the TCA, between the aircraft and the ATC facility, and the pilot provides position, altitude, and proposed flight path prior to entry.

91.91 TEMPORARY FLIGHT RESTRICTIONS ✈

(a) Whenever the Administrator determines it to be necessary in order to prevent an unsafe congestion of sight-seeing aircraft above an incident or event which may generate a high degree of public interest, or to provide a safe environment for the operation of disaster relief aircraft, a Notice to Airmen will be issued designating an area within which temporary flight restrictions apply.

(b) When a Notice to Airmen has been issued under this Section, no person may operate an aircraft within the designated area unless:

    (1) That aircraft is participating in disaster relief activities and is being operated under the

direction of the agency responsible for relief activities;

(2) That aircraft is being operated to or from an airport within the area and is operated so as not to hamper or endanger relief activities;

(3) That operation is specifically authorized under an IFR ATC clearance;

(4) VFR flight around or above the area is impracticable due to weather, terrain, or other considerations, prior notice is given to the Air Traffic Service facility specified in the Notice to Airmen, and en route operation through the area is conducted so as not to hamper or endanger relief activities; or

(5) That aircraft is carrying properly accredited news representatives, or persons on official business concerning the incident or event which generated the Notice to Airmen; the operation is conducted in accordance with 91.79; the operation is conducted above the altitudes being used by relief aircraft unless otherwise authorized by the agency responsible for relief activities; and further, in connection with this type of operation, prior to entering the area the operator has filed with the Air Traffic Service facility specified in the Notice to Airmen a flight plan which includes the following information:

(i) Aircraft identification, type, and color;

(ii) Radio communications frequencies to be used;

(iii) Proposed times of entry and exit of the designated area;

(iv) Name of news media or purpose of flight; and

(v) Any other information deemed necessary by ATC.

91.93 FLIGHT TEST AREAS

No person may flight test an aircraft except over open water, or sparsely populated areas, having light air traffic.

91.104 FLIGHT RESTRICTIONS IN THE PROXIMITY OF THE PRESIDENTIAL AND OTHER PARTIES

No person may operate an aircraft over or in the vicinity of any area to be visited or traveled by the President, the Vice President, or other public figures contrary to the restrictions established by the Administrator and published in Notice to Airmen (NOTAM).

VISUAL FLIGHT RULES

91.105 BASIC VFR WEATHER MINIMUMS

(a) Except as provided in 91.107, no person may operate an aircraft under VFR when the flight visibility is less, or at a distance from clouds that is less, than that prescribed for the corresponding altitude shown in figure 6-5.

(b) When the visiblity is less than one mile, a helicopter may be operated outside controlled airspace at 1,200 feet or less above the surface if operated at a speed that allows the pilot adequate opportunity to see any air traffic or other obstructions in time to avoid a collision.

(c) Except as provided in 91.107, no person may operate an aircraft, under VFR, within a control zone beneath the ceiling when the ceiling is less than 1,000 feet.

(d) Except as provided in 91.107, no person may take off or land an aircraft, or enter the traffic pattern of an airport, under VFR, within a control zone:

(1) Unless ground visibility at that airport is at least 3 statute miles; and

Altitude	Flight visibility	Distance from clouds
1,200 feet or less above the surface (regardless of MSL altitude)		
Within controlled airspace	3 statute miles	500 feet below 1,000 feet above 2,000 feet horizontal
Outside controlled airspace	1 statute mile except as provided in 91.105(b)	Clear of clouds
More than 1,200 feet above the surface but less than 10,000 feet MSL		
Within controlled airspace	3 statute miles	500 feet below 1,000 feet above 2,000 feet horizontal
Outside controlled airspace	1 statute mile	500 feet below 1,000 feet above 2,000 feet horizontal
More than 1,200 feet above the surface and at or above 10,000 feet MSL.	5 statute miles	1,000 feet below 1,000 feet above 1 mile horizontal

Fig. 6-5. Weather Minimum Table

(2) If ground visibility is not reported at that airport, unless flight visibility during landing or takeoff, or while operating in the traffic pattern, is at least 3 statute miles.

(e) For the purposes of this section, an aircraft operating at the base altitude of a transition area or control area is considered to be within the airspace directly below that area.

91.107 SPECIAL VFR WEATHER MINIMUMS

(a) Except as provided in 93.113 of this chapter, when a person has received an appropriate ATC clearance, the special weather minimums of this section instead of those contained in 91.105 apply to the operation of an aircraft by that person in a control zone under VFR.

(b) No person may operate an aircraft in a control zone under VFR except clear of clouds.

(c) No person may operate an aircraft (other than a helicopter) in a control zone under VFR unless flight visibility is at least one statute mile.

(d) No person may take off or land an aircraft (other than a helicopter) at any airport in a control zone under VFR:

(1) Unless ground visibility at that airport is at least 1 statute mile; or

(2) If ground visibility is not reported at that airport, unless flight visibility during landing or takeoff is at least 1 statute mile.

(e) No person may operate an aircraft (other than a helicopter) in a control zone under the special weather minimums of this section, between sunset and sunrise (or in Alaska, when the sun is more than 6 degrees below the horizon) unless:

(1) That person meets the applicable requirements for instru-

ment flight under Part 61 of this chapter; and

(2) The aircraft is equipped as required in 91.33(d).

Special VFR weather minimums allow non-instrument rated pilots and airplanes not on IFR flight plans to takeoff with ceiling and visibility below basic VFR minimums. An ATC clearance must be received, however, and the pilot must follow all ATC instructions. In addition, to receive such a clearance, a visibility of at least one mile must exist and the pilot must be able to remain clear of clouds. At many major air terminals, special VFR is not authorized, and this information can be obtained from aeronautical charts, the AIM or from FAR Part 93.

91.109 VFR CRUISING ALTITUDE OR FLIGHT LEVEL ✈

Except while holding in a holding pattern of two minutes or less, or while turning, each person operating an aircraft under VFR in level cruising flight at an altitude of more than 3,000 feet above the surface shall maintain the appropriate altitude prescribed below:

(a) When operating below 18,000 feet MSL and:
(1) On a magnetic course of zero degrees through 179 degrees, any odd thousand foot MSL altitude plus 500 feet (such as 3,500, 5,500 or 7,500); or
(2) On a magnetic course of 180 degrees through 359 degrees, any even thousand foot MSL altitude plus 500 feet (such as 4,500, 6,500, and 8,500).

(b) When operating above 18,000 feet MSL to flight level 290 (inclusive) and:
(1) On a magnetic course of zero degrees through 179 degrees, any odd flight level plus 500 feet (such as 195, 215, or 235); or

(2) On a magnetic course of 180 degrees through 359 degrees, any even flight level plus 500 feet (such as 185, 205, or 225).

(c) When operating above flight level 290 and:
(1) On a magnetic course of zero degrees through 179 degrees, any flight level, at 4,000 foot intervals beginning at and including flight level 300 (such as flight level 300, 340, or 380) or
(2) On a magnetic course of 180 degrees through 359 degrees, any flight level, at 4,000 foot intervals, beginning at and including flight level 320 (such as flight level 320, 360, or 400).

SUBPART C—MAINTENANCE, PREVENTATIVE MAINTENANCE, AND ALTERATIONS

91.161 APPLICABILITY ✈

(a) This subpart prescribes rules governing the maintenance, preventive maintenace, and alteration of U.S. registered civil aircraft operating within or without the United States.

91.163 GENERAL ✈

(a) The owner or operator of an aircraft is primarily responsible for maintaining that aircraft in an airworthy condition, including compliance with Part 39 of this chapter.

(b) No person may perform maintenance, preventive maintenance, or alterations on an aircraft other than as prescribed in this subpart and other applicable regulations, including Part 43 of this chapter.

Part 43 is entitled "Maintenance, Preventive Maintenance, Rebuilding, and Alteration."

(c) No person may operate a rotorcraft for which a Rotorcraft Maintenance Manual containing an "Airworthiness Limitations" section has been issued unless the replacement times, inspection intervals, and related procedures specified in that section of the manual are complied with.

91.165 MAINTENANCE REQUIRED ✈

Each owner or operator of an aircraft shall have that aircraft inspected as prescribed in 91.169 and 91.170 of this chapter and shall, between required inspections, have defects repaired as prescribed in Part 43 of this chapter. In addition, he shall insure that maintenance personnel make appropriate entries in the aircraft and maintenance records indicating the aircraft has been released to service.

91.169 INSPECTIONS ✈

(a) Except as provided in paragraph (c) of this section, no person may operate an aircraft, unless, within the preceding 12 calendar months, it has had:

(1) An annual inspection in accordance with Part 43 of this chapter and has been approved for return to service by a person authorized by 43.7 of this chapter; or

(2) An inspection for the issue of an airworthiness certificate.

No inspection performed under paragraph (b) of this section may be substitutued for any inspection required by this paragraph unless it is performed by a person authorized to perform annual inspections, and is entered as an "annual" inspection in the required maintenance records.

(b) Except as provided in paragraph (c) of this section, no person may operate an aircraft carrying any person (other than a crewmember) for hire, and no person may give flight instruction for hire in an aircraft which that person provides, unless within the preceding 100 hours of time in service it

has received an annual or 100-hour inspection and been approved for return to service in accordance with Part 43 of this chapter, or received an inspection for the issuance of an airworthiness certificate in accordance with Part 21 of this chapter. The 100-hour limitation may be exceeded by not more than 10 hours if necessary to reach a place at which the inspection can be done. The excess time, however, is included in computing the next 100 hours of time in service.

(c) Paragraphs (a) and (b) of this section do not apply to:

(1) Any aircraft for which its registered owner or operator complies with the progressive inspection requirements of 91.171 and Part 43 of this chapter;

(2) An aircraft that carries a special flight permit or a current experimental or provisional certificate;

(3) Any airplane operated by an airtravel club that is inspected in accordance with Part 123 of this chapter and the operator's manual and operations specifications; or

---

Part 123 governs the certification and operation of air travel clubs using large aircraft.

---

(4) Any aircraft inspected in accordance with an approved aircraft inspection program under Part 135 of this chapter and is identified, by registration number, in the operations specifications of the certificate holder having the approved inspection program.

91.173 MAINTENANCE RECORDS ✈

(a) Except for work performed in accordance with 91.170, each registered owner or operator shall keep the following records for the periods specified in paragraph (b) of this section:

(1) Records of the maintenance and alteration, and records of the 100-hour annual; progressive, and other required or approved inspections, as appropriate, for each aircraft (including the airframe, and each engine, propeller, rotor, and appliance of the aircraft. The records must include—

   (i) A description (or reference to data acceptable to the Administrator) of the work performed;

   (ii) The date of completion of the work performed; and

   (iii) The signature and certificate number of the person approving the aircraft for return to service.

(2) Records containing the following information:

   (i) The total time in service of the airframe.

   (ii) The current status of life-limited parts of each airframe, engine, propeller, rotor, and appliance.

   (iii) The time since last overhaul of all items installed on the aircraft which are required to be overhauled on a specified time basis.

   (iv) The identification of the current inspection status of the aircraft, including the times since the last inspections required by the inspection program under which the aircraft and its appliances are maintained.

   (v) The current status of applicable airworthiness directives, including the method of compliance.

   (vi) A list of current major alterations to each airframe, engine, propeller, rotor, and compliance.

(b) The owner or operator shall retain the records required to be kept by this section for the following periods:

(1) The records specified in paragraph (a)(1) of this section shall be retained until the work is repeated or superseded by other work or for 1 year after the work is performed.

(2) The records specified in paragraph (a)(2) of this section shall be retained and transferred with the aircraft at the time the aircraft is sold.

(c) The owner or operator shall make all maintenance records required to be kept by this section available for inspection by the Administrator or any authorized representative of the National Transportation Safety Board (NTSB).

SECTION E -
NTSB PART 830

RULES PERTAINING TO THE NOTIFICATION
AND REPORTING OF AIRCRAFT ACCIDENTS OR INCIDENTS AND
OVERDUE AIRCRAFT, AND PRESERVATION OF AIRCRAFT
WRECKAGE, MAIL, CARGO, AND RECORDS

SUBPART A — GENERAL

830.1 APPLICABILITY

This part contains rules pertaining to:

(a) Providing notice of and reporting, aircraft accidents and incidents and certain other occurrences in the operation of aircraft when they involve civil aircraft of the United States wherever they occur, or foreign civil aircraft when such events occur in the United States, its territories or possessions.

(b) Preservation of aircraft wreckage, mail, cargo, and records involving all civil aircraft in the United States, its territories or possessions.

830.2 DEFINITIONS

As used in this part the following words or phrases are defined as follows:

AIRCRAFT ACCIDENT means an occurrence associated with the operation of an aircraft which takes place between the time any person boards the aircraft with the intention of flight until such time as all such persons have disembarked, and in which any person suffers death or serious injury as a result of being in or upon the aircraft or by direct contact with the aircraft or anything attached thereto, or in which the aircraft receives substantial damage.

FATAL INJURY means any injury which results in death within 7 days of the accident.

OPERATOR means any person who causes or authorizes the operation of an aircraft, such as the owner, lessee, or bailee of an aircraft.

SERIOUS INJURY means any injury which (1) requires hospitalization for more than 48 hours, commencing within 7 days from the date the injury was received; (2) Results in a fracture of any bone (except simple fractures of fingers, toes, or nose); (3) involves lacerations which cause severe hemorrhages, nerve, muscle, or tendon damage; (4) involves injury to any internal organ; or (5) involves second or third degree burns, or any burns affecting more than 5 percent of the body surface.

SUBSTANTIAL DAMAGE:

(1) Except as provided in subparagraph (2) of this paragraph, substantial damage means damage or structural failure which adversely affects the structural strength, performance, or flight characteristics of the aircraft, and which would normally require major repair or replacement of the affected component.

SUBPART B — INITIAL NOTIFICATION OF AIRCRAFT ACCIDENTS, INCIDENTS, AND OVERDUE AIRCRAFT

830.5 IMMEDIATE NOTIFICATION

The operator of an aircraft shall immediately, and by the most expeditious means available, notify the nearest National Transportation Safety Board (Board), Bureau of Aviation Safety field office when:

(a) An aircraft accident or any of the following listed incidents occur:

(1) Flight control system malfunction or failure;

(2) Inability of any required flight crewmember to perform his normal flight duties as a result of injury or illness;

(3) Turbine engine rotor failures excluding compressor blades and turbine buckets;

(4) In-flight fire; or

(5) Aircraft collide in flight.

(b) An aircraft is overdue and is believed to have been involved in an accident.

830.6 INFORMATION TO BE GIVEN IN NOTIFICATION ✈

The notification required in 830.5 shall contain the following information, if available:

(a) Type, nationality, and registration marks of the aircraft;

(b) Name of owner, and operator of the aircraft;

(c) Name of the pilot-in-command;

(d) Date and time of the accident;

(e) Last point of departure and point of intended landing of the aircraft;

(f) Position of the aircraft with reference to some easily defined geographical point;

(g) Number of persons aboard, number killed, and number seriously injured;

(h) Nature of the accident, the weather and the extent of damage to the aircraft, so far as is known; and

(i) A description of any explosives, radioactive materials, or other dangerous articles carried.

SUBPART C — PRESERVATION OF AIRCRAFT WRECKAGE, MAIL, CARGO, AND RECORDS

830.10 PRESERVATION OF AIRCRAFT WRECKAGE, MAIL, CARGO, AND RECORDS ✈

(a) The operator of an aircraft is responsible for preserving to the extent possible any aircraft wreckage, cargo, and mail aboard the aircraft, and all records, including tapes of flight recorders and voice recorders, pertaining to the operation and maintenance of the aircraft and to the airmen involved in an accident or incident for which notification must be given until the Board takes custody thereof or a release is granted pursuant to 831.17.

(b) Prior to the time the Board or its authorized representative takes custody of aircraft wreckage, mail, or cargo, such wreckage, mail or cargo may not be disturbed or moved except to the extent necessary:

(1) To remove persons injured or trapped;

(2) To protect the wreckage from further damage, or

(3) To protect the public from injury.

(c) Where it is necessary to disturb or move aircraft wreckage, mail or cargo, sketches, descriptive notes, and photographs shall be made, if possible, of the accident locale including original position and condition of the wreckage and any significant impact marks.

(d) The operator of an aircraft involved in an accident or incident as defined in this part, shall retain all records and reports, including all internal documents and memoranda dealing with the accident or incident, until authorized by the Board to the contrary.

SUBPART D — REPORTING OF AIRCRAFT ACCIDENTS, INCIDENTS, AND OVERDUE AIRCRAFT

830.15 REPORTS AND STATEMENTS TO BE FILED ✈

(a) *Reports.* The operator of an aircraft shall file a report as provided in paragraph (c) of this section on Board Form 6120.1 or Board Form 6120.2 within 10 days after an accident, or after 7 days if an overdue aircraft is still missing. A report on an incident for which notification is required by 830.5(a) shall be filed only as requested by an authorized representative on the Board.

(b) *Crewmember statement.* Each crewmember, if physically able at the time the report is submitted, shall attach thereto a statement setting forth the facts, conditions, and circumstances relating to the accident or incident as they appear to him to the best of his knowledge and belief. If the crewmember is incapacitated, he shall submit the statement as soon as he is physically able.

(c) *Where to file the reports.* The operator of an aircraft shall file with the field office of the Board nearest the accident or incident any report required by this section.

Forms are obtainable from the Board field offices, the National Transportation Safety Board, Washington, D. C. 20594, and the Federal Aviation Administration, Flight Standards District Office. The National Transportation Safety Board field offices are listed under U.S. Government in the telephone directories in the following cities: Anchorage, Alaska; Chicago, Ill.; Denver, Colo.; Fort Worth, Tex.; Kansas City, Mo.; Los Angeles, Calif.; Miami, Fla.; New York, N.Y.; Oakland, Calif.; Seattle, Wash.; Washington, D. C.

chapter 7

INTRODUCTION

Pilot procedures and knowledge requirements are constantly changing. New airports are opening, old ones are closing, new regulations replace old ones, some radio facilities are temporarily closed for repairs, and others are shut down permanently. New factors affecting safety of flight are uncovered while new flight routes and techniques are established. These are just a few reasons why it is not only important that a pilot read the *Airman's Information Manual*, but why it is essential that he have an up-to-date copy.

Aviation is a dynamic and changing industry, and since the *Airman's Information Manual* contains up-to-date official information, each new issue should be checked to acquaint the pilot with the latest procedures and data available. The knowledge gained from a study of the AIM will provide the pilot with confidence, understanding and the ability to cope with the complexities of today's aviation environment.

The *Airman's Information Manual* is valuable in assisting preflight planning. Regulations require that, for a flight away from the vicinity of an airport, preflight planning must include information on available weather reports and forecasts, fuel requirements, and alternatives available if the flight cannot be completed as planned.

Three methods are used to provide aeronautical information for the National Airspace System. The aeronautical chart is the primary method, the *Airman's Information Manual* is the secondary method, and the Notices to Airmen is the third. The three methods complement and supplement each other.

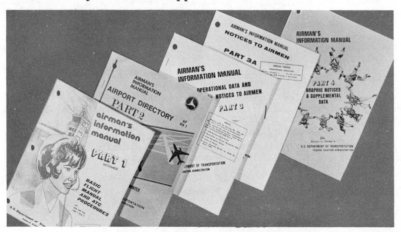

Fig. 7-1. The Airman's Information Manual

AIRMAN'S INFORMATION MANUAL (AIM)

The Airman's Information Manual has been designed primarily as a pilot's operational and information manual for use in the National Airspace System of the United States (unless otherwise indicated). It is divided into five basic parts, each of which may be purchased separately. Frequency of issuance, area of coverage, annual subscription costs and highlights of the contents of each part follow.

Part 1—Basic Flight Manual and ATC Procedures

Issued: Quarterly (Feb., May, Aug., Nov). *Coverage:* Entire U.S. unless otherwise indicated.

This part contains the basic fundamentals required to fly in the U.S. National Airspace System. Among other data it also contains adverse factors affecting Safety of Flight; Health and Medical Facts of interest to pilots; ATC information affecting rules, regulations and procedures; a Glossary of Aeronautical terms; Air Defense Identification Zones (ADIZ); Designated Mountainous Areas; and Emergency Procedures.

Part 2—Airport Directory

Issued: Semiannually (Mar. and Sept.). *Coverage:* Conterminous U.S., Puerto Rico and Virgin Islands (Note: similar information for Alaska and Hawaii appears in Alaska Supplement and Pacific Chart Supplement, respectively–See Special Notice Section, Part 3 for availability.)

Part 2 contains a Directory of all airports, seaplane bases, and heliports available for civil use. It includes all their services, *except communications*, in codified form. (Those airports with communications are also listed in Part 3.) A list of new and permanently closed airports which updates Part 2 is contained in Part 3. Also included in Part 2 are U.S. Entry and Departure Procedures, including Airports of Entry and Landing Rights Airports; and a listing of Flight Service Station and National Weather Service Telephone Numbers.

Parts 3 and 3A—Operational Data and Notices to Airmen

Issued: Part 3, every 56 days and Part 3A, every 14 days. *Coverage:* Part 3, Conterminous U.S., Puerto Rico and Virgin Islands (Note: Similar information for Alaska and Hawaii appears in Alaska Supplement and Pacific Chart Supplement, respectively—(For sale by National Ocean Survey, Distribution Division, C44, Riverdale, Md. 20840). Part 3A coverage is the same as Part 3 except that Notice-to-Airmen data for Puerto Rico and Virgin Islands appears in the International NOTAMS publication).

Part 3 contains an Airport-Facility Directory of all major airports with control towers and/or instrument landing systems; a tabulation of Air Navigation Radio Aids; Special, General, and Area Notices; a tabulation of New and Permanently Closed Airports (which updates Part 2); North Atlantic Routes; Locations of VOR Receiver Check Points (both ground and airborne); Restrictions to Enroute Navigation Aids; Preferred Routes; Area Navigation Routes, and supplemental data to Part 4.

Part 3A contains current Notices to Airmen considered essential to the safety of flight as well as supplemental data to Parts 3 and 4.

Part 4—Graphic Notices and Supplemental Data

Issued: Quarterly (Jan., April, July, Oct.). *Coverage:* Conterminous U.S., Puerto Rico and Virgin Islands (Note: similar information for Alaska and Hawaii appears in Alaska Supplement and Pacific Chart Supplement, respectively—(For sale by National Ocean Survey, Distribution Division, C44, Riverdale, Md. 20840).

Part 4 contains a list of abbreviations used in the AIM; a tabulation of Parachute Jump Areas; Special Notice—Area Graphics; Terminal Area Graphics: Olive Branch Routes and other data not requiring frequent change.

Where to Purchase AIM

The four basic parts described above are available from the Superintendent of Documents, Government Printing Office, Washington, D.C. 20402. Orders should be accompanied by check or money order made payable to the Superintendent of Documents.

Errors, Omissions, or Changes

Errors, omissions, or suggested changes should be forwarded to the Federal Aviation Administration, Flight Services Division, AAT–430, Washington, D.C. 20591.

For sale by the Superintendent of Documents, U.S. Government Printing Office, Washington, D.C. 20402

Fig. 7-2. General Subject Areas Page

THE FOUR PARTS OF AIM

The *Airman's Information Manual* is prepared in four basic parts, plus a supplement. Each of these parts is important for some phase of flight or preflight planning. The purpose of this chapter is to show the types of information that the AIM contains and how to locate and use this information. (See Fig. 7-1.)

Part 1 of the *Airman's Information Manual* contains the general flight information of a textbook nature and is called the *Basic Flight Manual and ATC Procedures.* It is revised and reissued every three months, or quarterly.

Part 2 of the AIM contains the *Airport Directory*, and the telephone numbers for flight service stations and Weather Services. Entry and departure procedures at United States international borders are also included in Part 2. This section is revised and reissued semiannually.

Part 3 contains *Operational Data and Notices to Airmen* and is issued every 56 days. Notices to Airmen, or NOTAMs, contained in a supplemental section of Part 3, referred to as Part 3A, which is issued every 14 days.

Graphic Notices and Supplemental Data make up the remainder of the *Airman's Information Manual* in a separate publication labeled Part 4. Part 4 is revised and reissued quarterly.

USE OF THE AIM

Each part of the AIM begins with the page shown in figure 7-2. This page lists the general subject areas covered in each part of the AIM in order to assist the pilot in determining which part to use to find specific information.

When the pilot has determined whether the desired information is located in Part 1, 2, 3, 3A, or 4, he should examine the table of contents for that part,(See Fig. 7-3.)

For instance, it is assumed that the pilot wishes to determine how to perform a VOR receiver check. Examining the first page of any part (see Fig. 7-2), it seems advisable to look first in Part 1. The desired information may be found in this part, since it contains *basic fundamentals required to fly in the National Airspace System.*

Turning to the table of contents for Part 1, the pilot finds that VOR re-

---

New or amended textual material is indicated by a solid black dot (●) prefixing the heading, paragraph, or line.

TABLE OF CONTENTS

Page

Chapter 1. GENERAL

Fig. 7-3. Part I, Alphabetical Index

AIR NAVIGATION RADIO AIDS

VOR RECEIVER CHECK

1. Periodic VOR receiver calibration is most important. If a receiver's Automatic Gain Control or modulation circuit deteriorates, it is possible for it to display acceptable accuracy and sensitivity close in to the VOR or VOT and display out-of-tolerance readings when located at greater distances where weaker signal areas exist. The likelihood of this deterioration varies between receivers, and is generally considered a function of time. The best assurance of having an accurate receiver is periodic calibration. Yearly intervals are recommended at which time an authorized repair facility should recalibrate the receiver to the manufacturer's specifications.

2. Part 91.25 of the Federal Aviation Regulations provides for certain VOR equipment accuracy checks prior to flight under instrument flight rules. To comply with this requirement and to ensure satisfactory operation of the airborne system, the FAA has provided pilots with the following means of checking VOR receiver accuracy: (1) VOR test facility (VOT), (2) certified airborne check points, and (3) certified check points on the airport surface.

a. The VOR test facility (VOT) transmits a test signal for VOR receivers which provides users of VOR a convenient and accurate means to determine the operational status of their receivers. The facility is designed to provide a means of checking the accuracy of a VOR receiver while the aircraft is on the ground. The radiated test signal is used by tuning the receiver to the published frequency of the test facility. With the Flight Path Deviation Indicator (FPDI) centered the omnibearing selector should read 0° with the to-from indication being "from" or the omnibearing selector should read 180° with the to-from indication reading "to." Should the VOR receiver be of the automatic indicating type, the indication should be 180°. Two means of identification are used with the VOR radiated test signal. In some cases a continuous series of dots is used while in others a continuous 1020 cycletone will identify the test signal. Information concerning an individual test signal can be obtained from the local Flight Service Station.

b. Airborne and ground check points consist of certified radials that should be received at specific points on the airport surface, or over specific landmarks while airborne in the immediate vicinity of the airport.

c. Should an error in excess of ±4° be indicated through use of a ground check, or ±6° using the airborne check, IFR flight shall not be attempted without first correcting the source of the error. CAUTION: no correction other than the "correction card" figures supplied by the manufacturer should be applied in making these VOR receiver checks.

d. The list of airborne check points and ground check points is published in Part 4. VOT's are included with the airport information in Part 3.

e. If dual system VOR (units independent of each other except for the antenna) is installed in the aircraft, the person checking the equipment may check one system against the other. He shall tune both systems to the same VOR ground facility and note the indicated bearings to that station. The maximum permissible variations between the two indicated bearings is 4°.

Fig. 7-4. VOR Receiver Checkpoints

ceiver checks are covered on page 1-8. (See Fig. 7-3.) Page 1-8 in the AIM provides the pilot with a general discussion of VOR receiver checks and describes the mechanics of using airborne checks, ground checks, and the VOR test facility. (See Fig. 7-4.)

SECTION A

PART 1 OF THE AIM

Chapter 1, entitled "General," includes a Glossary of Aeronautical Terms. (See Fig. 7-6.) In the Glossary of Aeronautical Terms, pilots find explanations of terminology used in the AIM. For example, the term *air traffic* is defined as *aircraft operating in the air or on an airport surface exclusive of loading ramps or parking areas.*

Fig. 7-5. Part I of the Airman's Information Manual

NAVIGATION AIDS

Chapter 2 is entitled "Navigational Aids," and begins with a section describing the various types of air navigation radio aids. In this section is found the VOR receiver check subject area, as noted previously. The section continues with a general discussion of distance measuring equipment or DME, its operating principles, and frequency range, plus an explanation of the instrument landing system or ILS. General characteristics as well as details of the localizer, glide path, marker beacons, and compass locator are presented.

The next subject covered in Chapter 2 is the VHF/UHF direction finder. The VHF/UHF direction finder display indicates the magnetic direction of the aircraft from the ground station each time the aircraft transmitter is activated. (See Fig. 7-7.) Direction finding equipment is of particular value in locating disoriented aircraft.

Chapter 1.   GENERAL

GLOSSARY OF AERONAUTICAL TERMS

ADVISORY SERVICE—Advice and information provided by a facility to assist pilots in the safe conduct of flight and aircraft movement.

AIR DEFENSE IDENTIFICATION ZONE (ADIZ)—The area of airspace over land or water within which the ready identification, the location, and the control of aircraft are required in the interest of national security. For operating details see ADIZ procedures.

AIR NAVIGATION FACILITY (NAVAID)—Any facility used in, available for use in, or designed for use in aid of air navigation, including landing areas, lights, any apparatus or equipment for disseminating weather information, for signaling, for radio direction-finding, or for radio or other electronic communication, and any other structure or mechanism having a similar purpose for guiding or controlling flight in the air or the landing or takeoff of aircraft.

AIRPORT ADVISORY AREA—The area within five statute miles of an uncontrolled airport on which is located a Flight Service Station so depicted on the appropriate Sectional Aeronautical Chart.

AIRPORT ADVISORY SERVICE—A terminal service provided by a Flight Service Station located at an airport where a control tower is not operating.

AIRPORT TRAFFIC AREA—Unless otherwise specifically designated (FAR Part 93), that airspace with a horizontal radius of five statute miles from the geographical center of any airport at which a control tower is operating, extending from the surface up to, but not including, an altitude of 3,000 feet above the elevation of the airport.

AIR ROUTE SURVEILLANCE RADAR (ARSR)—Long range radar which increases the capability of ATC for handling heavy en route traffic. An ARSR site is usually located at some distance from the ARTCC it serves. Range, approximately 200 NM.

AIR ROUTE TRAFFIC CONTROL CENTER (CENTER)—A facility established to provide air traffic control service to aircraft operating on an IFR flight plan within controlled airspace and principally during the en route phase of flight.

AIR TRAFFIC—Aircraft operating in the air or on an airport surface, exclusive of loading ramps and parking areas.

AIR TRAFFIC CLEARANCE (CLEARANCE)—An authorization by air traffic control for the purpose of preventing collision between known aircraft, for an aircraft to proceed under specified traffic conditions within controlled airspace.

AIR TRAFFIC CONTROL RADAR BEACON SYSTEM (ATCRBS)—See RADAR.

Fig. 7-6. Glossary of Aeronautical Terms

The pilot of an aircraft equipped with a transponder would find the section on air traffic control radar beacon systems of interest. In addition, Chapter 2 contains a section entitled "Airport, Air Navigation Lighting, and Marking Aids." A diagrammatical explanation of the different types of runway markings is also included in this section. (See Fig. 7-8.)

A black dot in front of a subject in the table of contents or on a chapter page indicates new or amended information. In figure 7-7, the black dot indicates that there is new information or a change in the information associated with the radar capabilities.

THE AIRSPACE

Because airspace utilization requirements are continuously changing, Chapter 3, entitled "The Airspace," is a valuable reference for pilots. This chapter includes a review of VFR cruising altitudes and a description of control areas, positive control areas, control zones, airport traffic areas, and related subjects. An example of this type of information is shown in figure 7-9..

AIR NAVIGATION RADIO AIDS

VHF/UHF DIRECTION FINDER

1. The VHF/UHF Direction Finder (VHF/UHF/DF) is one of the Common System equipments that helps the pilot without his being aware of its operation. The VHF/UHF/DF is a ground-based radio receiver used by the operator of the ground station where it is located.

2. The equipment consists of a directional antenna system, a VHF and a UHF radio receiver. At a radar-equipped tower or center, the cathode-ray tube indications may be superimposed on the radarscope.

3. The VHF/UHF/DF display indicates the magnetic direction of the aircraft from the station each time the aircraft transmits. Where DF equipment is tied into radar, a strobe of light is flashed from the center of the radarscope in the direction of the transmitting aircraft.

4. DF equipment is of particular value in locating lost aircraft and in helping to identify aircraft on radar. (See VHF/UHF Direction Finding Instrument Approach Procedure in Chapter 4.)

RADAR

1. **Capabilities.**

a. A method whereby radio waves are transmitted into the air and are then received when they have been reflected by an object in the path of the beam. *Range* is determined by measuring the time it takes (at the speed of light) for the radio wave to go out to the object and then return to the receiving antenna. The *direction* of a detected object from a radar site is determined by the position of the rotating antenna when the reflected portion of the radio wave is received.

b. More reliable maintenance and improved equipment have reduced radar system failures to a negligible factor. Most facilities actually have some components duplicated—one operating and another which immediately takes over when a malfunction occurs to the primary component.

AIR TRAFFIC CONTROL RADAR BEACON SYSTEM (ATCRBS)

1. The Air Traffic Control Radar Beacon System (ATCRBS), sometimes referred to as secondary surveillance radar, consists of three main components:

a. **Interrogator.** Primary radar relies on a signal being transmitted from the radar antenna site and for this signal to be reflected or "bounced back" from an object (such as an aircraft). This reflected signal is then displayed as a "target" on the controller's radarscope. In the ATCRBS, the *Interrogator*, a ground based radar beacon transmitter-receiver, scans in synchronism with the primary radar and transmits discrete radio signals which repetitiously requests all transponders, on the mode being used, to reply. The replies received are then mixed with the primary returns and both are displayed on the same radarscope.

b. **Transponder.** This airborne radar beacon transmitter-receiver automatically receives the signals from the interrogator and selectively replies with a specific pulse group (code) only to those interrogations being received on the mode to which it is set. These replies are independent of, and much stronger than a primary radar return.

c. **Radarscope.** The radarscope used by the controller displays returns from both the primary radar system and the ATCRBS. These returns, called targets, are what the controller refers to in the control and separation of traffic.

● 2. The job of identifying and maintaining identification of primary radar targets is a long and tedious task for the controller. Some of the advantages of ATCRBS over primary radar are:

a. Reinforcement of radar targets.

b. Rapid target identification.

c. Unique display of selected codes.

Fig. 7-7. Air Navigation Radio Guide

AIRPORT, AIR NAVIGATION LIGHTING AND MARKING AIDS

5. In haze or dust conditions or when the approach is made into the sun, the white lights may appear yellowish. This is also true at night when the VASI is operated at a low intensity. Certain atmospheric debris may give the white lights an orange or brownish tint; however, the red lights are not affected and the principle of color differentiation is still applicable.

RUNWAY END IDENTIFIER LIGHTS (REIL)

1. Runway End Identifier Lights are installed at many airfields to provide rapid and positive identification of the approach end of a particular runway. The system consists of a pair of synchronized flashing lights, one of which is located laterally on each side of the runway threshold facing the approach area. They are effective for:

a. Identification of a runway surrounded by a preponderance of other lighting.

b. Identification of a runway which lacks contrast with surrounding terrain.

c. Identification of a runway during reduced visibility.

AIRCRAFT ARRESTING DEVICES

Certain airports are equipped with a means of rapidly stopping military aircraft on a runway. This equipment, normally referred to as EMERGENCY ARRESTING GEAR, generally consists of pendant cables supported over the runway surface by rubber "donuts". Although most devices are located in the overrun areas, a few of these arresting systems have cables stretched over the operational areas near the ends of a runway. Arresting cables which cross over a runway require special markings on the runway to identify the cable location. Aircraft operations on the runway are NOT restricted by such installations.

MARKING

1. In the interest of safety, regularity, or efficiency of aircraft operations, the FAA has recommended for the guidance of the public the following airport marking. (Runway numbers and letters are determined from the approach direction. The number is the whole number nearest one-tenth the magnetic azimuth of the centerline of the runway, measured clockwise from the magnetic north.) The letter or letters differentiate between parallel runways:

    For two parallel runways "L" "R"
    For three parallel runways "L" "C" "R"

a. Basic Runway Marking—markings used for operations under Visual Flight Rules: centerline marking and runway direction numbers.

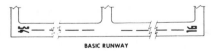

BASIC RUNWAY

b. Non-Precision Instrument Runway Marking—markings on runways served by a nonvisual navigation aid and intended for landings under instrument weather conditions: basic runway markings plus threshold marking.

NON-PRECISION INSTRUMENT RUNWAY

c. Precision Instrument Runway Marking—markings on runways served by non-visual precision approach aids and on runways having special operational requirements, non-precision instrument runway marking, touchdown zone marking, fixed distance marking, plus side stripes.

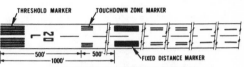

PRECISION INSTRUMENT RUNWAY

d. Threshold—A line perpendicular to the runway centerline designating the beginning of that portion of a runway usable for landing.

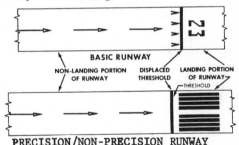

e. Displaced Threshold—A threshold that is not at the beginning of the full strength runway pavement.

f. Closed or Overrun/Stopway Areas—Any surface or area which appears usable but which, due to the nature of its structure, is usable only for taxiing aircraft.

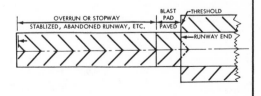

1. OVERRUN/STOPWAY AND BLAST PAD AREA

Fig. 7-8. Airport, Air Navigation Lighting, and Marking Aids

AIR TRAFFIC CONTROL

Chapter 4 details the many air traffic control services available to pilots. For example, automatic terminal information service and radar assistance to VFR aircraft are described in detail. Also included are radio phraseology and techniques used in communications with air traffic controllers. Recommended methods, standards, and procedures are clearly presented in this chapter.

Another example of the kind of information found in Chapter 4 of Part 1

pertains to airport operations. This section deals with procedures used at both tower controlled and uncontrolled airports. The standard FAA light signals used in controlling non-radio equipped aircraft and procedures for aircraft experiencing radio failure at controlled airports are explained. The airport operations section concludes with an illustration of the hand signals used to assist taxiing aircraft.

The section entitled "Preflight" serves as a guide to VFR and IFR flight planning.

THE AIRSPACE

certain hours of the day, this fact will also be noted on the charts. A typical control zone is depicted below.

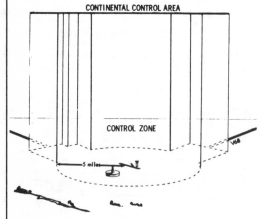

CONTINENTAL CONTROL AREA

CONTROL ZONE

5 miles

VFR REQUIREMENTS

Minimum flight visibility and distance from clouds have been prescribed for VFR operation in controlled airspace. In addition, appropriate altitudes/flight levels for VFR flight in controlled, as well as in uncontrolled airspace have been prescribed in FAR 91.105. The ever increasing speeds of aircraft results in increasing closure rates for opposite direction aircraft. This means that there is less time for pilots to see each other and react to avoid each other. By adhering to the altitude/flight level appropriate for the direction of flight, a "built-in" vertical separation is available for the pilots.

IFR REQUIREMENTS

Federal Aviation Regulations prescribe the pilot and aircraft requirements for IFR flight.

IFR ALTITUDES/FLIGHT LEVELS

Pilots operating IFR within controlled airspace will fly at an altitude/flight level assigned by ATC. When operating IFR within controlled airspace with an altitude assignment of "VFR-ON-TOP", an altitude/flight level appropriate for *VFR flight* is to be flown. (FAR 91.121)

VFR ALTITUDES/FLIGHT LEVELS—CONTROLLED AND UNCONTROLLED AIRSPACE

Under Visual Flight Rules (VFR)
More than 3,000 feet above the surface.

COURSES ARE MAGNETIC

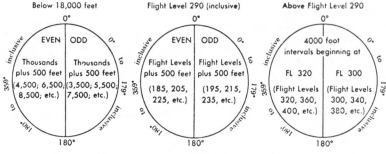

Below 18,000 feet

Above 18,000 feet MSL to
Flight Level 290 (inclusive)

Above Flight Level 290

Fig. 7-9. Airspace Utilization

The next three major subject areas covered, Departure, En Route, and Arrival are concerned with IFR procedures. The instrument pilot must take particular note of this portion of the AIM.

A subject of interest to all pilots is emergency procedures. The Emergency Procedures section in Chapter 4 of the AIM is principally concerned with communications procedures recommended for pilots in distress. Pilots in distress are urged to remember the four "C's." Summarized, they are: *confess* your predicament to any ground radio station; *communicate* with a ground station and pass along as much of the distress message on the first transmission as possible; *climb* if possible for better radar and direction finding detection; and comply—especially comply with advice and instructions received if help is really desired.

Included in the Emergency Procedures section is a discussion of the different types of locator beacons that are used as a means of locating downed aircraft and their occupants. These electronic, battery-operated beacons can be installed in an aircraft and are ideally suited for general aviation. (See Fig. 7-10.)

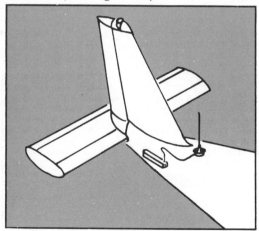

Fig. 7-10. Locator Beacon

Search and rescue is a lifesaving service provided through the combined efforts of the FAA, Air Force, Coast Guard, state aeronautics commissions, Civil Air Patrol, and assisted by other organizations such as the sheriff's department.

The system is so designed that if a pilot fails to cancel his flight plan within one-half hour after his estimated time of arrival, a search will be automatically activated.

The last part of Chapter 4 is devoted to national security. This section contains a chart which shows the different air defense identification zones. The requirements for penetrating the zones are also included in text form. Every pilot planning a flight outside the borders of the United States should consult this portion of the AIM.

SAFETY OF FLIGHT

Chapter 5, the last chapter in Part 1, concerns flight safety. Here, the pilot can find valuable hints and information to insure the safe operation of his aircraft. The first section of this chapter is devoted to weather, and contains important information on weather broadcasts, pilot weather reports, and different types of weather phenomena such as turbulence. A table is given which should be used as a guide by pilots when they make radio reports of turbulence. (See Fig. 7-11.)

Within the weather section of Chapter 5 are a few paragraphs concerning altimetry, which provide general information regarding factors affecting aircraft altimeters and guidelines for altimeter settings.

A full-page key to aviation weather reports and aviation forecasts is clearly illustrated in the AIM to simplify the interpretation of these reports. Other subjects covered in the weather section are visibility observations, weather radar, thunderstorms, detouring thunderstorms, airframe icing, and in-flight weather advisory test programs.

Pilots operating in the vicinity of large transport-type aircraft must be familiar with the phenomena called *wake turbulence.* An explanation of wingtip vortices is given, and the action that pilots should take when flying in the vicinity of a large aircraft is also explained. (See Fig. 7-12.)

Another example of the type of information found under the "Safety of Flight" section in Chapter 5 is medical facts for pilots. This coverage includes the effects of alcohol and drugs, vertigo, carbon monoxide, vision and middle ear problems, panic, and underwater diving, and how they all relate to the pilot.

TURBULENCE REPORTING CRITERIA TABLE

Intensity	Aircraft Reaction	Reaction Inside Aircraft	Reporting Term Definition
LIGHT	Turbulence that momentarily causes slight, erratic changes in altitude and/or attitude (pitch, roll, yaw). Report as *Light Turbulence*;* or Turbulence that causes slight, rapid and somewhat rhythmic bumpiness without appreciable changes in altitude or attitude. Report as *Light Chop.*	Occupants may feel a slight strain against seat belts or shoulder straps. Unsecured objects may be displaced slightly. Food service may be conducted and little or no difficulty is encountered in walking.	Occasional—Less than 1/3 of the time. Intermittent—1/3 to 2/3. Continuous—More than 2/3.
MODERATE	Turbulence that is similar to Light Turbulence but of greater intensity. Changes in altitude and/or attitude occur but the aircraft remains in positive control at all times. It usually causes variations in indicated airspeed. Report as *Moderate Turbulence*;* or Turbulence that is similar to Light Chop but of greater intensity. It causes rapid bumps or jolts without appreciable changes in aircraft altitude or attitude. Report as *Moderate Chop.*	Occupants feel definite strains against seat belts or shoulder straps. Unsecured objects are dislodged. Food service and walking are difficult.	NOTE—Pilots should report location(s), time (GMT), intensity, whether in or near clouds, altitude, type of aircraft and, when applicable, duration of turbulence. Duration may be based on time between two locations or over a single location. All locations should be readily identifiable. Example: a. Over Omaha, 1232Z, Moderate Turbulence, in cloud, Flight Level 310, B707. b. From 50 miles south of Albuquerque to 30 miles north of Phoennx, 1210Z to 1250Z, occasional Moderate Chop, Flight Level 330, DC8.
SEVERE	Turbulence that causes large, abrupt changes in altitude and/or attitude. It usualy causes large variations in indicated airspeed. Aircraft may be momentarily out of control. Report as *Severe Turbulence.**	Occupants are forced violently against seat belts or shoulder straps. Unsecured objects are tossed about. Food service and walking are impossible.	
EXTREME	Turbulence in which the aircraft is violently tossed about and is practically impossible to control. It may cause structural damage. Report as *Extreme Turbulence.**		

* High level turbulence (normally above 15,000 feet ASL) not associated with cumuliform cloudiness, including thunderstorms, should be reported as CAT (clear air turbulence) preceded by the appropriate intensity, or light or moderate chop.

Fig. 7-11. Turbulence Classification Table

Fig. 7-12. Wake Turbulence

On a cross-country flight during the migration season, it is conceivable that a pilot's course could cross bird flight corridors. In the chapter "Safety of Flight," a major section called Bird Hazards points out the danger of flying close to flocks of large migratory birds.

Procedures and practices that greatly enhance the safety of every flight are covered in the section entitled "Good Operating Practices." This information is valuable to all pilots.

SECTION B

PART 2 OF THE AIM

Part 2 of the AIM, the *Airport Directory* (see Fig. 7-13), is issued semi-annually and contains a directory of all airports, seaplane bases, and heliports which can be used by general aviation aircraft in the contiguous 48 states, Puerto Rico, and the Virgin Islands. Also available in Part 2 is a list of flight service station and Weather Service telephone numbers. Information on U.S. Customs and a list of the U.S. Airports of Entry and Landing

Rights Airports make up the balance of Part 2.

AIRPORT DIRECTORY

Subject matter in Part 2 is listed alphabetically in the table of contents. The most prominent listing is the Airport Directory (see Fig. 7-14) which covers nearly 250 pages. An airport legend is provided to explain how to read the Airport Directory listings. (See Fig. 7-15.) Detailed explanations of the symbols and abbreviations used in this listing are explained in the pages preceding the sample excerpt. Cross reference numbers have been added between the pages of figure 7-15 to aid in the location of this information.

Since airports within the directory, are arranged alphabetically by cities within states, it is easy to locate and interpret the airport listings. For example, Chisholm-Hibbing Field at Hibbing, Minnesota, has the following facilities: (See Fig. 7-16.)

Fig. 7-13. Part II of the Airman's Information Manual

AIRPORT DIRECTORY

PART 2

TABLE OF CONTENTS

Fig. 7-14. Airport Directory Table of Contents

(1) NOTAM service is available at the airport in Hibbing.

(2) An approved instrument approach procedure is available for instrument pilots.

(3) The letters "HIB" are used as the airport identifier.

(4) The airport is located five nautical miles southeast of the city of Hibbing.

(5) The geographical coordinates for Hibbing Airport are 47°23'18"N, 92°50'19"W.

(6) The Hibbing Flight Service Station is located on the airport.

(7) The field elevation at Hibbing is 1,353 feet and the longest runway, which is runway 13-31, is 6,600 feet long. The "3" indicates there are three hard-surfaced runways available at Hibbing.

(8) Runway end identifier lights are located on both ends of runway 13-31.

(9) Radar traffic advisories are available by contacting the Duluth International Approach Control.

(10) Very high frequency/direction finding equipment is available by contacting the flight service station.

(11) The remarks indicate that the runway 13 threshold is displaced 162 feet. Also, runway 4-22 is closed to wheeled aircraft during the winter months. This runway can be used in the winter months by airplanes which are equipped with skiis.

(12) The runways at Hibbing can support a 60,000-pound aircraft equipped with single-wheel type landing gear The other letters and numbers indicate maximum aircraft weights with twin-wheel type landing gear and twin-tandem type landing gear.

(13) The letters "BL" followed by the numbers "5" and "11" indicate the presence of a rotating beacon, medium-intensity runway lights, and runway end identifier lights.

(14) Aircraft storage, major airframe, and major powerplant repairs are available as indicated by the "S5."

(15) The fuel available at Hibbing is 80/87, 100/130 octane, and special fuel is available for turbine aircraft.

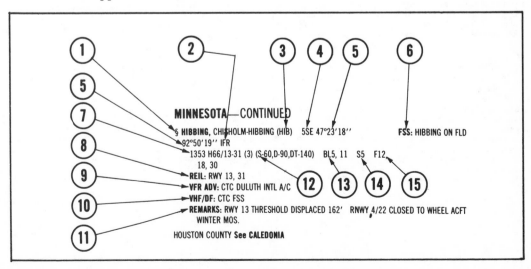

Fig. 7-15. Sample Excerpt From Airport Directory

Airport Directory

LEGEND

> The listing of an airport in this directory merely indicates the airport operator's willingness to accommodate transient aircraft, and does not represent that the facility conforms with any Federal or local standard, or that it has been approved for use on the part of the general public.

(1) LOCATION

Location Identifier—A three character code assigned to airports with a significant amount of activity (shown immediately after the airport name). These codes are used by ATC in lieu of the airport name in flight plans, flight strips and other written records.

The airport location is given in nautical miles (to the nearest mile), direction from center of referenced city, and geographical coordinates.

(2) ELEVATION

Elevation is given in feet above mean sea level and is based on highest usable portion of the landing area. When elevation is sea level, elevation will be indicated as "00." When elevation is below sea level, a minus sign (−) will precede the figure.

(3) RUNWAYS

Runway length and weight bearing capacity are listed for the longest instrument runway, or, if no instrument runway exists, for the longest runway. However, if an airport has hard surfaced and non-hard surfaced runways, the longest hard surfaced runway will be shown. Runway length is given to the nearest hundred feet using 70 feet as the division point, i.e., 1469 feet would be shown as "14"; 1470 feet would be shown as "15".

Runway lengths prefixed by the letter "H" indicates that the runway is hard surfaced (concrete or asphalt). If the runway length is not prefixed, the surface is turf, gravel, etc. The total number of runways available is shown in parenthesis. (However, only hard surfaced runways are counted at airfields with both hard surfaced and non-hard surfaced runways.) The full length and width of helipads are shown in the Heliport Directory, i.e., 50 x 50. For computer purposes, helipads are identified by the letter "H" and a number.

(4) RUNWAY WEIGHT BEARING CAPACITY

Runway strength data shown in this publication is derived from available information and is a realistic estimate of capability at an average level of activity. It is not intended as a maximum allowable weight or as an operating limitation. Many airport pavements are capable of supporting limited operations with gross weights of 25–50% in excess of the published figures. Permissible operating weights, insofar as runway strengths are concerned, are a matter of agreement between the owner and user. When desiring to operate into any airport at weights in excess of those published

in this publication, users should contact the airport management for permission.

Add 000 to figure following S, D, and DT for gross weight capacity, e.g., (S–10=10,000 lbs.)

S–Runway weight bearing capacity for aircraft with single-wheel type landing gear.

D–Runway weight bearing capacity for aircraft with twin-wheel type landing gear.

DT–Runway weight bearing capacity for aircraft with twin-tandem type landing gear.

Quadricycle and twin-tandem are considered virtually equal for runway weight bearing considerations, as are single-tandem and twin-wheel.

Omission of weight bearing capacity indicates information unknown. Footnote remarks are used to indicate a runway with a weight bearing greater than the longest runway.

(5) LIGHTING

B: Rotating Beacon. (Green and white, split-beam and other types.)

L: Field Lighting. An asterisk (*) may precede an element to indicate that it operates on prior request only (by phone call).

  4—Low Intensity Runway
  5—Medium Intensity Runway
  6—High Intensity Runway
  7—Instrument Approach (neon)
  7A—Medium Intensity Approach Lights (MALS)
  8—High Intensity Instrument Approach (ALS)
  10—Visual Approach Slope Indicator (VASI)
  11—Runway end identifier lights (threshold strobe) (REIL)
  12—Short approach light systems (SALS)
  13—Runway alignment lights (RAIL)
  14—Runway centerline
  15—Touchdown zone

Because the obstructions on virtually all lighted fields are lighted, obstruction lights have not been included in the codification.

(6) SERVICING

S2: Storage, minor airframe repairs.

S3: Storage, minor airframe and minor powerplant repairs.

S4: Storage, major airframe and minor powerplant repairs.

S5: Storage, major airframe and major powerplant repairs.

Fig. 7-16. (Page 1 of 3) Airport Directory Legend

AIRPORT DIRECTORY

FUEL (7)

Fuel Data Includes Each Grade Available

Code	Grade
F12	80/87
F15	91/98
F18	100/130
F22	115/145
F30	Kerosene, freeze point −40° F.
F34	Kerosene, freeze point −58° F.
F40	Wide-cut gasoline, freeze point −60° F.
F45	Wide-cut gasoline without icing inhibitor, freeze point −60° F.

OXYGEN (8)

Ox1	High pressure
Ox2	Low pressure
Ox3	High pressure—replacement bottles
Ox4	Low pressure—replacement bottles

OTHER (9)

§—Notam Service is provided. Applicable only to airports with established instrument procedures or high volume VFR activity.

AOE—Airport of Entry. (10)

LRA—Landing Rights Airport. (11)

VASI—Visual Approach Slope Indicator, applicable runway provided. (12)

IFR—Airport with approved FAA Standard Instrument Approach Procedure. (13)

RVV—Runway Visibility Value, applicable runway provided. (14)

RVR—Runway Visual Range, applicable runway provided.

TPA—Traffic Pattern Altitude—This information is provided for only those airports without a 24-hour operating control tower or FSS. Directions of turns are indicated only when turns of the pattern(s) are to the right (non-standard). TPA data are related to the runway listed under the tabulated airport information. Generally, only one altitude is listed; however, separate altitudes may be shown for aircraft of different performance or size. (16)

(17) FSS—The name of the associated FSS is shown in all instances. When the FSS is located on the named airport, "ON FLD" is shown following the FSS name. When the FSS can be called through the local telephone exchange, (Foreign Exchange) at the cost of a local call, it is indicated by "(LC)" (local call) with the phone number immediately below the FSS name. When an Interphone line exists between the field and the FSS, it is indicated by "(DL)" (direct line) below the FSS name.

(18) The availability of a VHF/DF at a FSS is indicated by the letters VHF/DF. For service, contact FSS on standard frequencies.

(19) AIRPORT REMARKS

(†)—Indicates that an air traffic control tower and/or an instrument landing system are associated with the airport. For specific details see the Airport/Facility Directory in Part 3 of the Airman's Information Manual.

"Fee" indicates landing charges for private or non-revenue producing aircraft. In addition, fees may be charged for planes that remain over a couple of hours and buy no services, or at major airline terminals for all aircraft.

"Rgt tfc rwy 13" indicates right turns should be made on landings and takeoffs on runway 13.

Obstructions.—Because of space limitations only the more dangerous obstructions are indicated. Natural obstructions, such as trees, clearly discernible for contact operations, are frequently omitted. On the other hand, all pole lines within at least 15:1 glide angle are indicated.

Remarks—data is confined to operational items affecting the status and usability of the airport, traffic patterns and departure procedures.

(15)

UNICOM (20)

A private aeronautical advisory communications facility operated for purposes other than air traffic control, transmits and receives on one of the following frequencies:

U1—122.8 MHz for Landing Areas (except heliports) without an ATC Tower or FSS;

U2—123.0 MHz for Landing Areas (except heliports) with an ATC Tower or FSS;

U3—123.05 MHz for heliports.

Fig. 7-16. (Page 2 of 3) Airport Directory Legend

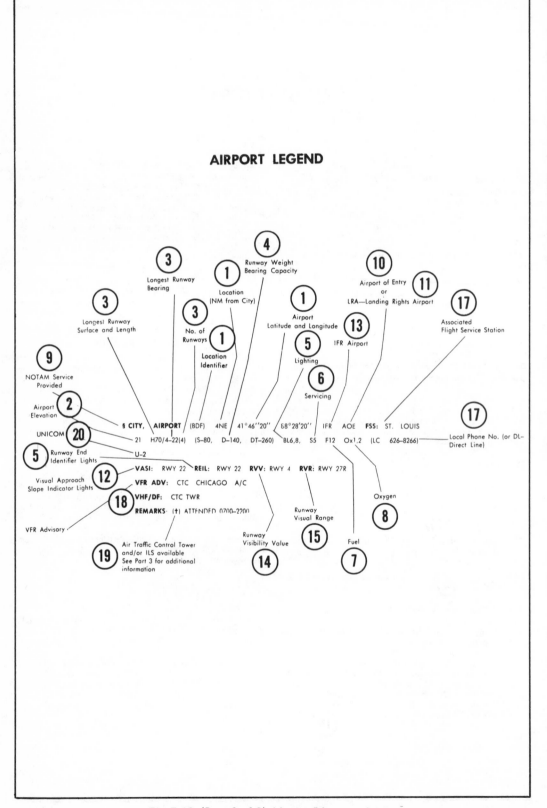

Fig. 7-16. (Page 3 of 3) Airport Directory Legend

OTHER SECTIONS OF PART 2

The Seaplane Base and Heliport Directory follow the Airport Directory and present information in a similar manner.

Customs, Immigration, and public health information as well as a list of airports of entry and landing rights airports are included in Part 2. (See Figs. 7-17 through 7-22.)

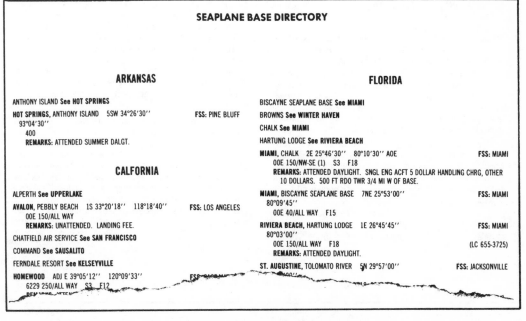

AIRPORT DIRECTORY

MONTANA—CONTINUED

BIG TIMBER (6S0) 3SW 45°48'15'' 109°58'45''   FSS: LIVINGSTON
4482 H46/6-24 (1) (S-12) BL5 F12, 18
REMARKS: UNATTENDED.

§ BILLINGS, LOGAN FIELD (BIL) 2NW 45°28'26''   FSS: BILLINGS ON FLD
108°32'13'' IFR
3606 H86/9-27 (3) (S-60,D-85,DT-135) BL5, 8, 11 S5 F12,
18, 30, 40 OX1,2,3,4
REIL: RWY 27RVV: RWY 9 **RVR:** RWY 9
VHF/DF: CTC FSS
REMARKS: (†) .

BOULDER 2S 46°12'42'' 112°06'18''   FSS: BUTTE
5068 36/11-29 (1)
REMARKS: UNATTENDED.

§ BOZEMAN, GALLATIN FLD (BZN) 8NW 45°46'46''   FSS: BOZEMAN ON FLD
111°09'24'' IFR
4461 H90/12-30 (2) (S-110,D-200,DT-370) BL5, 10 S5 F12,
18, 34 OX1,2,3,4 U-2
VASI: RWY 12, 30

BRADY, WOODS 23E 48°03'45'' 111°20'00''   FSS: GREAT FALLS
3308 20/7-25 (1) S5 F12
REMARKS: P-LINE IN RWY 25 APCH.

BRIDGER, MUNI (6S1) ADJ W 45°17'30'' 108°55'30''   FSS: BILLINGS
3720 H34/16-34 (1) BL4 F12, 18 U-1
REMARKS: ATTENDED ON REQ FONE 662-3382. RWY 16 THRESHOLD DISPLACED 100'
RWY 34 THRESHOLD DISPLACED 400'. P-LINE IN RWY 16 APCH. P-LINE IN RWY 34
APCH. RGT TFC RWY 34.

BRIGGS See **DELL**

BROADUS (BDX) 1S 45°26'00'' 105°25'00''
2992 H31/11-29 (1)
REMARKS

CHINOOK, MUNI (S71) 1W 48°35'15'' 109°14'30''   FSS: GREAT FALLS
2410 H39/8-25 (1) BL4 F12 U-1
REMARKS: ATTENDED 0700-1900. RWY 8 THRESHOLD DISPLACED 175'. RWY 26
THRESHOLD DISPLACED 380'. ANTENNA IN RWY 26 APCH.

CHINOOK, HEBBELMAN 10SE 48°31'00'' 109°04'00''   FSS: GREAT FALLS
2576 29/4-22 (2) S3 F12, 18

CHOTEAU 1NE 47°49'30'' 112°09'45''   FSS: GREAT FALLS
3945 H38/5-23 (1) (S-4) BL5 S5 F12, 18 U-1
REMARKS: ATTENDED MON-SAT 0800-1800.

CHRISTIANSEN FARM See **DAGMAR**

CIRCLE, MCCONE COUNTY 1E 47°25'00'' 105°34'00''   FSS: MILES CITY
2426 H30/12-30 (1) BL4 F12
REMARKS: UNATTENDED.

CLARK See **GRASS RANGE**

CLINTON, ELIOTT FIELD (25S) 3E 46°44'00'' 113°40'12''   FSS: MISSOULA
3540 38/7-25 (1)
REMARKS: UNATTENDED.

COLUMBUS (6S3) ADJ SE 45°37'45'' 109°15'00''   FSS: BILLINGS
3575 H26/9-27 (1) (S-4) *BL4 F12, 18 U-1
REMARKS: ATTENDED ON CALL FONE 322-4471. RWY 9 THRESHOLD DISPLACED 200'.

CONDON, USFS (S04) 1NE 47°32'30'' 113°42'30''   FSS: MISSOULA
3686 26/13-31 (1) U-1

CONNER, SHOOK MOUNTAIN 6SW 45°51'45'' 114°12'50''   FSS: MISSOULA
4247 45/NE-SW (1) F12, 18
REMARKS: CLSD WINTER MOS.

CONRAD (S01) 1W 48°10'15'' 111°58'15''   FSS: CUT BANK
H36/5-23 (1) (S-4) BL5 F12, 18 U-1
OCT-APR 0800-1800.

FSS: MILES CITY

Fig. 7-17. Airport Directory

SEAPLANE BASE DIRECTORY

ARKANSAS

ANTHONY ISLAND See **HOT SPRINGS**

HOT SPRINGS, ANTHONY ISLAND 5SW 34°26'30''   FSS: PINE BLUFF
93°04'30''
400
REMARKS: ATTENDED SUMMER DALGT.

CALFORNIA

ALPERTH See **UPPERLAKE**

AVALON, PEBBLY BEACH 1S 33°20'18'' 118°18'40''   FSS: LOS ANGELES
00E 150/ALL WAY
REMARKS: UNATTENDED. LANDING FEE.

CHATFIELD AIR SERVICE See **SAN FRANCISCO**

COMMAND See **SAUSALITO**

FERNDALE RESORT See **KELSEYVILLE**

HOMEWOOD ADJ E 39°05'12'' 120°09'33''   FSS
6229 250/ALL WAY S3 F12

FLORIDA

BISCAYNE SEAPLANE BASE See **MIAMI**

BROWNS See **WINTER HAVEN**

CHALK See **MIAMI**

HARTUNG LODGE See **RIVIERA BEACH**

MIAMI, CHALK 2E 25°46'30'' 80°10'30'' AOE   FSS: MIAMI
00E 150/NW-SE (1) S3 F18
REMARKS: ATTENDED DAYLIGHT. SNGL ENG ACFT 5 DOLLAR HANDLING CHRG, OTHER
10 DOLLARS. 500 FT RDO TWR 3/4 MI W OF BASE.

MIAMI, BISCAYNE SEAPLANE BASE 7NE 25°53'00''   FSS: MIAMI
80°09'45''
00E 40/ALL WAY F15

RIVIERA BEACH, HARTUNG LODGE 1E 26°45'45''   FSS: MIAMI
80°03'00''
00E 150/ALL WAY F18   (LC 655-3725)
REMARKS: ATTENDED DAYLIGHT.

ST. AUGUSTINE, TOLOMATO RIVER 5N 29°57'00''   FSS: JACKSONVILLE

Fig. 7-18. Seaplane Base Directory

HELIPORT DIRECTORY

ARIZONA

CAREFREE ADJ S 33°49'20'' 111°55'20'' **FSS: PHOENIX**
2300 H95X95
REMARKS: UNATTENDED. FUEL & UNICOM SVC AVBL AT CAREFREE ARPT 1 MI E.

FREEWAY See TUCSON

TUCSON, MEDICAL CENTER ADJ NE 32°15'04'' **FSS: TUCSON**
110°52'37''
2470 H97X75
REMARKS: UNATTENDED. P-LINE IN RWY H1 APCH.

TUCSON, FREEWAY 4NW 32°16'45'' 111°00'35'' **FSS: TUCSON**
2290 H40X40
REMARKS: APCH & DEPART TO NORTH.

WINDOW ROCK ADJ S 35°39'20'' 109°03'45'' **FSS: ZUNI**
6755 H100X100
REMARKS: ATTENDED 0800-1700.

CALIFORNIA

BAKERSFIELD, AIRPARK 3S 35°19'43'' 118°59'48'' **FSS: BAKERSFIELD**
378 H80X80 (S-12.5) (LC 399-1787)

BAKERSFIELD, MEADOWS FIELD 4NW 35°25'46'' **FSS: BAKERSFIELD**
119°03'05''
492 H150X150 (S-12.5)

§ **BERKELEY**, MUNI (JBK) 2W 37°52'02'' 122°18'11'' **FSS: OAKLAND**
14 H215X170 (S-12)
REMARKS: ATTENDED 0515-2300.

BISHOP 2E 37°22'24'' 118°21'54''

QUARTZ HILL 5W 34°39'03'' 118°12'18'' **FSS: PALMDALE**
2469 H50X50 (S-4)
REMARKS: HELIPAD LCTD BTWN RWYS 5 & 9 FRNT OF APRN.

RAMONA 2W 33°02'15'' 116°54'30'' **FSS: SAN DIEGO**
1393

SAN DIEGO, MONTGOMERY FLD 7N 32°48'57'' **FSS: SAN DIEGO**
117°08'23''
423 H100X132
REMARKS: UNATTENDED.

SAN DIEGO, BROWN FIELD MUNI 15SE 32°34'22'' **FSS: SAN DIEGO**
116°58'47''
510 H100X100

SAN DIEGO /SANTEE/, GILLESPIE FLD /HELIPAD/ (SEE) 12NE **FSS: SAN DIEGO**
32°49'33'' 116°58'19''
385

§ **SAUSALITO**, MARIN COUNTY (JMC) 2NW 37°52'45'' **FSS: OAKLAND**
122°30'45''
2 H100X100 (S-19) L5 F18
REMARKS: ATTENDED SUN-FRI 0530-2230 SAT 0630-2145.

TRI COUNTY COPTERS See MONTE RIO

VAN NUYS See LOS ANGELES/VAN NUYS/

COLORADO

ASPEN, VALLEY HOSPITAL 1N 39°11'48'' 106°49'02'' **FSS: DENVER**
7880 H100X80
REMARKS: ATTENDED. RED ROTG BCN. BCN & PAD FLOOD LGTS AVBL BY PHONE

FSS: DENVER ON FLD

Fig. 7-19. Heliport Directory

CUSTOMS, IMMIGRATION, AND NATURALIZATION, PUBLIC HEALTH, AND AGRICULTURE DEPARTMENT REQUIREMENTS

If not carrying passengers for hire, or cargo, a private aircraft departing from the United States on purely a business or pleasure flight does not require a U.S. Customs clearance of any type, although modified military-type privately owned aircraft are subject to certain restrictions under the regulations of the Office of Munitions Control of the Department of State even on a business or pleasure flight.

Any aircraft departing from the United States carrying passengers for hire or merchandise, or which will take on board or discharge passengers anywhere outside the U.S. is required to obtain clearance at the customs port of entry at or nearest the last place of take-off from the United States.

Any aircraft being exported from the United States must obtain U.S. Customs clearance and also meet appropriate Export Control requirements of the Department of Commerce for commercial aircraft, and the Department of State for privately owned military-type aircraft.

1. All aircraft entering the United States must land at designated airport of entry unless prior ap~~proval~~ ~~has~~ been obtained. In th~~...~~

should specify the following: (a) Type of aircraft; (b) Identification (NC number); (c) Name of pilot; (d) Place of last departure; (e) Airport of Entry; (f) Number of alien and citizen passengers, and estimated time of arrival. (Indicating whether C.S.T, E.S.T., etc.) Private aircraft arriving from Canada or Mexico may request that advance notice of arrival to customs officers be included in the *flight plan* filed in those countries if destined to an airport in the U.S. where flight notification service is available. At a landing rights airport such notices will then be treated as an application for permission to land.

Aircraft may use the following method of notifying customs when departing from a country or remote area where a predeparture flight plan cannot be filed or an "advise customs" message cannot be included in a predeparture flight plan: Call the nearest domestic or international FAA flight service station as soon as it is estimated that radio communications can be established and file a VFR (DVFR) flight plan and include as the last item "advise customs." The ~~station with which~~ such a flight plan is filed will ~~...~~ FAA station who will

Fig. 7-20. Customs and Public Health Information for Pilots

UNITED STATES INTERNATIONAL AIRPORTS
(Airports of Entry)

Any aircraft may land at one of the following International Airports and permission to land from U.S. Customs is not required. *At least one hour advance notice* of arrival must be furnished to U.S. Customs. This may be included in your *flight plan* filed in Canada or Mexico if destined to an airport where flight notification service is available.

Alaska

Juneau/Juneau Mun
Ketchikan/Ketchikan Port Area
*Wrangell/Wrangell SPB

Arizona

Douglas/Bisbee-Douglas Intl
Nogales/Nogales Intl
Tucson/Tucson Intl
Yuma/Yuma MCAS Intl

California

Calexico/Calexico Intl (Direct radio contact with Imperial FSS required for Customs notification.)
San Diego/San Diego Intl Lindbergh Fld

Florida

Fort Lauderdale/Ft. Lauderdale-Hollywood Intl
Key West/Key West Intl
Miami/Miami Intl
*Miami/Chalk SPB
Tampa/Tampa Intl
West Palm Beach/Palm Beach Intl

New York

Albany/Albany County (2 hours advance notice required)
Buffalo/Greater Buffalo Intl
†Massena/Richards Fld
Ogdensburg/Ogdensburg Mun
Rochester/Monroe County
*Rouses Point/Rouses Point SPB
†Watertown/Watertown Mun

North Dakota

Grand Forks/Grand Forks Intl
Minot/Minot Intl
*Pembina/Pembina Mun
*Portal/Portal Mun
Williston/Sloulin Fld Intl

Ohio

Akron/Mun
Cleveland/Cleveland Hopkins Intl Arpt
Put-in-Bay/Put-in

Fig. 7-21. Airports of Entry

UNITED STATES LANDING RIGHTS AIRPORTS

At the following airports an application for permission to land must be submitted in advance to U.S. Customs. *At least one hour advance notice* of arrival must also be furnished to U.S. Customs, unless otherwise noted. Advance notice of arrival may be included in your flight plan filed in Canada or Mexico if destined to an airport where flight notification service is available and this notice will be treated as an application for permission to land.

Alabama

Birmingham/Birmingham Mun (Not manned by customs personnel. Advance arrangements must be made for inspection by customs.)
Mobile/Bates Fld (Not manned by customs personnel. Advance arrangements must be made for inspection by customs.)
Montgomery/Dannelly Arpt (Not manned by customs personnel. Advance arrangements must be made for inspection by customs.)

Alaska

Anchorage/Anchorage Intl
†Annette/Annette Island Arpt
Cold Bay/Cold Bay Arpt
*Eagle/Eagle Mun
Fairbanks/Fairbanks Intl
*†Haines/Mun Arpt
*†Hyder/Hyder SPB (No communications. Report on arrival.)
Kodiak/Kodiak SPB

New Haven/Tweed-New Haven
Windsor Locks/Bradley Intl (Direct telephone notification prior to departure from Canada required to insure adequate prior notice.)

Delaware

Wilmington/Greater Wilmington (2 hours advance notice required.)

District of Columbia

Washington/Dulles Intl

Florida

Jacksonville/Jacksonville Intl
Panama City/Bay County Arpt
Pensacola/Pensacola Mun

Georgia

Atlanta/Atlanta Arpt (At least 2 hours advance notice

Fig. 7-22. Landing Rights Airports

TELEPHONE NUMBERS

Part 2 of the AIM concludes with a listing of flight service station and Weather Services telephone numbers. (See Fig. 7-23.) These telephone numbers are arranged alphabetically by cities within states. For example, the listing for Omaha, Nebraska, provides the following information:

(1) The designation "OMA" following Omaha is the three letter identifier.

(2) The name in parentheses indicates that the listed numbers connect to Eppley Field.

(3) "FSS" means the listed numbers are flight service station numbers.

(4) The area code, 402, is in parentheses.

(5) Following the first listed number, a black rectangle is noted. The legend at the beginning of this listing indicates that this number is a "one call" flight service station/Weather Service briefing service number.

(6) The star beneath the rectangle is explained as a "pilot's automatic telephone weather answering service" (PATWAS), and is connected to the transcribed weather broadcast.

(7) A diamond indicates that the telephone number is restricted to aviation weather information.

FSS—CS/T AND NATIONAL WEATHER SERVICE TELEPHONE NUMBERS

Flight Service Stations (FSS) and Combined Station/Tower (CS/T) provide information on airport conditions, radio aids and other facilities, and process flight plans. CS/T personnel are not certificated pilot weather briefers; however, they provide factual data from weather reports and forecasts. Airport Advisory Serviice is provided at the pilot's request on 123.6 by FSSs located at airports where there are not control towers in operation. (See Part 1 ARRIVALS.)

The telephone area code number is shown in parentheses. Each number given is the preferred telephone number to obtain flight weather information. Automatic answering devices are sometimes used on listed lines to given general local weather information during peak workloads. To avoid getting the recorded general weather announcement, use the selected telephone number listed.

★ Indicates Pilot's Automatic Telephone Weather Answering Service (PATWAS) or telephone connected to the Transcribed Weather Broadcast (TWEB) providing transcribed aviation weather information.

◆ Indicates a restricted number, use for aviation weather information

■ Call FSS for "one call" FSS/WSO briefing service.

✗ Automatic Aviation Weather Service (AAWS).

Location and Identifier		Area Code	Telephone
MISSISSIPPI (Con't.)			
McComb MCB (Pike County)	FSS	(601)	684-7070
Meridian MEI (Key Field)	FSS	(601)	482-1243
	WS	(601)	483-3725 ◆
MISSOURI			
Cape Girardeau CGI	FSS	(314)	ED 4-2803
Columbia COU	FSS	(314)	449-1136 ■
Joplin JLN	FSS	(417)	MA 3-6868

Location and Identifier		Area Code	Telephone
NEBRASKA			
Chadron CDR	FSS	(308)	HE 2-3153
Grand Island GRI	FSS	(308)	382-5196 ■
Imperial IML	FSS	(308)	TU 2-4887
Lincoln LNK (Muni/AFB)	FSS	(402)	477-3929 ■
Norfolk	WS	(402)	371-3386
			(0530–2130)
North Platte LBF (Lee Bird)	FSS	(308)	532-4034 ■
Omaha OMA (Eppley)	FSS	(402)	341-6178 ■
	FSS	(402)	342-3603 ★
Scottsbluff BFF	FSS	(308)	635-2615 ■
Sidney SNY	FSS	(308)	254-3130
Valentine	WS	(402)	376-8442

Fig. 7-23. Weather Service Telephone Numbers

SECTION C

PARTS 3 AND 3A OF THE AIM

Parts 3 and 3A of the *Airman's Information Manual* contain Operational Data and Notices to Airmen. (See Fig. 7-24.) Part 3 is issued every 56 days; Part 3A is published every 14 days between issues of Part 3. Notices to Airmen make up the entire content of Part 3A. All subjects contained in Part 3 are listed in the table of contents in alphabetical order. (See Fig. 7-25.)

AIRPORT/FACILITY DIRECTORY

In addition to the airport information listed in Part 2 of the AIM, Part 3 lists considerable supplemental information including all of the communications and navigation frequencies and facilities. Only those airports which have an operating control tower are listed in the Airport/Facility Directory of Part 3.

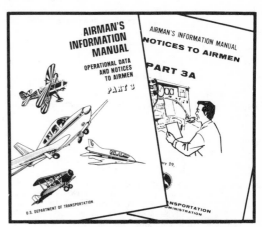

Fig. 7-24. Parts 3 and 3A of the AIM

As was the case with the Airport Directory of Part 2, a legend is provided at the beginning of the Airport/Facility Directory of Part 3. (See Fig. 7-26.) Cross-reference numbers have been added between the pages of figure 7-26 to aid in the understanding of the legend example.

TABLE OF CONTENTS

Part 3

OPERATIONAL DATA

> New or amended textual or tabulated information except in the Airport/Facility Directory, is indicated by a solid dot ● prefixing the heading, paragraph, or line.

Fig. 7-25. Part 3 Table of Contents

AIRPORT/FACILITY DIRECTORY

> The Airport Directory in this publication is limited to airports with control towers and/or instrument landing systems. See Part 2 for a complete listing of all public use airports.

 LOCATION

The airport location is given in nautical miles (to the nearest mile) and direction from center of referenced city.

 ELEVATION

Elevation is given in feet above mean sea level and is based on highest usable portion of the landing area. When elevation is sea level, elevation will be indicated as "00." When elevation is below sea level, a minus sign (−) will precede the figure.

 RUNWAYS

The runway surface length, and weight bearing capacity are listed for the longest instrument runway or sealane, or the longest active landing portion of the runway or strip, given to the nearest hundred feet, using 70 feet as the division point, i.e., 1469 feet would be shown as "14"; 1470 feet would be shown as "15". Runway lengths prefixed by the letter "H" indicates that runways are hard surfaced (concrete; asphalt; bitumen, or macadam with a seal coat). If the runway length is not prefixed, the surface is sod, clay, etc. The total number of runways available is shown in parenthesis. (However, only hard surfaced runways are counted at airfields with both hard surfaced and sod runways.)

 RUNWAY WEIGHT BEARING CAPACITY

Runway strength data shown in this publication is derived from available information and is a realistic estimate of capability at an average level of activity. It is not intended as a maximum allowable weight or as an operating limitation. Many airport pavements are capable of supporting limited operations with gross weights of 25–50% in excess of the published figures. Permissible operating weights, insofar as runway strengths are concerned, are a matter of agreement between the owner and user. When desiring to operate into any airport at weights in excess of those published in this publication, users should contact the airport management for permission.

Add 000 to figure following S, D, DT and MAX for gross weight capacity, e.g., (S–000).

S–Runway weight bearing capacity for aircraft with single-wheel type landing gear. (DC–3), etc.

D–Runway weight bearing capacity for aircraft with dual-wheel type landing gear. (DC–6), etc.

DT–Runway weight bearing capacity for aircraft with dual-tandem type landing gear. (707), etc.

Quadricycle and dual-tandem are considered virtually equal for runway weight bearing considerations, as are single-tandem and dual-wheel. Omission of weight bearing capacity indicates information unknown. Footnote remarks are used to indicate a runway with a weight bearing greater than the longest runway.

 LIGHTING

B: Rotating Beacon. Green and white, split-beam and other types.

L: Field Lighting. An asterisk (*) may precede an element to indicate that it operates on prior request only (by phone call).

    4—Low Intensity Runway
    5—Medium Intensity Runway
    6—High Intensity Runway
    7—Instrument Approach (neon)
    7A—Medium Intensity Approach Lights (MALS)
    8—High Intensity Instrument Approach (ALS)
    10—Visual Approach Slope Indicator (VASI)
    11—Runway end identifier lights (threshold strobe) (REIL)
    12—Short approach light systems (SALS)
    13—Runway alignment lights (RAIL)
    14—Runway centerline
    15—Touchdown zone

Because the obstructions on virtually all lighted fields are lighted, obstruction lights have not been included in the codification.

 SERVICING

S2: Minor airframe repairs.
S3: Minor airframe and minor powerplant repairs.
S4: Major airframe and minor powerplant repairs.
S5: Major airframe and major powerplant repairs.

 FUEL

(Fuel data includes each grade available.)

Code	Grade
F12	80/87
F15	91/96
F18	100/130
F22	115/145
F30	Kerosene, freeze point −40°F
F34	Kerosene, freeze point −58°F
F40	Wide-cut gasoline, freeze point −60°F
F45	Wide-cut gasoline without icing inhibitor, freeze point −60°F

 OXYGEN

Ox1	High Pressure
Ox2	Low Pressure
Ox3	High Pressure—Replacement Bottles
Ox4	Low Pressure—Replacement Bottles

Fig. 7-26. (Page 1 of 4) Airport/Facility Directory Legend

AIRPORT/FACILITY DIRECTORY

 OTHER

!—NOTAM Service is provided. Applicable only to airports with established instrument approach procedures, or high volume VFR activity.

AOE—Airport of Entry—A customs Airport of Entry where permission from U.S. Customs is not required, however, at least one hour advance notice of arrival must be furnished.

AVASI—Abbreviated Visual Approach Slope Indicator—2 boxes.

FSS—The name of the associated FSS is shown in all instances. When the FSS is located on the named airport, "on fld" is shown following the FSS name. When the FSS can be called through the local telephone exchange, (Foreign Exchange) at the cost of a local call, it is indicated by "(LC)" (local call) with the phone number immediately following the name of the FSS, i.e., "FSS: WICHITA (LC481–5867)." When an Interphone line exists between the field and the FSS, it is indicated by "(DL)" (direct line) immediately following the name of the FSS, i.e., "FSS: OTTO (DL)."

IFR—Airport with approved FAA Standard Instrument Approach Procedure.

LRA—Landing Rights Airport—Application for permission to land must be submitted in advance to U.S. Customs. At least one hour advance notice of arrival must also be furnished.

REIL—Runway end identifier lights (threshold strobe).

RVV—Runway Visibility Values, applicable runway provided.

RVR—Runway Visual Range, applicable runway provided.

TPA—Traffic Pattern Altitude—This information is provided for only those airports without a 24-hour operating control tower or FSS.

VASI—Visual Approach Slope Indicator, applicable runway provided.

  TCH—Threshold Crossing Height.

  RRP—Runway Reference Point.

 AIRPORT REMARKS

Aircraft Categories—*Category I*—Light-weight, single-engine, personal-type propeller driven aircraft, (Does not include higher performance single-engine aircraft such as the T–28.)

Category II—Light-weight, twin engine, propeller driven aircraft weighing 12,500 pounds or less such as the Aero Commander, Twin Beechcraft, DeHavilland Dove, Twin Cessna. (Does not include such aircraft as a Lodestar, Learstar, DC–3).

Category III—All other aircraft such as the higher performance single-engine, heavy twin-engine, four engine and turbojet aircraft.

"FEE" indicates landing charges for private or non-revenue producing aircraft. In addition, fees may be charged for planes that remain over a couple of hours and buy no services, or at major airline terminals for all aircraft.

"Rgt tfc 13–31" indicates right turns should be made on landings and takeoffs on runways 13 and 31.

Remarks data are confined to operational items affecting the status and usability of the airport, traffic patterns and departure procedures.

Obstructions.—Because of space limitations only the more dangerous obstructions are indicated. Natural obstructions, such as trees, clearly discernible for contact operations, are frequently omitted. On the other hand, all pole lines within at least 15:1 glide angle are indicated.

 FLIGHT SERVICE STATIONS

Flight Service Station (FSSs) and Combined Station/Tower (CS/Ts) are listed alphabetically by state in the Airport/Facility Directory. At certain locations the preflight briefing and flight plan processing responsibilities of the CS/T have been reassigned to an adjacent FSS. At these locations the adjacent FSS will be listed as the 'Associated FSS,' otherwise, the CS/T will be listed. Limited Remote Communications Outlet (LRCO) and Remote Communications Outlet (RCO), where available at the facility, are shown following the three letter identifier. If located at other than a facility site they are listed alphabetically.

FSSs and CS/Ts provide information on airport conditions, radio aids and other facilities, and process flight plans. Airport Advisory Service is provided at the pilot's request on 123.6 by FSSs located at non-tower airports or when the tower is not in operation. (See Part 1, ADVISORIES AT NON TOWER AIRPORTS.)

Aviation weather briefing service is provided by FSSs and CS/Ts; however, CS/T personnel are not certified weather briefers and therefore provide only factual data from weather reports and forecasts. Flight and weather briefing services are also available by calling the telephone numbers listed in the chapter entitled 'FSS-CS/T Information and Weather Service Office Telephone Numbers,' located in Part 2."

● **Civil communications frequencies used in the FSS air/ground system are now operated simplex on 122.0, 122.2, 122.3, 122.4, 122.6, 122.7, 123.6; emergency 121.5; plus receive-only on 122.05, 122.1, 122.15 and 123.6.**

a. 122.0 is assigned to selected FSSs as a weather channel for both general aviation and air carrier.

b. 122.2 is assigned to all FSSs as a common en route simplex service.

c. 123.6 is assigned as the airport advisory channel at non-tower FSS locations, however, it is still in commission at some FSSs collocated with towers to provide part-time Airport Advisory Service.

d. 122.1 is the primary receive-only frequency at VORs. 122.05, 122.15 and 123.6 are assigned at selected VORs meeting certain criteria.

e. Some FSSs are assigned 50KHz channels for simplex operation in the 122–123 MHz band (e.g. 122.35).

Pilots using the FSS A/G system should refer to this directory or appropriate charts to determine frequencies available at the FSS or remoted facility through which they wish to communicate.

Part time FSS hours of operation are shown in remarks under facility name.

 COMMUNICATIONS

Clearance is required prior to taxiing on a runway, taking off, or landing at a tower controlled airport.

When operating at an airport where the control tower is operated by the U.S. Government, two-way radio communication is required unless otherwise authorized by the tower. (When the tower is operated by someone other than the U.S. Government, two-way radio communication is required if the aircraft has the necessary equipment.

Frequencies transmit and receive unless specified as: T—Transmit only, R—Receive only, X—On request. Primary frequencies are listed first in each frequency grouping, i.e., **VHF, LF.** Emergency frequency 121.5 is available at all TOWER, APPROACH CONTROL and RADAR facilities, unless indicated as not available in remarks.

Fig. 7-26. (Page 2 of 4) Airport/Facility Directory Legend

AIRPORT/FACILITY DIRECTORY

 COMMUNICATIONS REMARKS

Remarks data are confined to operational items affecting the status and usability of navigational aids, such as: ILS component restrictions, part time hours of operation, frequency sectorization, VOT frequencies.

VOICE CALL

The voice call for contact with the air traffic control tower is listed at each airport assigned such a facility.

 SERVICES AVAILABLE

TOWER

Pre-Taxi Clearance Procedure
Clearance Delivery (CLRNC DEL).
Approach Control (App Con) Radar and Non-Radar.
Departure Control (Dep Con) Radar and Non-Radar.
VFR Advisory Service (VFR Adv) Service provided by Non Radar Approach Control.
Radar Advisory Service for VFR Acft (Stage I).
Radar Advisory and Sequencing Service for VFR Acft (Stage II).
Radar Sequencing and Separation Service for VFR Acft Terminal Area Control (TCA).
Radar vectoring and sequencing on a full time basis of all IFR and VFR acft, (Stage III—Terminal Radar Service Area–TRSA.)
Ground Control (GND CON).
VHF Direction Finding (VHF/DF).

RADIO NAVIGATION AIDS

Included in this section is a tabulation of all Air Navigation Radio Aids in the National Airspace System and those upon which the FAA has approved an instrument approach. Private or military Navigation Radio Aids not in the National Airspace System are not tabulated.

AUTOMATIC TERMINAL INFORMATION SERVICE (ATIS)

ATIS is continuous broadcast of recorded non-control information in selected areas of high activity. See Part 1.

FLIGHT SERVICE STATION (FSS)

Airport Advisory Service (AAS).
En Route Weather Advisory Service (EWAS).
Island, Mountain and Lake Reporting Service.
Remote Weather Radar Display (WR).
VHF Direction Finding (DF).

UNICOM

A private aeronautical advisory communications facility operated for purposes other than air traffic control, transmits and receives on one of the following frequencies:

U–1—122.8 MHz for Landing Areas (except heliports) without an ATC Tower or FSS;
U–2—123.0 MHz for Landing Areas (except heliports with an ATC Tower or FSS;
U–3—123.05 MHz for heliports with or without ATC Tower or FSS;
U–4—122.85 MHz for landing areas not open to the public;
U–5—122.95 MHz for landing areas not open to the public.

NOTE.—UNICOM used for communications must be licensed by the Federal Communication Commission in order to be listed in this publication.

 RADIO CLASS DESIGNATIONS

Identification of VOR/VORTAC/TACAN Stations by Class (Operational Limitations):

Normal Usable Altitudes and Radius Distances

Class	Altitudes	Distance (miles)
T	12,000′ and below	25
L	Below 18,000′	40
H	Below 18,000′	40
H	14,500′ — 17,999′	100*
H	18,000′ — FL 450	130
H	Above FL 450	100

*Applicable only within the contiguous 48 States.

(H) = High   (L) = Low   (T) = Terminal

NOTE: An H facility is capable of providing L and T service volume and an L facility additionally provides T service volume.

The term VOR is, operationally, a general term covering the VHF omnidirectional bearing type of facility without regard to the fact that the power, the frequency-protected service volume, the equipment configuration, and operational requirements may vary between facilities at different locations.

AB .... Automatic Weather Broadcast (also shown with ■ following frequency).

B .... Scheduled Broadcast Station (broadcasts weather at 15 minutes after the hour.

DME .... UHF standard (TACAN compatible) distance measuring equipment.

H .... Non-directional radio beacon (homing), power 50 watts to less than 2,000 watts.

HH .... Non-directional radio beacon (homing), power 2,000 watts or more.

H–SAB ..... Non-directional radio beacons providing automatic transcribed weather service.

ILS .... Instrument Landing System (voice, where available, on localizer channel).

LDA .... Localizer Directional Aid.

LMM .... Compass locator station when installed at middle marker site.

LOM .... Compass locator station when installed at outer marker site.

MH .... Non-directional radio beacon (homing) power less than 50 watts.

S .... Simultaneous range, homing signal and/or voice.

SABH ...... Non-directional radio beacon not authorized for IFR or ATC. Provides automatic weather broadcasts.

SDF .... Simplified Direction Facility.

TACAN ..... UHF navigational facility—omnidirectional course and distance information.

VOR .... VHF navigational facility—omnidirectional, course only.

VOR/DME .. Collocated VOR navigational facility and UHF standard distance measuring equipment.

VORTAC ... Collocated VOR and TACAN navigational facilities.

W .... Without voice on radio facility frequency.

Z .... VHF station location marker at a LF radio facility.

Fig. 7-26. (Page 3 of 4) Airport/Facility Directory Legend

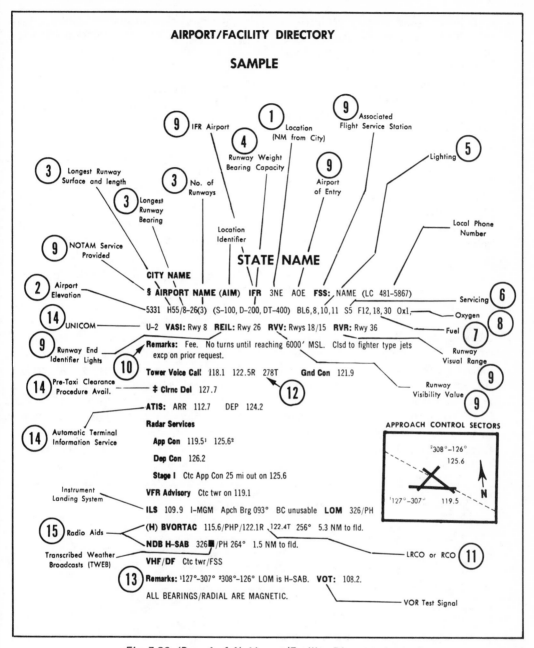

Fig. 7-26. (Page 4 of 4) Airport/Facility Directory Legend

An example of a typical listing in the Airport/Facility Directory is Kansas City Municipal Airport. (See Fig. 7-27.) It is interpreted as:

(1) Kansas City Municipal Airport has NOTAM service available and it has an approved instrument approach procedure. It is located four nautical miles north of the center of the city. The Kansas City Flight Service Station is located on the field.

(2) The field elevation is 758 feet. The longest instrument hard-surfaced runway, 18-36, is 7,000 feet long and there are two hard-surfaced runways. The runway weight bearing capacity is 100,000 pounds for aircraft with single-wheel type landing gear, 185,000 pounds for aircraft with twin-wheel type landing gear, and 350,000 pounds for aircraft with twin-tandem type landing gear. Lighting consists of: rotating beacon, low- and high-intensity runway lights, high-intensity instrument approach lights, sequence flashing lights, runway end identifier lights, and runway alignment lights.

(3) Service facilities on the field include: storage, major airframe and powerplane repairs, 100/130 octane and kerosene fuel, and high and low pressure oxygen service and replacement bottles.

(4) Runway visibility values are provided for runway 18.

(5) Overrun areas are at each end of runway 18-36. A tower, 1,163 feet high (2,049 feet MSL), is located three nautical miles south of the airport. Another tower, 1,042 feet high (2,049 feet MSL) is located 3.5 nautical miles south of the airport. A third tower, 1,023 feet high (1,946 feet MSL) is located 4.5 nautical

miles east southeast of the airport. Lead-in lights are provided for runway 36.

(6) The tower transmits and receives on 118.3, 126.6, and 121.1 MHz and the tower receives only on 122.7 MHz.

(7) Ground control transmits and receives on 121.9 MHz.

(8) Pre-taxi clearance procedure is available to receive IFR clearance.

(9) The field is equipped with radar beacon equipment. Approach control transmits and receives on 118.9 MHz in the east sector and 121.1 MHz in the west sector and transmits only on 112.6 and 109.9 which are local navaid frequencies. Departure control transmits and receives on 118.1 MHz. For "Stage 1" radar service (VFR traffic advisories), a pilot can contact approach control on 118.9 MHz in the east sector and 121.1 MHz in the west sector.

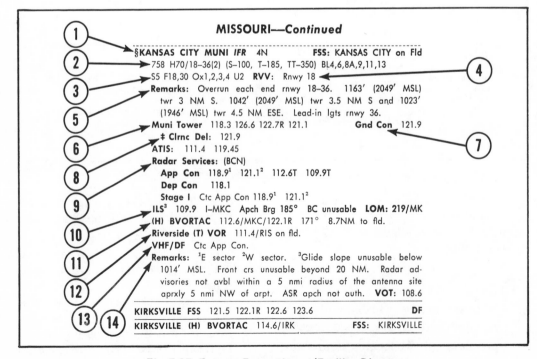

Fig. 7-27. Excerpt From Airport/Facility Directory

(10) The ILS glide slope is unusable below 1,014 feet MSL. The front course is unusable beyond 20 nautical miles. I-MKC approach bearing is 185°. Back course is unusable. Locator outer marker frequency is 219 kHz, and station identification letters are "MK."

(11) The Kansas City VORTAC station is a class "H" station with scheduled weather broadcasts. The VORTAC frequency is 112.6 MHz, and the call letters are "MKC." The VORTAC station receives only on 122.1 MHz. The airport is located on the 171° radial 8.7 nautical miles from the station.

(12) Riverside VOR, which is a terminal (T) class station, is located on the field and operates on a frequency of 111.4 MHz with station call letters of "RIS."

(13) For VHF direction finding service, a pilot can contact approach control.

(14) Radar advisories are not available within a five nautical mile radius of the antenna site which is located approximately five nautical miles northwest of the airport. Airport surveillance radar approach is not authorized. VOR test signal is available on 108.6 MHz.

RADIO NAVIGATION FACILITIES

The Airport/Facility Directory section of Part 3 also lists the radio navigation and flight service station facilities for the 48 conterminous states. For example, the Priest VOR transmitting frequency is 110.0 MHz. (See Fig. 7-28, item 1.) The FSS that handles communications through the Priest VOR is the Paso Robles Flight Service Station. The numbers, "122.1R," indicate that the flight service station can receive a pilot's transmissions on 122.1 MHz.

Another example is the Red Bluff Flight Service Station. (See Fig. 7-28, item 2.) It can transmit and receive on the emergency frequency of 121.5 in addition to 122.3, 122.6, and 123.6 MHz. Direction finding assistance is available through the Red Bluff Flight Service Station. The letter "B," which is the prefix to the word VORTAC, indicates the Red Bluff Flight Service Station broadcasts the scheduled weather at 15 minutes past the hour.

The Red Bluff VORTAC station transmits on 115.7 MHz which is the navigation frequency. The station identifier for the Red Bluff VORTAC is "RBL." All transmitting and receiving done through facilities located at the Red Bluff VORTAC are handled by the Red Bluff Flight Service Station.

A black square following a frequency in this section indicates that automatic

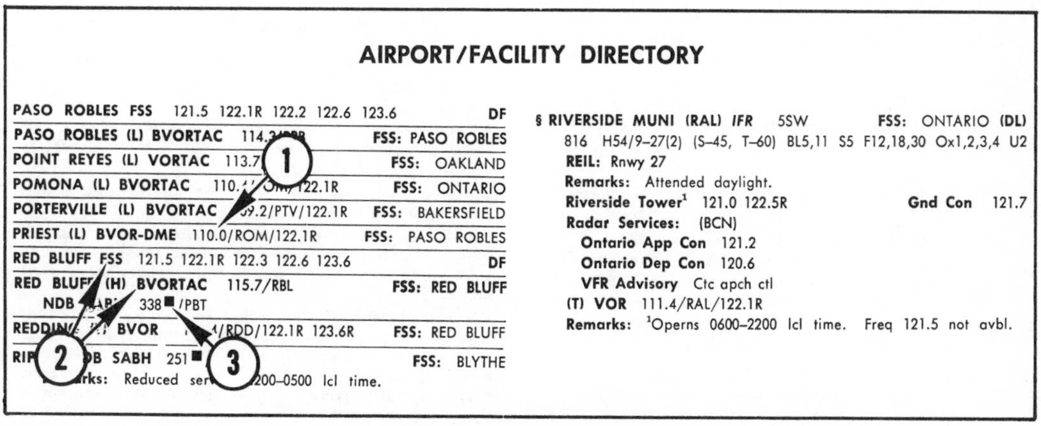

AIRPORT/FACILITY DIRECTORY

PASO ROBLES FSS	121.5 122.1R 122.2 122.6 123.6	DF
PASO ROBLES (L) BVORTAC	114.3	FSS: PASO ROBLES
POINT REYES (L) VORTAC	113.7	FSS: OAKLAND
POMONA (L) BVORTAC	110.4/OM/122.1R	FSS: ONTARIO
PORTERVILLE (L) BVORTAC	109.2/PTV/122.1R	FSS: BAKERSFIELD
PRIEST (L) BVOR-DME	110.0/ROM/122.1R	FSS: PASO ROBLES
RED BLUFF FSS	121.5 122.1R 122.3 122.6 123.6	DF
RED BLUFF (H) BVORTAC	115.7/RBL	FSS: RED BLUFF
NDB	338■/PBT	
REDDING BVOR	4/RDD/122.1R 123.6R	FSS: RED BLUFF
RIPON DB SABH	251■	FSS: BLYTHE
Remarks: Reduced ser 200–0500 lcl time.		

§ RIVERSIDE MUNI (RAL) IFR 5SW   FSS: ONTARIO (DL)
816 H54/9–27(2) (S–45, T–60) BL5,11 S5 F12,18,30 Ox1,2,3,4 U2
REIL: Rnwy 27
Remarks: Attended daylight.
Riverside Tower[1] 121.0 122.5R   Gnd Con 121.7
Radar Services: (BCN)
  Ontario App Con 121.2
  Ontario Dep Con 120.6
  VFR Advisory Ctc apch ctl
(T) VOR 111.4/RAL/122.1R
Remarks: [1]Operns 0600–2200 lcl time. Freq 121.5 not avbl.

Fig. 7-28. Radio Navigation Facilities

weather broadcasts, or "TWEBs," are transmitted through that facility. (See Fig. 7-28, item 3.) The initials "NDB" stand for the Red Bluff low-frequency non-directional radio beacon which transmits on the frequency of 338 kiloHertz.

SPECIAL NOTICES

The Special Notices section of AIM Part 3 contains notices of general interest to pilots across the country, such as Civil

Use of Military Fields. (See Fig. 7-29.) A solid dot is used to prefix new or revised data in this section.

Most control zones are in operation continuously. However, where traffic conditions are light, a control zone may be in operation for a limited time period. When a note is placed on the sectional chart, the pilot should refer to the Special Notices section of AIM Part 3 to determine the hours of operation. (See Fig. 7-30.)

SPECIAL NOTICES

Special Notices of a general nature or universal application, and other than for a specific geographical location, are grouped together under General Notices. Special Notices pertinent to a specific geographic area are grouped together under Area Notices by state, then city, airport, or location within the state. The month and year the notice is initially inserted into the manual is provided at the conclusion of each Special Notice. A solid dot ● prefixes new or revised Special Notices.

GENERAL NOTICES

FLIGHT INFORMATION PUBLICATION POLICY

The following is, in essence, the statement issued by the FAA Administrator and published in the December 10, 1964, issue of the Federal Register, concerning the FAA policy as pertaining to the type of information that will be published as NOTAMs and in the Airman's Information Manual.

It is a pilot's inherent responsibility that he be alert at all times for and in anticipation of all circumstances, situations and conditions which affect the safe operation of his aircraft. For example, a pilot should expect to find air traffic at any time or place. At or near both civil and military airports and in the vicinity of known training areas, a pilot should expect concentrated air traffic although he should realize concentrations of air traffic are not limited to these places.

It is the general practice of the agency to advertise by NOTAM or other flight information publications such information it may deem appropriate; information which the agency may from time to time make available to pilots is solely for the purpose of assisting them in executing their regulatory responsibilities. Such information serves the aviation community as a whole and not pilots individually.

The fact that the agency under one particular situation or another may or may not furnish information does not serve as a precedent of the agency's responsibility to the aviation community; neither does it give assurance that other information of the same or similar nature will be advertised nor does it guarantee that any and all information known to the agency will be advertised.

Consistent with the foregoing, it shall be the policy of the Federal Aviation Administration to furnish information only when, in the opinion of the agency, a unique situation should be advertised and not to furnish routine information such as concentrations of air traffic, either civil or military. The Airman's Information Manual will not contain informative items concerning everyday circumstances that pilots should, either by good practice or regulation, expect to encounter or avoid.

July 19

CIVIL USE OF MILITARY FIELDS:

U.S. Army, Air Force, Navy and Coast Guard Fields are open to civil fliers only in emergency or with prior permission.

Army Installations, prior permission authorized by the Commanding Officer of the installation.

For Air Force installations, prior permission should be requested at least 30 days prior to first intended landing from either Headquarters USAF (PRPOC) or the Commander of the installation concerned (who has authority to approve landing rights for certain categories of civil aircraft). For use of more than one Air Force installation, requests should be forwarded direct to Hq USAF (PRPOC), Washington, D.C. 20330. Use of USAF installations must be specifically justified.

For Navy and Marine Corps installations, prior permission should be requested at least 30 days prior to first intended landing. An Aviation Facility License must be approved and *executed* by the Navy prior to any landing by civil aircraft. Applications must include the following:

a. Application for Aviation Facility License (OPNAV Form 3770/1 (Rev. 7-70)) in quadruplicate.

b. Certificate of Insurance (NAVFAC Form 7-11011/36 (7-70)) in duplicate, signed by an official of the insurance company.

Forms may be obtained from the nearest U.S. Navy or Marine Corps aviation activity.

With minor exceptions, authority to use Navy and Marine Corps fields is granted only to aircraft on government business, or when no suitable civil airport is available in the vicinity.

Applications should be forwarded to the appropriate one of the following:

a. Use of one airfield, only: to the Commanding Officer of the field concerned (who has the authority to approve landing rights for certain categories of civil aircraft).

b. Use of Naval Station Adak, Alaska; Naval Air Station, Agana, Guam; and/or Naval Station Midway for trans-Pacific ferry operations making refueling and crew rest stops ("technical stops") wherein crew members, only, (no passengers) are embarked: to Commander

Fig. 7-29. General Notices.

SPECIAL NOTICES—GENERAL

in Chief, U.S. Pacific Fleet, Fleet Post Office, San Francisco 96610, with an additional copy of the application via air mail to the Commanding Officer of each of the foregoing facilities at which a technical stop is desired.

c. All others: to the Chief of Naval Operations (OP–513), Washington, D.C. 20350.

For Coast Guard fields prior permission should be requested from the Commandant, U.S. Coast Guard via the Commanding Officer of the field.

Use of Coast Guard fields is limited to persons on government business only when there is no suitable civil airport in the vicinity.

When instrument approaches are conducted by civil aircraft at military airports, they shall be conducted in accordance with the procedures and minimums approved by the military agency having jurisdiction over the airport. ——————————— **December 1964**

FEDERAL AVIATION REGULATION 91.103

The provisions of FAR 91.103 will apply as follows:

Air traffic clearances to aircraft of Cuban registry not engaged in scheduled International Air Service in U.S. airspace will require that the flight plan be filed with appropriate authorities at least five days prior to the proposed departure time. Route changes while en route will normally not be authorized. The procedures set forth herein do not apply at this time to overflights by aircraft of Cuban registry engaged in scheduled International Air Service.

——————————— **August 1967**

PART–TIME CONTROLLED AIRSPACE DESIGNATION SUBJECT TO IRREGULAR EFFECTIVE HOURS/DATES

Listed below are those control zones and transition areas whose hours are irregular and subject to change by Notam (as provided for in FAR 71) Note: These are not to be confused with other part-time controlled airspace with fixed hours, as specified in FAR 71, and which can only be changed by subsequent rule making action. All times are local and seven days a week unless otherwise noted.

The changes to Standard Time will affect the hours of operation of most Part Time Control Zones. Check NOTAMS for latest changes.

Control Zones

ALABAMA
Troy: 0600–1800 Mon-Fri except holidays.

ARIZONA
Fort Huachuca, Libby AAF: Continuous.
Phoenix-Litchfield: 0600–2300.

ARKANSAS
Jonesboro: 0600–2200.

CALIFORNIA
Anaheim (Disneyland Hel:port): Not in effect.
Camp Pendleton MCALF: 0800–1800 Mon-Fri.
Chico: 0600–2230.
Crescent City: 0600–2200.
El Monte: 0700–2300.
Fort Ord: 0600–2100 Mon-Fri., 0600–1700 Sat, Sun
Fullerton: 0600–2200.
Lake Tahoe: 0700–2300.
Los Angeles (Hawthorne Muni): 0700–2300.
LaVerne-Brackett Fld: 0700–2300.
Modesto: 0600–2200.

Oxnard (Ventura Co): 0600–2200.
Palmdale: Continuously.
Palm Springs: 0600–2200.
Palo Alto: 0700–2300.
Redding: 0530–2200.
Riverside (Muni): 0600–2200.
San Carlos: 0700–2300.
San Diego (Brown Fld Muni): 0700–2300.
San Diego (Gillespie Fld): 0700–2300.
San Diego (Montgomery Fld): 0700–2300.
San Jose (Reid Hillview): 0700–2300.
Santa Ana (Orange Co): 0600–2200
Santa Ana (MCAS El Toro): 0600–2300.
Santa Ana (MCAS, Santa Ana): 0600–2300.
Santa Monica: 0700–2300.
Santa Rosa: 0700–2300.
Torrance: 0600–2200.
Visalia: 0700–1700.

COLORADO
Alamosa: 0700–2000.
Aspen: 0700 til 30 min after official SS.
Broomfield: 0600–2200.
Cortez: 1000–2100 daily, except holidays.
Durango: 0800–2000 daily, except holidays.
Fort Carson (Butts AAF): 0600–2200 daily.
Greenwood Village (Arapahoe Co. Arpt): Not in effect.
Montrose: 1300–1900 daily, except holidays.

CONNECTICUT
Groton: 0530–2200 Mon-Sat, 0900–2200 Sun.
New Haven: 0600–2400.

FLORIDA
Fort Lauderdale (Executive): 0700–2300.
Hollywood: 0600–2200.
Miami (Dade Collier Training and Transition): Continuously.
Miami (New Tamiami) Continuous.
Miami (Opa Locka): 0700–2300.
Milton NAS Whiting Field (North): 0500–2300 Mon-Fri; 0800–1100 Sat; 1500–2100 Sun.
Panama City: 0600–2300.
Pensacola NAS (Saufley Field): 0600–2200.
St. Petersburg (Albert Whitted): 0600–2200.
Sarasota: 0600–2400.
Titusville: 0800–2300.

GEORGIA
Moultrie: 0700–2245.
Valdosta (Moody AFB): 0700–2300 Mon thru Thurs; 0700–2130 Fri; 1000–1800 Sat; 1200–1800 Sun. Not designated on holidays.

IDAHO
Lewiston: 0500–1930.
Twin Falls: 0630–2200.

ILLINOIS
Alton: 0700–2300.
Bloomington: 0615–2030 Mon-Fri; 0800–1800 Sat; 0900–2030 Sun.
Carbondale: 0700–1900.
Galesburg: 0530–2200.
Marion (Williamson County): 0600–2200.
Mattoon: 0615–2030 Mon-Fri. 0830–1630 Sat. 0830–2030 Sun.
Mt. Vernon: 0615–2100 Mon-Fri; 0830–1630 Sat; 0830–2100 Sun.
Rantoul: Not in effect.

●INDIANA
Bloomington: 0600–2100 Mon-Fri; 0600–1830 Sat; 1230–2100 Sun.
Marion: Not in effect.
Muncie: 0700–2300.

Fig. 7-30. Part-Time Controlled Airspace

NEW AND ABANDONED/CLOSED TO PUBLIC USE AIRPORTS

When an airport is newly opened or when it becomes closed or abandoned, it is listed in AIM Part 3 under a section entitled "New And Abandoned/Closed To Public Use Airports." It will be listed in this section until the next edition of AIM Part 2 Airport Directory.

New airports are listed as shown in figure 7-31. The name of the nearest city, the name of the airport, its distance from the city, and the associated Flight Service Station are tabulated. For example, in Arizona, Montezuma Airport has been added. It is located three miles north of Camp Verde, Arizona and its associated Flight Service Station is Prescott.

Likewise, abandoned or closed airports are listed as shown in figure 7-32. For example, in Arizona, there are two airports that have been abandoned since the last issue of AIM Part 2 Airport Directory.

```
              NEW AND ABANDONED/CLOSED TO PUBLIC USE AIRPORTS

                    (INCLUDING HELIPORTS AND SEAPLANE BASES)

                          *** NEW AIRPORTS ***

    THE FOLLOWING AIRPORTS WILL BE ADDED TO THE NEXT EDITION OF THE AIM PART 2 AIRPORT DIRECTORY:

       CITY, AIRPORT NAME              ASSOCIATED        CITY, AIRPORT NAME              ASSOCIATED
    DISTANCE & DIRECTION FROM CITY        FSS          DISTANCE & DIRECTION FROM CITY        FSS

              ALABAMA                                            KENTUCKY

FORT DEPOSIT, FORT DEPOSIT-LOWNDES COUNTY,           COLUMBIA, LUDOT, 2 SW--------------BOWLING GREEN
   1 SW-----------------------------MONTGOMERY       DAWSON SPRINGS, TRADEWATER, 2 NE-------BOWLING GREEN
                                                     RUSSELLVILLE, ABBOTT, 9 SW------------BOWLING GREEN
              ARIZONA
                                                              LOUISIANA
.CAMP VERDE, MONTEZUMA, 3 N-----------PRESCOTT
                                                     BATON ROUGE, OUR LADY OF THE LAKE HOSPITAL
              ARKANSAS                                  HELIPORT,-------------------------NEW ORLEANS
                                                     HOMER, HOMER MUNI, 3 E-----------------MONROE
MOUNTAIN HOME, LAHAR FIELD, 2 SW---------HARRISON    THIBODAUX, THIBODAUX MUNI, 3 S---------NEW ORLEANS

              CALIFORNIA                                        MAINE

ALTURAS, CALIFORNIA PINES, 9 SW---------KLAMATH FALLS .LINCOLN, LINCOLN REGIONAL, 2 SW--------UNKNOWN
BAKERSFIELD, FASEIMCO INC, 7 NW---------BAKERSFIELD  .SEBOOMOOK, DEPOT CAMP, 35 SE----------HOULTON
LOYALTON, LOYALTON, 2 N-------------------RENO
SAN MARTIN, SOUTH COUNTY ARPT OF SANTA CLARA                   MARYLAND
   COUNTY, 1 E-------------------------SALINAS
                                                     GERMANTOWN, AEC .HELIPORT, 1 NE--------WASHINGTON
              COLORADO
                                                              MICHIGAN
.DENVER, CAPRI HELIPORT, 3 N------------DENVER
                                                     GOBLES, WIDNERS FIELD, 14 W-----------SOUTH BEND
              DELAWARE                                MECOSTA, CANADIAN LAKES DIV CO, 5 SW-----SAGINAW
                                                     OVID, KOSHT, 6 NE---------------------LANSING
.DOVER, CAESAR RODNEY HELISTOP HELIPORT,------MILLVILLE SHERIDAN, CLARK, 3 W------------------SAGINAW

              GEORGIA                                          MINNESOTA

FAYETTEVILLE, MCCOMBS FLD, 4 S----------ATLANTA      .MOTLEY, MOREY FISH CO, ADJ S----------ALEXANDRIA
.JACKSON, FLYING H RANCH INC, 10 SW----------ATLANTA WALKER, SPORTS CRAFT SKY HARBOUR SEAPLANE BASE,
                                                        1 N-------------------------------HIBBING
              IDAHO                                   .WHEATON, TRAVERSE AIR, 14 NE---------ALEXANDRIA

CHALLIS, UPPER LOON CREEK, 30 NW----------IDAHO FALLS           MISSISSIPPI
HOME, HOWE, 5 N-------------------------IDAHO FALLS
MUD LAKE, MUD LAKE, 1 W-----------------IDAHO FALLS  .BRUCE, CALHOUN COUNTY, 4 S------------GREENWOOD
                                                     OLIVE BRANCH, OLIVE BRANCH, 3 NE-------MEMPHIS
              ILLINOIS
                                                              MONTANA
ALLERTON, VILLAGE OF ALLERTON, 1 SE--------TERRE HAUTE
BELVIDERE, BELVIDERE LTD, 2 N-----------CHICAGO      .CONRAD, CONRAD, 1 W-------------------CUT BANK
CENTRALIA, MATHIS, 4 SW-----------------ST. LOUIS
COLUMBIA, QUINT/KING HELIPORT, 3 SE----------ST. LOUIS         NEVADA
HINCKLEY, HINCKLEY, 3 W-----------------CHICAGO
NEWTON, JASPER COUNTY FLYING CLUB, 3 S--------VANDALIA CARSON CITY, PARKER GYROPORT, 4 E-------RENO
.PLAINFIELD, CLOW INTL, 7 NE------------CHICAGO
YATES CITY, TRI COUNTY FLYING CLUB ARPT, 3 W---QUINCY          NEW JERSEY

              INDIANA                                WALPACK, WALPACK LANDING FIELD, 4 SW--------TETERBORO

CROWN POINT, KLINEDORF, 2 NW-----------CHICAGO                 NEW YORK
ELKHART, MIDWAY, 3 SE-------------------SOUTH BEND
.INWOOD, TRI STATE HELIPORT, 1 N--------SOUTH BEND   .CAPE VINCENT, KENT BAY SEAPLANE BASE, 3 S----WATERTOWN
                                                     .COMSTANTIA, SHERWOOD AIRPARK,----------UTICA
              IOWA                                    .NEW YORK, EAST 34TH STREET HELIPORT, 5 NE----TETERBORO
                                                     .WESTPORT, WESTPORT, ADJ NE------------MONTPELIER
MONTEZUMA, MONTEZUMA EAST, 4 E---------OTTUMWA
MONTEZUMA, SIG FIELD, 2 S--------------OTTUMWA                 NORTH DAKOTA

              KANSAS                                  ALEXANDER, TAYLOR FIELD, 5 NW---------DICKINSON
                                                     CHAFFEE, POULSON, 3 E-----------------JAMESTOWN
FRANKLIN, GALICHIA, 2 S----------------JOPLIN        GARDNER, WOITZEL FIELD, 5 NW----------JAMESTOWN
HALSTEAD, HALSTEAD, 1 N----------------WICHITA       GARDNER, SWENSON, 6 NE----------------JAMESTOWN
.LENEXA, SHAWNEE MISSION, ADJ S--------KANSAS CITY   GLENBURN, GLENBURN MARPT, 1 SW--------MINOT
.SALINA, SALINA MUNICIPAL HELIPORT, 5 SW-------SALINA WYNOMERE, SANDEN, 6 NE----------------JAMESTOWN
```

Fig. 7-31. New Airports

NEW AND ABANDONED/CLOSED TO PUBLIC USE AIRPORTS

OHIO

DELPHOS, DELPHOS, 2 SE------------------------FINDLAY
FOSTORIA, FOSTORIA METROPOLITAN, 2 NE---------FINDLAY
.GREENVILLE, HOBBS FLD, 3 SE------------------DAYTON
LONG BOTTOM, BELLVILLE DAM LANDING STRIP,
 2 E------------------------------------PARKERSBURG
WILMINGTON, WILMINGTON INDUSTRIAL AIRPARK, 2 SE-DAYTON

OKLAHOMA

MOORELAND, MOORELAND MUNI, 3 N-----------------GAGE

PENNSYLVANIA

.HADLEY, LEAP,-----------------------------ALTOONA
MARS, LAKEHILL, 1 NW-----------------------PITTSBURGH
PITTSFIELD, BROKENSTRAW, 1 E----------------BRADFORD
.PORT ALLEGANY, JOHNSONS PRIVATE, 2 S--------BRADFORD
SEVEN SPRINGS BOROUGH, SEVEN SPRINGS, 2 SW--PITTSBURGH

SOUTH CAROLINA

LITTLE RIVER, CYPRESS BAY, 1 W--------------UNKNOWN

TENNESSEE

JOHNSON CITY, JOHNSON CITY, 4 NE-------------TRI CITY
MANCHESTER, SPENCER FLD, 9 SE----------------NASHVILLE

TEXAS

ABILENE, ZIMMERLE, 16 SE--------------------ABILENE
CORPUS CHRISTI, MEMORIAL MEDICAL CENTER HELIPORT,
 ADJ N----------------------------------ALICE
DELL CITY, DELL CITY MUNI, ADJ N------------EL PASO
FOLLETT, FOLLETT/LIPSCOMB COUNTY, 1 E--------GAGE
.GROVETON, GROVETON-TRINITY COUNTY, 3 NW-----LUFKIN
HOUSTON, SKYHAVEN, 14 E---------------------HOUSTON
.JUSTIN, SAGEBRUSH, 7 NW--------------------FORT WORTH
.PANHANDLE, STAMPS FLD, ADJ SW--------------AMARILLO
PORT OCONNOR, PORT OCONNOR, 1 W-------------PALACIOS
POWDERLY, POWDERLY, 1 S---------------------DALLAS
SMITHVILLE, SMITHVILLE, 1 N-----------------AUSTIN
SPICEWOOD, WINDEMERE, 2 E-------------------AUSTIN
WICHITA FALLS, TOM DANAHER, 6 SW-----WICHITA FALLS

VIRGINIA

.AMELIA, EASTER FIELD, 5 SW-----------------RICHMOND
.SAXE, WALTHALL,---------------------------DANVILLE

WEST VIRGINIA

NEW MARTINSVILLE, CIVIL AIR PATROL, 3 N----PARKERSBURG

WISCONSIN

.COTTAGE GROVE, FLYING HOOF, 3 NE-----------LONE ROCK
.TOMAHAWK, TOMAHAWK REGIONAL, 3 W-----------WAUSAU

*** ABANDONED OR CLOSED TO PUBLIC AIRPORTS ***

THE FOLLOWING AIRPORTS HAVE BEEN ABANDONED (A) OR CLOSED TO PUBLIC USE (C) AND WILL BE DELETED FROM THE NEXT EDITION OF THE AIM PART 2 AIRPORT DIRECTORY:

ARIZONA

BONITA, EUREKA RANCH, 12 NW (A)
PARKER CANYON, PARKER CANYON LAKE, 2 S (A)

ARKANSAS

WRIGHT, ARCHER, ADJ NE (C)

CALIFORNIA

CALIENTE, W-BAR-B RANCH, 11 NE (C)
.CAMBRIA, RANCHO SAN SIMEON, 3 NW (C)
MAMMOTH LAKES, ARCULARIUS RANCH, 8 NE (A)
PAYNES CREEK, PONDEROSA SKY RANCH, 7 E (C)
SANTA MARIA, NORTHSIDE AIRPARK, 3 NW (C)
WINDSOR, FLYING K RANCH, 1 N (C)

COLORADO

.ELIZABETH, PINE VIEW, 1 S (C)

CONNECTICUT

EAST HAMPTON, LINDQUIST, 2 NW (A)

GEORGIA

ASHBURN, OLSENS AIR PARK, 6 NE (C)
.TIFTON, EAGLEHEAD, 7 SW (C)

ILLINOIS

.EAST ST LOUIS, SPRING LAKE FARM, 4 E (C)
EAST ST LOUIS/COLLINSVILLE, ST LOUIS DOWNTOWN AIRPARK
 4 W (A)
.IRVING, MCNEELY, 2 NE (C)
.LEXINGTON, CORN BELT HELIPORT, 2 NW (A)
MARSHALL, LINCOLN TRAILS STATE PARK, 3 SW (C)
MOKENA, RICHARDS HELIPORT, 1 S (C)
NASHVILLE, DIVISION OF PARKS, 5 SE (C)
.NORTHBROOK, SKY HARBOR, 2 NW (C)
PLYMOUTH, GILES GRISWOLD, 1 SE (C)
SAN JOSE, REED, 3 SW (C)
STRASBURG, WITTENBURG, ADJ E (C)
WEST BROOKLYN, MARVIN BERNARDIN, 4 NW (C)

INDIANA

RICHMOND, EAST RICHMOND, 4 E (A)
SPRINGPORT, CRANDALL FIELD, 1 E (A)

IOWA

.ANITA, DRESSLER, 7 SE (A)
.DE WITT, GREEN MEADOWS, 1 W (A)
MANSON, ANDERSON PVT, 1 N (C)
RUDD, EHLEBRACHT FLD, 1 S (C)

KANSAS

.EDNA, EDNA MUNI, 5 N (C)
.LAKIN, DIENST RANCH STRIP, 13 N (A)
.MANHATTAN, ROESENER, 8 E (A)
NORCATUR, LAWSON FLD, 9 N (A)

LOUISIANA

.AMITE, HAYDEN, 6 E (C)

MAINE

BELGRADE, HUTCHINSON FLD, 1 S (A)
.WAYNE, WAYNE FLYERS SEAPLANE BASE, ADJ SW (A)

MICHIGAN

BRIGHTON, MILLER, 4 NW (C)
FOWLERVILLE, NEWTON FIELD, 5 SW (A)

MINNESOTA

GARDEN CITY, BERGEMANN, 3 S (C)

MISSISSIPPI

MENDENHALL, MENDENHALL MUNI, 1 S (A)
MONTICELLO, CLAY, 2 N (C)

MISSOURI

.CASSVILLE, TIMBER LINE AIRPARK, 4 SE (C)
.KIMBERLING, KIMBERLING SEAPLANE BASE, 1 E (A)
.SUMMERSVILLE, STEELMAN, 2 NW (A)
.VIOLA, TOMAHAWK, 1 NE (A)

MONTANA

.CONRAD, CONRAD HELIPORT, 1 W (A)

Fig. 7-32. Abandoned Or Closed To Public Airports

VOR RECEIVER CHECKS

Good operating practice dictates that the VOR receivers in the airplane be checked periodically for accuracy. Part 3 of the AIM includes a list of the airborne and ground VOR receiver checkpoints described in Part 1 as well as the VOR test facilities located on larger airports.

Figure 7-33 is an example of a ground checkpoint. The top of this illustration is the AIM excerpt which lists the information on the checkpoint. The point on the airport where the check should be made is shown with a painted, segmented circle and a small sign. The sign gives such information as the VOR name and the radial from the VOR station. With the aircraft parked in the segmented circle and the VOR properly tuned, the VOR receiver's accuracy can

be checked with the omni bearing selector set to the designated radial.

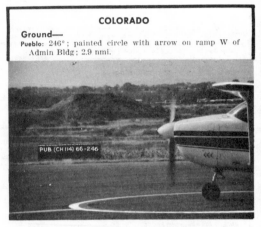

Fig. 7-33. VOR Ground Checkpoint

The location of the airborne VOR checkpoints listed in Part 3 can be found by

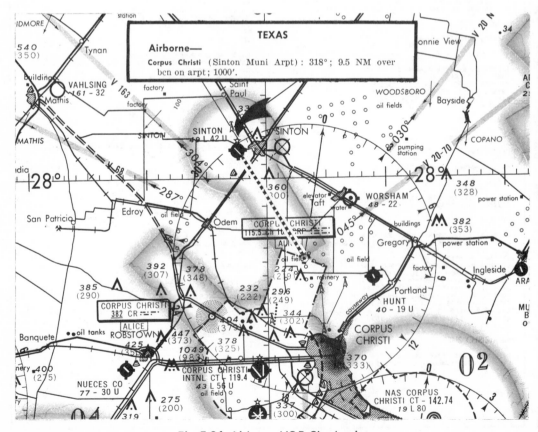

Fig. 7-34. Airborne VOR Checkpoint

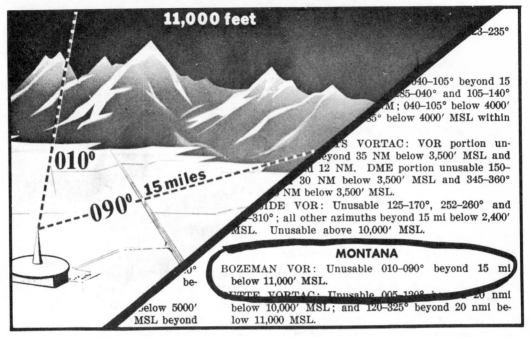

11,000 feet

3-235°

040-105° beyond 15
285-040° and 105-140°
M; 040-105° below 4000'
5° below 4000' MSL within

010°

090° 15 miles

TS VORTAC: VOR portion un-
eyond 35 NM below 3,500' MSL and
d 12 NM. DME portion unusable 150-
30 NM below 3,500' MSL and 345-360°
NM below 3,500' MSL.
DE VOR: Unusable 125-170°, 252-260° and
310°; all other azimuths beyond 15 mi below 2,400'
MSL. Unusable above 10,000' MSL.

MONTANA

BOZEMAN VOR: Unusable 010-090° beyond 15 mi
below 11,000' MSL.

be-

TE VORTAC: Unusable 005-120° b___ __ 20 nmi
below 10,000' MSL; and 120-325° beyond 20 nmi below 11,000 MSL.

elow 5000'
MSL beyond

Fig. 7-35. Restriction to Navigation Aids

referring to the appropriate sectional chart. As shown in figure 7-34, the VOR can be checked for accuracy near Corpus Christi, Texas, when flying over the rotating beacon located on the Sinton Municipal Airport. The flight should be made at 1,000 feet MSL.

In addition to the ground and VOR receiver checkpoints listed in Part 3 of the AIM, special VOR test facilities are available at many major terminals to test the VOR accuracy. A VOR test facility is referred to simply as a VOT. The availability and frequency of the VOT is included in the remarks section of the Airport/Facility Directory information found in Part 3 of the AIM.

To use the VOT facility, the omni bearing selector should be set to zero which will cause a FROM indication and the CDI should center. The zero degree setting on the OBS is commonly referred to as 360° in radio communications. A VOT indication will be the same regardless of the aircraft's position on the airport.

RESTRICTIONS TO NAVIGATIONAL AIDS

Because of obstructions and various kinds of interference, not all of the radials of every VOR are usable. The list of unusable radials is found in the Restrictions to En Route Navigation Aids in Part 3. As illustrated in figure 7-35, a range of mountains northeast of the Bozeman, Montana, VOR makes the radials between 010° and 090° unusable below 11,000 feet MSL when beyond 15 miles from the station.

AREA NAVIGATION

A new type of radio navigation has recently been presented to the aviation industry. With the proper radio equipment, a pilot can navigate by radio direct to any location without flying directly toward or away from a navigational aid. This new method is called "area navigation." Area navigation routes, which are approved for instrument navigation, are listed in Part 3 of the AIM.

PART 3A-NOTICES TO AIRMEN

The *Airman's Information Manual* Part 3A contains Notices to Airmen considered essential to the safety of flight as well as supplemental data to AIM Parts 3 and 4. Since the Part 3A section is revised between issues of Part 3, it frequently contains information that must be distributed to airmen rapidly. (See Fig. 7-36.)

Part 3A is published every 14 days between issues of AIM Part 3. It contains appropriate notices from the daily NOTAM Summary and other items considered essential to flight safety. This section contains Notices to Airmen that are expected to remain in effect for at least seven days. Temporary notices without published duration dates are normally carried twice unless resubmitted. (See Fig. 7-37.)

The data in Part 3A is arranged alphabetically by state and within the state by city or locality. New or revised data are indicated by underlining the first line of the affected item. The new information is not necessarily limited to the underlined portion.

An example of a NOTAM for an airport is shown in figure 7-37 for the Blythe Airport. This NOTAM reads as follows:

"Blythe Airport: Intensive airline jet aircraft training in progress 24 hours daily. Inbound aircraft report 20 miles out on 123.6 and guard 123.6 for airport advisory service. Use other frequencies for other purposes. Unicom is not for airport advisory use."

Fig. 7-36. Part 3A of the Airman's Information Manual

AIRMAN'S INFORMATION MANUAL—PART 3A

INFORMATION CURRENT AS OF DECEMBER 17, 19

THIS SECTION CONTAINS NOTICES TO AIRMEN THAT ARE EXPECTED TO REMAIN IN EFFECT FOR AT LEAST SEVEN DAYS.
NOTE: NOTICES ARE ARRANGED IN ALPHABETICAL ORDER BY STATE (AND WITHIN STATE BY CITY OR LOCALITY).
NEW OR REVISED DATA: NEW OR REVISED DATA ARE INDICATED BY UNDERLINING THE AIRPORT NAME.
NOTE: ALL TIMES ARE LOCAL UNLESS OTHERWISE INDICATED.
NOTE: SEE AIM PART 3 FOR PILOT CONTROLLED LIGHTING LEGEND.

ALABAMA

ALABASTER, SHELBY COUNTY ARPT: Open ditch 100 ft from threshold rwy 15 until Jun. (12/)
ANDALUSIA OPP ARPT: ARPT closed. (7/)
BREWTON MUNI ARPT: Rwy lights now medium intensity. (12/)
DEMOPOLIS MUNI ARPT: Rwy lights dusk-dawn. (12/)
DOTHAN WHELLESS ARPT: Rwy lights after 2400 on request. (11/)
HUNTSVILLE-MADISON COUNTY JETPORT: LOC, GS, MM and OM OTS. (12/) t) (
STEVENSON BRIDGEPORT ARPT: Strobe lighted stack 1623 MSL 1500 ft E.

ALASKA

For complete information on Alaska consult the Alaska Supplement.

ARIZONA

FLAGSTAFF, PILLIAM ARPT: ATCT hours 0600-2200. (1/76-3)
GRAND CANYON AND PETRIFIED FOREST NATIONAL PARKS: All pilots are requested to avoid flying below the canyon rim and to maintain a distance 1500ft above and horizontally from all scenic overlooks, parks, and trails. (10/)
PHOENIX DEER VALLEY MUNI ARPT: Rwy lights north east 2700 ft rwy 7R-25L OTS until Apr 1.

ARKANSAS

BATESVILLE REGIONAL ARPT: SDF LOC rwy 7 cmsnd. Batesville NDB now Independence County ident "INY" freq 317 kHz. (12/)
EL DORADO, GOODWIN FIELD: Rwy 17-35 closed air carrier. (12/)
FAYETTEVILLE DRAKE FLD: MALS rwy 16 OTS. (10/)
HOPE MUNI ARPT: Threshold rwy 22 dsplcd 250 ft. Threshold rwy 28 dsplcd 250 ft. (11/)
PARAGOULD MUNI ARPT: Rwy 5-23 S -11,000 lbs. (1/)
PINE BLUFF, GRIDER FIELD: S 350 ft rwy 17-35 closed. Non-Radar A/C and departure control cmsnd freq 118.4, hours 0700-2300. From 2300-0700 ctc Little Rock A/C freq 124.2. (1/)
ROGERS MUNI ARPT: Arpt closed nights. Rwy 1-19 now 6000 ft. (10/)

CALIFORNIA

SPECIAL NOTICE: Do not mistake dirt strip on large island, Lake Berryessa, lctd lat 38-34 long 122-13 for airport. Strip is unauthorized and unsafe. (10/)

SPECIAL NOTICE: The MELONES RESERVOIR/DAM AREA, lctd about 5 to 12 miles SW of Columbia Arpt and about 8 miles S of Frog Town Arpt at Angels Camps:acft are cautioned to avoid the area, if necessary to fly over the area maintain a minimum alt of 4000 ft, due to possible damage from flying rock or turbulence caused by continuous blasting for dam construction until 1979.
SPECIAL NOTICE: Avoid flying below 1000 ft over a 400 acre Wildlife Refuge along the south side of the mouth of the Salinas River and ocean shoreline eastward. (1/)
ANO NUEVO ISLAND: Avoid low flying in the vicinity and over island. Biological research of wild life in progress. (10/)
AVALON CATALINA ARPT: MIRL rwy 4-22 cmsnd. (12/)
BLYTHE ARPT: Intensive airline jet acft training in progress 24 hrs daily. Inbound acft report 20 miles out on 123.6 and guard 123.6 for arpt advisory service. Use other freqs for other purposes. UNICOM is not for arpt advisory use. (10/)
CHICO MUNI ARPT: ATCT ground control freq 121.9. (1/)
CHINO ARPT: ATCT ground control freq 121.6. (1/)
DIXON MAINE PRAIRIE ARPT: TPA 800 ft. (12/)
LONG BEACH-DAUGHERTY FIELD ARPT: Oil drilling rig 150 ft AGL 11/8 MI W apch end rwy 7R lighted thru Mar 17. No turbo-jet operations between 2300-0700. (12/)
OAKLAND, METROPOLITAN OAKLAND INTL ARPT: All turbo-jet, all turbo-prop over 12500 lbs GWT, and all 4 engine reciprocating acft are prohibited from landing rwys 9L/R and takeoff from rwys 27L/R, 2200-0700. Same acft prohibited landing rwys 9L/R or takeoff rwys 27L/R 0700-2200 unless certificated under Part 36 or are operd at or below applicable Part 36, Appendix C noise limits for type and weight of acft. Restriction waived if rwy 11-29 closed or if required by acft for operational safety. (10/)
PALO ALTO: Crane 75 ft 1/2 mile east Palo Alto and Santa Clara County arpts. Operating in harbor days thru Feb 15. (1/)
PORTERVILLE MUNI ARPT: Rwy 12-30 S-40000 lbs D-60000 lbs DT-100000 lbs. Rwy 7-25 permly closed. (12/)
SAN DIEGO INTL-LINDBERGH FLD ARPT: Rwy 9-27 closed 0015-0630 except Sat, Sun, Holidays. Rwy 9-27 closed 0015-0630 except Sat-Sun and Holidays. (1/)
SAN DIEGO-SANTEE GILLESPIE FLD ARPT: Threshold 17 dsplcd 470 ft. (12/)
TULARE AIRPARK: Arpt closed night, rwy lights OTS. (10/)

COLORADO

ALAMOSA: NDB "ALS" dcmsnd. (1/1-3)
CRAIG-MOFFAT ARPT: Rwy 7-25 unmarked (7/)
GRAND JUNCTION WALKER FIELD: MALSR rwy 11 cmsnd. (11/)
HAYDEN YAMPA VALLEY ARPT: Apch/departure service on 128.5 and 327.8 provided by Denver ARTCC. (12/)
HAYDEN: Hayden VOR "CHE" freq 115.4 cmsn. (12/)
STEAMBOAT SPRINGS: LRCO freq 122.2 cmsnd. (10/)

CONNECTICUT

WATERBURY AIRPORT INC: Extensive glider soaring activity. (5/)

Fig. 7-37. Notices to Airmen

SECTION D

PART 4 OF THE AIM

Part 4 of the AIM is entitled "Graphic Notices and Supplemental Data." (See Fig. 7-38.) Part 4 is issued quarterly and contains abbreviations, lists of parachute jump areas, maps showing *olive branch* routes, and terminal area graphics. The list of FAA official abbreviations used throughout the *Airman's Information Manual* is included to insure the proper interpretations of abbreviations used throughout the AIM.

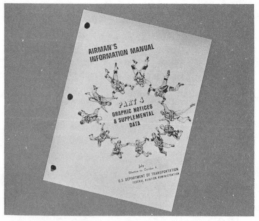

Fig. 7-38. Part 4 of the Airman's Information Manual

PARACHUTE JUMPING AREAS

A potential hazard to flight is a parachute jumping operation. The established parachute jumping areas are listed in Part 4 of the *Airman's Information Manual.* The parachute jumping areas are depicted with the miniature parachutes on the sectional chart. (See Fig. 7-39.) When these symbols are illustrated near the intended route of flight, it is advisable to check the listing of the parachute jumping areas in Part 4. The AIM listing contains the parachute jumping altitudes and any pertinent remarks such as the time of operation.

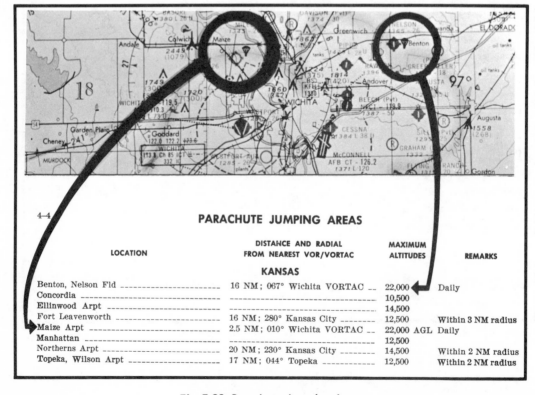

4-4

PARACHUTE JUMPING AREAS

LOCATION	DISTANCE AND RADIAL FROM NEAREST VOR/VORTAC	MAXIMUM ALTITUDES	REMARKS
KANSAS			
Benton, Nelson Fld	16 NM; 067° Wichita VORTAC	22,000	Daily
Concordia		10,500	
Ellinwood Arpt		14,500	
Fort Leavenworth	16 NM; 280° Kansas City	12,500	Within 3 NM radius
Maize Arpt	2.5 NM; 010° Wichita VORTAC	22,000 AGL	Daily
Manhattan		12,500	
Northerns Arpt	20 NM; 230° Kansas City	14,500	Within 2 NM radius
Topeka, Wilson Arpt	17 NM; 044° Topeka	12,500	Within 2 NM radius

Fig. 7-39. Parachute Jumping Areas

MILITARY OPERATIONS

AIM Part 4 also describes low-altitude military operations which can represent hazards to civil VFR aircraft. An awareness of these operations is essential for safe flight. The types of operations covered in Part 4 are described in the following paragraphs.

VFR LOW-ALTITUDE TRAINING ROUTES

The military services have a continuing requirement to conduct VFR low-altitude training flights at or below 1,500 feet above the surface in excess of 250 knots indicated airspeed. These flights are conducted only when weather conditions are equal to or better than a 3,000-foot ceiling and five miles visibility. The routes used by these training flights are selected to avoid control zones, control zone extensions, airport traffic areas, and, to the extent possible, uncontrolled airports by three statute miles.

ALL-WEATHER LOW-ALTITUDE (OLIVE BRANCH) ROUTES

The United States Air Force and the United States Navy conduct low-level navigation/bombing training flights in jet aircraft in both VFR and IFR weather conditions along the routes shown on graphic depictions in AIM Part 4. Operations, both IFR and VFR, are contained within four nautical miles on either side of the centerline. An exception to this is the maneuvering area, where the IFR route width will not exceed nine nautical miles and the VFR route width will not exceed seven nautical miles on either side of centerline. When the ceiling is at least 3,000 feet and visibility five miles, aircraft may be operating anywhere between the altitude published on the olive branch route charts and the surface

while within the confines of the low-level route or maneuvering area. All VFR operations will be conducted in accordance with FAR Part 91.79. Flights are conducted only during the times specified for each route. The current operational status of a particular route may be obtained by calling a flight service station near that route.

Indexes of olive branch routes, military aerial refueling tracks, and low-level training areas are provided at the beginning of the Special Operations section. To determine whether the intended flight route passes through any of these areas or routes, the pilot should consult the AIM during preflight planning. A typical olive branch route is shown in figure 7-40.

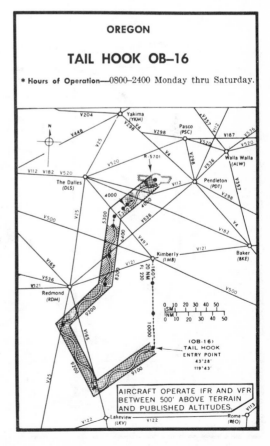

Fig. 7-40. Olive Branch Route

GRAPHIC NOTICES

AIM Part 4 contains graphic depictions of three types of airport terminal areas. These terminal maps are designed to provide procedural information for VFR pilots operating at the busier terminals throughout the country. These three types of graphic notices include the following:

1. Terminal control areas
2. Terminal area graphic notices
3. Terminal radar service areas

TERMINAL CONTROL AREAS

This section contains graphic charts of terminal control areas (TCAs) in the United States and provides information on operational requirements in TCAs. Terminal control area configurations are tailored to the operational needs of the individual area so all TCA depictions in AIM Part 4 are distinctive to a particular airport. Figure 7-41 shows a typical AIM Part 4 TCA chart. These charts are for procedural information and are not intended for navigation. VFR pilots should use *VFR terminal area charts* for navigation, not the AIM depictions.

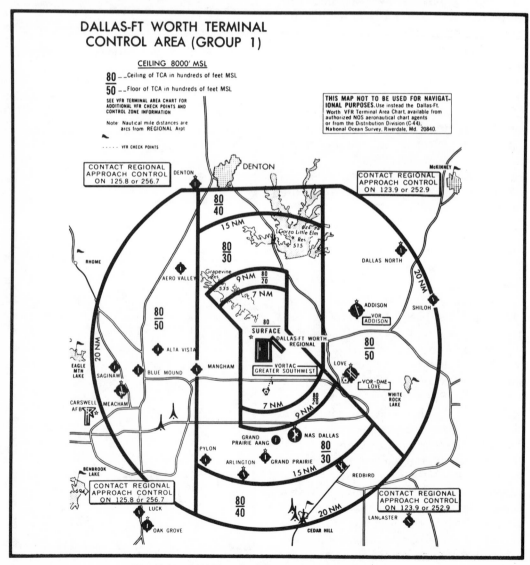

Fig. 7-41. AIM Terminal Control Area Depiction

Terminal control areas are established to separate all arriving and departing traffic, both VFR and IFR, at large airports. A TCA may be designated as Group I, Group II, or Group III. Each area is designed to facilitate traffic separation at a particular terminal and, therefore, is different from other terminal control areas. Within the expanding controlled airspace of the terminal control area, all aircraft are subject to operating rules and pilot and equipment requirements, as specified by FAR Parts 91.24 and 91.90.

Pilot participation is mandatory and an ATC clearance must be received before entering a Group I TCA. To operate within a Group I TCA, the pilot also must possess at least a private pilot certificate. The airplane must be equipped with a VOR or TACAN receiver, two-way radio, a transponder with 4096-code capability, and automatic pressure altitude reporting equipment.

Within Group II TCAs, there are no minimum pilot certification requirements. The aircraft must be equipped as in a Group I TCA, except that automatic pressure altitude reporting equipment is not required. In addition, a transponder is not required for those IFR flights operating to or from an airport outside of, but in close proximity to the TCA when the commonly used approach, departure, or transition procedures require flight within the TCA. A transponder is required for other operations within the TCA. However, no person may operate an aircraft within a Group II terminal control area unless he has received an appropriate authorization from ATC prior to operation of that aircraft in that area. Finally, two-way radio communications must be maintained within that area between that aircraft and the ATC facility.

The requirements for the Group III TCA are the least restrictive of the three groups. There are no minimum pilot

certification requirements. A transponder and automatic pressure altitude reporting equipment are not required at all *if* the pilot is capable of maintaining two-way communications with ATC while in the TCA boundaries and provides altitude, position, and proposed route of flight information to ATC prior to entering the TCA.

TERMINAL AREA— GRAPHIC NOTICES

Following the TCA charts, Part 4 provides additional charts termed "Terminal Area—Graphic Notices." The charts show concentrated IFR traffic routes and preferred VFR routes around terminal airports that do not provide Stage III radar service (TRSAs). However, pilots are still requested to contact approach or departure control when operating in these terminal areas or to operate above or below the airspace segment occupied by IFR traffic, as depicted on the charts. Since Part 4 is published on a quarterly revision cycle, new or revised TRSA charts or terminal area graphic charts are usually published first in the AIM Part 3A.

TERMINAL RADAR SERVICE AREAS

Terminal radar service areas are established wherever Stage III radar service is implemented. Stage III radar service provides *vectoring*, *sequencing*, and *separation*, on a full-time basis, of both VFR and IFR air traffic landing at the primary airport. Separation service and advisories concerning unknown aircraft are also provided to all participating aircraft not landing at the primary airport but operating within the TRSA. *Participating traffic* means that an aircraft is in radar and radio contact with approach or departure control.

AIM Part 4 provides a description of TRSA operational procedures for both pilots and air traffic controllers. In addition, Part 4 contains charts of each terminal radar service area in the United States. The charts are prepared for the FAA by the National Ocean Survey for the benefit of pilots conducting VFR operations in the terminal radar service area. When VFR pilots utilize the information contained on the TRSA chart and request Stage III radar service, they will be vectored around the heaviest concentrations of IFR traffic within the TRSA.

ADVISORY CIRCULARS

Advisory circulars are issued by the FAA to provide aviation information of a non-regularatory nature. Although the contents of most advisory circulars are not binding on the public, they contain information and accepted procedures necessary for good operating practices.

Each of the subjects covered is identified by a general subject number, followed by the number of the specific subject matter, as shown in the following list.

00	General
10	Procedural
20	Aircraft
60	Airmen
70	Airspace
90	Air Traffic Control and General Operations
140	Schools and Other Certified Agencies
150	Airports
170	Air Navigational Facilities

When the volume of circulars in a general series warrants a further breakdown, the general number is followed by a slash and a specific subject number. For example, some of the Airports series (150) is issued under the following numbers.

150/4000	Resource Management
150/5000	Airport Planning
150/5200	Airport Safety—General
150/5210	Airport Safety Operations (Recommended Training, Standards, Manning)

150/5220	Airport Safety Equipment and Facilities
150/5230	Airport Ground Safety System

Each circular has a subject number followed by a dash and a sequential number identifying the individual circular (150/4000-1). Changes to circulars have the notation "CHG 1," "CHG 2," etc. after the identification number on pages that have been changed. The date on a revised page reflects the effective date of that change.

The subjects covered by advisory circulars and the availability of each are contained in the Advisory Circular Checklist, which is printed as part of the *Federal Register*. Generally, this checklist is issued three times each year and contains a listing of all current circulars, those circulars which have been canceled, and any additions since the last checklist was printed. This checklist can be obtained free of charge by writing to the following address.

Department of Transportation
Distribution Unit, TAD 443.1
Washington, D.C. 20590

JEPPESEN J-AID

The *Jeppesen Airport and Information Directory (J-AID)* is a comprehensive collection of current airport and aeronautical information. This looseleaf reference book, shown in figure 7-42, is the same size as the Jeppesen approach chart

Fig. 7-42. J-AID

binders, and contains information applicable to both VFR and IFR pilots.

The *J-AID* includes the material covered in the *Airman's Information Manual*, plus information on many other aeronautical knowledge areas. For this reason, FAR Part 135.39 authorizes it as a substitute for the AIM during air taxi and commercial opeations. Because it contains more information and is revised more often than the AIM, many airlines and commercial operators prefer its use.

There are two basic revision cycles employed with the *Jeppesen J-AID*. First, revisions to information concerning air traffic control are distributed approximately 10 times a year. Second, airport diagrams and related information are revised and disseminated every four months.

J-AID SECTIONS

The *J-AID* is divided into seven basic sections—introduction, radio facilities, airport directory, meteorology, tables and codes, air traffic control, and regulations. All material relating to a specific subject is included in the appropriate section.

INTRODUCTION

The introduction section contains a table of contents, a revision record, and any recently issued Jeppesen briefing bulletins. The briefing bulletins deal primarily with new material or subjects that must be specifically brought to the attention of the pilot.

RADIO FACILITIES

Information contained in the radio facilities section pertains to frequency allocations, characteristics of navigation facilities, location of radar units, direction finding procedures, and the location of direction finding stations. In addition, it lists locations and frequencies for commercial broadcast stations and provides information concerning VOR receiver checkpoints and test signals. Unlike the AIM, all published VOR receiver tests are listed in this one section. Information also is presented concerning the availability of automatic terminal information service and the locations and frequencies of navigation aids within that region.

AIRPORT DIRECTORY

The airport directory section is divided into three geographic regions. The airports within each region are listed alphabetically by state, city, and airport name. The airport listings of each state are preceded by a map of the state showing all of the locations under which airports are indexed. The state maps can be used with sectional and enroute charts for selection of the airport nearest the destination.

An airport diagram and tabular airport information are provided for each airport, as shown in figure 7-43. The tabular information includes the airport name, elevation, geographic coordinates, manager's name and telephone number, hours of operation, service facilities available, runway information, communications facilities, and the distance and direction from a local navaid.

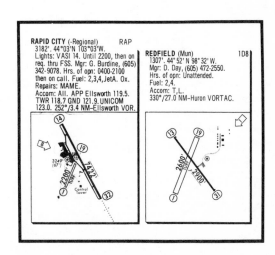

Fig. 7-43. J-AID Airport Diagrams

METEOROLOGY

The meteorology section contains general information concerning the weather services available and the meanings of weather station models, symbols, and codes used on various weather charts and forecasts. In addition, it lists all continuous automatic transcribed weather broadcast (TWEB) stations and National Weather Service and flight service station telephone numbers.

TABLES AND CODES

A comprehensive collection of conversion tables and data on commonly used codes is included in the tables and codes portion of the *J-AID*. For example, there are tables for converting temperatures from Celsius to Fahrenheit and making various metric conversions. The phonetic alphabet, Morse code, wind component table, and sunrise, sunset, and twilight tables also are provided.

AIR TRAFFIC CONTROL

The air traffic control section consists of the basic information contained in AIM Part 1. This includes data pertaining to airspace, services available to pilots, radiotelephone phraseology and technique, clearances, emergency procedures, and departure, enroute, and arrival procedures. It also provides data concerning wake turbulence avoidance, bird hazards, safety of flight, and good operating practices.

REGULATIONS

The last section of the *J-AID* provides a current set of Federal Aviation Regulations reprinted from Parts 1, 61, 91, 103, 121, 135, 141, and NTSB 830. The regulations are updated for changes on a timely basis. The regulations also may be purchased separately from the *J-AID* with update service.

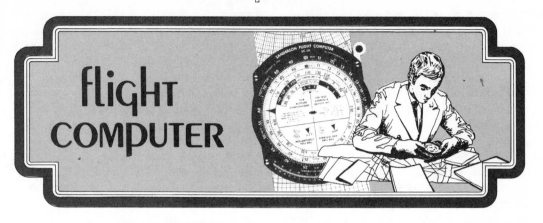

flight COMPUTER

SECTION A - CALCULATOR SIDE

THE SCALES

Mathematicians have long used a slide rule as the tool to speed solutions to their problems. The calculator side of the flight computer is basically a circular slide rule with special indices for solving

flight problems. The wind side of the computer is merely a device which makes it easy to draw a wind vector. The slide rule part of the calculator side of the computer consists of three scales referred to as A, B, and C as shown in figure 8-1, items 1, 2, and 3.

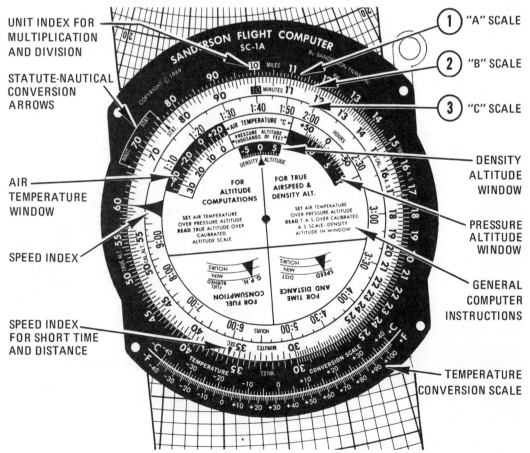

Fig. 8-1. Flight Computer

The outer scale (A) is fixed to the computer. Scales B and C are printed together on the disc that pivots in the center of the computer. With these three scales, it is a simple matter to establish mathematical ratios which are the basis of problem solving on this side of the computer.

The A scale is used to represent *miles, gallons,* or *true airspeed.* When used to represent miles, the scale provides the distance traveled or speed of the aircraft in miles per hour. Gallons on this scale can represent two values: fuel consumption (gallons per hour), or total quantity of fuel used by the aircraft. TAS on the A scale is an abbreviation for *true airspeed.*

The graduations on the B scale are used for *time* and *minutes* or for *IAS (indicated airspeed)* in either miles per hour or knots. The C scale graduations represent *hours* and *minutes* only.

CHANGING VALUES

In order to accurately read the scales on the calculator side of the computer, it is necessary that the changing values of the graduations of these scales be understood.

When using the A, B, and C scales to solve problems, common sense must be used to determine the value of the number. If a short distance is involved, 25 on the A scale might be read as 2.5 miles. (See Fig. 8-2.) If a long distance is involved, zeros are added to the 25 to get the proper answer. For example, 25 might be 250 or 2,500 miles. If the number 14 is used as 14, each graduation between

14 and 15 is equal to 0.1. If these numbers are used as 140 and 150, each graduation is equal to one. If used as 1,400 and 1,500, they represent 10.

Between the numbers 15 and 16, there are only five graduations as compared to the 10 graduations between 14 and 15. (See Fig. 8-3.) When the numbers are used as 15 and 16 respectively, each graduation between them is equal to 0.2; when used as 150 and 160, each unit represents two.

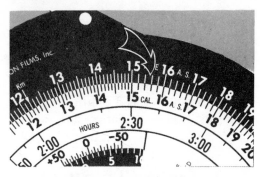

Fig. 8-3. Scale Graduations

The changing values on the C scale are somewhat different than those on the A and B scales. (See Fig. 8-4.) These graduations always represent minutes and are equal to five or ten minutes as shown by the arrows. For example, between 1:50 and 2:00, the graduations are equal to 5 minutes; between 2:00 and 2:30, the graduations represent 10 minutes.

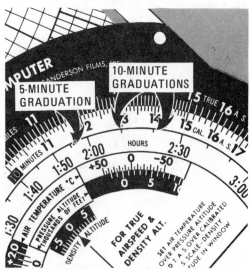

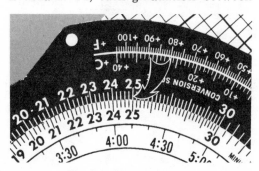

Fig. 8-2. Changing Values

Fig. 8-4. Minute Graduations

TIME, SPEED, AND DISTANCE

The *speed index* is the large black triangle symbol on the B and C scales and is used as a reference in time and distance problems. (See Fig. 8-5.) The speed index always represents 60 minutes or one hour. In time and distance problems, there are three items for which to solve: time, speed, and distance. Two of these items must be known to work the problem.

Fig. 8-5. Speed Index

SOLVING FOR TIME

If an aircraft is flying at a speed of 100 miles per hour (m.p.h.), how long will it take to fly 132 miles? The steps involved in solving this problem are as follows: (See Fig. 8-6.)

1. Rotate the computer disc until the speed index is located directly under 10, which represents 100 m.p.h.

Fig. 8-6. Time to Fly

2. Look clockwise on the A scale to 132, which is two graduations past 13. Look directly below 132 and find the answer to the problem.
3. It takes 79 minutes, as noted on the B scale, or 1 hour and 19 minutes as noted on the C scale, to travel 132 miles at 100 m.p.h.

SOLVING FOR DISTANCE

In order to solve for distance, another hypothetical problem is shown below. If the aircraft flies at 136 m.p.h. for a two-hour period, how many miles will it fly? (See Fig. 8-7.)

1. Place the speed index under 136, which is six graduations to the right of 13 on the A scale. As noted before, this setting denotes the fact that the airplane is flying at 136 miles in one hour.
2. Move clockwise on the C scale to 2:00 (two hours).
3. Look directly above 2:00 and find

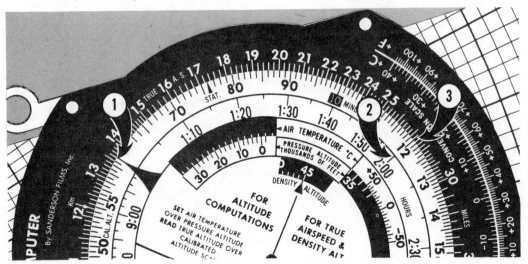

Fig. 8-7. Finding Distance

272 on the A scale. The 272 is interpreted by noting that there are five major graduations between 25 and 30 on the A scale. Each one of the major graduations in this case will represent 10 miles. Between each of the major graduations there are five minor graduations which, in this case, will have a value of two miles per graduation. Therefore, the answer is 272 miles.

SOLVING FOR SPEED

In order to find the speed, consider the next hypothetical problem. If an airplane flies 180 miles in one hour and ten minutes, what is the speed? To solve this problem, proceed as follows: (See Fig. 8-8.)

1. Position 180 miles on the A scale directly over 1:10 on the C scale.
2. Directly over the speed index, read the answer: 154 m.p.h.

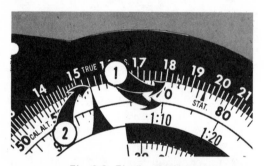

Fig. 8-8. Finding Speed

FUEL CONSUMPTION

Fuel consumption problems are solved in the same manner as time and distance problems except gallons per hour and gallons are used in lieu of miles per hour and miles.

FINDING ENDURANCE

If an aircraft burns fuel at the rate of eight gallons per hour (g.p.h.) and has 18 gallons of usable fuel on board, the flight endurance of the airplane may be computed by the following steps: (See Fig. 8-9.)

1. Rotate the computer disc until the speed index is directly under 80, which represents eight gallons per hour in this problem.
2. Look clockwise on the A scale and locate 18 which represents the 18 gallons of usable fuel.
3. Directly under this, combine the B and C scales and interpret the answer as two hours and 15 minutes.

COMPUTING AMOUNT OF FUEL

If an aircraft burns 11 gallons per hour for a period of one hour and 30 minutes, the number of gallons of fuel burned can be computed by the following steps: (See Fig. 8-10.)

1. Place the speed index under 11, the rate of fuel consumption in 60 minutes, or one hour.

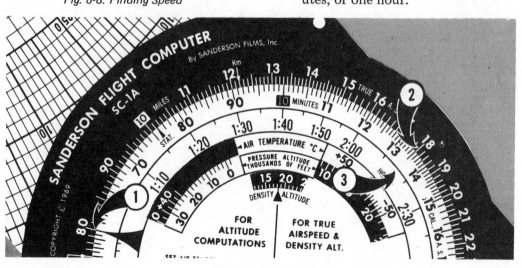

Fig. 8-9. Endurance

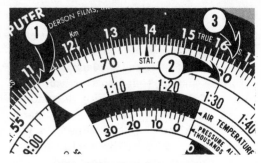

Fig. 8-10. Fuel Consumed

2. Locate one hour and 30 minutes on the C scale.
3. Look directly over one hour 30 minutes and find the number of gallons burned as computed on the A scale. The answer is 16.5 gallons.

FINDING FUEL CONSUMPTION RATE

The rate of fuel consumption can be determined if the gallons of fuel burned and the total time are known. (See Fig. 8-11.) For example, if an aircraft burns 45 gallons of fuel in two hours and 20 minutes, the rate of fuel consumption per hour may be determined by the following steps:

1. Rotate the calculator disc until 45 on the A scale is directly over two hours 20 minutes on the C scale.
2. Rotate the complete computer and look for the answer on the A scale above the speed index. The answer to the problem is 19.2 gallons per hour.

AIRSPEED CORRECTIONS

An error in the airspeed instrument occurs in some airplanes due to inherent errors in the position of the static source on the fuselage. The airplane manual will have a calibration table to show this error if it is significant. The result of the correction is called *calibrated airspeed.* For purposes of this discussion, however, *indicated airspeed*, which is read directly off the airspeed indicator, and calibrated airspeed will be considered synonymous.

TRUE AIRSPEED

The airspeed indicator is a pressure instrument and as such will indicate different values with changes in pressure and temperature. The correction of indicated airspeed to allow for pressure, altitude, and temperature variations results in a value called *true airspeed.* In a no-wind condition, true airspeed, and the actual speed across the ground, or ground speed, are equal.

The computer has a means of converting indicated airspeed to true airspeed. The steps involved are as follows:

1. In the small window labeled "PRESSURE ALTITUDE THOUSANDS OF FEET," place the pressure altitude for the flight under the temperature for the flight on the scale labeled "AIR TEMPERATURE °C" (Celsius). (See Fig. 8-12.)

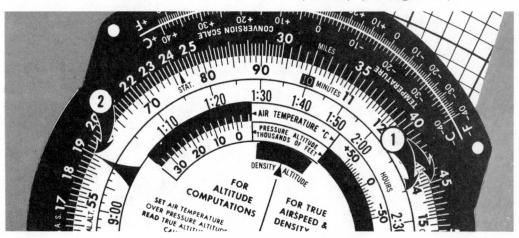

Fig. 8-11. Rate of Fuel Consumption

Fig. 8-12. Airspeed Scales

2. Without moving any scales, locate the indicated airspeed on the B scale. The B scale is labeled "CAL. A.S." (calibrated airspeed) which is considered synonymous with indicated airspeed.

3. Find the true airspeed on the A scale immediately above the indicated airspeed value.

For a hypothetical problem, assume that the airplane is flying at an altitude of 9,000 feet, the temperature is +10° Celsius, and the indicated airspeed is 150 m.p.h. (See Fig. 8-13.)

1. Rotate the computer disc until +10° Celsius appears over 9,000 feet.

2. Opposite the indicated airspeed of 150 m.p.h., read the true airspeed which is 176 m.p.h.

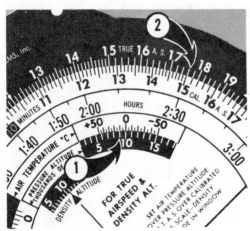

Fig. 8-13. True Airspeed

DENSITY ALTITUDE

The density altitude is an important consideration in determining the performance of an aircraft. If an airplane is taking off at an airport that has a pressure altitude of 6,000 feet (the pilot determines pressure altitude by setting his altimeter to 29.92 and reading the altimeter) and a temperature of 86° Fahrenheit, the density altitude could be determined as follows:

1. Since the temperature is given in degrees Fahrenheit and the computer is scaled in degrees Celsius, the conversion scale on the computer may be consulted to determine the temperature in degrees Celsius. Looking opposite 86° F, find +30° C. (See Fig. 8-14.)

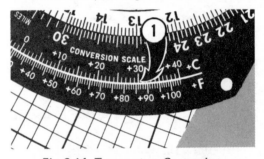

Fig. 8-14. Temperature Conversion

2. Position +30° Celsius opposite the pressure altitude of 6,000 feet as shown in figure 8-15.

3. The density altitude is read in the density altitude window. In this case, the density altitude is approximately 9,000 feet.

In this example, the pilot knows that the performance of his airplane will be equivalent to flying at an altitude of 9,000 feet under standard conditions.

Fig. 8-15. Density Altitude

CONVERSION INDEXES

Another use for the calculator side of the computer is for converting nautical miles to statute miles, or knots to miles per hour. This type of problem is made very simple by a small conversion scale consisting of two arrows labeled "NAUT" (nautical) and "STAT" (statute) respectively. These arrows are located on the A scale and point toward the B scale as shown in figure 8-16.

Fig. 8-16. Knots to Miles Per Hour

An example of conversion from nautical to statute miles would be converting 100 nautical miles to 115 statute miles as shown in figure 8-16. Of course, this could also be interpreted as 10 nautical miles equals 11.5 statute miles, or 1,000 nautical miles equal 1,150 statute miles, etc. With these indexes, statute miles may also be converted back to nautical miles.

SECTION B - WIND SIDE

In order to understand the wind side of the computer a brief explanation of how wind affects an airplane in flight is presented along with related terms and their definitions.

WIND EFFECTS

After a true course is plotted, the effects of wind speed and direction at the planned flight altitude must be considered. Forecast wind information for winds aloft is obtained from the National Weather Service and flight service stations.

Wind is the movement of an air mass over the earth. Its motion is made apparent through blowing trees, choppy water, or the tugging of clothing. If it were a balloon, the airplane would simply move with the air mass and drift above the terrain. Since the airplane moves under its own power, however, the speed and direction it travels over the terrain is the result of combining aircraft speed, wind speed, aircraft heading, and wind direction.

WIND SPEED

The National Weather Service reports the speed of the wind in knots, and wind direction is always the direction *from* which it is blowing. Wind speed is the rate or progress of an air mass over the earth. The term *knots*, when referring to speed, means nautical miles per hour. Remember that a nautical mile, which is 6,076 feet long, is equal to one minute of latitude. The term *miles per hour* means statute miles per hour. A statute mile is 5,280 feet long.

AIRSPEED

INDICATED AIRSPEED

An airplane's speed through the air mass is shown on the airspeed indicator. This indicated airspeed has no direct correlation with the speed over the ground. Only at sea level on a standard day are the in-dicated airspeed and the aircraft's actual speed through the air the same.

TRUE AIRSPEED

In the thin air of high altitudes, compensations must be made for variances in pressure and temperature to find the actual speed through the air mass. When calculated corrections for temperature and pressure altitude are applied to indicated airspeed, the adjusted airspeed is called *true airspeed* and is expressed in miles per hour or knots.

GROUND SPEED

Another type of speed that affects the aircraft's time in flight is called *ground speed*, because it relates to the aircraft's progress over the ground. If there is no wind, the airplane's true airspeed and its ground speed will be the same. However, although true airspeed may be constant, the headwind or tailwind affects the ground speed.

For example, an airplane has a true airspeed of 100 miles per hour. It is flying in an air mass which is moving in the same direction at 20 miles per hour.

The aircraft is considered to have a tailwind, which is assisting its progress over the ground. The airplane's ground speed is then 120 miles per hour; however, the true airspeed is unaffected by wind. (See Fig. 8-17.)

Reversing this example, assume the airplane is moving against a 20 mile-per-hour air mass. Its progress over the ground is determined by subtracting the 20 mile-per-hour air mass speed from its 100 mile-per-hour true airspeed. The ground speed is now 80 miles per hour. (See Fig. 8-17.)

DRIFT

An airplane, in figure 8-18, is heading east in an air mass moving from the south at 20 miles per hour, so the aircraft has a

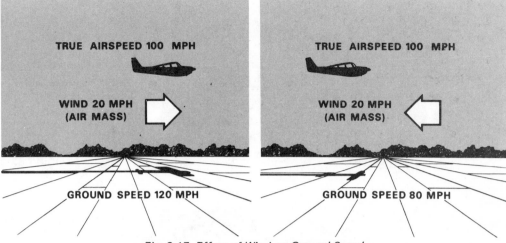

Fig. 8-17. Effect of Wind on Ground Speed

crosswind. At the end of one hour, the plane will have drifted 20 miles north of the true course line because of the air mass movement.

When flying from one point to another, the airplane must not be allowed to drift off course, nor can slow ground speed be permitted to extend time en route beyond fuel endurance. Wind corrections and fuel stops must be determined by careful preflight planning.

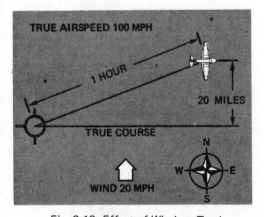

Fig. 8-18. Effect of Wind on Track

TRUE HEADING

The direction that an aircraft is actually heading, as measured in degrees from true north, is called *true heading*. The path it makes over the ground is called *track*. A specific heading must be maintained to keep the track directly over the planned true course line.

PROBLEM SOLVING

The solution of the effect of wind direction and velocity on the path and speed over the ground is a wind vector triangle. The wind side of the flight computer simplifies the drawing of the vectors of the wind triangle.

The components of the wind side are the *rotating azimuth* and the *sliding grid*. (See Fig. 8-19.) The sliding grid is a section of a graduated circle. The straight lines represent degrees right or left of the centerline and the curved lines represent distance. The sliding grid of this computer has two sides, a high speed side and a low speed side.

The *azimuth circle* rotates freely and is graduated into 360°. The transparent portion is frosted so that it can be written on with a pencil. In the center is a small circle called the *grommet* that is moved directly over the centerline of the grid.

To maintain a specific course, the pilot must determine the *wind correction angle* and adjust the aircraft heading accordingly. The time en route, an important factor in any flight, is influenced by wind as it affects the ground speed. To determine the total effect of wind on a flight, the *true course, true airspeed, wind direction,* and *velocity* must be known.

FLIGHT COMPUTER

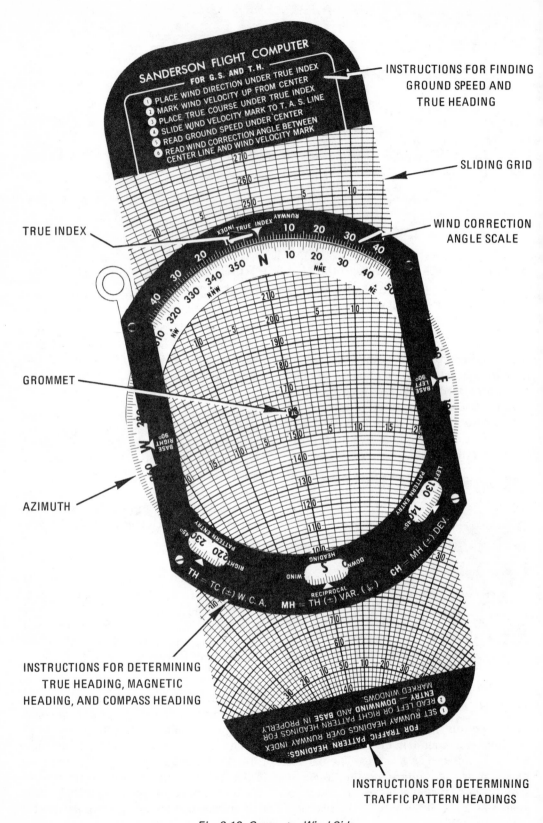

Fig. 8-19. Computer Wind Side

FINDING TRUE HEADING AND GROUND SPEED

An explanation of how the wind side works is provided with the following example:

Given:

    True course . . . . 084°
    True airspeed . . . . 90 m.p.h.
    Wind Direction and
       Velocity . . . .
    . . . . . . 330° (true at 23 m.p.h.)

Determine:

    Wind Correction Angle
    True Heading
    Ground Speed

1. Rotate the azimuth until the wind direction of 330° is located directly under the true index. (See Fig. 8-20.)
2. Next, slide the grid through the computer until the grommet is on any one of the heavy curved lines extending from right to left on the grid. For the problem, the 100-mile grid has been arbitrarily chosen.
3. Then place the wind velocity on the azimuth by placing a pencil mark 23 miles (m.p.h.) above the grommet of the azimuth.
4. Now, place the true course of 084° under the true index. (See Fig. 8-21.)

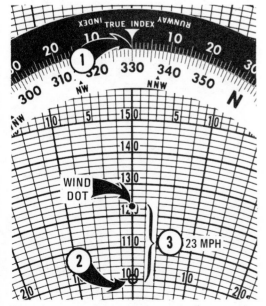

Fig. 8-20. Plotting the Wind

5. With the true course under the true index, slide the grid until the pencil mark, or wind dot, is positioned on the true airspeed line of 90 m.p.h.

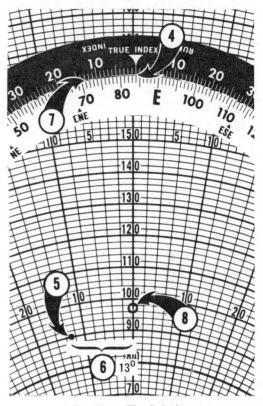

Fig. 8-21. The Solution

6. Find the wind correction angle by checking the number of vertical lines (degrees) to the right or left (in this case to the left), between the grid center line and the wind dot. In this problem, the wind correction angle is 13° to the left.
7. To find true heading, the scales on either side of the true index are used. Since the wind correction angle is 13° left, start at the true index symbol and count 13 units to the left and find the true heading of 071° directly below this graduation. (See Fig. 8-21.) If the wind dot is to the left, count left for true heading; if the wind dot is to the right, count right for true heading.
8. Without moving the computer setting, read the ground speed under the grommet. The ground speed in this problem is 97 m.p.h.

OTHER COMPUTER USES

This chapter has covered the functions of the computer most commonly used by the student pilot. In addition, a flight computer can be used for solving such problems as:

1. short time and distance.
2. finding drift angle (off course).
3. time and distance to a VOR station.
4. multiplication and division.
5. finding unknown winds.
6. finding altitude for most favorable winds.
7. radius of action.
8. metric and other conversions.

basic navigation

SECTION A - SECTIONAL AND WAC CHARTS

Just as a motorist uses roadmaps when making a long trip, the pilot of an airplane uses aeronautical charts to assist him in finding his way. Knowledge of charts and the ability to use charts, therefore, is another of the skills of an experienced, competent pilot.

PROJECTIONS

If a pilot wanted to carry with him a completely accurate small-scale representation of the surface over which he was flying, he would have to carry a globe or at least a portion of a globe. This procedure, however, would be impractical. Therefore, for the sake of portability and convenience, the representation of the spherical surface of the globe is *flattened* and placed on a chart. In the process of flattening, however, there is a certain amount of *distortion* of the surface represented. This can be illustrated using an orange peel as an example. (See Fig. 9-1.) An orange peel, like the earth, cannot be flattened or mapped without distortion.

The techniques used in converting a portion of the earth's spherical surface to a flat, small-scale representation are known as *projections*. Each projection technique attempts to minimize distortion; however, each has its limitations and advantages. The two principal map projections used in air navigation are the *Mercator* and

Fig. 9-1. Flattening of an Orange Peel Produces Distortion

Lambert Conformal Conic Projections. (See Fig. 9-2.)

The Lambert Conformal Conic is used to a greater extent than the Mercator, especially for flight in the middle latitudes. The sectional and the world aeronautical charts, which are studied and explained in this manual, are Lambert Conformal Conic projections.

Two distinct advantages of the Lambert projection over the Mercator are:

1. A straight line on the chart approximately represents a *great circle* (a great circle is the shortest distance between two points over the earth's surface). This means that a straight-line course drawn on a Lambert chart nearly represents the shortest distance between those two points.

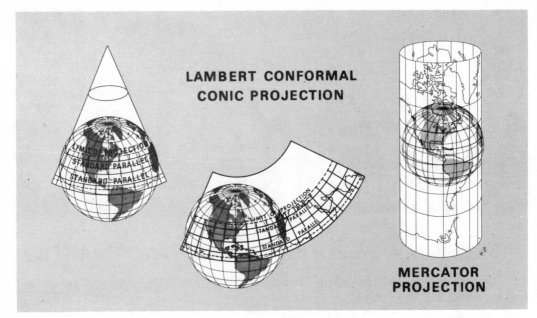

Fig. 9-2. Chart Projections

2. Scale inaccuracies (the ratio of miles on the earth to inches on the chart) are small, so the scale error may be considered negligible over a single chart.

LATITUDE AND LONGITUDE

In order to locate places on the earth's surface, or on any representation of the earth's surface, such as a chart or globe, some method of referencing an exact point must be used. The method accepted worldwide utilizes a system based on lines of *latitude* and *longitude*.

The *Equator* is a line running east and west around the earth midway between the poles. Other lines around the earth parallel to the Equator are called *parallels* or *lines of latitude*. (See Fig. 9-3.) The Equator is labeled as 0° of latitude and the other lines of latitude are numbered in degrees from 0 to 90 north of the Equator. Similarly, parallels of latitude in the Southern Hemisphere are labeled from 0° to 90° south of the Equator. (See Fig. 9-3.)

Other lines called *meridians*, or *lines of longitude*, are drawn from pole to pole; therefore, the direction along these lines

is always true north or south. These lines are also numbered starting at the *Prime Meridian*, which is the line of longitude passing through the observatory at Greenwich, England. The Prime Meridian is labeled as 0° of longitude and the other meridians are numbered in degrees east and west of the Prime Meridian.

Fig. 9-3. Parallels and Meridians (Latitude and Longitude)

There are 360° of longitude; therefore, 180° are labeled east longitude and the remaining 180° are labeled west longitude. Pilots who take their training and fly in the Western Hemisphere will always be dealing with west longitude. The line of longitude labeled 180° is called the *International Date Line* and is on the opposite side of the world from the Prime Meridian.

All charts are laid out with lines of longitude and latitude extending up, down, and across the chart. The lines of longitude and latitude are printed and labeled on the chart for each degree interval. Each degree is divided into 60 equal segments called *minutes*. The minutes, indicated by the symbol ('), are marked along the lines of longitude and latitude in 30 minute, 5 minute, and 1 minute segments. (See Fig. 9-4.)

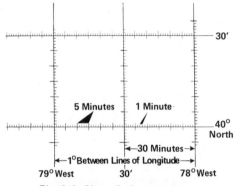

Fig. 9-4. Chart Reference System

With the system of lines just discussed, the location of objects such as airports and cities may be designated by *geographical coordinates*, or in other words, the intersection of the longitude and latitude lines on the chart. In figure 9-5, the coordinates of the airport are given as 40°15'N, 78°18'W. In this example, the airport is located 40° and 15' north of the zero line of latitude, or Equator, and 78° and 18' west of the Prime Meridian.

The pilot would first locate the parallel labeled 40° north on his chart, then count upward 15' from that line and draw a light line on the chart from the 15' mark parallel to the 40° north latitude line. (See Fig. 9-5, item 1.) Next, he would locate the 78° west longitude line and count 18' to the left, or west, of that line. He would then lightly draw a second

line through the 18' mark and parallel to the 78° west longitude line (item 2). The point where the two lines intersect is the location of the airport. It is not always necessary to actually draw the lines as suggested in this example; however, the procedure outlined should be used in locating places on aeronautical charts when the geographical coordinates are given.

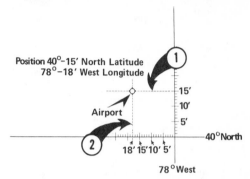

Fig. 9-5. Geographical Coordinates

THE EARTH'S TRUE NORTH POLE AND MAGNETIC NORTH POLE

The position of the earth's True North Pole, or point upon which it would appear to rotate if viewed from space, and the position of the earth's Magnetic North Pole are not the same. In fact, the Magnetic North Pole is located on Prince Wales Island, Canada, approximately 1,000 miles south of the geographic, or True North Pole. The problem introduced into a pilot's navigational planning and piloting technique is simply this: the charts which he uses are oriented toward *true north* and the compass needle in his airplane points to the *Magnetic North Pole*. This angular difference between the direction to true north and the direction to magnetic north is called *magnetic variation*, or simply *variation*. (See Fig. 9-6.)

Figure 9-6 shows that the position of the aircraft relative to the earth's surface will determine whether the compass variation is to the east of true north or to the west of true north. Also, when the aircraft's position on the earth is in alignment with

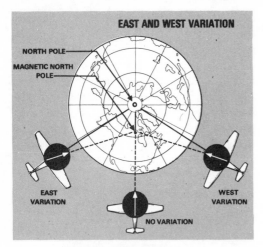

Fig. 9-6. Magnetic Variation

true north and magnetic north, no variation exists.

Most charts show variation with *isogonic lines*, which are lines connecting points of equal magnetic variation. When flying on a true north heading over the panhandle of Nebraska as seen in figure 9-7, the aircraft compass would read 348° rather than 360° because the magnetic variation is 12° east (of true north). To fly south at the same point, the aircraft compass would read 168° rather than 180° because the variation at any point is always constant regardless of the airplane heading. Figure 9-7 also shows the line of 0° magnetic variation which is termed an *agonic line*.

MEASURING DIRECTION IN DEGREES OF THE COMPASS

Direction is measured in degrees clockwise from north. The *compass rose* in figure 9-8 shows the relationship of the basic points of the compass and direction in degrees. For example, north is 360° (0°), east is 090°, south is 180°, northwest is 315°, and so on. If 0°, or north, is oriented with the True North Pole, then *true* direction is the result. If 0° is oriented with magnetic north, then *magnetic* direction results.

CHARTS

Aeronautical charts are a necessary part of a pilot's equipment for any flight away from the home base or local area. There are two basic types:

1. Charts that show all significant

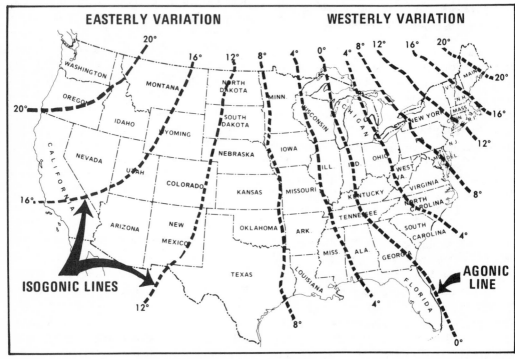

Fig. 9-7. Magnetic Agonic and Isogonic Lines

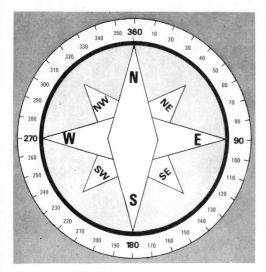

Fig. 9-8. Compass Rose

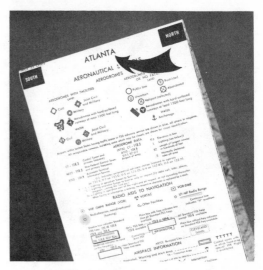

Fig. 9-9. Sectional Chart

topography (terrain, natural and man-made features) and most aeronautical information.

2. Charts that show all radio navigation facilities and routes and very little topographical information.

The beginning pilot generally flies "low and slow" and is primarily interested in topographical and some aeronautical information, while pilots of higher performance aircraft flying on airways (the highways of the sky) at high altitudes or in instrument flight conditions are interested in radio navigation and aeronautical information.

Every pilot should be familiar with the charts described in this section. The charts listed here are discussed in the relative order of their importance to the beginning pilot. A detailed discussion of the chart used with this manual appears later in this section.

Sectional charts are the charts most commonly used by private pilots. Each sectional chart shows a portion of the United States and is identified by a name of a principal city. (See Fig. 9-9.) The scale on the sectional chart is 1:500,000, or about eight miles to one inch.

The sectional chart contains topographical information which features the por-

trayal of the terrain relief and also includes visual checkpoints, such as cities, towns, villages, drainages, roads, railroads, and other distinctive landmarks used in VFR flight. The aeronautical information on the sectional chart includes visual and radio aids to navigation, aerodromes (airports, etc.), controlled airspace, special use airspace, obstructions, and related data. The sectional charts are revised and issued every six months.

Local charts are large-scale charts which show many details of highly populated areas around selected major cities and major air terminals. The local chart's scale is 1:250,000, or about four miles to one inch. These charts identify and emphasize checkpoints used for local control of VFR traffic with small flags placed prominently on the chart. (See Fig. 9-10.)

When air traffic controllers and pilots have the same references, they can better communicate and "fix" the position of an aircraft according to local checkpoints. Where local charts are available, they are included as an inset on one side of the sectional chart which covers the same area.

VFR terminal area charts are also large-scale charts (1:250,000) published to assist the private pilot in navigating within or through the terminal control areas, or

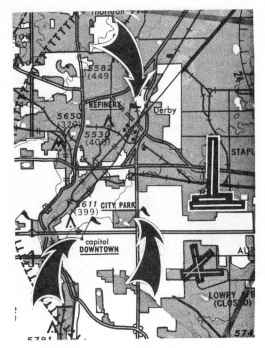

Fig. 9-10. Local Chart Excerpt

TCA's, presently established at certain of the nation's airports. (See Fig. 9-11.) As with the local chart, considerable detail is shown and VFR reporting points are also designated with small flags. The VFR terminal area chart is revised and reissued every six months like the sectional chart.

World Aeronautical Charts (WAC) are similar to sectional charts except that WAC charts are drawn to a scale which is one-half that of a sectional chart. WAC charts use a 1:1,000,000 scale, which is

about 16 miles to 1 inch and, therefore, omit some detail. WAC charts are usually used for navigation of higher performance airplanes or for long trips where frequent change of charts en route is bothersome. Each WAC chart is identified by a number. (See Fig. 9-12.) The WAC charts are revised and issued once a year.

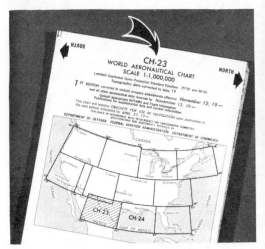

Fig. 9-12. World Aeronautical Chart

VFR/IFR planning charts are large charts designed to fulfill the requirements of preflight planning for VFR and IFR operations. The planning chart uses a scale of 1:2,333,232 in which 1 inch equals about 36 statute miles. This chart is printed in two parts in such a manner that, when assembled, it forms a composite VFR planning chart on one side and IFR planning chart on the other.

This large chart is often attached to the wall of a ground school classroom, preflight planning area at the local fixed base operator, or in the flight planning area of a flight service station. Information on the VFR chart side includes selected populated areas, large bodies of water, major drainage areas, shaded relief, navigation facilities, airports, special use airspace, and related data.

IFR avigation charts (en route) are radio facility charts. (See Fig. 9-13.) The IFR en route charts include all the information necessary for radio navigation and omit all terrain features except for large

Fig. 9-11. VFR Terminal Area Chart

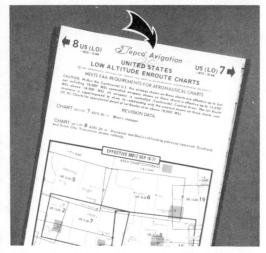

Fig. 9-13. IFR Avigation Chart

bodies of water. They are drawn to various scales, depending on the needs of the particular chart. Much of the radio information found on these charts is not available on any other chart series. The en route charts are on a four-week revision cycle; therefore, they are the most current charts available for radio navigation, radio facilities, and communication frequencies.

SECTIONAL CHARTS

The private pilot must be thoroughly familiar with sectional charts as they are the most popular chart used for VFR flying. Sufficient detail for navigation by pilotage is included on these charts, but much of the aeronautical information is represented by symbols and colors. Their meanings should be memorized because dependence on the legend makes navigation by pilotage awkward.

The name of each sectional chart, derived from a principal city on the chart, is displayed prominently at the top of *both* sides of the chart legend panel. (See Fig. 9-14, item 1.) Both the date at which the chart becomes effective, and also the date at which the chart is considered obsolete for use in navigation are also found on the same panel. (See Fig. 9-14, item 2.) It is important that the dates on the charts be checked so that flight will be made with the most current publication.

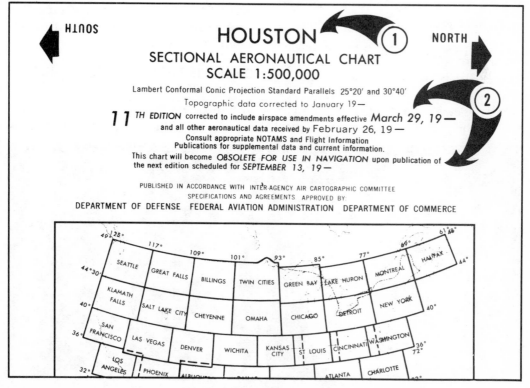

Fig. 9-14. Chart Name and Date

The convenient size of the folded sectional chart provides for easy handling, storage, and identification, especially while in flight. The sectionals are printed back to back, making it possible to cover the 48 contiguous (touching or adjacent) states with 37 charts. The two labels marked "NORTH" on the legend margin indicate that the northern half is shown when the chart is extended to the right from the legend margin. (See Fig. 9-15.)

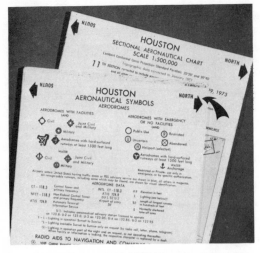

Fig. 9-15. North Half of Chart Indicators

The arrow pointing left and labeled "SOUTH" alerts the pilot to turn the chart so that the legend on the margin is *upside down*. (See Fig. 9-16.) Then, when the chart is extended to the right, the southern half of the sectional chart

is shown. The margin along the bottom is used to list such information as control tower frequencies and data concerning special use airspace.

In the right portion of the south side bottom margin are plotting instructions using the overlap feature to plot the chart to a point on the other side of the chart. (See Fig. 9-17.) By using this method, a direct course is easily plotted between points on the north and south panels.

SELECTING CHARTS

On the front panel is a box showing the names of the adjoining charts. (See Fig. 9-18.) To fly west of the area represented on the Houston Chart, the San Antonio Chart is used. It is wise for the pilot to carry additional charts of the surrounding area during flight, particularly if the course is near the edge of the chart; otherwise, in the event of an unplanned course change, deteriorating weather, adverse winds, or low fuel supply, the pilot might find himself flying over unfamiliar terrain without the necessary charts needed for safe navigation.

CHART INTERPRETATION AND USE OF SYMBOLS

Sectional charts are printed in several colors, depending on the range of eleva-

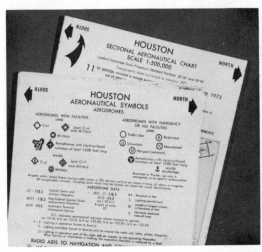

Fig. 9-16. Using South Half of Chart Indicators

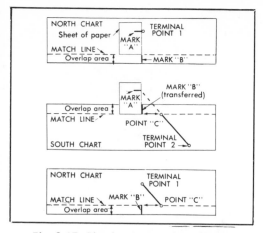

Fig. 9-17. Plotting Instructions (Excerpt)

tions encompassed by the chart. Terrain elevations below 1,000 feet are shown by light green; 1,000 to 2,000 feet by darker green; 2,000 to 3,000 feet by yellow; and elevations above 3,000 feet are shown by various shades of brown. (See Fig. 9-19, item 1.) The highest terrain elevation on a particular chart is shown at the top of the color band legend (item 2). The legend for certain elevations is also specified (item 3), and contour intervals are shown (item 4). The elevation legend for the Houston Sectional Chart, which is similar to figure 9-19, is shown on the legend panel.

Water (lakes, rivers, etc.) is shown in blue. Roads are shown by brown lines; heavy magenta lines indicate principal highways, and narrower lines indicate less heavily traveled roads. The relative size and shape of cities is indicated by a yellow area outlined in black. Yellow squares represent towns, and small towns, villages, and hamlets are represented by small circles.

Control tower frequencies are listed directly below the mileage scale on the south half of the chart. To the right of the frequency group, the special use airspace areas, which are explained later in this chapter, are tabulated and defined.

SYMBOLS USED ON THE HOUSTON SECTIONAL

To obtain the greatest benefit from the following explanation of sectional chart symbols, it is recommended that the student refer to the chart excerpt in figure 9-20 each time a new encircled number is encountered. This should help reinforce the meaning of each symbol represented.

① The information enclosed within the magenta box describes the name, frequency, three-letter identifier,

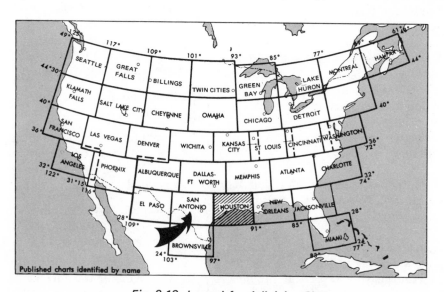

Fig. 9-18. Legend for Adjoining Chart

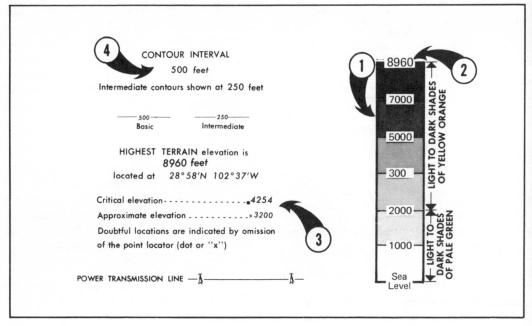

Fig. 9-19. Elevation Legend

and Morse code for the DeRidder nondirectional radio beacon (NDB). The magenta color, as used here, is used *for all* low-frequency radio information.

(2) The magenta dot pattern surrounding the Beauregard Parish Airport symbol represents the location of the DeRidder nondirectional radio beacon.

(3) Maximum terrain elevation figures centered in the areas completely bounded by *ticked* (subdivided into minutes) lines of latitude and longitude are represented in hundreds and thousands of feet, but do not include elevations of vertical obstructions. The numbers "0_5" are used to represent the elevation of 500 feet.

(4) A power transmission line and its supporting towers are represented by this symbol.

(5) A shaded light blue color represents a body of water. The black line on the southeast end of the Bundick Lake represents a man-made dam.

(6) The terrain elevation at the location of the dot is 175 feet mean sea level (MSL). This is the highest elevation in the immediate area.

(7) The inverted V represents an obstruction whose height at the top is less than 1,000 feet above ground level (AGL). The number 360 represents the elevation of this obstruction above mean sea level, and the number 250 in parentheses represents the elevation above ground level. The student should refer to the chart legend for the symbol used to represent obstructions 1,000 feet and *higher* above ground level.

(8) Shaded blue lines represent rivers or streams.

(9) The number "100" represents the MSL altitude of the adjacent contour line, or line connecting points of equal elevation.

(10) A lookout tower which is air marked for identification is shown here. The site number is 90 and the base of the tower is 89 feet MSL.

(11) The open-air theater symbol is representative of those symbols which

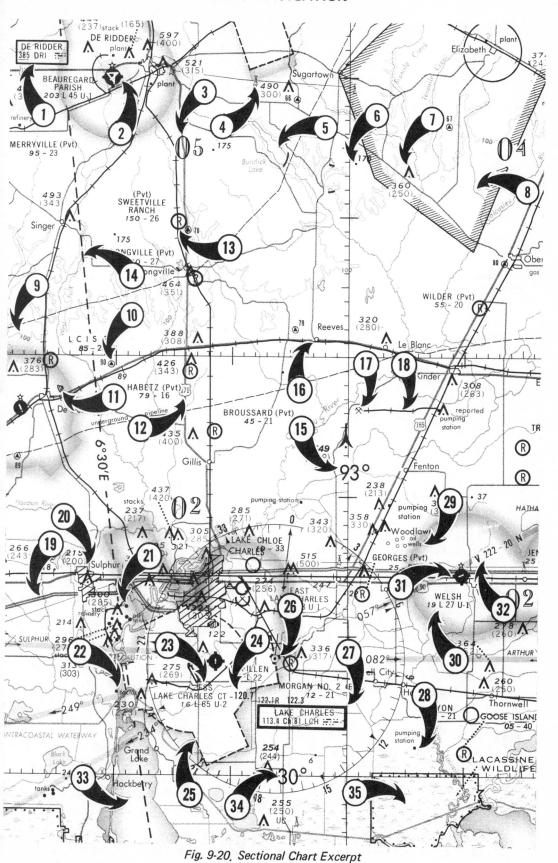

Fig. 9-20. Sectional Chart Excerpt

actually look like the objects they represent.

(12) Identifying markers, such as this one for U.S. Highway 171, are shown.

(13) This is a *restricted* airport not open to the general public. Notice in the legend by the airport symbol that Sweetville Ranch is a private field.

(14) This symbol represents an isogonic line (line of equal magnetic variation) which is 6°30' east on this chart.

(15) This line of longitude is 93° west of the Prime Meridian in Greenwich, England.

(16) The location of small towns or villages, such as Reeves, is shown with a circle.

(17) Shaft mines or quarries are represented by a symbol that looks like two crossed pick axes.

(18) This line with spaced single crossbars represents a single-track railroad. Abandoned railroads are represented by broken single-track symbols. This type of symbol may be seen on the Houston Chart from approximately 30° to 31° north latitude, running approximately parallel to the 97° west longitude line.

(19) Interstate freeways and tollways are depicted with a double magenta line.

(20) The shape and relative size of the town of Sulphur is represented by the area outlined in yellow.

(21) The black dots represent storage tanks which contain oil in this case.

(22) The small solid black square is used to represent a variety of prominent landmarks. In this case, a factory is designated.

(23) The letters "FSS" indicate that a flight service station is available on the field at the Lake Charles Airport.

(24) The Lake Charles Control Tower frequency is 120.7 MHz. When FAA communication frequencies are available from either a flight service station or a control tower, the

name of the airport will be printed in *blue* as in this example.

(25) The dashed blue line surrounding the Lake Charles Airport defines the boundaries of the control zone. A control zone surrounds airports where there are special weather (ceiling and visibility) requirements. (The details of control zones are discussed in Section B of this chapter.)

(26) This symbol represents a VORTAC station, a type of radio navigation station discussed fully in Chapter 8.

(27) The dark blue box encloses information regarding the Lake Charles VORTAC such as the VORTAC frequency which is 113.4 MHz. This station has a Flight Service Station which transmits and receives on 122.2 and 122.3. In addition, it receives on 122.1 for two-way radio communications.

(28) In this case, the solid black square represents a pumping station and is so labeled.

(29) Oil wells are represented by the closely-spaced group of circles.

(30) The Welsh Airport is 19 feet above sea level, has lighting for night operations, and has a runway that is 2,700 feet long. "U-1" representing UNICOM, a non-FAA advisory facility, is available on the frequency 122.8 MHz. The airport information and symbol are printed in *magenta* indicating that FAA communication facilities *are not* available.

(31) An airport which is illustrated with a pictorial runway representation has at least one runway which is more than 1,500 feet long. The star above the airport symbol indicates the presence of a rotating beacon. The three extensions on the sides and bottom of the airport diagram indicate that facilities such as fuel, maintenance, etc. *are* available.

(32) The airway extending northeast from the Lake Charles VORTAC on

the 057° radial is V-222/20N. This VHF airway connects two navigational facilities.

(33) This large body of water would provide an excellent VFR checkpoint.

(34) The numerals indicate that this east-west line of latitude is 30° north of the Equator.

(35) Swamps and marshes are depicted by horizontal blue lines interspersed with short vertical dashes suggesting the clumps of marsh grass common in such areas.

TABLE p. 6-50

SECTION B - AIRSPACE UTILIZATION

Certainly when the Wright brothers made their first flight, and for some time thereafter, all of the airspace was available to all of the users. As aircraft, and particularly airplanes, became more complex and their numbers and types of operations increased, the necessity for implementing regulation of the use of airspace became apparent.

When a portion of the airplanes of those early days of aviation were equipped to operate in what is now known as "instrument conditions," it was necessary to establish basic flight rules and to set aside certain portions of the airspace in which aircraft flying by instrument reference could be controlled and separated from other aircraft. It was also necessary to provide for those aircraft *not* under air traffic control (control by radio). These requirements have generated the necessity for many of the various divisions of airspace existing today.

A study of *uncontrolled* and *controlled* airspace is important to the private pilot so that he may learn the segments of controlled airspace in which he will be subject to air traffic control. A second reason for studying airspace utilization is that different weather minimums (visibility, ceiling, and distance from clouds) apply in the various types of controlled and uncontrolled airspace.

UNCONTROLLED AIRSPACE

Uncontrolled airspace is defined as that portion of airspace within which FAA Air Traffic Control (ATC) does not have authority nor the responsibility to exercise any control over air traffic.

CONTROLLED AIRSPACE

Certain elements of airspace were originally controlled only to prevent collisions of aircraft operating in instrument conditions and to exclude non-instrument equipped aircraft and non-instrument

rated pilots. However, the complexity and density of aircraft movements today have generated the necessity for establishing control functions in other airspace segments; for example, to provide for the orderly and safe flow of traffic around air terminals.

Controlled airspace now consists of those areas designated as positive control areas, continental control areas, control areas, control zones, terminal control areas, and transition areas within which some or all aircraft may be subject to air traffic control.

POSITIVE CONTROL AREA

Positive control area airspace extends from 18,000 feet MSL to flight level (FL) 600 (approximately 60,000 feet MSL), and aircraft within this airspace are required to be operated under instrument flight rules (IFR). Positive control airspace is not shown on sectional charts; however, it is depicted on en route high altitude charts. For operations within positive control areas, aircraft must be:

1. operated under IFR at specified flight levels assigned by air traffic control;
2. equipped with instruments and equipment required for IFR operations and flown by a pilot rated for instrument flight;
3. equipped with a radar beacon transponder; and
4. radio equipped to provide direct pilot-to-controller communications.

Even though non-instrument rated pilots cannot legally operate in positive control airspace, the topic is presented to better acquaint the student with the airspace utilization plan.

CONTINENTAL CONTROL AREA

The continental control area consists of the airspace at and above 14,500 feet MSL over the 48 contiguous states, the

District of Columbia, and that portion of Alaska east of longitude 160° west. The continental control area does not include the airspace less than 1,500 feet above the surface of the earth or prohibited and restricted areas. (See Fig. 9-21.) The definition of the continental control area given in the FAR's implies that there is no upward limit to this airspace.

The continental control area was established to increase the visibility and cloud clearance minimums at the higher altitudes at which high performance airplanes fly. For example, one cannot fly VFR in this airspace unless there is five miles visibility. Also, there must be one mile horizontal separation and 1,000 feet vertical separation from all clouds.

CONTROL AREAS

Control areas consist of the airspace designated as VOR federal airways, additional control areas, and control area extensions which include transition areas. These are depicted on sectional charts as shown in figure 9-22.

VOR FEDERAL AIRWAYS

Each airway is based on a centerline that extends from one radio navigational aid or intersection to another navigational aid specified for that airway. An airway is depicted in figure 9-23. Notice that the airway begins at 1,200 feet above ground level (AGL), intersects the base of the continental control area, and extends up to, but not including, 18,000 feet MSL. The centerline is shown in blue on sectional charts and the magnetic course and the airway identification number are indicated.

Each airway includes airspace within parallel boundary lines, normally four nautical miles each side of the airway centerline. Where airways pass through otherwise uncontrolled airspace, the outer limit of the airway is shown by the darker and more definite edges of the blue tinted boundaries, and the vanishing

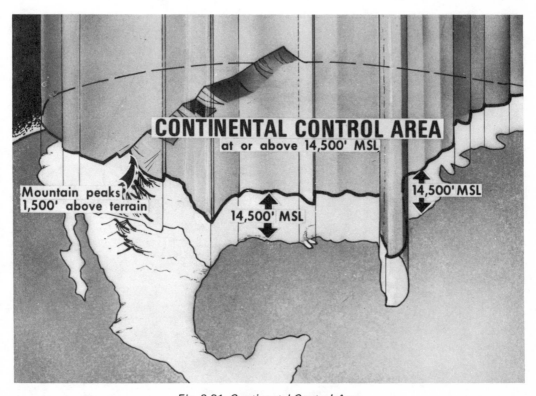

Fig. 9-21. Continental Control Area

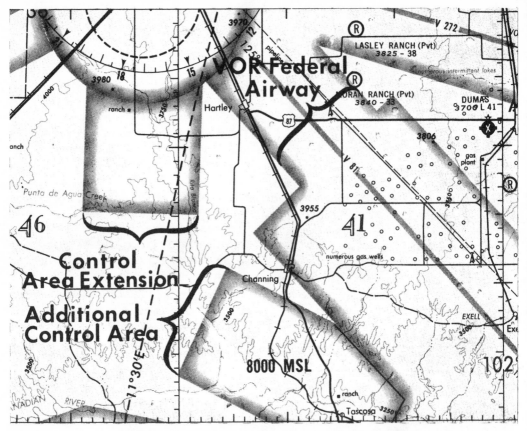

Fig. 9-22. Typical Control Areas Shown on Sectional Charts

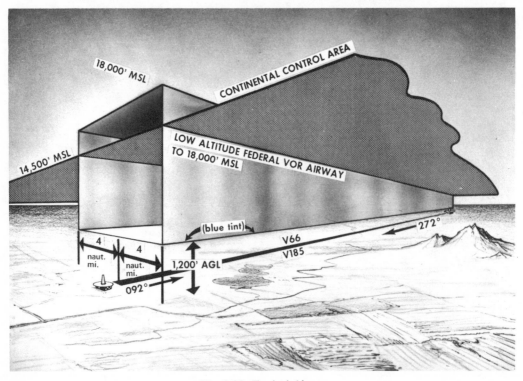

Fig. 9-23. Typical Airway

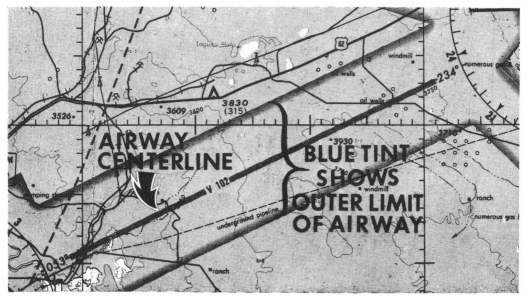

Fig. 9-24. Airway Boundaries

airspace within the airway. (See Fig. 9-24.)

TRANSITION AREAS

Transition areas from airways to airports for which an instrument approach procedure has been prescribed have been established. These transition areas extend upward from 700 feet or 1,200 feet above the surface and enable an instrument pilot to remain within controlled airspace as he descends for an instrument approach to an airport. Transition areas are shown in magenta or in blue and, as with other controlled airspace, the outer limit of the transition area is shown by the darker and more definite edge, and the feathered edges show the direction within the controlled airspace. (See Fig. 9-25.)

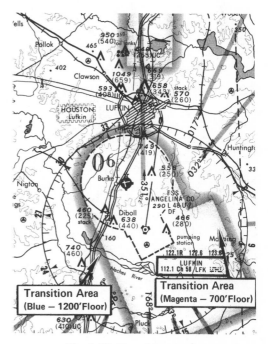

Fig. 9-25. Transition Areas

CONTROL ZONES

Controlled airspace which extends upward from the surface and terminates at the base of the continental control area is called a *control zone*. Control zones that do not underlie the continental control area have no upper limit. A control zone may include one or more airports and is normally a circular area with a radius of five statute miles and any extensions necessary to include instrument departure and arrival paths. (See Fig. 9-26.)

The control zone was established in order that air traffic control could be exercised on aircraft taking off or landing when instrument conditions exist within the control zone, and also to exclude possible conflicting VFR traffic when weather conditions are below VFR minimums.

Fig. 9-26. Control Zone

Control zones are depicted on sectional charts with a broken blue line (see Fig. 9-27), and if a control zone is effective only during certain hours of the day, this fact will also be noted on the chart. Student pilots often erroneously assume that a control zone exists only where there is a control tower; however, such is not the case. Many control zones, such as the one shown in figure 9-27, exist at airports without control towers.

In order to operate within a control zone, the ceiling must be not less than 1,000 feet and the visibility must be at least three miles at that airport. However, a special VFR clearance may be obtained from the nearest tower, ATC center, or FSS if there is one mile visibility and the aircraft can be maintained clear of clouds.

Some major terminals have experienced such high traffic density that special VFR

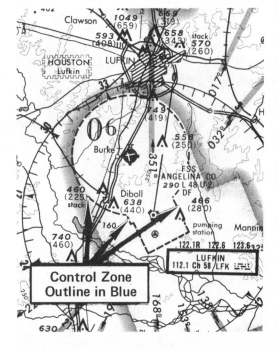

Fig. 9-27. Typical Control Zone

clearances were determined to be incompatible with maximum safety and separation of aircraft. Therefore, there are a number of terminals where special VFR clearances are not issued. The control zone outline where special VFR is not authorized is depicted as a series of the letter "T."

AIRPORT TRAFFIC AREA

In order to provide for the separation and orderly flow of traffic around many of the nation's larger airports, control towers and mandatory two-way radio communications were established. The *airport traffic area* is the airspace within which the control of all air traffic is exercised and is normally defined as the airspace within five statute miles of an airport at which a control tower is in operation.

This airspace extends from the ground up to, but not including, 3,000 feet above ground level. (See Fig. 9-28.) It is important for the student to recognize that the boundaries are not depicted on aeronautical charts, but the presence of an airport traffic area is indicated when the letters "CT" and control tower radio frequencies are shown in the airport data listing. (See Fig. 9-29.)

AIRPORT ADVISORY AREA

The area within five statute miles of an airport where a control tower is *not* oper-

Fig. 9-28. Airport Traffic Area

ating but where a flight service station *is* located is called an *airport advisory area.*

At such locations, flight service stations provide advisory service to arriving and departing aircraft. On sectional charts, the letters FSS above the other airport data indicates the availability of airport advisory service.

TERMINAL CONTROL AREAS

Terminal control areas have been established to separate all arriving traffic, both VFR and IFR, at large airports. Although each terminal control area is designed to facilitate traffic separation at a particular terminal and, therefore, is different from other terminal control areas, the TCA is generally shaped like an upside-down wedding cake. (See Fig. 9-30.)

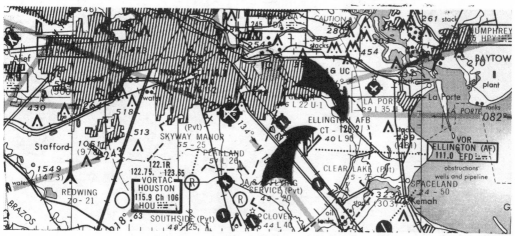

Fig. 9-29. Indicators of an Airport Traffic Area

BASIC NAVIGATION

Fig. 9-30. Terminal Control Area

Within the expanding controlled airspace of the terminal control area, all aircraft are subject to operating rules and pilot

and equipment requirements as specified in Part 91.90 of the FAR's. Each such location includes at least one primary airport around which the TCA is located.

A plan view of the Atlanta Terminal Control Area is shown in figure 9-31. Notice that the controlled airspace is divided into ceilings and floors. For instance, notice that the ceiling for the entire terminal control area is 8,000 feet and that area B is controlled from a floor of 2,500 feet up to this elevation. A pilot wishing to operate through this area without meeting the necessary equipment and pilot requirements could do so if he were either below 2,500 feet or above 8,000 feet.

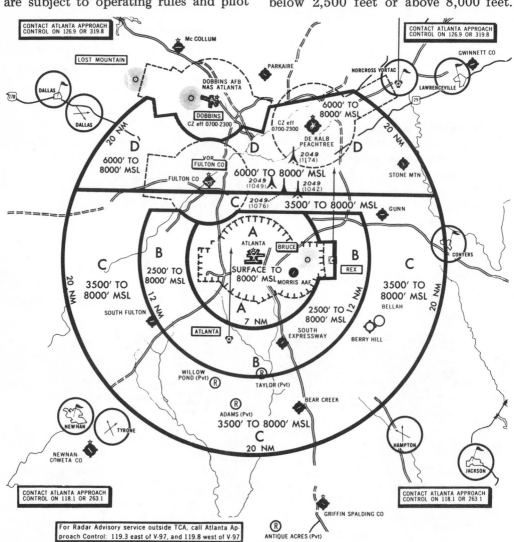

Fig. 9-31. Atlanta TCA Planview

TCA's are outlined with a blue band (1/8 in. wide) on sectional charts and detailed, expanded scale terminal area charts are available for each TCA.

SPECIAL USE AIRSPACE

Special use airspace consists of that airspace where the nature of certain activities within the airspace, or in proximity to the airspace, requires limitations upon aircraft operation. These areas are shown on sectional charts and are listed by type as follows:

1. Prohibited area.
2. Restricted area.
3. Warning area.
4. Intensive student jet training area.
5. Alert area.

PROHIBITED AREAS

Prohibited areas are portions of the airspace within which the flight of aircraft is not permitted. Such areas are generally established for reasons of national security or national welfare. An example of a prohibited area is the area surrounding the Rocky Mountain Arsenal near Stapleton International Airport, Denver, Colorado. (See Fig. 9-32.)

Fig. 9-32. Prohibited Area

The same type of symbol is used to show prohibited, restricted, warning, intensive student jet training areas, and alert areas; however, the number of each special use airspace area is preceded by a letter designating its type; i.e., P indicates prohibited; R — restricted; W — warning; and A — alert.

A listing of prohibited, restricted, warning, and alert areas appearing on the Houston Sectional Chart is found at the bottom of the south panel. Notice that the last entry at the right indicates the appropriate authority to contact if transition through a restricted or warning area is desired.

RESTRICTED AREA

Restricted areas include the airspace within which the flight of aircraft, while not wholly prohibited, is subject to certain limitations. Restricted areas denote the existance of unusual, but often invisible, hazards to aircraft, such as artillery firing, aerial gunnery, or the flight of guided missiles. Penetration of restricted areas without authorization from the controlling agency may be extremely hazardous to the aircraft and its occupants.

WARNING AREA

A *warning area* is that airspace which may contain hazards to non-participating aircraft in international airspace; therefore, warning areas are established beyond the three-mile limit. Although the activities conducted within warning areas may be as hazardous as those in restricted areas, warning areas cannot be legally designated as restricted because they are over international waters.

INTENSIVE STUDENT JET TRAINING AREAS

The VFR pilot should be aware of *intensive student jet training areas* and the potential hazards created by numerous high-speed jet aircraft. These areas are shown on sectional charts and information may be obtained from any flight service station within 100 miles of the area.

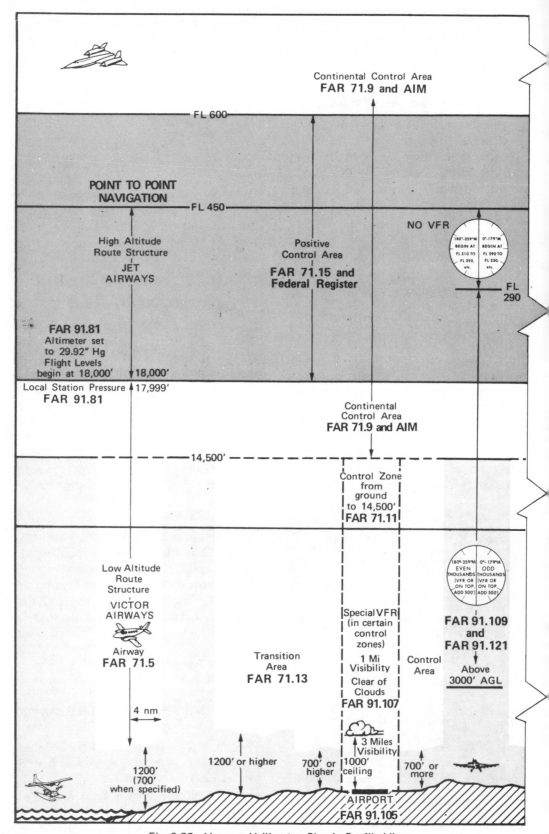

Fig. 9-33. Airspace Utilization Plan in Profile View

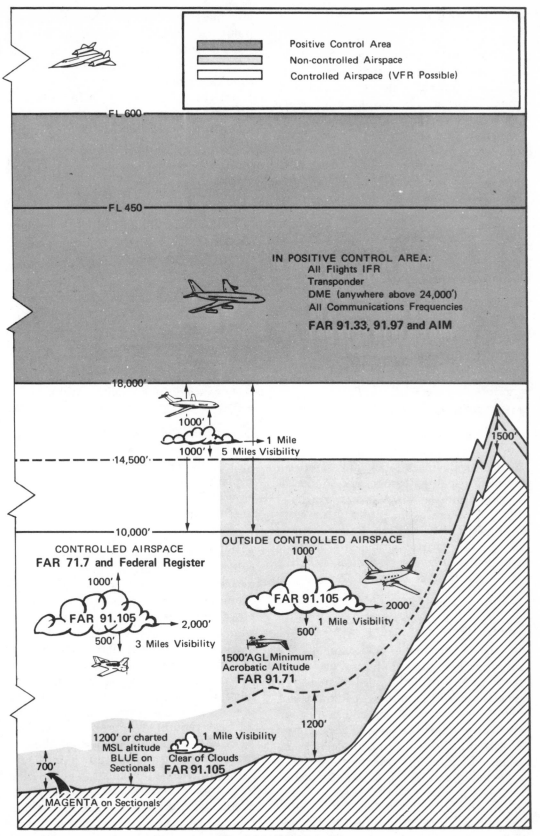

Positive Control Area
Non-controlled Airspace
Controlled Airspace (VFR Possible)

FL 600

FL 450

IN POSITIVE CONTROL AREA:
All Flights IFR
Transponder
DME (anywhere above 24,000')
All Communications Frequencies
FAR 91.33, 91.97 and AIM

18,000'

1000'
1 Mile
1000'   5 Miles Visibility

14,500'

1500'

10,000'

CONTROLLED AIRSPACE
FAR 71.7 and Federal Register

OUTSIDE CONTROLLED AIRSPACE
1000'

1000'
FAR 91.105
FAR 91.105   2,000'
500'   3 Miles Visibility

2000'
1 Mile Visibility
500'

1500' AGL Minimum
Acrobatic Altitude
FAR 91.71

1200'

1200' or charted
MSL altitude
BLUE on
Sectionals   **FAR 91.105**
1 Mile Visibility

Clear of Clouds

700'

MAGENTA on Sectionals

Fig. 9-33. (Continued)

ALERT AREA

Alert areas contain airspace which is depicted on sectional charts to inform transient or cross-country pilots of areas that may contain a high volume of pilot training or an unusual type of aerial activity. The example in figure 9-34 shows an alert area where a high volume of Army helicopter and airplane pilot training activity is occurring. Flight within alert areas is not restricted, but pilots are urged to exercise extreme caution.

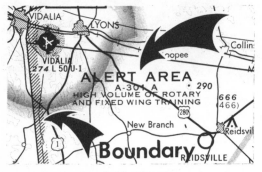

Fig. 9-34. Alert Area

MILITARY OPERATIONS AREAS

ISJTAs and alert areas are being replaced by military operation areas (MOAs). On aeronautical charts, they are identified by names such as Tarheel and Moody 1. Flight within an MOA is not restricted for VFR aircraft, but pilots are urged to exercise extreme caution. Both the pilots of aircraft participating in activities within the area and pilots transiting the MOA are fully responsible for collision avoidance.

COLLISION AVOIDANCE

In the VFR flight environment, the responsibility for collision avoidance rests entirely with the pilot. Visual scanning is one of the pilot's most important techniques for practicing collision avoidance. It requires the pilot to concentrate his attention *outside* the airplane, using head and eye movements to make a systematic sweep of the entire area.

Effective collision avoidance techniques are especially important in airport operations, where the volume of traffic usu-

ally increases. The pilot should make a point of checking that both the approach and departure paths are clear prior to takeoff. During climbout, he should accelerate the aircraft to cruise climb speed as soon as a safe altitude is reached in order to reduce pitch and increase forward visibility. During operations from an airport without an operating control tower, the pilot should broadcast his intentions on the appropriate frequency, as indicated in the *Airman's Information Manual.* In addition, the pilot should utilize such aids as airport advisory service and radar traffic information service, when available.

An FAA program called "Operation—Lights On" provides an added margin of safety for operations in the vicinity of airports. Under this program, pilots are requested to turn on their landing/taxi lights whenever they are operating within 10 miles of an airport, and to operate their anticollision lights anytime the engine is running.

During all flight operations, the pilot must maintain an awareness of the blind spots inherent in the design of the airplane. Clearing turns must be employed to expose the airspace hidden by the fuselage and wings.

Another excellent collision avoidance aid is the spot method for determining the collision potential of another aircraft within the pilot's field of vision. For example, if the other aircraft appears to move up in relation to a spot on the windshield, door post, or engine cowling, it will pass above the observer's aircraft. If it moves down, the aircraft will pass below. If there is no movement in relation to the spot, the flight paths of the two aircraft are on converging courses requiring appropriate action. A careful study and review of the airspace utilization plan shown in figure 9-33 will assist the student in correlating the various divisions of airspace. Compliance with the visual flight rules applicable to each airspace segment is a basic collision avoidance precaution.

SECTION C - PLOTTER AND WIND TRIANGLES

CONVENTIONAL NAVIGATION PLOTTER

The navigation plotter, developed to assist the pilot in flight planning, combines the protractor, straightedge, and scales into one convenient and compact device to serve the needs of pilots. The most conventional style of navigation plotter is shown in figure 9-36. It is a precision instrument made of clear plastic so that chart details can be viewed through the plotter. Since it is made of plastic, care should be taken to keep it from being exposed to high temperatures or prolonged exposure to direct sunlight, such as might be encountered if left on top of the airplane instrument panel. Such abuse could cause the plotter to warp out of shape and to shrink in size.

The direction of a course is measured by the protractor or semicircular scale of the plotter. It is labeled from 0° to 180° on the outside scale and 180° to 360° on the inside scale. (See Fig. 9-35.)

The straightedge portion of the plotter is used to measure distance and draw the course line from the center of the departure airport to the center of the

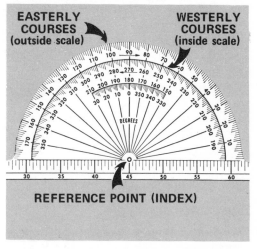

Fig. 9-35. Protractor Portion of Plotter

arrival airport. This line, when measured from true north, is called the *true course*. The plotter shown in figure 9-36 incorporates the following features in the straightedge portion of the plotter:

1. On one side of the straightedge portion of the plotter, the mileage scale matches that of the sectional chart which has a 1:500,000 scale. (See item 1.)

   As shown by item 2, the scale along the bottom of the straightedge is

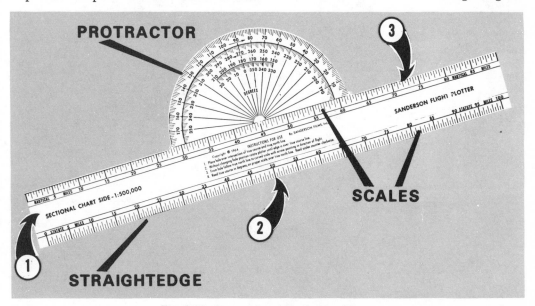

Fig. 9-36. Conventional Navigation Plotter

graduated in statute miles. The scale along the top of the straight-edge is graduated in nautical miles. (See item 3.)

2. The opposite side of the plotter displays both statute and nautical mile scales for the WAC, or world aeronautical charts, which have a scale of 1:1,000,000.

3. Concise instructions on how to use this device are provided in the center area of the plotter. For convenience, these instructions appear on both sides.

USING THE PLOTTER

The plotter is used to measure the true course angle as follows: (See Fig. 9-37.)

1. Place the small hole in the center of the plotter directly over an intersection of the true course line and one of the longitude lines printed on the chart.

2. Next, **align the edge of the plotter** that is adjacent to the protractor part of the plotter with the true course line, being careful to keep the hole in the plotter over the intersection of the longitude and the true course lines.

3. Read from the scale graduation of the protractor that lies directly over the longitude line. The outside scale is used to measure easterly courses, and the inner scale is used to measure westerly courses. In this case, the true course line is easterly and measures 097°.

The plotter is used to measure the course distance as follows: (See Fig. 9-38.)

1. Determine the correct side of the plotter (in this case the sectional chart side) to use for the measurement.

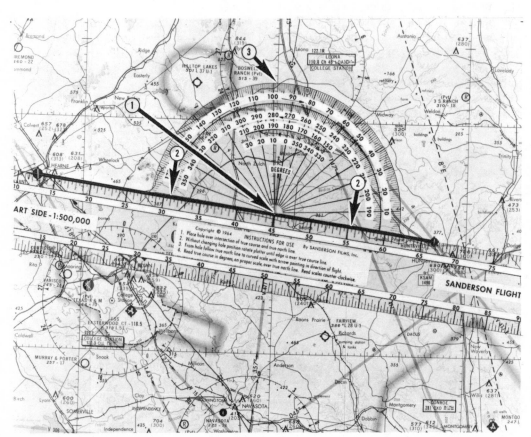

Fig. 9-37. Measuring the True Course

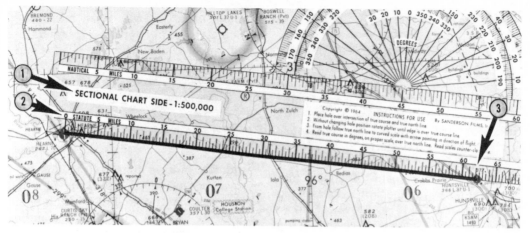

Fig. 9-38. Measuring Course Distance

(2) Place the statute miles scale parallel to the true course line and align the zero mile mark with the center of the airport from which the departure is being made.

(3) Read the statute mile mark closest to the center of the airport of arrival. In this example, the distance from Hearne Airport to Huntsville Airport is 62 statute miles.

USING THE NORTH-SOUTH SCALES

Sometimes in the process of flight planning, a true course is laid out and is so directly northerly or southerly that the course does not cross a meridian. Rather than extend the course until it intersects a meridian, the pilot may use a parallel of latitude instead and apply the special innermost north-south scales provided on the plotter. With a northerly course, as shown in figure 9-39 the edge of the plotter is aligned with the course line. The intersection of the course line and the latitude line is directly under the small index hole of the plotter. The true course of 004° is read at the point where the north scale directly overlies the latitude line. It should be remembered that the outer special scale is used to measure the direction of southerly courses, and the inner special scale in this example is used to determine the direction of northerly courses. Each reading should be checked to be certain that it coincides with the direction of flight.

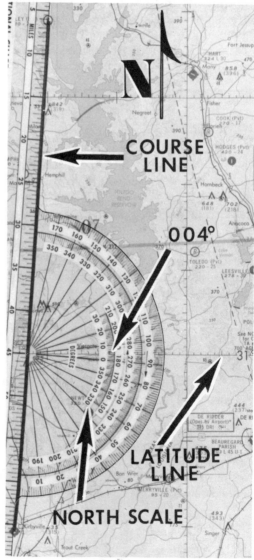

Fig. 9-39. Measuring a North-South True Course Line

WIND TRIANGLES

WIND EFFECTS

After a true course is plotted, the effects of wind speed and direction at the planned flight altitude must be considered. Forecast wind information for winds aloft is obtained from the Weather Bureau and flight service stations.

Wind is the movement of an air mass over the Earth. Its motion is made apparent through blowing trees, choppy water, or the tugging of clothing. If it were a balloon, the airplane would simply move with the air mass and drift above the terrain. Since the airplane moves under its own power however, the speed and direction it travels over the terrain is the result of combining aircraft speed, wind speed, aircraft heading, and wind direction.

WIND SPEED

The Weather Bureau reports the speed of the wind in knots, and wind direction is always the direction *from* which it is blowing. Wind speed is the rate of progress of an air mass over the Earth. The term *knots*, when referring to speed, means nautical miles per hour. Remember that a nautical mile, which is 6,076 feet long, is equal to one minute of latitude. The term *miles per hour* means statute miles per hour. A statute mile is 5,280 feet long.

AIRSPEED

Indicated Airspeed

An airplane's speed through the air mass is shown on the airspeed indicator. This indicated airspeed has no direct correlation with the speed over the ground. Only at sea level on a standard day are the indicated airspeed and the aircraft's actual speed through the air the same.

True Airspeed

In the thin air of high altitudes, compensations must be made for variances in pressure and temperature to find the actual speed through the air mass. When calculated corrections for temperature and pressure altitude are applied to indicated airspeed, the adjusted airspeed is called *true airspeed* and is expressed in miles per hour, or knots.

Ground Speed

Another type of speed that affects the aircraft's time in flight is called *ground speed*, because it relates to the aircraft's progress over the ground. If there is no wind, the airplane's true airspeed and its ground speed will be the same. However, although true airspeed may be constant, the headwind or tailwind affects the ground speed.

For example, an airplane has a true airspeed of 100 miles per hour. It is flying in an air mass which is moving in the same direction at 20 miles per hour. The aircraft is considered to have a tailwind, which is assisting its progress over the ground. The airplane's ground speed then is 120 miles per hour; however, the true airspeed is unaffected by wind. (See Fig. 9-40.)

Reversing this example, assume the airplane is moving against a 20 mile-per-hour air mass. Its progress over the ground is determined by subtracting the 20 mile-per-hour air mass speed from its 100 mile-per-hour true airspeed. The ground speed is now 80 miles per hour. (See Fig. 9-40.)

DRIFT

An airplane, in figure 9-41, is heading east in an air mass moving from the south at 20 miles per hour, so the aircraft has a crosswind. At the end of one hour, the plane will have drifted 20 miles north of the true course line because of the air mass movement.

When flying from point to another the airplane must not be allowed to drift off course, nor can slow ground speed be permitted to extend time en route be-

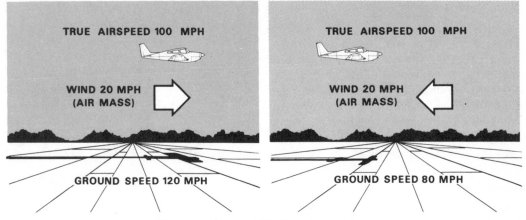

Fig. 9-40. Effect of Wind on Ground Speed

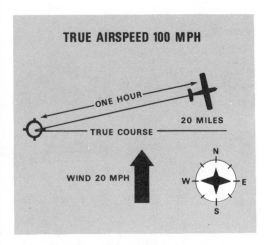

Fig. 9-41. Effect of Wind on Track

yond fuel endurance. Wind corrections and fuel stops must be determined by careful preflight planning.

TRUE HEADING

The direction that an aircraft is actually heading, as measured in degrees from true north, is called *true heading*. The path it makes over the ground is called *track*. A specific heading must be maintained to keep the track directly over the planned true course line.

WIND TRIANGLE SOLUTION

Preflight planning, to determine the effect of the wind, requires solving a wind triangle which is simply a pictorial display of the wind, true course, true heading, true airspeed, and ground speed. A wind triangle uses vectors to represent directions and magnitude of those values. The flight plotter may be used to construct a wind triangle and to determine the effects of wind. (See Fig. 9-42.) A flight computer is used most often for the same purpose.

WIND TRIANGLE CONSTRUCTION

For the construction of the sample wind triangle, the following values will be used:

True Course . . . . .043°
Distance. . . . . 66 nautical miles
True Airspeed . . . . .114 knots
Wind . . . . 090° (true) at 20 knots

By using nautical distances and speed, conversion of wind speed to miles per hour is unnecessary to work the problem since the wind is always reported in knots.

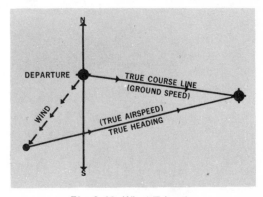

Fig. 9-42. Wind Triangle

North-South Line

In order to solve this wind problem, a vertical or north-south line is first drawn on a sheet of paper. This represents a meridian, or true north line, and is the basis from which all directions will be plotted. Then, a small mark is placed on the line as a point from which to begin drawing the wind triangle. This represents the point of departure.

Course Line

Next, the plotter is positioned so that the hole in the protractor is centered over the mark, or point of departure, with the 43° mark on the protractor aligned with the north-south line. Now, a short reference line is drawn along the top edge of the plotter. (See Fig. 9-43.)

Using the straightedge as a guide, a line is drawn so that it passes through the departure point and the short reference line. This represents the true course or route to be followed between the airports. This true course is labeled "TC 043°." (See Fig. 9-44.)

Wind Line

Next, the plotter is rotated until 090°, which is the wind direction, is on the

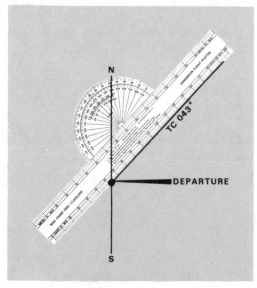

Fig. 9-44. True Course Line

north-south line. Then, a short 090° reference line is drawn along the straightedge to locate the wind direction. (See Fig. 9-45.)

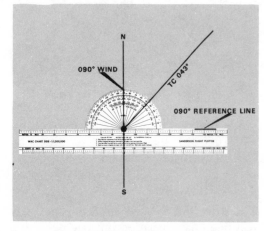

Fig. 9-45. Wind Reference Line

With the straightedge aligned on the wind mark, a line is drawn beginning at the point of departure to the left to the 20 nautical mile mark. Remember, this is the direction *from* which the wind is blowing. (The WAC chart side of the plotter is normally used because the distances are scaled to smaller increments which are adaptable for use on convenient-sized paper.) After measuring the wind line, an arrow is drawn at the end

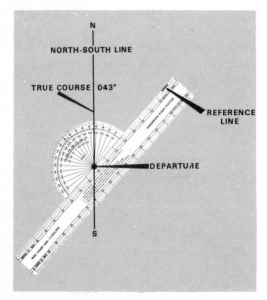

Fig. 9-43. Drawing Reference Line

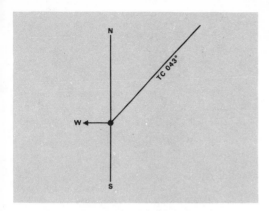

Fig. 9-46. Wind Line

of it and the letter "W" is added to label this line as representing wind. (See Fig. 9-46.)

PLOTTING AIRSPEED

From pre-trip planning it is known that the aircraft will have a true airspeed of 114 knots at the cruising altitude. Using the WAC side of the plotter, the zero mark of the nautical scale is placed over the plotter so the 114 mile mark intersects the true course line. (See Fig. 9-47.)

A line connecting those two points is drawn and is labeled "true airspeed, 114 knots." The intersection of the true airspeed and true course lines represent the position of the aircraft at the end of one hour. This point is labeled "P:: for position. (See Fig. 9-47.)

MEASURING THE GROUND SPEED

The ground speed, or distance flown in one hour, is measured along the true course line from the point of departure to the symbol "P." In this case, the ground speed measures 99 knots. Remember, the true course and ground speed are always represented by the same line in wind triangle solutions. (See Fig. 9-48.)

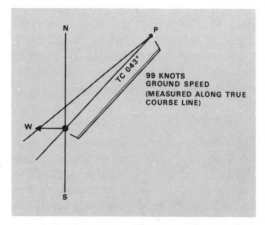

Fig. 9-48. Ground Speed Line

MEASURING THE TRUE HEADING

The true heading is measured by placing the plotter with the protractor hole centered over the intersection of the airspeed and north-south lines with the straightedge aligned with the true airspeed line. (See Fig. 9-49.) The north-south line crosses the curved edge of the

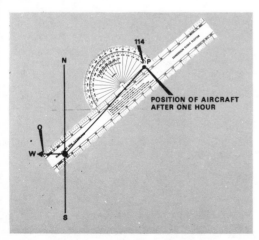

Fig. 9-47. True Airspeed Line

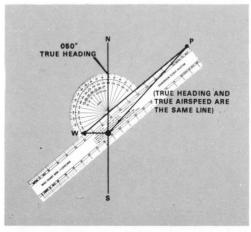

Fig. 9-49. Finding True Heading

protractor at 050°, which is the true heading. In wind triangle problems, the true heading and true airspeed are on the same line. The direction of the line is true heading and the length of the line is the true airspeed.

True heading could also be determined by measuring the number of degrees between the true course and the true heading lines. With the protractor centered over point "P," the angle is seven degrees, which is called the *wind correction angle*. It is the number of degrees the airplane must be turned into the wind to maintain a planned track over the ground. (See Fig. 9-50.)

WIND CORRECTION ANGLE

The wind correction angle must be applied to the true course to find true heading. If wind is from the right of the true course, the angle is added to the true course. If wind is from the left, the wind correction angle is subtracted from the true course. In the previous problem,

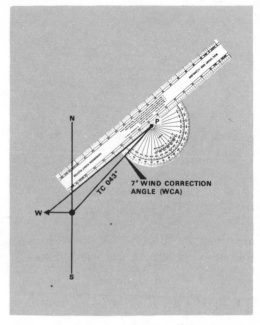

Fig. 9-50. Measuring Wind Correction Angle

the wind is from the right, so the seven degree wind correction angle must be *added* to the true course of 043°, resulting in a true heading of 050°.

SECTION D- DEAD RECKONING NAVIGATION (FLIGHT PLANNING)

SELECTING THE CHARTS

The first step for a pilot in planning a trip is to select his course. In this instance, it is assumed that a pilot desires to fly from Jefferson County Airport, southeast of Beaumont, Texas, to Rusk County Airport near Henderson, Texas.

With this information in mind, the pilot refers to the Houston Sectional Aeronautical Chart on which his airport of departure is located. He finds that the point of departure and his destination are both on the same side of the chart. If

this were not the case, he would use the procedure printed on the chart's lower margin to plot the course from one side of the chart to the other. (See Fig. 9-51.)

Since Rusk County Airport is very close to the northern edge of the Houston Sectional Chart, the pilot refers to the sectional chart diagram printed on the index margin for the names of bordering charts. (See Fig. 9-52.) He finds that the Dallas-Fort Worth and the Memphis Sectional Charts border the northwest area of the Houston Sectional and will be

---

PLOTTING DIRECT ROUTE ON NORTH/SOUTH CHART

The dashed lines in the sketches below represent the match line for the NORTH and SOUTH chart. This is indicated on the printed charts by the termination of the elevation tints. The overlap area, which is void of tints, is the area band between the match line and the papers' edge.

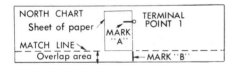

Step No 1
On the side of the chart having a terminal point nearer the match line, place a sheet of paper (or chart) so that one edge corresponds to the match line while another edge intersects Terminal Point 1. Make a mark on the edge of the paper at Terminal Point 1 (Mark "A" as illustrated) and also another mark on the chart extending from match line to the edge of the chart (Mark "B" as illustrated)

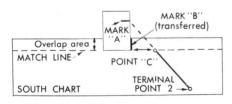

Step No. 2
Roll the chart over and transfer Mark "B" from the north chart to the south chart, extending the mark to the match line. Align the sheet of paper to the match line with the corner of the sheet at the transferred Mark "B". With a straight edge draw a line from Terminal Point 2 to Mark "A" to intersect the match line forming Point "C".

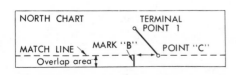

Step No. 3
Turn chart over and transfer Point "C" from the south chart to the north chart. (This can be done by measuring the distance from Mark "B" to Point "C" along the match line from the south chart). With a straight edge draw a line from Point "C" to Terminal Point 1. This line is the continuation of the direct route drawn on the south chart.

Fig. 9-51. Continuing Course Line

valuable in addition to the Houston Sectional Chart. It is wise to carry bordering charts when flying over an area near the edge of a chart because the additional charts may be needed to navigate to another airport in the event of bad weather.

The next step is to unfold the Houston Sectional Chart and locate the departure and destination airports. The chart is folded in accordion-pleated fashion for convenient reference to any point on one side, and also makes it easy for the pilot to refer back to the margin containing the legend.

DRAWING THE COURSE

Now, the pilot draws the true course from Jefferson County Airport in a color that is easily seen. (See Fig. 9-53.) If the trip is over 100 miles, as in this example, a long straightedge is needed to draw the true course. A yardstick is very good for this purpose; however, if one is not available, the edge of another sectional chart can be used.

CHECKPOINTS

After the true course has been drawn, prominent checkpoints are selected along the course and circled with a color that can be easily seen. These checkpoints will be needed during the flight to verify positions and to make checks on groundspeed. The reasons for choosing these checkpoints will be discussed and explained as the student "flies" the trip in chapter 12 of this publication.

To help in planning a trip, it is recommended that the pilot use a navigation log to record all of the data that is pertinent to his trip. An example of such a log is shown in figure 9-54. Once the checkpoints are selected, they are recorded in a navigation log so that they can be referred to during pre-trip planning and in flight. (See Fig. 9-55, item 1.)

The next element in planning is the measurement of distances using the navigation plotter. Then the distance between each checkpoint is recorded in the navigation log. (See Fig. 9-55, item 2.) The total distance is found to be 159 statute miles which is entered in the navigation log. (See Fig. 9-55, item 3.) Now the "distance remaining" column can be filled in. This will allow the pilot to

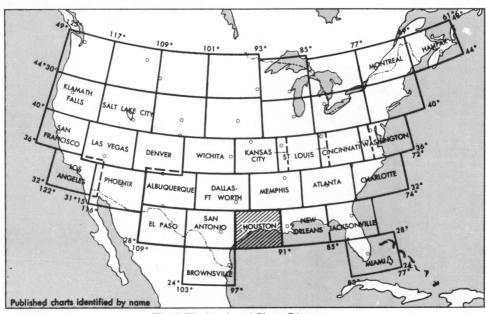

Fig. 9-52. Sectional Chart Diagram

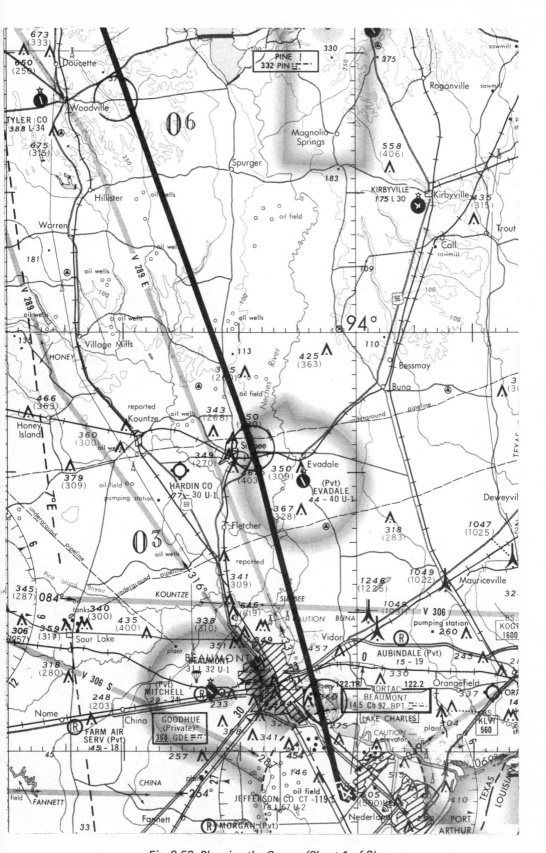

Fig. 9-53. Planning the Course (Sheet 1 of 2)

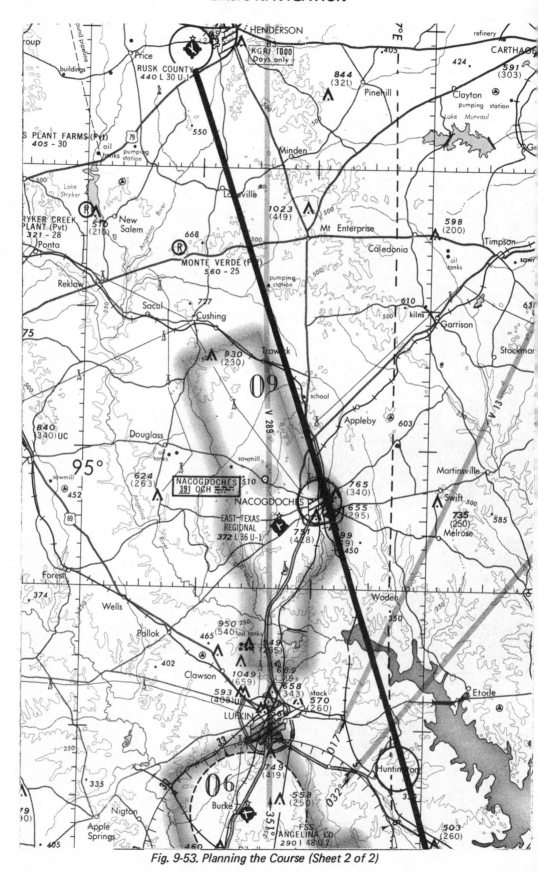

Fig. 9-53. Planning the Course (Sheet 2 of 2)

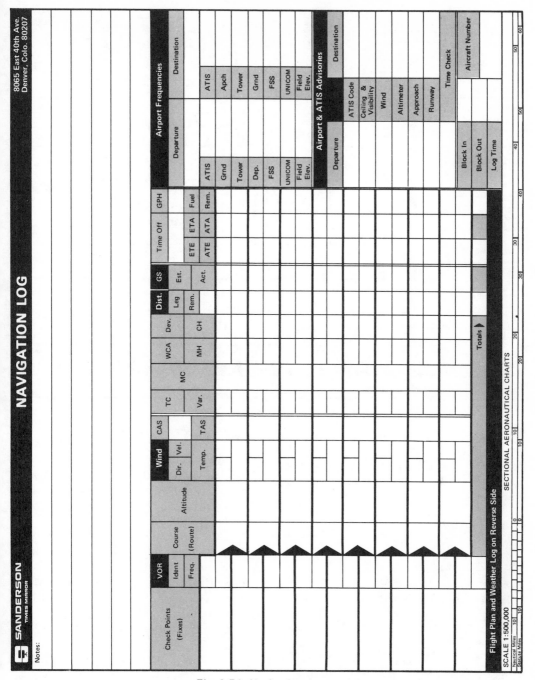

Fig. 9-54. Navigation Log

recompute his destination ETA over any checkpoint in the event of a *change* in ground speed.

MEASURING THE TRUE COURSE

With the plotter positioned so that the small hole in the plotter is centered over the intersection of the course line and a line of longitude, the true course can be read directly from a semicircular scale. The correct semicircular scale is the one that has the arrow pointing in the same direction as the course. (See Fig. 9-56, item 1.) In this case, the inner scale is the correct one. The true course is 342°. (See Fig. 9-56, item 2.) This means that

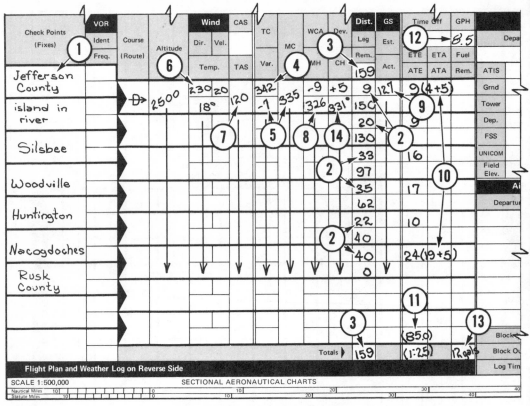

Fig. 9-55. Navigation Log Excerpt

Check Points (Fixes)	VOR Ident Freq.	Course (Route)	Altitude	Wind Dir. Vel. / Temp.	CAS / TAS	TC / MC / Var.	WCA / MH	Dev. / CH	Dist. Leg / Rem.	GS Est. / Act.	Time Off ETE / ATE	GPH ETA / ATA	Fuel / Rem.	
Jefferson County		⑥ ⊕→ 2500		230 20 18°	120	342 / 335	-9	+5	159 9	127	⑫ 8.5 9(4+5)			Depa / ATIS / Grnd
island in river						326	331°	150		⑨ 9			Tower
								20					Dep.
Silsbee								130	②				FSS / UNICOM
								33	16				Field Elev.
Woodville								97		⑩			Ai
								35	17				Departu
Huntington								62					
								22	10				
Nacogdoches								40					
								40	24(19+5)				
Rusk County								0					
										⑪			
								③	(85.0)			⑬	Block
			Totals ▶					159	(1:25)	12 gals			Block Ou
Flight Plan and Weather Log on Reverse Side													Log Tim

SCALE 1:500,000   SECTIONAL AERONAUTICAL CHARTS
Nautical Miles
Statute Miles

the true course is 342° clockwise from a line projected through the true North Pole. The pilot now records the true course in the navigation log. (See Fig. 9-55, item 4.)

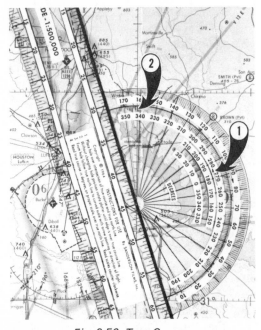

Fig. 9-56. True Course

MAGNETIC VARIATION

The next item to be determined is the *magnetic course.* Actually, the magnetic course is nothing more than the true course corrected for *magnetic variation.*

For the pilot to clearly understand magnetic variation, he must know the reason for magnetic variation and how it affects his *preflight* planning. As noted previously, variation is the angular difference between the True North Pole and the Magnetic North Pole. The True North Pole is the geographic North Pole of the earth where all the lines of longitude converge. The Magnetic North Pole is where the lines of force of the earth's magnetic field converge at a point approximately 1,000 miles south of the True North Pole. Lines drawn from a position on the earth's surface to the True North Pole and to the Magnetic North Pole can show an angular difference in direction.

In review, figure 9-57 shows three aircraft with their *noses* pointing toward true north. The aircraft to the left has its *compass* pointing 10° to the east. The center aircraft is flying on a line of no variation, called the agonic line. True north and magnetic north are in direct alignment in this area. The aircraft on the right has its compass pointing 10° to the west of true north. This means that there is a 10° westerly variation in this area. It is this variation for which the pilot compensates when he corrects the true course to a magnetic course.

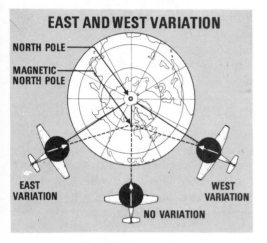

EAST AND WEST VARIATION

Fig. 9-57. Variation

There is an old saying that has been in aviation circles for years that *east is least and west is best*. This means that east variation is subtracted and west variation is added to the true course to obtain the magnetic course. This saying makes it easy to remember the magnetic variation rule.

The magnetic course is determined by the following formula:

TC - East Var. = MC

TC + West Var. = MC

MAGNETIC COURSE

The first step in solving the magnetic variation problem is to refer to the sectional chart and look for a magnetic variation line along the course. The sectional chart shows a line of equal variation labeled 7° east. (See Fig. 9-58.) As noted in a previous chapter, this line is known as an *isogonic line*. If the course crosses several lines of variation, as on long trips flying

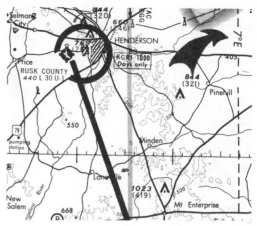

Fig. 9-58. Magnetic Variation

east or west, several values for variation must be used. However, on a short flight, the one used in the magnetic heading calculation would be the one *halfway* between the departure and destination.

The magnetic course is then determined by taking the true course of 342° and subtracting the 7° easterly magnetic variation in this area. This gives a magnetic course of 335°, which must be flown to make good a true course of 342°. Now, the magnetic course of 335° is recorded in the navigation log. (See Fig. 9-55, item 5.)

OBTAINING WEATHER INFORMATION

For preflight planning, flight service stations and National Weather Service Offices stand ready 24 hours a day to make their wealth of weather information available to the pilot.

It is best to obtain a weather briefing in person. This procedure gives the pilot the opportunity to examine the weather reports, forecasts, and charts himself and to receive a complete weather picture for his proposed flight. If, however, this is impossible, a telephone briefing is the next best method. The telephone numbers of flight service stations and National Weather Service Offices for weather briefings are found in Part 2 of the *Airman's Information Manual.*

When telephoning for a weather briefing, the individual should identify himself as a pilot, give his aircraft identification number, state his intended route, and advise the briefer if he is qualified to fly using instrument flight rules (IFR). After gathering this information, the weather briefer will normally supply all the essential weather information concerning the flight. The pilot should insure, however, that he secures the following information:

1. Weather at destination (derived from terminal forecasts and hourly weather reports).
2. En route weather (mainly gathered from area forecasts). Terminal forecasts and hourly weather reports along the proposed route help to indicate surface weather.
3. Weather for a proposed alternate route, if necessary.
4. Precipitation, sky cover, and visibility for all points along the route and alternate route.
5. Frontal activity (from area forecasts and surface weather charts).
6. Winds aloft forecasts.
7. Any teletype NOTAMs.

While en route, the pilot has several sources of weather information available. First, using his VHF radio, he may request information at any time from any flight service station within radio range. In addition, at 15 minutes past each hour, flight service stations regularly broadcast hourly weather reports for their station, and for reporting points within approximately 150 miles from the broadcast station. In addition, SIGMETs affecting the area are broadcast at 15-minute intervals (on the hour, 15, 30, and 45 minutes past the hour), and AIRMETs are broadcast at 30-minute intervals (at 15 and 45 minutes past the hour).

Transcribed weather broadcasts (TWEBs) recorded on tape are prepared by selected flight service stations and broadcast continuously over specified low frequency navigational aids (200-450 kHz) and VOR's. This broadcast identifies the station, gives general weather forecast conditions in the area, PIREPs available, and weather reports at selected locations within a 400-mile radius of the broadcasting station. As changes occur, they are transcribed onto the tape.

With all this available weather information and a minimum of effort, the pilot can become well briefed for his flight. With a basic knowledge of meteorology and an ability to obtain a weather briefing, the pilot can make the decision to continue with his flight or try another day.

PILOT'S WEATHER BRIEFING

To obtain weather information for this cross-country flight, the pilot uses the telephone number listed in the *Airman's Information Manual*, calls the Flight Service Station at Lake Charles, and obtains the following data: (See Fig. 9-59.)

1. Lufkin Flight Service Station is reporting high scattered clouds with visibility 10 to 15 miles.
2. There are a few isolated thunderstorms to the east of Lufkin, moving northeast, then clear weather north, including Rusk County Airport.
3. Winds aloft forecasts are:

   a. surface to 3,000 feet — 230° at 17 knots.
   b. 6,000 feet — 270° at 22 knots.
   c. 9,000 feet — 270° at 23 knots.

After receiving the wind information, the pilot can choose his altitude for the flight. It should be remembered that both winds aloft forecasts and aircraft altimeters are read in feet above sea level. Since the wind is reported as the same from the surface to 3,000 feet, the flight altitude can be chosen anywhere within this range with the same headwind effect.

The pilot arbitrarily chooses 2,500 feet mean sea level as his flight altitude, which

```
        FDUS2 KWBC 170545
DATA BASED ON 170000Z

VALID 171800Z  FOR USE 1500-2100Z.  TEMPS NEG ABV 24000

FT   3000    6000    9000    12000   18000   24000   30000   34000   39000

DAL 2211 2414+10 2714+02 2820-06 2832-14 2940-25 294535 295545 295055
HOU 2010 2015+09 2115+01 2120-07 2220-15 2429-25 253532 264240 265045
LFK 2317 2722+12 2723+04 2725-02 2835-16 2840-30 285444 287050 288059
MSY 2405 2408+07 2412+01 2518-07 2523-12 2530-18 263734 274540 275049
SHV 2305 2308+08 2312-01 2420-09 2422-15 2530-25 263530 264035 264540


        SA23   171500
LFK 150SCT 10+ 146/75/58/2215/998/FEW ISOLD TSTMS MVG NE
BPT 100SCT 15+ 145/86/60/2015/005 → BPT↘6/4


→NOSUM171502
→BPT 6/4 BPT 16-34 CLSD UFN

TX 171505
GGG CLR15+ 145/89/60/2110/000/FEW CU SE
LCH 120BKN 12 143/78/65/1805/987
TYR 200SCT 12 140/82/65/1710/980


        BPT   Beaumont, TX (Jefferson County Airport)
        DAL   Dallas, TX
        HOU   Houston, TX
        LCH   Lake Charles, LA (Lake Charles Municipal Airport)
        GGG   Longview, TX (Gregg County Airport)
        LFK   Lufkin, TX (Angelina County Airport)
        MSY   New Orleans, LA
        SHV   Shreveport, LA
        TYR   Tyler, TX (Pounds Field)
```

Fig. 9-59. Weather Information

would result in the aircraft flying about 2,000 feet above the ground for the trip. This altitude is low enough to easily observe checkpoints and high enough to be well above any terrain or obstacles on his course.

Having analyzed the weather, the pilot now records the applicable parts of the weather report in the log. (See Fig. 9-60.)

Since the wind is given in knots, it must be converted to miles per hour by using the conversion scale on the flight computer. The wind at the flight altitude is 230° at 17 knots. So, using the computer, the pilot notes that 17 knots is

WEATHER LOG						
	Ceiling and Visibility		Winds Aloft	Icing and Turbulence	Freezing Level and Cloud Tops	NOTAMs
	Reported	Forecast				
LFK Departure	HIGH SCATTERED CLOUDS VIS. 10-15 MI.		3000' 230°-17 KTS			NONE
En Route	TSTMS EAST MOVING NE					NONE
Destination	015+ (GGG FSS)		↓			NONE
Alternate						

Fig. 9-60. Entering Weather Conditions

19.6 miles per hour and rounds off this value to 20 miles per hour.

The temperature at 3,000 feet AGL is not shown in the winds aloft forecast but may be found by using the standard lapse rate of 2^O C/1,000 feet. Since the surface temperature at Lufkin is 75° F (See Fig. 9-59) or 24^OC, the temperature at 3,000' AGL will be 18^OC (24° - 6° = 18°). The wind and temperature information is now entered in the navigation log. (See Fig. 9-55, item 6.)

Before deciding upon the altitude, it is wise to check the *Airman's Information Manual* in regard to the direction of flight. There the pilot finds that under visual flight rules (VFR) the cruising altitude regulation is applicable for flights that are more than 3,000 feet above the earth's surface. Since the flight will be less than 3,000 feet above the surface, the VFR cruising altitude regulation does not apply to this trip.

DETERMINING TRUE AIRSPEED

For VFR planning purposes, true airspeed can be obtained by two methods: (1) by referring to the airplane operation manual and performance charts; or (2) by computer calculations. The values obtained from these sources are based on standard atmospheric conditions at various pressure altitudes.

Many factors, such as outside air temperature, atmospheric pressure, and power settings affect true airspeed; however, true airspeed, based on standard conditions, is accurate enough for VFR pre-flight planning. Referring to the cruise performance table in the airplane owner's manual, the pilot finds that his aircraft has a TAS of 120 m.p.h. at the altitude and power setting that he plans to use.

Using an alternate method of calculating true airspeed, the pilot uses the 18^OC temperature already calculated for 3,000 feet, a calibrated airspeed of 113 m.p.h. (based on previous experience), and a pressure altitude of 3,000 feet. With the flight computer and the above data, the pilot is able to compute a TAS of 120 m.p.h. This value is now entered in the navigation log for use in determining magnetic heading and ground speed. (See Fig. 9-55, item 7.)

FINDING MAGNETIC HEADING AND GROUND SPEED

With true course, true airspeed, and wind direction and velocity known, the pilot has the three items which are needed to solve the groundspeed and magnetic heading problems on the wind side of the computer.

Compute magnetic heading as follows:

1. Place 230°, the wind direction, under the true index on the wind side of the flight computer. (See Fig. 9-61, item 1.)
2. Plot up from the grommet 20 m.p.h., which is the wind velocity, and place a dot on the centerline. It should be remembered that each minor graduation up and down the scale is equal to 2 m.p.h. (See Fig. 9-61, item 2.)
3. Rotate the azimuth until the true course of 342° is directly under the true index. (See Fig. 9-62, item 3.)
4. Now, slide the grid through the computer until the wind dot rests on the 120 mile-per-hour line on the grid. This would be 120 m.p.h., the true airspeed. (See Fig. 9-62, item 4.)
5. Note the wind correction angle of 9° left. (See Fig. 9-62, item 5.) Each graduation to the left and right of the centerline below 150 m.p.h. is equal to 2^O.
6. The magnetic course is then determined by taking the true course of 342° and subtracting the 7° easterly magnetic variation. This gives a magnetic course of 335° which must be flown to make good a true course of 342°. Now, the magnetic course of 335° is recorded in the navigation log.

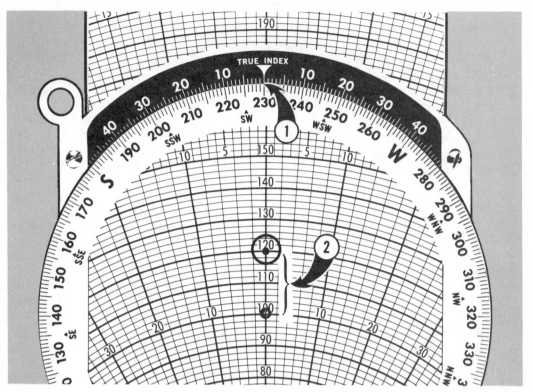

Fig. 9-61. Setting Wind Direction and Velocity

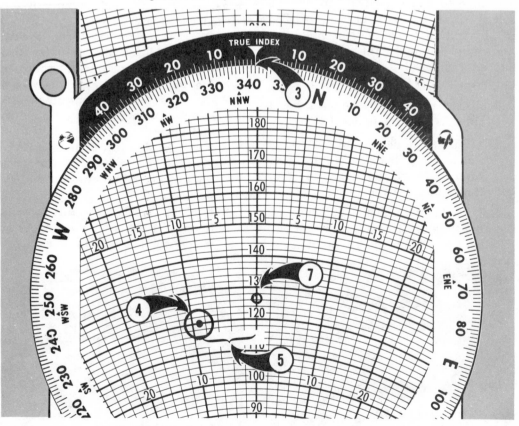

Fig. 9-62. Finding Groundspeed and Magnetic Heading

7. After noting that the wind correction angle is to the left 9°, subtract 9° from the magnetic course of 335° and arrive at a magnetic heading of 326°, which is recorded in the magnetic heading column of the log. (See Fig. 9-55, item 8.) Always remember that if the wind correction is to the left, subtract from the magnetic course. If it is to the right, add to the magnetic course.

8. The groundspeed can be determined by referring to the wind side of the computer and checking the grommet position. In this problem, the grommet is on 127 m.p.h., which is the groundspeed of the aircraft. (See Fig. 9-62, item 7.) The 127 m.p.h. is now recorded in the navigation log. (See Fig. 9-55, item 9.)

TIME REQUIRED TO FLY THE TRIP

The time to make the trip can now be solved since the groundspeed of 127 m.p.h. and distance of 159 miles are known. The time for the trip is worked out on the calculator side of the computer as follows: (See Fig. 12-13.)

1. Rotate the disc until the black speed index is directly under 127 m.p.h.

2. Then, move along the "A" scale and locate 159 miles, the distance from Jefferson County Airport to Rusk County Airport.

3. Look directly under "159" and find the time of 75 minutes on the "B" scale, or 1 hour and 15 minutes.

Normally, additional time is allowed for departing the traffic pattern and for the climb to cruising altitude. Time also should be alotted for entry into the traffic pattern at the destination airport.

For this cross-country, 5 minutes will be allowed for climb and 5 minutes for descent and landing, or an additional 10 minutes for the entire trip. (See Fig. 9-55, item 10.)

The calculated time between checkpoints, arrival and departure allowance, and the total time for flying the trip are entered in the flight log. (See Fig. 9-55, item 11.)

In referring to the aircraft owner's manual, the pilot finds that the fuel consumption of the aircraft is 8-1/2 gallons per hour. He records this in the aircraft log. (See Fig. 9-55, item 12.)

Now, with this information and the time required to fly the trip, he uses the computer to find the total amount of fuel burned: (See Fig. 9-64.)

1. First, he rotates the disc on the calculator side of the computer until the speed index is directly under the fuel consumption rate of 8-1/2 gallons per hour. (Remember, the speed index represents 60 minutes, or 1 hour.)

2. Next, he locates 1 hour and 25 minutes on the "C" scale, which is the same as 85 minutes on the "B" scale.

3. Looking directly over this time, he finds 12 gallons on the "A" scale, which will be the fuel burned.

Fig. 9-63. Time for the Trip

Fig. 9-64. Fuel Consumption

The pilot now records 12 gallons in the fuel column of the navigation log. (See Fig. 9-55, item 13.) However, this does not take into consideration any reserve fuel. It is wise to figure at least 45 minutes fuel reserve so that if any bad weather is encountered, the pilot would have enough fuel to take him around the weather, or to an alternate airport.

Forty-five minutes of fuel reserve is equal to approximately 6½ gallons. The owner's manual recommends that 2 gallons of fuel should be allowed for engine warmup and climb to 2,000 feet. Therefore, the aircraft should have at least 21 gallons (to the nearest gallon) of fuel on board at takeoff. Since the aircraft will be carrying 38 gallons of fuel, there will be plenty of fuel for the trip.

COMPASS HEADING

One of the last flight planning steps is the determination of the compass heading.

The information needed to determine the compass heading is found on the compass correction card, which is mounted in the aircraft. The compass correction card is a means of recording compass deviation that is caused by the magnetic interference of the aircraft parts, such as radio, engine, instruments, or anything in the aircraft that would disturb the compass.

Figure 9-65 shows the compass correction card for the aircraft used in this flight. Since the magnetic heading is 326°, the pilot can locate the *nearest* value on the card which is 330°. The pilot finds that for a 330° magnetic heading, he must steer a compass heading of 335°, which is a +5° deviation.

Adding the +5° deviation to the magnetic heading of 326°, he obtains a compass heading of 331° for the trip. The deviation and compass heading of 331° are recorded in the navigation log. (See Fig. 9-55, item 14.)

FOR	STEER
N	002
30	031
60	060
E	090
120	120
150	150
S	180
210	210
240	240
W	272
300	303
330	335

RADIO ON ⊠
NO RADIO ☐

Fig. 9-65. Compass Correction Card

It is wise for the pilot to remember that any metal (such as notebooks, ring binders, tools, or other objects) that would affect the compass should be kept well clear of the instrument panel area, because this can contribute to erroneous compass readings.

The formula for compass heading is:

MH ± Deviation = Compass Heading

THE OTHER FLIGHT LOG COLUMNS

The frequency box on the navigation log should also be filled in prior to departing on the cross-country flight. The departure and destination airport frequencies may be found in the AIM. Since Jefferson County Airport has a control tower located on the field, airport facility information is found in AIM Part 3. (See Fig. 9-66.) Information concerning the destination airport (Rusk County) is found in the AIM Part 2, as shown in figure 9-67. This pertinent information is entered in the frequency box. (See Fig. 9-68.)

The navigation log is used for two distinctly different operations. Its first use is to organize all the preflight planning

AIRPORT/FACILITY DIRECTORY

TEXAS—Continued

BEAUMONT LRCO 122.2   FSS: LAKE CHARLES

BEAUMONT-PORT ARTHUR
§ JEFFERSON CO (BPT) *IFR* 9SE LRA   FSS: LAKE CHARLES
(LC 722-0288)
  16   H67/11–29(3) (S–70, D–95, DT–150) BL5,6,8,11 S5 F12,18,30 U2
  REIL: Rwy 16 **RVR–**1: Rwy 11
  Remarks: Rwy 16–34 (S–83, D–118, DT–212). Cert.-FAR 139,
     CFR Index A.
  Jefferson Co. Tower 119.5   **Gnd Con** 121.9
  Radar Services:
     Beaumont App Con 121.3 (West) 124.85 (East)
     Beaumont Dep Con 121.3 (West) 124.85 (East)
     Stage I Ctc App Con within 30 mi
  ILS 109.5 I–BPT Rwy 11   **LOM:** 257/BP
  Beaumont (L) **BVORTAC** 114.5/BPT on fld/122.1R
  VHF/DF Ctc twr

Fig. 9-66. Airport/Facility Directory

Time Off	GPH	AIRPORT FREQUENCIES		
			Departure	Destination
ETE	ETA	Fuel	JEFFERSON Co.	RUSK Co.
ATE	ATA	Rem.	ATIS ———	ATIS ———
			Grnd 121.9	Apch ———
			Tower 119.5	Tower ———
			Dep. ———	Grnd ———
			FSS 122.2	FSS
			UNICOM 123.0	UNICOM 122.8
			Field Elev. 16'	Field Elev. 440'

Fig. 9-68. Frequency Box

information which includes the course, distance, weather briefing, estimated time en route, and the fuel required for the flight. The content of this section has been aimed primarily at explaining the steps toward completion of the planning necessary to insure a safe flight. The columns and blocks of information on the navigation log that were not explained were left blank since these spaces are used during the actual flight. Chapter 12 will explain the second use of the flight log — its function during the course of a flight.

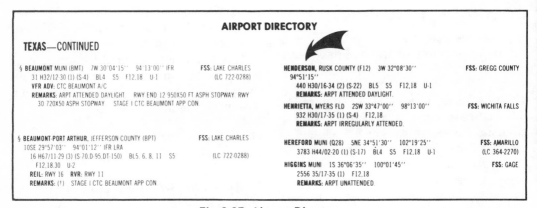

AIRPORT DIRECTORY

TEXAS—CONTINUED

§ **BEAUMONT** MUNI (BMT)   7W 30°04'15''   94°13'00'' IFR   FSS: LAKE CHARLES
   31 H32/12-30 (1) (S-4)   BL4   S5   F12,18   U-1   (LC 722-0288)
   VFR ADV: CTC BEAUMONT A/C
   REMARKS: ARPT ATTENDED DAYLIGHT   RWY END 12 950X50 FT ASPH STOPWAY RWY
   30 720X50 ASPH STOPWAY   STAGE I CTC BEAUMONT APP CON

§ **BEAUMONT-PORT ARTHUR,** JEFFERSON COUNTY (BPT)   FSS: LAKE CHARLES
   10SE 29°57'03''   94°01'12'' IFR LRA
   16 H67/11-29 (3) (S 70,D-95,DT-150)   BL5, 6, 8, 11   S5   (LC 722-0288)
   F12,18,30   U-2
   REIL: RWY 16   **RVR:** RWY 11
   REMARKS: (†)   STAGE I CTC BEAUMONT APP CON

HENDERSON, RUSK COUNTY (F12)   3W 32°08'30''   FSS: GREGG COUNTY
   94°51'15''
   440 H30/16-34 (2) (S-22)   BL5   S5   F12,18   U-1
   REMARKS: ARPT ATTENDED DAYLIGHT.

HENRIETTA, MYERS FLD   2SW 33°47'00''   98°13'00''   FSS: WICHITA FALLS
   932 H30/17-35 (1) (S-4)   F12,18
   REMARKS: ARPT IRREGULARLY ATTENDED.

HEREFORD MUNI (Q28)   5NE 34°51'30''   102°19'25''   FSS: AMARILLO
   3783 H44/02-20 (1) (S-17)   BL4   S5   F12,18   U-1   (LC 364-2270)

HIGGINS MUNI   1S 36°06'35''   100°01'45''   FSS: GAGE
   2556 35/17-35 (1)   F12,18
   REMARKS: ARPT UNATTENDED.

Fig. 9-67. Airport Directory

chapter 10

RADIO NAVIGATION

INTRODUCTION

In the early days of flying, it became apparent that a system to assist the pilot in navigation would be necessary if the airplane was really going to be useful as a transportational tool. One of the most primitive navigational aids consisted of bonfires built by farmers and ranchers at predetermined times. (See Fig. 10-1.) The pilots would fly from bonfire to bonfire across the country. Obviously, in today's airspace this would be an impossible task.

Another system used in the early days of navigation was a cross-country system of light beacons and emergency airfields. In 1926, these light beacons covered approximately 2,000 miles; by 1929, over 10,000 miles. The biggest disadvantage of this system was that it was only effective at night and only when visibility was good. (See Fig. 10-2)

Early in 1925, experimental work was begun utilizing a radio beacon which would produce a radio beam for navigational purposes. When put into use, these became known as radio range stations, and some are still in use in Alaska, Canada, Mexico, and many other countries.

The old radio range stations, however, were destined to give way to the new, more sophisticated VHF omnidirectional range which is the primary mode of navigation in the United States today.

Fig. 10-1. The Bonfire Airway System

10-2. Airway Light Beacons

SECTION A - VOR NAVIGATION SYSTEM

In the late 1940's, a system of navigation was developed called the *very high frequency omnidirectional range* (VOR). Approximately 400 VOR stations had been installed by 1953 as the VOR began to take the place of the low frequency airway system. Today, there are approximately 900 VOR stations in the United States operated by the Federal Aviation Administration, state agencies, and private operators.

VOR RADIALS

A VOR station transmits beams called *radials* outward in every direction, as shown in figure 10-3. An airborne VOR receiver in the airplane is used to detect these radiated signals and then to indicate on which radial an aircraft is located, thus enabling a pilot to follow a radial to or from a VOR.

Each radial is numbered according to its *bearing* from the station. A compass rose is drawn around each VOR on a chart, with the 0° radial pointed toward magnetic north. Thus, the 090° radial of a VOR represents a magnetic course of

090° away *from* the station, and is located 90° clockwise from the 0° radial. The radials are all magnetic courses and many are used to form the *Victor airways* that connect most VOR's.

TACAN, which stands for tactical air navigation, is a military navigational aid. However, civil aviation can and does use a certain portion of the TACAN system. That portion deals with distance measuring equipment (DME) which will be discussed later. Therefore, since TACAN was developed for the military and VOR was developed for civil aviation, it was determined practical to co-locate the civil VOR and the military TACAN at the same site. These integrated facilities are called VORTACs.

A VORTAC facility, being comprised of VOR and TACAN, provides three individual services: VOR azimuth for civil aviation, TACAN azimuth for military aviation, and TACAN distance (DME) which is used by both civil and military aviation. When the term "VOR" is used in this section, "VORTAC" is also implied.

Technically, the VOR station broadcasts an infinite number of radials; however, the airborne receiver is generally calibrated to reflect only 360 separate radials. The VOR is accurate to a tolerance of one full degree — hence 360 radials are assumed. Radials are like spokes of a wheel, as depicted by figure 10-3, and are considered to point in a direction *from* the station.

Remember, each radial is a magnetic direction from a VOR station.

In figure 10-4, a representative VOR station azimuth from a sectional chart is reproduced. A radial of 056° FROM the station is indicated (item 1); the other radials shown are 128°, 238°, 263°, and 315° (items 2 through 5). These are only

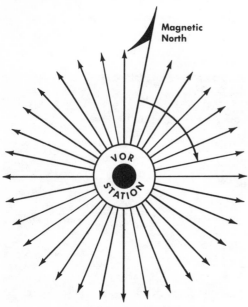

Fig. 10-3. VOR Radials

Magnetic North

VOR STATION

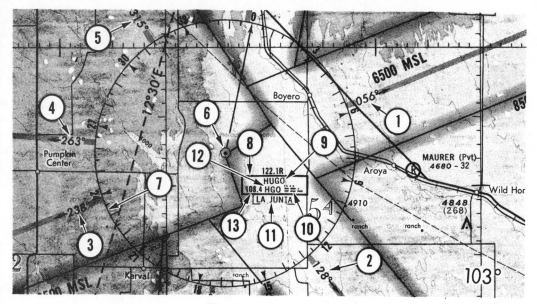

Fig. 10-4. VOR Information

five of the possible 360 radials available to the pilot from this station.

The airplane VOR receiver is able to detect each radial, and therefore can identify the one on which the airplane is currently located. Thus, if within receiving distance, the pilot is able to determine a bearing or line of position *at any direction from the VOR station.*

VOR NAVIGATION

The VOR is the most popular and easiest to use of the various radio navigational aids principally because it has the following advantages:

1. MANY AVAILABLE COURSES — Omnidirectional means in all directions. The pilot can navigate on any of 360 courses to or away from an omni station.
2. FREEDOM FROM INTERFERENCE — The VOR frequencies are virtually free from precipitation static and annoying interference caused by electrical storms.
3. ACCURACY — Course accuracy of plus or minus 2° is possible when flying VOR.
4. STRAIGHT LINE COURSE — A pilot can fly a straight line, and wind drift is compensated for automatically.
5. MAGNETIC COURSES — All VOR courses are related to magnetic north. Therefore, under a no-wind condition, the magnetic course will be the same as the magnetic heading. When correction for wind is taken into consideration, the difference between the VOR course and the magnetic heading is the *wind correction angle.*

On a sectional or world aeronautical chart, a VOR station symbol (see Fig. 10-4, item 6) is shown in blue and is surrounded by an azimuth symbol (item 7). The calibrated azimuth surrounding the station is a built-in protractor corrected for local magnetic variation. Therefore, all that is needed to plot a course is a straightedge.

Information about the VOR station is given in the small box (see Fig. 10-4, item 8) within the VOR azimuth symbol. This box contains the following information:

1. Name of station (item 9).
2. Station frequency in MegaHertz (item 13). A station without voice facilities will have the station frequency underlined.

3. Station code letters (item 12).
4. If the voice over the VOR station is handled by a remote flight service station, brackets (item 11) are placed just under the frequency box.
5. Morse code identification of the station (item 10).

There is only one positive method of identifying a VOR station. That is by listening to the Morse code identification or the recorded automatic voice identification. Automatic voice identification has been added to numerous VOR stations. When used, it is always indicated by the word "VOR" following the station name. The transmission consists of a voice announcement "Denver VOR or VORTAC" alternating with the usual Morse code identification. If no air/ground communication facility is associated with the omni range "Denver unattended VOR or VORTAC" will be heard.

VOR identification should never be determined by listening to the voice transmissions aired by flight service stations or any other radio facility. Many FSS stations remotely operate several VOR's which have different names, and in some cases none have the name of the "parent" flight service station. Also, it should be remembered that when maintenance is being performed on a VOR station, the coded identification is removed.

The VOR stations *with voice facilities* broadcast weather reports and notice-to-airmen information at 15 minutes past each hour *from reporting points within approximately 150 miles of the broadcast station.* In addition, unscheduled broadcasts are made at random times. Consequently, pilots often monitor the VOR stations as they fly across the country in order to get up-to-date weather information.

Victor airways are shown on charts with a light blue line. (See Fig. 10-5.) The airway centerlines are VOR radials which

are listed on the chart near the VOR station azimuth. Each airway has a certain magnetic direction that is determined by the VOR radial being used. Victor airways are designated with a "V" followed by a number; for example, V222.

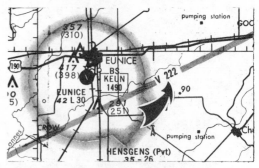

Fig. 10-5. Victor (VHF) Airway

THE VOR RECEIVER

Figure 10-6 shows a close-up view of a modern navigation/communication (nav/com) radio and VOR indicator which will be used in the remainder of this section to show how a pilot uses VOR to navigate. The radio is crystal-tuned and has separate volume controls and positive-action squelch.

The communications transmitter, which is the left half of this radio, has 360 frequencies or channels, including a frequency range of 118.0 to 135.95 MHz with 50 kHz spacing. The navigation receiver, which occupies the right half of the radio, has 200 channels with a frequency range of 108.00 to 117.95 MHz and 50 kHz spacing.

THE VOR INDICATOR

The VOR indicator has three basic components which the pilot uses to interpret the radio signals. Figure 10-7 shows a diagrammatical presentation of these components. The three basic components are:

1. COURSE DEVIATION INDICATOR (CDI) — This indicator is shown in the left portion of figure 10-7. The vertical needle, which swings to the left or right, indicates whether the

Fig. 10-6. VOR Receiver and Indicator

Course Deviation Indicator (C.D.I.)   TO/FROM Indicator   Course Selector

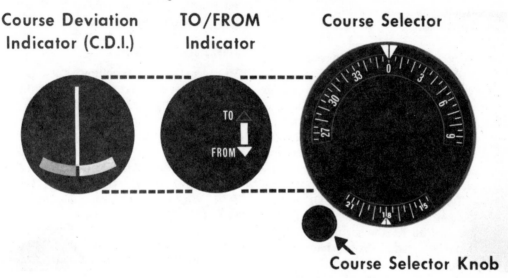

Course Selector Knob

Fig. 10-7. VOR Indicator

aircraft is to the left of course, the right of course, or exactly on course. The CDI is commonly called the "left-right" needle.

2. TO-FROM INDICATOR — The TO-FROM indicator is shown in the center of figure 10-7. This instrument indicates whether the course selected by the pilot will take him TO the station or FROM the station. It also provides positive indications of station passage.

3. COURSE SELECTOR — The right side of figure 10-7 shows a diagram of a course selector, also called the omni bearing selector (OBS). The course selector has an azimuth grad-

uated in 360° which can be rotated to select the desired radial, or to find the aircraft's position. The course is shown at the top of the dial, while the reciprocal or opposite course is on the bottom index.

VOR NAVIGATION PROCEDURES

The procedure for using the VOR receiver is very simple. For example, to fly to a VOR station, a pilot would use the following four steps in sequence:

1. Tune the receiver to the frequency printed on the chart and *identify the station by listening to the Morse*

code identifier. (The pilot may select either voice or the Morse code identifier by the use of the inside right knob shown in figure 10-6.)

2. Turn the course selector (OBS) until the TO-FROM window indicates a TO.

3. Continue turning the course selector until the course deviation indicator (CDI) is centered.

4. The reading on the course selector is the magnetic course *to the station.* Turn the airplane until the directional gyro and the course selector correspond; then with the CDI still centered, the airplane will proceed directly TO the station on the selected course.

Some very important things to remember:

1. Directions indicated by the course selector are in relation to the station. That is, the heading of the airplane has no relationship to the direction indicated by the course selector. Figure 10-8 shows three airplanes on the same radial, but with different headings. The CDI indications would be the same in each case because all of the airplanes are on the 120° radial from the station. The VOR receiver in the airplane reacts only to transmissions from a VOR station, and *not* to the heading of the airplane.

2. When the course selected is toward a station, be sure that the TO-FROM indicator reads TO and the course selector and the directional gyro are in general agreement. When

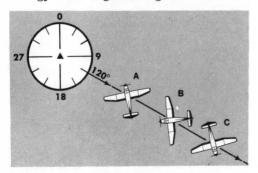

Fig. 10-8. VOR — Airplane Positions

the course selected is from a station, be sure that the TO-FROM indicator reads FROM and that the course selector is in general agreement with the directional gyro.

The TO-FROM indicator has two positive indications: TO and FROM. The area between the words TO and FROM is colored red and may be printed with the word OFF. When the red area is shown in the TO-FROM window, either the signal is too weak to be used for reliable navigation, or the airplane may be in a position abeam or 90° from the selected course. On some equipment, a red warning flag is used to indicate a poor signal or the abeam position.

VOR INDICATIONS

An airplane approaching the Leona VORTAC station from the west with intentions of flying to the station on the 270° radial, should tune the VOR receiver to 110.8 MHz which is the frequency printed on the sectional chart. (See Fig. 10-9.) The code letters LOA, − ·· / − − − / · −, the Morse code identification of Leona VOR, can be received on the *voice facilities* of the VORTAC station.

The course selector should be rotated to 090° which is the magnetic course *to* the station. The TO-FROM indicator will read TO. Assuming that the airplane in figure 10-9 is on the 270° radial and the course selector is set to 090°, the CDI needle will be centered. This indicates that the airplane is on the selected course. Since the aircraft heading and selected course are in general agreement, if the airplane deviates from the course northward, the CDI needle will be deflected to the right, telling the pilot that the course is to the right. If the pilot deviates to the south of the course, the CDI needle will deflect to the left, indicating that the course is to the left.

Assuming a no-wind condition, while maintaining a heading of 090°, the airplane will fly to the Leona VORTAC on the 270° radial and the CDI needle will

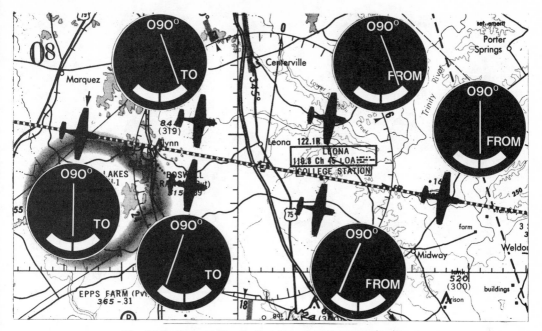

Fig. 10-9. Correct Needle Sensing and TO-FROM Indications

be centered. The TO-FROM indicator will continually display a TO indication until the aircraft passes the Leona VORTAC.

After crossing the Leona VORTAC, the aircraft is now outbound *from* the VORTAC station on the 090° radial. With the course selector still set to 090° (as it should be), the TO-FROM indicator will change to a FROM indication since the 090° course will take the pilot away *from* the VORTAC station.

The airplane, proceeding away from the Leona VORTAC in figure 10-9, will experience the same CDI deflections as the other aircraft. If, while tracking away from the VORTAC station on the 090° radial, the aircraft deviated north of the course, the CDI would be deflected to the right, indicating that the course was to the right of the aircraft. If the airplane deviated south of the course, the CDI needle would deviate to the left, indicating that the course was to the left of the aircraft.

VOR PRACTICE PROBLEMS

To further explain and help visualize how the VOR works, a series of situa-

tions is illustrated in figure 10-10. The Mountain VOR station is tuned and identified by the airplane at position one in figure 10-10. To identify the aircraft's position, the course selector is turned until the TO-FROM indicator indicates FROM and the CDI is centered. The course selector is checked, and because of the aircraft's position, 180° is noted. The pilot knows from this information that he is on the 180° radial from the Mountain VOR.

Assuming that the pilot wants to fly to the station, the course selector should be rotated until the TO-FROM indicator indicates TO and the CDI is centered. This will cause the course selector to read 0° to fly to the station, as shown at position two in figure 10-10. As the aircraft is flown to the station, the pilot makes course corrections by noting the deflection of the CDI. If the needle were to move to the right, the aircraft would be turned to the right to recenter the needle.

As the aircraft gets very close to the station, as shown in position three, the CDI needle becomes very sensitive. Because the VOR courses are *degrees*, the *distance*

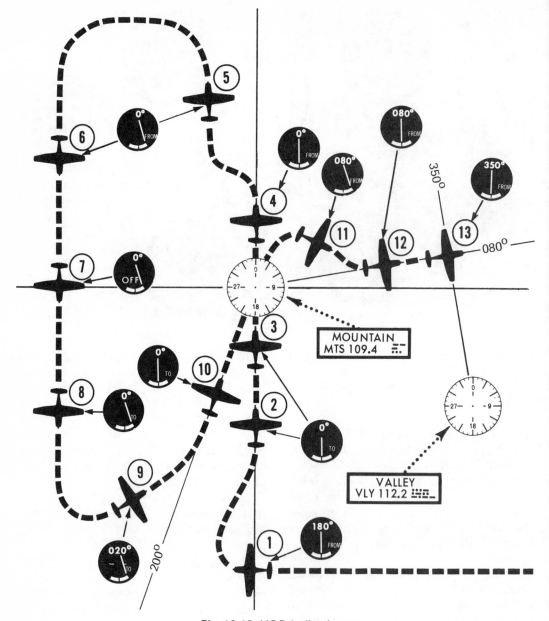

Fig. 10-10. VOR Indications

between courses is very small near the VOR station.

Station passage is determined by noting the change of the TO-FROM indicator from TO to FROM. At position two, the aircraft was flying inbound on the 180° radial. At position four, the aircraft is shown as flying outbound on the 360° radial. If the aircraft were flown to position five with the same settings on the receiver as in position four, the CDI needle

would be deflected to the right indicating to the pilot that a right heading correction would be necessary to bring the airplane back to the selected course.

REVERSE NEEDLE SENSING

If the aircraft made a 180° turn and flew to position six in figure 10-10, with the course selector still set on 0°, the CDI needle would still be to the right and the TO-FROM indicator would still indicate FROM.

The heading of the airplane has absolutely no effect on the VOR indications; however, if the airplane heading is the same as the course selector, a right CDI needle indicates the course is to the right. In the case of the airplane at position six, since the heading is 180° and the course selector is 0°, a 180° ambiguity exists. The CDI needle for airplane six will be to the right even though the selected course is to the pilot's left. This could be remedied by turning the course selector to 180°; however, for the sake of instruction, the course selector will remain at 0° for the airplane in positions six, seven, and eight.

With these same settings, the aircraft is flown to position seven where the TO-FROM indicator reflects neither TO nor FROM and the OFF flag is shown because the aircraft is in the area of ambiguity. The airplane at position seven is on a radial which is 90° from the selected course. This is called the *abeam* position. At position eight, with the course selector still set on 0°, the CDI needle still is to the right, and the TO-FROM indicator indicates TO.

INTERCEPTING AN INBOUND COURSE

Now, assuming that the pilot wants to fly inbound on the 200° radial, the inbound course of 020° (reciprocal of 200°) is selected, the TO-FROM indicator is noted to read TO, and the CDI indicator shows a needle deflection to the right. The aircraft, then, is turned to a heading of 050° which is used to intercept the 200° radial, as shown at position nine. When the needle on the CDI moves to center, the aircraft is turned to a heading of 020° as shown at position ten. The pilot then flies to the station. Station passage is noted by the sensitivity of the CDI and the changing of the TO-FROM indicator from TO to FROM.

Assuming that the pilot wants to fly away from the station on the 080° radial, the course selector is turned to 080° and the TO-FROM indicator is noted to read FROM. Since the CDI is to the right, the pilot turns to the right to an intercept heading of 120° as shown at position 11. When the CDI needle comes from the right to center, the aircraft is intercepting the course and the heading of the aircraft, as shown at position 12, should be changed to 080°.

USING THE VOR FOR A CROSSCHECK

If the position of the airplane along the 080° radial was desired, a second VOR station could be used to locate the position. For example, if the airplane at position 13 tuned in the Valley VOR station illustrated in figure 10-10, the course selector should be turned until the TO-FROM indicator indicated FROM and the CDI centered. In this case, the course selector would read 350° indicating that the aircraft was on the 350° radial from the Valley VOR. The 080° radial from the Mountain VOR and the 350° radial from the Valley VOR could be easily plotted on a chart thereby showing the pilot's exact position.

VOR ORIENTATION

To illustrate an example of the many possible combinations of aircraft headings, aircraft locations, and VOR indications, some typical samples are shown in figure 10-11. Airplane "A" is located on the 0° radial of the Cicero VOR and is flying a magnetic heading of 090°. By referring to the indications on the bottom of figure 10-11, it can be seen that with the course selector set to 0°, the VOR indicator in airplane A would have a centered CDI and a FROM indication. (Sample VOR indications are also shown for Airplanes B through I.)

A rule may be used to assist in orientation while solving the problems in figures 10-10 and 10-11: "*When orientation is in doubt, mentally or physically turn the airplane to a heading that coincides with the setting on the course selector; then, the CDI needle will point toward the*

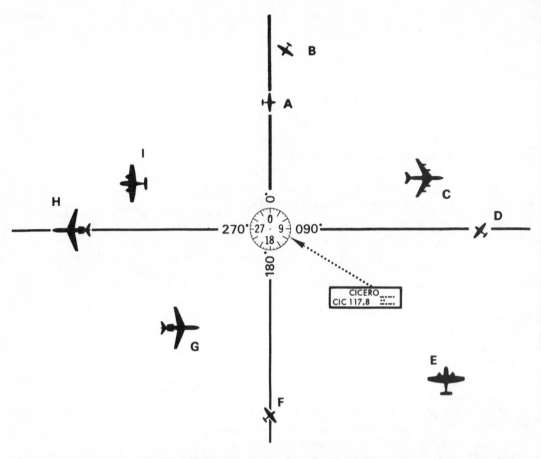

Airplane	A	B	C	D	E	F	G	H	I
Heading	090°	210°	090	310°	0°	045°	090°	270°	270°
Course Selector	0°	180°	270°	270°	130°	180°	180°	270°	270°
CDI									
To-From	FR	TO	TO	TO	FR	FR	FR	FR	FR

Fig. 10-11. VOR Orientation

selected course; the TO-FROM needle will read TO if the course will lead toward the station and FROM if the course will take the aircraft away from the station."

Since VOR equipment is VHF, it is subject to line-of-sight restriction. There is some "spillover," however, and reception at an altitude of 1,000 feet above the ground can be expected to be about 40 to 45 miles. This distance will increase with altitude.

VOR TEST SIGNALS

VOR test facilities have been established at selected locations in order to allow pilots to test the accuracy of their VOR receivers. The VOT will function anywhere on the airport. The frequencies for the radiated test signals (VOT) are listed in the *Airman's Information Manual* (AIM).

To use the VOT, the pilot first selects the frequency and notes that the aural signal is a steady series of dots. Then, when the course selector is set to 0°, the CDI should be centered and the TO-FROM indicator should read FROM. When 180° is selected on the course selector, the CDI should center and the TO-FROM indicator should read TO.

If the CDI needle does not center, the pilot can determine the exact amount of error. To determine the error, the course selector should be rotated until the CDI centers, and then the number of degrees that the course selector reads should be noted. The difference between the course selector indications and 180° or 360° is the error.

There are means other than the VOT of testing the accuracy of the VOR receiver. The AIM gives instructions for VOR checks that may be made at a point on an airport, or over a prominent landmark *that may be overflown.*

SECTION B - DME AND AREA NAVIGATION

DISTANCE MEASURING EQUIPMENT

Another radio that is commonly used in today's aircraft is *distance measuring equipment* (DME). (See Fig. 10-12.) The DME consists of ground equipment installed in VORTAC (co-located VOR and TACAN) stations and the airborne equipment installed in the airplane.

During operation, the airborne equipment originates a signal which the ground equipment receives. The ground equipment automatically returns a response to the airplane. The DME equipment in the airplane measures the elapsed time between the transmission of the original signal and the time that the response is received from the ground station. The resultant information is converted into an instrument indication that displays the *distance* from the station in *nautical miles*.

Most DME units also have a ground speed function which indicates the speed in *knots* toward or away from the station. This ground speed indication is accurate only when the aircraft is flying directly to or directly away from the VORTAC station.

The DME, in addition to being useful for en route navigational situations, is also useful in finding an airport, since it indicates the exact mileage from a VORTAC station.

Additionally, the combination of VOR and DME is useful in giving air traffic control position reports. For example, a pilot might report that he is on the 120° radial of a VORTAC station and showing 22 DME miles from the station, thereby giving a positive location of the airplane.

The DME does have an inherent error that must be taken into consideration particularly when flying at higher altitudes and close to the station. This error is due to the fact that the DME actually measures *slant distance*, not horizontal distance. For example, the airplane in figure 10-13 will be over the VORTAC station upon flying an additional 1.7 miles. However, the DME indicator, in this case, will

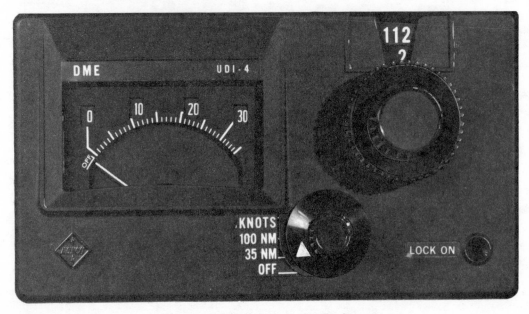

Fig. 10-12. Distance Measuring Equipment

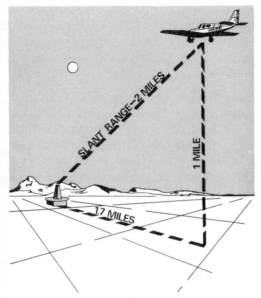

Fig. 10-13. DME Slant Range Error

read two miles from the station. The error diminishes as the distance from the station increases.

AREA NAVIGATION

Area navigation has been designed and developed to provide more lateral freedom and, thus, more complete use of available airspace. This mode of navigation does not require a track directly to or away from existing radio navigational aids. Therefore, it will prevent the tunneling effect of aircraft transitioning to or away from heavily used en route navigational aids.

Another advantage is the saving effected by flying direct routes. Figure 10-14 shows a typical comparison of a flight that is made on airways and a flight that uses area navigation. On a long flight, the time and cost advantages of flying with area navigation can be considerable.

Doppler radar, inertial navigation systems, and *course line computers* are typical of area navigation systems presently in use. The Doppler and inertial navigation systems are not dependent upon ground based navigational aids, while the course line computer utilizes VORTAC ground stations.

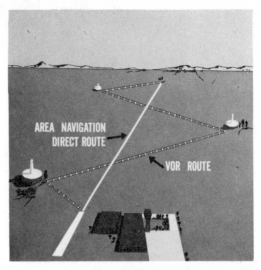

Fig. 10-14. Area Navigation

A characteristic common to all three types of area navigation is the use of *waypoints*. A waypoint may be described as a predetermined geographical location that may be a desired destination or a checkpoint along the route to a destination. The three systems differ in the manner in which the waypoints are determined.

DOPPLER RADAR SYSTEM

The Doppler system utilizes radar to detect direction and rate of movement of the aircraft across the ground. This information is sent to the computer. The computer in turn relates the aircraft's position and the desired course on the pilot's navigational display.

INERTIAL NAVIGATION SYSTEM

The inertial navigation system utilizes an inertial platform and a computer. The inertial platform contains two gyroscopes and three accelerometers. The gyroscopes maintain their orientation in space, while the accelerometers sense all direction and rate of movement. The information from the inertial platform is sent to the computer. The computer makes corrections in the information to allow for such things as the rotation of the earth, and then sends the information to the pilot's display unit. The inertial platform may also be used in conjunction with the air-

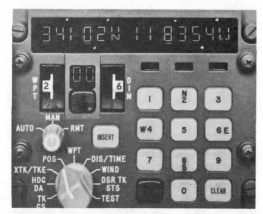

Fig. 10-15. Inertial Navigation Control Panel

craft's attitude instruments. Figure 10-15 shows the control display for a popular inertial navigation system.

The push buttons are used to "type" the geographical coordinates of a desired waypoint. The waypoint coordinates, shown in the window of the control unit in figure 10-15 would be $34^O10.2$' North latitude by $118^O35.4$' West longitude.

COURSE LINE COMPUTER

The course line computer utilizes ground-based VORTAC facilities and a vector analog computer to create a *phantom station* at the desired waypoint. Figure 10-17 shows an airplane using a course line computer to fly to a waypoint or phantom station.

The waypoint is established by setting 280° and 54 miles in the course line computer control illustrated in figure 10-16. The computer then creates the

phantom station 54 miles from the Midway VOR on the 280° radial. The pilot's VOR and DME indications are exactly the same as if the Midway VOR were actually at the waypoint.

The pilot may fly TO or FROM the waypoint on any desired radial, and the distance relative to the waypoint will be shown in the DME window on the area navigation control panel. (See Fig. 10-16.) The aircraft in this example is 61 nautical miles from the waypoint.

Figure 10-18 shows an excerpt from an area navigation chart. Note that the Dillon waypoint has both the VOR radial and DME distance (RLD 123°/ 30.0 Miles) for the course line computer and the geographical coordinates (39° 38.0'N, 106° 01.5'W) for the inertial navigation systems.

Fig. 10-16. Area Navigation Control Panel

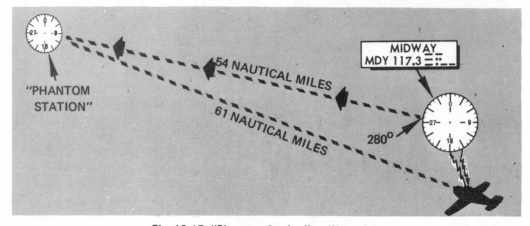

Fig. 10-17. "Phantom Station" or Waypoint

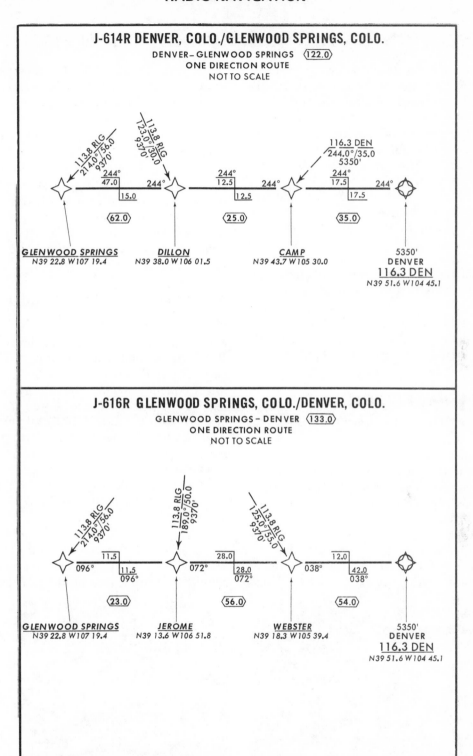

Fig. 10-18. Excerpt From Area Navigation Chart

SECTION C - AUTOMATIC DIRECTION FINDER

Automatic direction finding equipment is a very necessary and integral part of the radio navigation picture. Not only does ADF provide assistance in flying the old low/medium frequency range, but it also allows navigation to or from any non-directional low or medium frequency radio station. ADF is used with nondirectional radio beacons, compass locators used in the instrument landing systems, and standard (AM) broadcast stations.

An example of a nondirectional radio facility chart symbol and associated information box is shown in figure 10-19. An example of how a standard broadcast station is shown on a chart is depicted in figure 10-20. The station call letters and frequency appear in a box adjacent to the antenna location.

ADF RECEIVERS

The ADF receiver, shown in figure 10-21 has a frequency selector (or tuner), a volume control, and a function switch similar to many personal radios used in the home. This type of radio requires careful tuning to select the proper station.

The ADF receiver and bearing indicator shown in figure 10-21 give the pilot the necessary navigational information. Note that the bearing indicator has a 360° azimuth ring and a needle. On most ADF indicators, the azimuth is fixed and zero represents the nose of the airplane.

When a station is tuned and the function selector knob is set to the ADF position, the needle on the indicator will point to the station, and a bearing, relative to the nose of the airplane, is noted. If the magnetic heading of the airplane and the relative bearing on the ADF indicator are added together, the *magnetic bearing to the station* may be determined.

RELATIVE BEARING AND MAGNETIC BEARING

For example, if the aircraft in figure 10-22 was on a magnetic heading of 200° and the ADF was tuned to the Kedzie low frequency station, the *relative bearing* (angle between the nose of the aircraft and the L/MF station) on the ADF indicator would be 080°. The magnetic heading of the airplane (200°) is added to the relative bearing on the ADF indicator (080°), and the resultant *magnetic bearing* (280°) to the station is determined.

If the pilot of the aircraft in figure 10-22 wanted to fly to the station, the airplane should be turned to a heading of 280°.

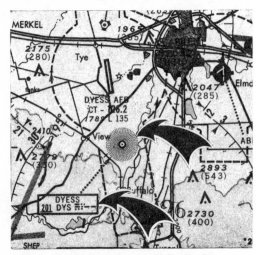

Fig. 10-19. Nondirectional Radio Beacon

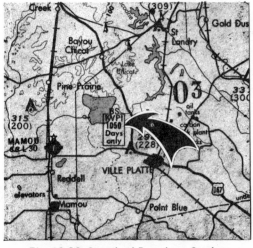

Fig. 10-20. Standard Broadcast Station

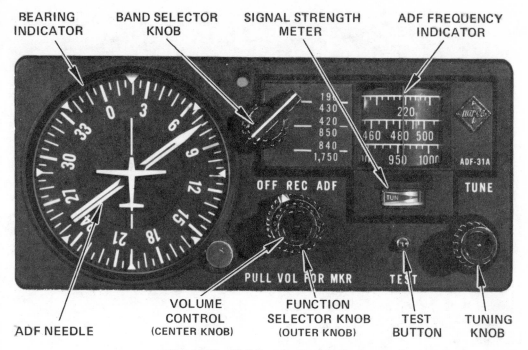

BEARING INDICATOR · BAND SELECTOR KNOB · SIGNAL STRENGTH METER · ADF FREQUENCY INDICATOR

OFF REC ADF · TUNE

PULL VOL FOR MKR · TEST

ADF NEEDLE · VOLUME CONTROL (CENTER KNOB) · FUNCTION SELECTOR KNOB (OUTER KNOB) · TEST BUTTON · TUNING KNOB

Fig. 10-21. ADF Receiver and Indicator

After the pilot turns the airplane, the ADF indicator needle would point to zero, indicating that the station is on the nose of the airplane. In a no-wind situation, the pilot could *home* to the station by keeping the ADF indicator needle on zero.

TRACKING

Through a procedure known as *tracking*, the aircraft may be flown to or away from a radio station on any desired course. This procedure involves a directed trial-and-error method of determining

the wind correction angle. The airplane in figure 10-23 is flying to a station on a 360° bearing and has established a wind correction angle of 15°. Note that the heading of the airplane is 015° and the ADF indicator is pointing to the station and showing the 15° wind correction angle on the ADF azimuth.

HOMING

Figure 10-24 illustrates a procedure known as *homing to a station*. This procedure is simple in that all the pilot has to do is continually change his heading so

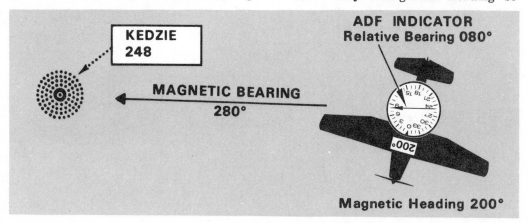

KEDZIE 248

ADF INDICATOR
Relative Bearing 080°

MAGNETIC BEARING 280°

Magnetic Heading 200°

Fig. 10-22. Determining Magnetic Bearing

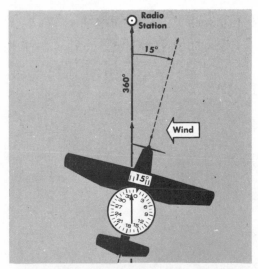

Fig. 10-23 "Inbound" ADF Wind Correction

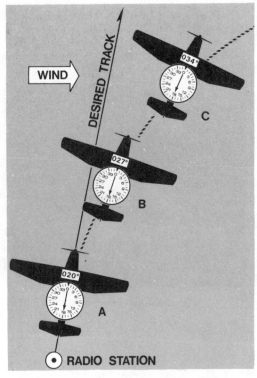

Fig. 10-25. "Outbound" ADF Without
Wind Correction

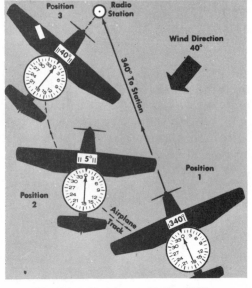

Fig. 10-24. Homing to Station Without
Compensations for Wind

that the ADF indicator remains on zero.
If there is a crosswind, however, the aircraft will actually drift slightly downwind
from the station rather than flying a direct
course to the station. *This procedure
should never be used for flying away from
a station.* If a pilot used the ADF to fly
away from the station and did not allow
for wind correction, the result would be
as shown in figure 10-25, positions B or
C. Position A shows the no-wind track.

SECTION D - RADAR AND TRANSPONDERS

RADAR

The term *radar* is derived from *radio detection and ranging*. Radar combines radio and television in that radio signals obtain data which is displayed on a scope similar to a television tube.

Radar works on an *echo* principle. (See Fig. 10-26.) The radar set transmits a high energy radio signal. The metal structure on aircraft reflects or *echoes* the radio signal back to the radar antenna. The object that reflects the signal is often referred to as a *target*.

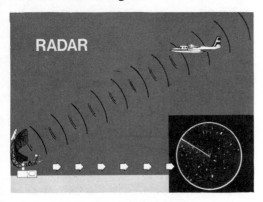

Fig. 10-26. Radar Echo

The distance to a target is determined by measuring the time required for the radar signal to reach the object or target and return to the radar antenna. The radar signal, traveling at the speed of light, makes the round trip in microseconds as shown in figure 10-27.

Normal radar reflects terrain features, buildings, traffic on a nearby highway, and clouds or precipitation, as well as aircraft. For use in air traffic control, a device known as a moving target indicator automatically eliminates stationary and very slow moving targets.

The use of radar in air traffic control has become a valuable tool because it serves two basic functions: separation of air traffic and pilot navigational assistance. The FAA provides radar coverage to most of the nation's airway system and provides traffic separation to en route traffic on instrument flight plans.

Radar is also used at terminal facilities by the tower, approach control, and departure control. Approach control normally uses *airport surveillance radar* (ASR) for the separation of traffic, but can assist a pilot in making an approach in instrument weather conditions if the pilot's own navigational equipment is inadequate or inoperative.

At selected locations, approach control also uses *precision approach radar* (PAR) which has very accurate range, azimuth, and glide path information used to bring the pilot right down to the touchdown zone area of the instrument runway. It is interesting to note that precision approach radar was first used during the famous Berlin Airlift after World War II.

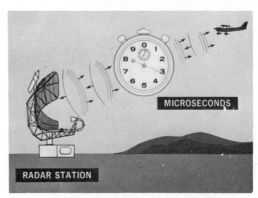

Fig. 10-27. Timing of Transmitted and Reflected Signal

TRANSPONDERS

The *transponder* is the airborne portion of the *air traffic control radar beacon system* (ATCRBS), otherwise known as *secondary radar*. Ordinary radar, or *primary* radar, operates by originating radio waves which are reflected back from a target such as an aircraft, to an antenna on the ground. Some of the disadvantages of primary radar are that these reflected waves cannot penetrate high terrain, heavy precipitation, or certain other atmospheric phenomena, and are not effective at great distances. The effectiveness of radar, therefore, has been greatly increased by placing companion equipment, referred to as the *transponder*, in the aircraft itself.

A coded interrogation signal is transmitted by the ground radar, and any aircraft within line of sight and equipped with a transponder that is set to the same code responds automatically by sending back a coded signal. This causes the aircraft to appear as a distinct pattern on the radar controller's scope. The main advantages of the secondary radar system are:

1. Radar target reinforcement — The beacon reinforces the normal radar return on the ground radarscope.
2. Radar target identification — By having the pilot set a specific code on his transponder, his airplane can be positively identified.
3. Extended radar coverage — Through use of special codes, more airplanes can receive radar guidance over a greater area, and their signals can be received at much greater distances than by primary radar.
4. Vertical separation of aircraft by altitude assignment is easier because special codes can be assigned to aircraft flying at specified altitudes.
5. Emergency codes can be used for rapid identification of airplanes in distress.
6. A transponder signal can penetrate extensive clouds or heavy precipitation.

In the United States, a common system known as Mode 3/A is now used by both civil and military aircraft. When the ground radar transmits a signal on 1030 MHz, all airborne transponders that are tuned to Mode 3/A will automatically respond on 1090 MHz.

The controller sees *beacon targets*, as illustrated in figure 10-28, on his scope only for aircraft which have been instructed

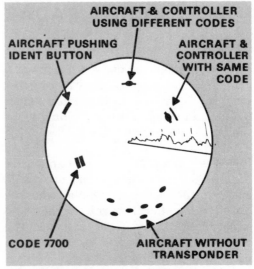

Fig. 10-28. "Targets" on Radar Screen

to reply on the code on which the controller is interrogating. Since it is important for the ATC personnel to know the transponder capability of all aircraft operating in the radar environment, a transponder notation is marked on the flight plan if the airplane is so equipped.

Figure 10-29 shows a typical transponder.
1. STANDBY — This position of the function switch allows the set to remain warmed up and ready for immediate use.
2. ON — This position will transmit the selected code at the normal power level.
3. LO SENS — This position, when requested by a controller, will make the signal more usable when the aircraft is close to the interrogating station.
4. MONITOR LIGHT — This light will blink each time the set is interrogated by the ground station.
5. MODE SWITCH — This switch is not provided on all equipment. Mode A is used in most areas, except where an automatic altitude reporting feature (Mode C) can be interrogated by the ground radar unit.

TRANSPONDER PHRASEOLOGY

Air traffic controllers use the following phraseology to indicate to the pilot the

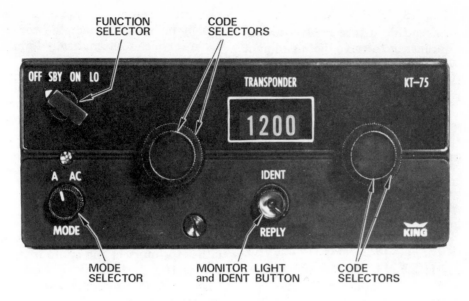

Fig. 10-29. Typical Transponder

type of radar beacon transponder operation required by ATC. Mode A/3 or Mode C operations are the only operations to which ATC refers. ATC will not specify other modes on the transponder.

SQUAWK (NUMBER) — Operate radar beacon transponder on designated code in Mode A/3.

IDENT — Push the ident button for identification purposes only upon controller's request. Release the ident button immediately after engaging since the ident signal transmits for 20 seconds after the button is released.

SQUAWK (NUMBER AND IDENT) — Operate transponder on specified code and Mode A/3 and engage the ident feature.

SQUAWK STANDBY — Switch transponder to standby position.

SQUAWK LOW/NORMAL — Operate transponder on low or normal sensitivity as specified. The transponder is operated in the normal position unless ATC specifies low. The word "ON" is used instead of "NORMAL" as a master control label on most types of transponders.

STOP SQUAWK (MODE IN USE) — Switch the specified mode off.

STOP SQUAWK — Switch transponder to the "OFF" position.

SQUAWK MAYDAY — Operate transponder in emergency position (Mode A Code 7700).

TRANSPONDER CODES

Under no circumstances are civil airplanes permitted to operate on code 0000 as this code is reserved for Air Defense Command use.

Some of the more common codes used by ATC are:

   1200 — VFR regardless of altitude.
   4000 — within restricted/warning areas.
* 7600 — two-way communication loss.
   7700 — emergency.

* Squawk 7700 for 1 minute; then 7600 for 15 minutes. Repeat cycle as required.

ATC will state the code which is needed for their operations if codes other than those listed above are required.

VFR RADAR PROGRAMS

Approach and departure control are intended primarily to serve the instrument pilot while he is transitioning from the enroute portion of a flight to the terminal area or vice versa. At many air terminals, the services of approach and departure control are also available to the VFR pilot. It is not mandatory that the VFR pilot use these services, but, when available, their use is highly recommended. The following services are explained as they would be used by the VFR pilot.

RADAR TRAFFIC INFORMATION SERVICE

Pilots receiving radar traffic information service are advised of any radar "targets" which are near the route of flight. The purpose of this service is very simple; it alerts the pilot to possible conflicting traffic so he is in a better position to take action. However, this service is not intended to relieve the pilot of his responsibility for continual vigilance to *see and avoid* other aircraft.

It should be noted that when operating in the terminal area, many times the altitude of the reported traffic is unknown. Therefore, the traffic may be well above or below the flight path of the aircraft.

The pilot operating under VFR must understand that this service is available on a "workload permitting" basis. The controller's first responsibility is to serve the needs of aircraft flying on IFR flight plans.

STAGE I SERVICE

Stage I service provides the pilot with *traffic information and limited vectoring* on a workload permitting basis. The pilot of an arriving aircraft should contact approach control on the frequency given over ATIS or in the AIM. He should provide his position, altitude, transponder code (if so equipped), and destination, followed by a request for traffic information. As the pilot continues toward the terminal, approach control specifies the time or place at which he is to contact the tower, and radar service is automatically terminated.

STAGE II SERVICE

As indicated by the title, when pilots use Stage II radar, the additional service of *sequencing* aircraft into the traffic pattern is provided. As the pilot of an arriving aircraft reaches the area in which Stage II service is provided, he should contact approach control and state, *"Request radar service."*

After radar contact is established, the pilot is directed to fly specific headings, either to intersect the appropriate leg of the traffic pattern or to position the flight behind preceding aircraft in the approach sequence. When the aircraft correctly positioned, and the pilot reports *"traffic in sight,"* he will be directed to follow the preceding aircraft. When the pilot is told to contact the tower, radar service is terminated automatically. A landing sequence number will be issued by the tower, unless previously issued by approach control.

The pilot of a *departing* VFR aircraft desiring Stage II service should state, *"Request radar traffic information,"* on his initial contact with ground control and advise the controller of the proposed direction of his flight. Following takeoff, the tower will advise him when to contact departure control and which frequency to use.

STAGE III SERVICE

In addition to radar vectoring and sequencing, *separation service* is provided with Stage III radar. It is designed to provide separation between participating VFR aircraft and all IFR aircraft operating within the airspace described as the *terminal radar service area* (TRSA). The availability of Stage III service is normally indicated by the designation TRSA under radar services in the AIM Part 3 Airport/Facility Directory. Each terminal where Stage III services are available also has the notation, "See Graphic in AIM Part 4."

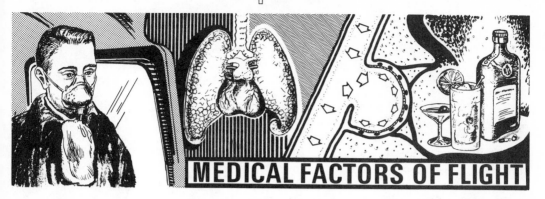

chapter 11

MEDICAL FACTORS OF FLIGHT

SECTION A - OXYGEN, ALTITUDE, AND THE BODY

THE ATMOSPHERE

Modern general aviation aircraft are capable of transporting man to altitudes where a knowledge and application of physiological principles is not only desirable, but necessary to sustain life; therefore, the pilot should understand the interrelationships between oxygen, altitude, and the body.

The atmosphere consists of 78% nitrogen, 20.9% oxygen, 1.1% carbon dioxide and other gases. These three main gases are important to the body physiologically. Due to the constant mixing of winds and to other weather factors, the percentages of each gas in the atmosphere are normally constant to 70,000 feet. (See Fig. 11-1.)

Nitrogen, present in a high percentage, is responsible for the major portion of the total atmospheric pressure or weight. Some nitrogen is dissolved in and is carried by the blood, but this gas does not

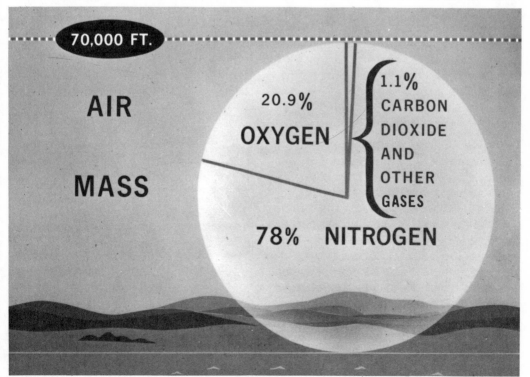

Fig. 11-1. Percentage of Gases in the Atmosphere

enter into chemical combination as it is carried throughout the human body. Each time we breathe, the same amount of nitrogen is exhaled as was inhaled.

Oxygen is a colorless, odorless, tasteless gas and is essential to life. Each time man breathes, approximately 20.9% of that breath is oxygen. In the lungs, a portion of this oxygen is absorbed into the bloodstream and is carried to all parts of the body. It is used to "burn" or oxidize food material and produce energy transformations in the body.

Man can live for weeks without food and for days without water, but only minutes if totally deprived of oxygen. Because he cannot store oxygen in his body, he lives a breath-to-breath existence. He continues to live only as long as he can continually replenish the oxygen consumed by his metabolic processes.

Air is heavy. It weighs 14.7 pounds per square inch at the earth's surface at sea level. Technically speaking, that is the pressure created by a column of air one inch square and about 100 miles high (the approximate thickness of the layer of free air or atmosphere covering the earth). (See Fig. 11-2.) A person normally does not notice this pressure because it presses upon him equally from all directions.

The weight of the atmosphere does not remain the same from top to bottom. Some people like to talk of the atmosphere as an ocean in which the diver finds that the pressure gets greater the deeper he goes. Others picture the air as a haystack because hay packs down at the bottom of the stack and is fairly loose at the top. In the same manner, air always remains the same in composition, but it is denser at the bottom because of all the weight on top.

The pilot should recognize that atmospheric pressure does not diminish at a

uniform rate with altitude. Although the atmosphere covers the earth to a height of 100 miles, three-fourths of the molecules of air are packed into a region not much taller than Mt. Everest. (See Fig. 11-3.)

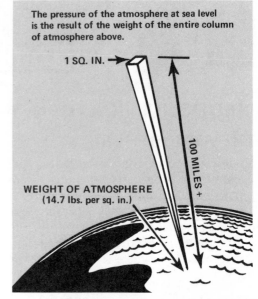

The pressure of the atmosphere at sea level is the result of the weight of the entire column of atmosphere above.

1 SQ. IN.

100 MILES +

WEIGHT OF ATMOSPHERE
(14.7 lbs. per sq. in.)

Fig. 11-2. Weight of the Atmosphere

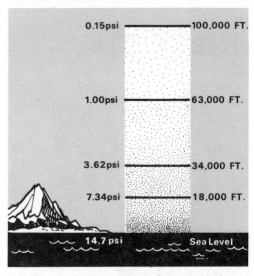

0.15psi — 100,000 FT.

1.00psi — 63,000 FT.

3.62psi — 34,000 FT.

7.34psi — 18,000 FT.

14.7 psi — Sea Level

Fig. 11-3. Atmospheric Pressure Varies with Altitude

At an altitude of 18,000 feet above sea level or about three and one-half miles up, the air pressure has been decreased by one-half and is only about seven pounds per square inch. At 34,000 feet, the pressure has been cut in half again to a little more than three and one-half pounds per square inch. At 63,000 feet, there is only one pound of pressure remaining. At 100,000 feet, or nearly 20 miles up, the atmospheric pressure is about 0.15 pounds per square inch. Beyond that, the atmosphere is largely a vacuum.

RESPIRATION

Respiration, which is the exchange of oxygen and carbon dioxide between an organism and its environment, takes place in two phases, internal and external. External respiration is the exchange of gases between the lungs and the surrounding atmosphere. Internal respiration is the exchange of gases across the tissue cell membranes.

The respiratory cycle begins with inhalation of air into the lungs. Inhalation is produced by contraction of the *diaphragm*, the large muscle separating the thoracic and the abdominal cavity.

Ordinarily, a person breathes 12 to 16 times a minute, although the rate will be slower when resting and faster when exercising. The average, quiet, resting man inhales a little more than a pint of air at a time, or from six to eight quarts per minute.

Oxygen used in the body is inhaled through the mouth or nose, passes through the trachea and bronchial tubes, and is directed into the lungs where it transfers to the blood. The blood then carries this oxygen to living cells where energy is obtained for all body processes and functions through the action of oxygen upon compounds in the cells. This transfer of energy produces a waste product gas known as carbon dioxide or CO_2. As carbon dioxide is produced, blood then carries it back to the lungs where it is released to the atmosphere by exhaling through the nose or mouth, completing the respiration cycle. (See Fig. 11-4.)

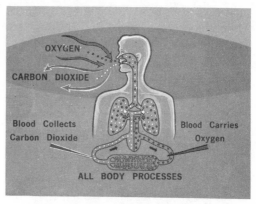

Fig. 11-4. Oxygen Used in the Body

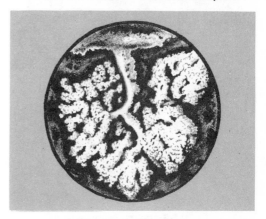

Fig. 11-5. The Alveoli

Within the lungs, there are millions of tiny air sacs called *alveoli* which inflate like tiny balloons. (See Fig. 11-5.) The number of air sacs in the lungs is estimated to be around 750 million with a surface area between 700 and 800 square feet, or about the size of a tennis court. It is a miracle of human body structure that this large area of gaseous exchange is available within the confines of the chest cavity.

Blood is pumped from the heart through arteries to microscopic *capillaries*, or tiny tubes, through which blood is constantly flowing. The walls of the air sac contain these tiny tubes in which the oxygen is diffused into the blood. The enriched blood then flows through veins back to the heart and ultimately to all parts of the body.

The wall of each air sac consists of a minutely thin membrane, 0.00002 of an inch thick, or about one cell in thickness. This membrane is moist and porous. It is dense enough to retain or hold liquid but is semipermeable enough to permit gas molecules to flow through it.

The tiny capillaries carrying blood through these membranes or walls are so small that the red blood cells are, for the most part, in single file. Thus, a microscopically thin film of blood is exposed to the air in the air sac through the semipermeable membrane. (See Fig. 11-6.)

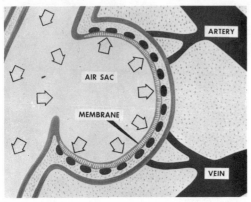

Fig. 11-6. Blood Exposed to Air in the Air Sac

GAS TRANSFER

In order to understand how the oxygen can pass through the air sac membrane and into the blood, certain basic information should be examined.

All gases tend to move from high to low pressure areas, and in so doing, are able to pass through thin membranes. These gases will flow from the high to the low pressure side of the membrane until the pressure on both sides equalizes.

For example, in the top portion of figure 11-7, the pressure within the mixture of gases on the left is gas A, 6 pounds per square inch (p.s.i.); gas B, 10 p.s.i.; and gas C, 2 p.s.i. On the right side of the membrane separating the two spaces, gas A has a pressure of 4 p.s.i.; gas B, 10 p.s.i.; and gas C, 4 p.s.i. As shown in the center

portion of figure 11-7, gas A will flow from the left to the right side of the membrane until both sides have equal pressures of gas A. There is no flow of gas B because the pressures were equal at the beginning. Gas C, however, will flow from right to left until the pressures have equalized. The bottom portion of figure 11-7 shows that on both sides of the membrane the pressure of gas A has stabilized at 5 p.s.i.; gas B at 10 p.s.i.; and gas C at 3 p.s.i.

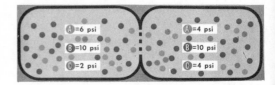

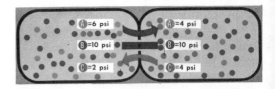

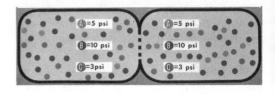

Fig. 11-7. Equalization of Pressures in a Mixture of Gases Across a Membrane

OXYGEN TRANSFER

This same principle of gas movement to equalize individual pressures on both sides of a membrane takes place in the human body.

Figure 11-8 shows a close-up view of the air sac wall. Since the water vapor and carbon dioxide are added to the air mixture entering the air sac, the percent volumes occupied by oxygen and the other gases are also changed. Less oxygen can enter the air sac due to the pressure of this water vapor and CO_2. The oxygen now occupies a smaller percent of the

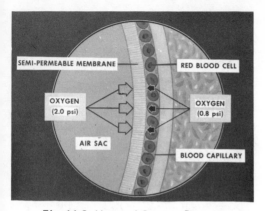

Fig. 11-8. Unequal Oxygen Pressure

total volume, so its pressure is reduced. The oxygen pressure in the air sacs becomes approximately two p.s.i. at sea level.

The pressure of oxygen in the blood as it enters the capillaries around the air sac is only 0.8 p.s.i. Since gases flow from high to low pressure areas, oxygen passes from the air sacs through the semipermeable membrane, through the blood capillary wall, and into the blood where it combines chemically with the *hemoglobin* of the red blood cells. This transfer is accomplished because the oxygen attempts to equalize the pressures on each side of the membrane.

The oxygen-rich blood is pumped by the heart to the cells throughout the body where an oxygen transfer takes place. The cells then use the oxygen to convert carbohydrates, fats, and proteins into new protoplasm or into energy for the life processes.

This energy conversion produces carbon dioxide as a waste product which is released by the body cells and transferred to the plasma of the blood. This exchange of oxygen for carbon dioxide takes place simultaneously as the blood passes through the tissues. The carbon dioxide-rich blood then returns to the heart. From the heart, the carbon dioxide-rich blood is pumped to the lungs where it flows through the tiny capillaries in the minute membranes of the air sacs. Since the air

in the air sacs has a lower carbon dioxide pressure than the carbon dioxide in the blood, the carbon dioxide flows through the blood vessel walls through the semipermeable membranes, and into the air sacs. Exhalation of the air in the lungs expels the carbon dioxide from the body, completing the respiratory cycle.

OXYGEN NEEDS AND AVAILABILITY

The amount of oxygen consumed by the body during the respiratory cycle depends primarily upon the degree of physical or mental activity of the individual. A person walking at a brisk pace will consume about four times as much oxygen as when resting. In the course of an average day, a normal adult male will consume about two and one-half pounds of oxygen. This is approximately equivalent to the weight of food consumed daily. An oxygen supply which might be adequate for a person at rest would be inadequate for the same individual when flying under severe weather conditions or when under mental stress.

Oxygen becomes harder to obtain with altitude because the air becomes less and less dense, and the total pressure decreases. As the total pressure decreases, so does the pressure of oxygen even though the percentage of oxygen in the air remains constant. For example, at 40,000 feet, the total pressure is only 2.7 p.s.i., and the oxygen pressure is 0.56 p.s.i.

As altitude increases and the pressure of oxygen is reduced, the amount of oxygen transfer in the lung air sacs is reduced which results in a decrease in the percentage of oxygen saturation in the blood. This causes a deficiency of oxygen throughout the body, and, for this reason, supplemental oxygen, pressurized cabins, and/or pressurized suits are required if the body is to receive adequate oxygen.

The total effect on an oxygen-deprived person is the result of both altitude and time of exposure. Every cell in the body is affected by the lack of oxygen, but the primary effects are on the brain and the body's nervous system. Above 10,000 feet, deterioration of physical and mental performance is a progressive condition. The curtailment becomes more severe with increased altitude as well as prolonged exposure.

HYPOXIA

The effects of an insufficient supply of oxygen on the body is called *hypoxia*. "Hypo" means "under," and "oxia" refers to oxygen. Some of the common symptoms of hypoxia are:

1. an increased breathing rate.
2. lightheaded or dizzy sensation.
3. tingling or warm sensation.
4. sweating.
5. reduced visual field.
6. sleepiness.
7. blue coloring of skin, fingernails, and lips.
8. behavior changes.

Subtle hypoxic effects begin at 5,000 feet at night. In the average individual, night vision will be blurred and narrowed. Also, dark adaptation will be affected, and at 8,000 feet, night vision is reduced as much as 25% without supplemental oxygen. Little or no effects will be noticed during daylight at these altitudes.

At 10,000 feet, the oxygen pressure in the atmosphere is approximately two p.s.i. Accounting for the dilution effect of water vapor and carbon dioxide in the air sacs, this is not enough to deliver a normal supply of oxygen into the lungs, but is enough to keep the blood about 90% saturated.

This mild deficiency is ordinarily of no great consequence. However, flying in the vicinity of 10,000 feet for over four hours may result in fatigue. Pilots report experiencing difficulty in concentrating, reasoning, solving problems, and making precise adjustments of aircraft controls under prolonged flight conditions at this altitude.

At 15,000 feet, drowsiness, a headache, weariness, fatigue, and a false sense of well-being will normally be experienced in two hours or less. Most important and less evident to the individual is the psychological impairment which could cause judgment errors, poor coordination, and difficulty in performing important tasks.

Proceeding up to 20,000 feet, the oxygen pressure in the air drops to 1.38 p.s.i., or less than half that at sea level. Oxygen saturation of the blood drops to 64% at this altitude. Collapse or convulsions will result within 15 minutes.

An atmospheric pressure reading at 34,000 feet altitude will show 3.6 p.s.i. The oxygen pressure is only 0.76 p.s.i. This is less than the constant pressures of water vapor and carbon dioxide that exist in the lung air sacs and about the same as the oxygen pressure in the blood entering the air sacs. Blood saturation with oxygen is only 30% at this altitude. Little or no oxygen will transfer from the air sacs to the blood because of the small pressure differential across the membrane. Unconsciousness will occur in less than four minutes.

It is true that susceptibility to hypoxia varies, and there are some who can tolerate altitudes well above 10,000 feet without any significant effects. It is equally true that there are those who develop hypoxic effects below 10,000 feet. As a general rule, individuals who do not exercise regularly or who are not in good physical condition will have less resistance to oxygen deficiency.

Also, those persons who have recently overindulged in alcohol, who are moderate to heavy smokers, or who take certain drugs will be more susceptible to hypoxia. Susceptibility can also vary in the same individual from day to day, or from morning to evening.

Because it affects the central nervous system, the general effects of hypoxia are, in many respects, very similar to alcoholic intoxication. A typical individual suffering from hypoxia, induced by exposure between 15,000 and 20,000 feet, will be comparable to an individual who has consumed five to six ounces of whiskey.

The most hazardous feature of hypoxia, as it may be encountered in general aviation aircraft, is its gradual and insidious onset. Its production of a false feeling of well-being called euphoria is particularly dangerous. Since it obscures a person's ability and desire to be critical of himself, he generally does not recognize the symptoms. (See Fig. 11-9.) The hypoxic individual commonly believes things are getting progressively better as he nears total collapse.

Fig. 11-9. Euphoria Often is Experienced by the Hypoxic Individual

While not all of the symptoms mentioned occur in each individual, any given person will develop the same symptoms in the same order each time he becomes hypoxic. For this reason, a person, having once experienced hypoxia under careful supervision is better prepared to recognize his condition if hypoxia occurs again. Such a controlled experience is available in what is known as an altitude chamber. Figure 11-10 shows the progressive hypoxia of a pilot from initial removal of his oxygen mask until total collapse occurred at simulated high altitudes.

There are some false indicators of a hypoxic condition which should be considered. The color of the fingernails has been suggested by some as a guide to determine the degree of hypoxia, depending upon their blueness, but this approach is invalid because when hypoxic, an individual should consider himself as "looking through the optimistic eyes of an alcoholic." Another unreliable clue is shortness of breath. As enlarged upon later, the breathing rate is controlled primarily by carbon dioxide accumulation. When carbon dioxide is exhausted adequately, the individual does not get this ordinarily useful clue to hypoxia.

TIME OF USEFUL CONSCIOUSNESS

The term "time of useful consciousness" refers to the maximum length of time an individual has to perform the purposeful tasks necessary for his survival, such as putting on an oxygen mask. Although the hypoxic individual may remain conscious for a longer period, he has a *limited* time in which his brain receives sufficient oxygen to make decisions and perform useful acts.

Fig. 11-10. Progressive Hypoxia

The table shown in figure 11-11 lists the times of useful consciousness for various altitudes. The times shown represent the average times for flying personnel. Note that the times vary from 10 minutes at 22,000 feet to 1 minute at 30,000 feet. For example, if the pressurization equipment fails in an aircraft flying at 28,000 feet, the pilot and passengers have only two and one-half to three minutes to get the oxygen mask on before they exceed their time of useful consciousness.

ALTITUDE	TIME OF USEFUL CONSCIOUSNESS
22,000 FT.	5 TO 10 MINUTES
25,000 FT.	3 TO 5 MINUTES
28,000 FT.	2½ TO 3 MINUTES
30,000 FT.	1 TO 2 MINUTES

Fig. 11-11. Time of Useful Consciousness vs. Altitude

RECOVERY FROM HYPOXIA

Recovery from hypoxia is rapid, usually within 15 seconds after oxygen is administered. Transient dizziness may occur during the recovery. The severely hypoxic individual recovering from moderate or severe hypoxia is usually quite fatigued and may suffer measurable deficiency in mental and physical performance for many hours.

PREVENTION OF HYPOXIA

Aircrew members have found that a good way to protect themselves from hypoxia is to be constantly aware of the problem and use the altimeter as the primary guide for the use of oxygen. It is recommended that oxygen be used by pilots when they fly at altitudes over 10,000 feet during the day and altitudes of over 5,000 feet at night. Night vision is reduced and dark adaptation is affected above 5,000 feet. Passengers are urged to use supplemental oxygen above 10,000 feet if the cabin is not pressurized. (See Fig. 11-12.) The pilot should also be aware of the Federal Aviation Regulations requiring use of supplemental oxygen.

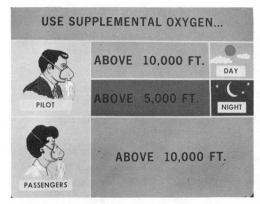

USE SUPPLEMENTAL OXYGEN...

PILOT — ABOVE 10,000 FT. DAY

ABOVE 5,000 FT. NIGHT

PASSENGERS — ABOVE 10,000 FT.

Fig. 11-12. Prevention of Hypoxia

Paraphrasing FAR 91.32: at cabin pressure altitudes above 12,500 feet MSL through 14,000 feet MSL, the pilot is required to use supplemental oxygen for any portion of his flight of over 30 minutes duration at that altitude. If he is above 14,000 feet MSL, he must use supplemental oxygen continuously; above 15,000 feet MSL, the passengers must have supplemental oxygen available.

The chart in figure 11-13 shows the percentage of oxygen needed to maintain sea level and 10,000 foot equivalents. Flying at 15,000 feet, 39% oxygen is needed to maintain a sea level atmosphere. If it is desired to maintain a 10,000 foot atmosphere while flying at 15,000 feet, only

INHALED AIR ··· %OXYGEN REQUIRED

I ALTITUDE	II S.L. EQUIV.	III 10,000 FT. EQUIV.
40,000 FT.		100
36,000 FT.		81
33,000 FT.	100	67
30,000 FT.	84	56
25,000 FT.	63	42
20,000 FT.	49	33
15,000 FT.	39	26
10,000 FT.	31	20.9
5,000 FT.	25	
S.L.	20.9	

Fig. 11-13. Percent of Oxygen Required vs. Altitudes

26% oxygen is required. Notice that above 33,000 feet, even 100% oxygen will no longer provide inhaled air with oxygen equal to that in sea level air. At 40,000 feet, the best that 100% oxygen will do is maintain a 10,000 foot atmosphere.

AIRCRAFT OXYGEN SYSTEMS

Figure 11-14 pictures one of the most common systems used in general aviation aircraft to supply supplemental oxygen. This type of equipment may be permanently installed or may be portable.

The operation of the re-breather continuous-flow oxygen mask, as it is called, is quite simple. (See Fig. 11-15.) First, oxygen flows into the accumulator bag at a continuous rate, filling the bag. A flow indicator in the tubing to the oxygen mask informs the pilot that oxygen is flowing. The rate can be adjusted for the flight altitude by opening or closing the flow control valve, or in most cases, atmospheric pressure is used to control an automatic regulator. (See Fig. 11-14.)

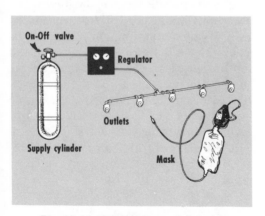

Fig. 11-14. Typical General Aviation Oxygen System

As the user inhales, the oxygen is sucked out of the bag and into the lungs. As the bag collapses, outside air is also drawn into the face mask through several tiny holes in the mask nose piece. This air then mixes with pure oxygen drawn from the bottle. The amount of outside air that is inhaled depends on the rate of oxygen flowing into the mask. The higher the flow rate, the more the bag and mask are

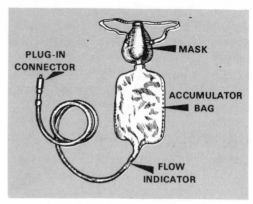

Fig. 11-15. Mask Assembly

pressurized resulting in less outside air entering the mask during inhalation.

The major portion of the oxygen used by the body is from the lower two-thirds of the inhaled air/oxygen volume in the lungs. The upper one-third of the gas mixture, trapped in the bronchial passages and throat cavities, remains relatively concentrated in oxygen.

The oxygen-rich top one-third of the air/oxygen volume in the lungs is exhaled first, passing into the mask and partially re-filling the accumulator bag. Because the bag becomes pressurized by the incoming oxygen from the oxygen system and the initial exhalation air from the lungs, the remaining two-thirds volume of exhaled air is forced out of the tiny holes into the outside air. The small opening, leading from the mask to the accumulator bag, provides a restriction that helps force the majority of the exhaled air out of the mask.

This cycle is repeated during each breath. The higher the altitude, the more the accumulator bag expands due to the lower outside air pressure around the bag. Thus, the inside of the bag receives an expanded volume of oxygen as the outside air pressure diminishes. Since the oxygen inside the bag has a larger volume, less outside air can be drawn into the mask during inhalation.

An advantage of this type of system is the conservation of oxygen by using the re-breather bag. However, at higher altitudes, systems which supply larger volumes of undiluted oxygen upon inhalation and which exhaust the exhaled air more effectively become necessary.

CARBON DIOXIDE

The respiratory center of the brain reacts to the amount of carbon dioxide found in the bloodstream. When a person is in a physically-relaxed state, the amount of carbon dioxide in the blood stimulates the respiratory center, and the breathing rate is stabilized at about 12 to 16 breaths per minute.

When physical activity occurs, the body cells use more oxygen, and more carbon dioxide is produced. Excessive carbon dioxide enters the blood, and through reflex action, the respiratory center responds to this excess. This center reacts, and breathing increases in depth and rate to remove the excess carbon dioxide. When the excess is removed, the respiratory center changes the breathing rate back to normal.

HYPERVENTILATION

"Hyper" means "over or above," and "ventilation" refers to breathing. Thus, *hyperventilation* is defined as excessive ventilation of the lungs from breathing too rapidly and too deeply with resulting loss of too much carbon dioxide from the body. This excessive loss of carbon dioxide causes the blood to become more alkaline.

Over-ventilation of the lungs in pilots or passengers is usually caused by anxiety or apprehension. When this condition occurs, the respiratory center would normally compensate by adjusting the breathing rate. However, persons who are anxious frequently override the respiratory signal, continue the forced breathing, and aggravate the carbon dioxide wash-out.

The resulting chemical imbalance may produce symptoms that are often mistaken for hypoxia. The symptoms associated with hyperventilation are:

1. dizziness.
2. tingling of the fingers and toes.
3. muscle spasms.
4. increased sensation of body heat.
5. nausea.
6. rapid heart rate.
7. blurred vision.
8. finally, loss of consciousness can occur.

A person may induce such symptoms experimentally by simply breathing deeply and rapidly over a period of time. Caution is advised, however, and such breathing should be limited to a period of a minute or two. By this time, a number of the symptoms will be experienced by the individual.

After becoming unconscious, the body's respiratory signals will take over, and the breathing rate will be exceedingly low until carbon dioxide is increased to a normal level. Hyperventilation symptoms can be relieved by voluntarily reducing rate and depth of breathing, allowing the body to build up a normal carbon dioxide level. Pilots are sometimes advised to rebreathe from a paper bag to hasten the return to normal carbon dioxide levels. This procedure artificially increases the percentage of carbon dioxide in the available air supply. (See Fig. 11-16.)

Fig. 11-16. Rebreathing Technique

SECTION B - VERTIGO AND VISION

In day-to-day life, the ability of the human body to function correctly and precisely depends partly upon a person's ability to correctly determine his position relative to the earth. This ability to correctly orient himself is especially important to the pilot. The sensations used by a person to maintain balance are primarily from:

1. the eyes.
2. the nerve endings in the muscles and about the joints (commonly called the "seat-of-the-pants" sensation).
3. certain tiny balance organs which are part of the inner ear structure.

A pilot who has entered clouds and can no longer perceive ground references will literally not be able to tell "which end is up" from his "seat-of-the-pants" sensations and from his balance organs' signals. This is because these nerve endings and balance organs depend, for the most part, upon a correct orientation to the pull of gravity and yet, these forces may not be oriented in the same direction as gravity. The brain struggles to decipher signals sent from the senses, but without the clue normally supplied by vision, incorrect or conflicting interpretations may result.

VERTIGO

The results of such sensory confusion is a dizzy, whirling sensation termed *vertigo*, sometimes referred to as *spacial disorientation*. Vertigo may take a variety of forms and may be produced by a number of different flight situations.

For example, a pilot conducting a night flight above an inclined cloud bank which is well-illuminated by moonlight might experience sensory confusion. The reason for this confusion is that most people naturally assume a cloud layer to be parallel to the surface of the earth. If the cloud bank, however, is inclined, the pi-

lot might attempt to align the aircraft with the cloud bank, thereby flying with a wing-low attitude or gradually increasing rate of climb or descent. (See Fig. 11-17.)

Fig. 11-17. Flying on Top of a Sloping Cloud Deck

In turbulence, if an aircraft is rolled abruptly right or left and then slowly resumes straight-and-level flight, the pilot may be aware of the roll, but not the recovery. On the other hand, if the aircraft rolls very slowly to the left, the pilot might still believe the aircraft to be in straight-and-level flight. A pitch change of 20° from the horizontal, if done slowly and without visual reference, may go unnoticed.

Also, in a banked turn, the acceleration tends to force the body into the seat much the same as when the aircraft is entering a climb or leveling off from a descent. Without visual reference, a banked turn may be interpreted as a climb. When completing a turn, the reduction in pressure gives the pilot the same sensation as going into a dive and may be interpreted this way.

INSTRUMENT CONDITIONS

A very serious, but common type of sensory illusion may occur when a non-instrument rated pilot continues flight into instrument conditions. When this occurs, the pilot has lost outside visual reference.

Often the aircraft will enter a very slight bank at a rate undetectable to the sense of balance. This bank will generally increase until there is a noticeable loss of altitude. The pilot, noting the decrease in altitude and still believing that he is in level flight, may pull back on the wheel and perhaps add power in an attempt to gain back the lost altitude. This maneuver only serves to tighten the spiral, unless he has the presence of mind to first correct the bank attitude of the aircraft.

Once the spiral has started, the pilot will suffer an illusion of turning in the opposite direction if he tries to stop the turning motion of the aircraft. Under these circumstances, it is unlikely that he would take the appropriate corrective action. Rather, he would continue tightening the spiral until a possible dangerous situation develops. (See Fig. 11-18.) This unfortunate situation all too often occurs to pilots without instrument training who mistakenly believe that they can maintain their orientation in clouds.

Fig. 11-18. Graveyard Spiral

GROUND LIGHT VS. STARS

A common problem associated with night flying is the confusion of ground lights with stars. (See Fig. 11-19.) Many incidents are recorded in which pilots have put their aircraft into very unusual attitudes in order to keep some ground lights above them, having mistaken them for stars. Some pilots have misinterpreted the lights along the seashore as the horizon, and have maneuvered their aircraft dangerously close to the sea while under the impression that they were flying straight and level.

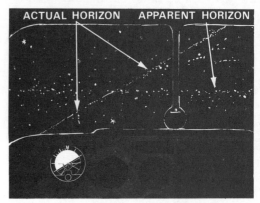

Fig. 11-19. Confusion of Ground Lights with Stars

FLICKER VERTIGO

There are other special causes of vertigo in addition to spacial disorientation (balance organ confusion). Among them is the imposition of a flickering light or shadow at a constant frequency. *Flicker vertigo* can result from a light flickering 14 to 20 times a second and may produce unpleasant and dangerous reactions in some persons. These reactions include dizziness, nausea, unconsciousness, or even reactions similar to an epileptic fit. They are especially insidious because the subject is often not aware of the cause of distress.

In a single-engine airplane flying toward the sun, the propeller can cause a vertigo-producing flickering effect, especially when the engine is throttled for a landing approach. The flickering shadows of helicopter blades have been known to cause flicker vertigo, as has been the bounce-back from rotating beacons or strobe lights when operated in or near the clouds at night. Pilots operating under instrument conditions are advised to turn off the rotating beacon or strobe lights in order to avoid this effect. Slight changes in propeller or rotor r.p.m. usually will produce relief when the effect cannot be avoided otherwise.

It should be clear from the previous discussion that instrument training is a valuable asset to any pilot, whether or not he actually flies in instrument conditions.

VISION IN FLIGHT

Since the earliest days of aviation, a pilot's vision has been regarded as a most vital part of his physical equipment for flight. In flying, constant demands are made on the vision in order to avoid obstacles, to judge distances or interpret air traffic control light signals, to read flight instruments inside the cockpit, and to study the terrain.

The pilot's vision may be adversely affected by such diverse factors as dietary deficiencies, hypoxia, and "G" forces.

NIGHT VISION

The ability to see at night can be greatly improved by understanding and applying certain techniques. The pilot's night visual ability can actually be increased by practice. If the eyes are exposed to strong light even briefly, a substance called visual purple is bleached from the nerve cells of the retina and night vision is temporarily destroyed. For this reason, avoidance of strong light must begin well in advance of a night flight, since several minutes are required to regain night vision after it is lost.

After the visual purple has been bleached by the exposure to light, the period during which the substance is resynthesized is known as the period of dark adaptation. A significant portion of dark adaptation is accomplished rather quickly, usually within a few minutes. When one enters a darkened motion picture theater, within a few minutes the seats or aisles may be seen more clearly. Dark adaptation is usually complete after approximately 30 minutes.

Physiologists have found that dark adaptation is destroyed most quickly and completely by exposure to white light while dark red light has been found to be the least detrimental.

DIET AND NIGHT VISION

Vitamin A is a chemical factor essential to good night vision. Each individual should know which foods contain Vitamin A and eat them regularly. Foods high in Vitamin A content are eggs, butter, cheese, liver, apricots, peaches, carrots, squash, spinach, peas, and all types of green vegetables. Cod liver oil and green vegetables are richest in Vitamin A. Persons who normally eat well-balanced meals should have no difficulty in receiving enough Vitamin A for normal night vision.

CARBON MONOXIDE AND NIGHT VISION

Hypoxia resulting from carbon monoxide poisoning affects visual sharpness, brightness, discrimination, and dark adaptation in the same way and extent as does hypoxia resulting from high altitudes. As an example, it has been shown that five percent saturation of carbon monoxide in the red blood cells has the same effect as an altitude increase of 8,000 to 10,000 feet. Smoking three cigarettes can cause a carbon monoxide saturation of four percent with an effect on visual sensitivity equal to that of an altitude increase of 8,000 feet. These facts stress the seriousness of minor concentrations of carbon monoxide in the cockpits of aircraft involved in night flying.

SECTION C - FLIGHT EFFECTS, DRUGS AND ALCOHOL

Once man lifts himself off the earth's surface, his body is suddenly subject to many motions and conditions to which it may not be accustomed. New references or lack of references may cause the body to grope for stabilization. Most persons adapt to new environments easily and quickly as evidenced by the world's many space flights which are now routine. It is the purpose of this section to explain how the body may react to the various conditions associated with flight.

For new flyers, flight means a new freedom. The body experiences sensations of movement which may be strange to those experienced in routine, earthbound existence. Some persons, however, are more sensitive to motion than others and are the ones most likely to get airsick. After repeated flights, these people usually become accustomed to their new environment and thereafter are not disturbed. Many times, ignorance and fear of the unknown are the main culprits, and a simple explanation of what is happening and what causes it will often cure the problem.

AIRSICKNESS

Airsickness, like seasickness or any other motion-induced disturbance, is primarily the result of the brain receiving conflicting messages from the eyes, organs of balance, and muscle receptors. Thus, a person gets dizzy and feels nauseated, and because the stomach is very sympathetic to nervous or mental fatigue, it reacts to these stimuli.

The airlines have found that the overall rate of motion discomfort in passengers is one percent or less, and that there are marked differences in sex and age. Adult men are the least susceptible. The incidence in women is five times as great as in men, and the incidence in children is nine times as great.

Many factors enter into the cause of motion discomfort and one or all can play a part in any given individual. There are three primary factors which can be involved: physical, environmental, and mental.

PHYSICAL FACTORS

The physical factors are those that are directly related to the organs in the body that perceive motion such as stimulation of the organs of balance in the inner ear, visual response to unusual motions, sensations in the internal organs, the position of the head, illness, and fatigue.

Motion discomfort can be caused in most people by continuous up and down motion, either slow or fast. Specific individuals also vary greatly in susceptibility. The up and down motion of short duration and frequent occurrence is the most common cause. This is true particularly in rough air and turbulence.

POSITION

The position of the head with respect to the direction of the motion is also important. Up-and-down motion with the head back is tolerated more readily than with the head erect. (See Fig. 11-20.) This is why people have less motion-induced distress when reclining or feel better after reclining. The same is true in ocean travel.

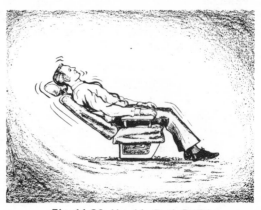

Fig. 11-20. Head in Back Position Tolerates Up and Down Motion

ILLNESS

Good sense dictates that a person who is ill should not fly except in emergencies. Illness makes the body less able to withstand discomfort or unusual experiences of any type. People who are ill should not act as pilot in command or accept responsibility as a required crewmember. For example, the common cold could mean trouble for a pilot. A cold may contribute to hypoxia, vertigo, impaired vision, or ear, tooth, or abdominal pain.

ENVIRONMENTAL FACTORS

The environment surrounding the person also contributes to the susceptibility of motion distress. Many things can cause a person to subconsciously tense up because they are uncomfortable or feel different. This feeling can then trigger a chain reaction that results in airsickness.

One of the external factors affecting the body is temperature. Excessive heat is frequently encountered in flying, particularly in light aircraft at low altitudes in the summer. In the winter, excessive cold is sometimes experienced. Either may increase sensitivity to motion.

Noise and vibration can contribute to increased discomfort, fatigue, and anxiety, particularly in people who are apprehensive and unfamiliar with flying.

MENTAL FACTORS

The mental factors associated with motion discomfort are hard to define, but one of these factors is fear. For people new to flying, the anticipation of flight may be enough to generate uncertainty that can develop motion discomfort. Also, some people have a definite fear of becoming airsick, and sure enough, they talk themselves into it. Impressionable people who have heard tall tales about air travel may be very susceptible.

PREVENTATIVE AIDS

What can be done to minimize the problem? If a pilot should become ill and does not have a co-pilot, his safest action is to land as soon as possible.

Student pilots who experience symptoms of motion discomfort are afforded special consideration in the beginning phase of flight instruction. The best approach emphasizes careful flight planning, provides a thorough briefing, restricts practice to gentle maneuvers, and avoids turbulence. Starting with extremely short flights and slowly increasing the length of each training flight sometimes overcomes the tension associated with these sessions. These measures will usually help the pilot to gain confidence and experience that will eliminate motion-induced disturbance after the first few flights.

Since motion distress may be encountered by pilots and passengers, it is helpful to know some of the things that can be done to avoid or relieve the causes and symptoms. (See Fig. 11-21.)

1. Look for distant landmarks and discuss their relative positions, distances, and features. This minimizes concentration on nearby landmarks which may emphasize motion.

2. An explanation of the trip before takeoff and keeping passengers advised during the flight helps to reduce apprehensions. Information concerning changes in direction, altitude, power changes, engine noise, time en route, and points of scenic interest tend to keep passengers from becoming bored. It also increases their understanding and reduces anxiety.

3. Secure medication prior to flight. Dramamine may be useful, especially in persons who are susceptible to motion discomfort. The advice of a physician, preferably an Aviation Medical Examiner, should be sought regarding dosage and desirability of the use of drugs in adults and children.

4. Take a position as near as possible to the center of gravity where there is a minimum of up-and-down or swaying motion.

MEDICAL FACTORS OF FLIGHT

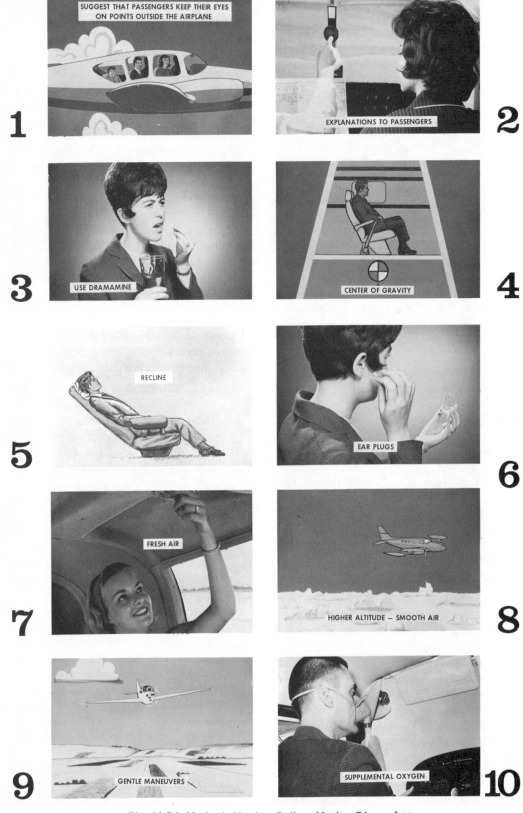

Fig. 11-21. Methods Used to Relieve Motion Discomfort

5. Lie back if possible. This keeps the head reclined and increases the body tolerance to up-and-down motion.

6. Use ear plugs to reduce noise.

7. Provide all the ventilation possible. This reduces cabin odors and keeps air fresh.

8. Flight at higher altitudes in smoother air is helpful. Flying at night may be very useful since it reduces visual stimulation and turbulence is usually at a minimum.

9. Avoid steep banks and sudden attitude changes. Gentle maneuvers insure greater passenger comfort.

10. Use oxygen at high altitudes. Oxygen may be useful when high altitudes are unavoidable. Of course, supplemental oxygen should be available for passengers and pilots for flights above 10,000 feet.

DRUGS

Pilots are normal people who have common, everyday ailments. They also get fatigued and may even be overweight and nervous. Dozens of drug prescriptions and medicines are available for each of these conditions. Many persons think that most commonly used remedies are harmless items which can be used routinely to add to daily well-being. This may not be so, and pilots should know the possible pitfalls and their tolerance to some common medications.

DRUG SIDE EFFECTS

Almost all drugs have side effects detrimental to the body, and the severity of the effects is often increased by flying. Major side effects of common medications can include drowsiness, mental depression, decreased coordination, reduced sharpness of vision, diminished function of the organs of balance, increased nervousness, decreased depth perception, and impaired judgment. Any medical condition for which drugs are necessary is sufficient cause for the pilot not to fly at all.

ASPIRIN

Most headache remedies contain aspirin, which is one of the few drugs normally accepted as having little or no adverse effect on people who are not allergic to it. However, many headache preparations contain pain relievers other than aspirin which reduce tolerance to altitude. Aspirin alone is considered safe for flyers.

NOSE DROPS AND INHALERS

In general, nose drops and inhalers cause little effect on the body other than the nose, if used on occasion. Excessive use of either or both could lead to rapid heartbeat and increased nervousness, either of which may decrease the performance of the flyer.

ANTIHISTAMINES

Antihistamines are used for almost everything, including colds, motion sickness, allergies, hay fever, and as sleeping pills. But they also may have dangerous side effects, such as drowsiness, decreased coordination, mental depression, reduced sense of balance, and diminished alertness. It is recommended that persons not pilot an aircraft for 24 hours after the administration of a usual dose of antihistamines.

COUGH MEDICINE

Cough medicines usually contain a substance which depresses the brain and cough center, a good-tasting syrup, an antihistamine, and a decongestant. It may decrease mental functions. The antihistamine decreases secretions and adds side effects. Decongestant side effects often cause increased heart rate and nervousness. In flight, all the side effects contribute to reduced pilot performance. Cold pills, which contain antihistamines, can also produce these side effects.

TRANQUILIZERS

The pilot who is taking tranquilizers will react abnormally and poorly when under stress, such as when flying on instruments, in marginal weather, at night, or in crowded traffic situations. A tranquilized pilot is a sluggish pilot. His alertness, judgment, reaction time, and perception are all affected adversely.

The FAA recommends that pilots not fly for a 24-hour period after taking a tranquilizer. Stomach soothers, heartburn preparations, and ulcer medications also have a depressant or tranquilizing effect. The same rule would apply to these drugs.

PEP PILLS

Pep pills are an artificial and harmful means of increasing energy and fighting fatigue. Pep pills may cause blurred vision, nervousness, irritability, impaired judgment, and loss of coordination. Reducing pills are essentially pep pills with sedatives or tranquilizers and are also undesirable for pilots.

MOTION SICKNESS DRUGS

Motion sickness drugs, while useful for passengers, should not be used by pilots unless prescribed for a student who experiences airsickness early in flight training. In this case, he should have a qualified instructor pilot at the dual controls, and the entire operation should be under the guidance of a physician and with the knowledge and consent of the instructor. Motion sickness drugs are basically antihistamines which have already been discussed.

Although the drugs which have just been discussed are the most common causes of impaired pilot performance, it should be pointed out that any form of medicine may have adverse effects if the pilot is sensitive to that particular drug. A pilot should not take any drugs or medicines before or during flight unless he is completely familiar with his reaction to the medication. If there is any doubt, he should consult his aviation medical examiner.

Probably the best general recommendation for flying personnel and others directly associated with flight control is abstinence from all drugs. However, some illnesses and symptoms may not preclude flying or ground traffic control work, but may be benefited by appropriate drugs. The question, therefore, arises as to whether use of the required drugs in such instances may be safe and, in fact, advantageous. This led to the preparation of *Drug Hazards in Aviation Medicine* for Aviation Medical Examiners. It lists drugs, the toxic effects relevant to aviation that they may produce, and a conservative estimate of their allowable use. This book may be purchased from the U.S. Government Printing Office. (See Fig. 11-22.)

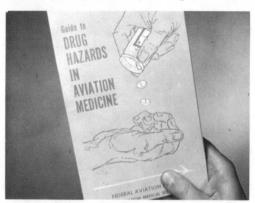

Fig. 11-22. FAA Booklet on Drug Hazards

ANESTHETICS

Following local and general dental and other anesthetics, a period of at least 48 hours should be spent on the ground. If any doubt remains concerning the right time to resume flying, seek appropriate medical counsel.

BLOOD DONATION

It should also be noted that blood donation and flying are not compatible. Blood circulation following a donation takes several weeks to return to normal and, although effects are slight at ground level, there are risks when flying during this period. It is recommended that pilots do not volunteer as blood donors when actively flying, but if blood has been given, an appropriate medical source should be consulted before flying.

CARBON MONOXIDE

The pilot must be aware of carbon monoxide as a threat, especially since most single-engine aircraft are heated by air which has passed over exhaust manifolds. Flying with a defective heat exchanger could be dangerous just as driving a car with a defective muffler is dangerous.

Carbon monoxide is a colorless, odorless, tasteless product of an internal combustion engine and is always present in exhaust fumes. The onset of carbon monoxide symptoms is insidious with "blurred thinking," a possible feeling of uneasiness, and subsequent dizziness. Later a headache may occur. For biochemical reasons, carbon monoxide has a greater affinity for the hemoglobin of the blood than oxygen. Furthermore, once carbon monoxide is absorbed in the blood, it "sticks like glue" to the hemoglobin and actually prevents the oxygen from attaching to the hemoglobin. (See Fig. 11-23.)

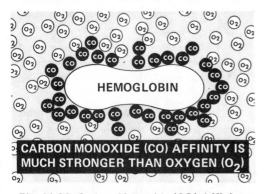

Fig. 11-23. Carbon Monoxide (CO) Affinity

The hemoglobin of the blood has 200 times more affinity for carbon monoxide than for oxygen. Thus, if there is only one part of carbon monoxide to 200 parts of oxygen in the cabin air, the hemoglobin will pick up about the same amount of each. More than this amount of carbon monoxide tends to block out the oxygen completely.

An individual exposed to hypoxia alone recovers quickly when oxygen is added to his inhaled air. In fact, recovery occurs usually within the circulation time of the blood, which is just a few minutes. However, an individual affected by carbon monoxide poisoning may be affected for 24 to 48 hours as carbon monoxide remains tenaciously attached to the hemoglobin molecule.

Inexpensive carbon monoxide detectors are available for cabin use. If carbon monoxide is suspected in the cabin, it is important that cabin heat be turned off immediately and a window be opened. An oxygen mask should be used with the oxygen turned to maximum if the aircraft is so equipped. Of course, a landing should be made as soon as a suitable airport is available.

ALCOHOL

Alcohol has effects similar to tranquilizers and sleeping tablets and may remain circulating in the blood for a considerable time, especially if taken with food. FAR's require that a pilot must allow eight hours between the consumption of alcoholic beverages and flying an aircraft. Common sense dictates that large amounts would require an even longer recovery period. A good rule is to allow 24 hours between the last drink and take-off time.

The following physiological and psychological effects of alcohol have been substantiated by research and are considered to be extremely detrimental to piloting an aircraft.

1. A dulling of critical judgment
2. A decreased sense of responsibility
3. Diminished skill reactions and co-ordination
4. Decreased speed and strength of muscular reflexes (even after one ounce of alcohol)
5. Efficiency of eye movements decreased 20% during reading (after one ounce of alcohol)
6. Significantly increased frequency of errors (after one ounce of alcohol)
7. Constriction of visual field
8. Decreased ability to see under dim illumination
9. Loss of efficiency of sense of touch

10. Decrease of memory and reasoning ability
11. Increased susceptibility to fatigue and decreased attention span
12. Decreased relevance of responses in free association test, with an increase in nonsensical reactions
13. Increased self-confidence with decreased insight into immediate capabilities and mental and physical status

PSYCHOLOGICAL CONSIDERATIONS

The psychological effect of desire may drive a person to be overmotivated to achieve a goal, perhaps to the extent judgment is impaired. A pilot may become so intent on getting home, he may attempt to fly through weather that demands skill beyond his experience. (See Fig. 11-24.)

Fig. 11-24. Negative Effect of Desire

Ironically, some pilots who are in professions requiring a high degree of knowledge sometimes place themselves in undesirable situations because of an overpowering desire to fulfill an ambition or goal. An ambitious or aggressive person who lives daily with challenge must continually examine his motives carefully and avoid flying beyond his experience and ability levels.

ANXIETY AND STRESS

Since a pilot in his aviation environment, is required to continually evaluate information, perform complex tasks, and make decisions, aviation psychology deals with a pilot's performance during various emotional states. Two of these states—stress and anxiety—may have adverse effects on a pilot's performance. The causes of emotional stress are many and varied. They can be divided into two categories. The first includes those situations not specifically related to flying such as family problems, financial or business considerations, or the demands of a pressing schedule.

The second category includes situations directly related to flight: for instance, apprehension about adverse weather conditions, malfunctioning equipment, or lack of confidence on the part of the pilot.

Fear is a normal, protective emotion which can build from stress or anxiety. Fear progressing to panic, however, is certainly undesirable. Panic can be avoided or overcome by forcibly maintaining or re-establishing self-control. A person who completely understands a situation can maintain control of his emotions, think more clearly, and reason properly. Once reason and logic are applied to the facts, the proper decision can be made and appropriate action taken.

If a pilot becomes lost or disoriented and panics, for instance, he may wander around aimlessly and compound his problem. If, however, he maintains self-control, he can follow a logical means of evaluating his situation, re-orienting himself, and taking proper corrective action.

As man has created more complex, faster, and higher flying aircraft, his physiological and psychological characteristics have become increasingly important. Although these characteristics create some limitations, man's adaptability and his capacity to design a protective environment have permitted aeronautical exploration undreamed of a few short years ago. It is expected that aviation physiology will become an even greater aspect of the total flight environment of the future.

chapter 12

FLYING A VFR TRIP

INTRODUCTION

The student has learned, up to this point, all types of necessary information concerning the airplane, weather, computer, communication, and regulations governing flight. In this chapter, these knowledge areas are put together in their proper sequence in order to proficiently plan and fly a cross-country trip using the dead reckoning method of navigation.

In order to have the proper mental perspective for a study of cross-country flight, the student must be aware that within this training exercise, *all* the elements of flight planning for a *long* cross-country flight are presented. He must also recognize that even though only *one* flight, encompassing the relatively short distance of 159 miles is made, the skills and techniques used by *many* experienced pilots on *numerous* cross-country flights are incorporated into this single learning experience.

Lest the student be overwhelmed, he should keep in mind the preceding thoughts, and also as with many other complex activities, once the "basics" are mastered, a seemingly complex task becomes "second nature." Furthermore, a student who is thoroughly prepared on the ground will approach cross-country flight with confidence and find the actual flight to be a comfortable, satisfying realization of a goal for which he has strived.

NAVIGATION LOG

The flight which will be taken in this chapter will be the same flight that was planned in the Dead Reckoning Section of Chapter 9. The navigation log as previously planned is shown in figure 12-1.

FEDERAL AVIATION ADMINISTRATION FLIGHT PLAN

After the completion of the navigation log, the FAA Flight Plan should be filled out and filed with the FSS, preferably in person. If the flight plan cannot be filed in person, the FSS will accept the flight plan either by telephone or, if en route, by radio. However, filing by telephone is desired in order to reduce frequency congestion. After the flight plan has been filed, it is not *activated* until the pilot is airborne and calls the flight service station to inform them of the actual time of departure. If the flight plan is not activated within an *hour* of the proposed departure time, the FSS will *cancel* the proposed flight plan. Any trip that is to be canceled after the flight plan is filed should be canceled by calling the FSS.

After the flight plan is activated, the flight service station sends a flight notification

NAVIGATION LOG

Notes:

Check Points (Fixes)	VOR Ident / Freq.	Course (Route)	Altitude	Wind Dir. / Vel. / Temp.	CAS / TAS	TC / Var.	MC	WCA / MH	Dev. / CH	Dist. Leg / Rem.	GS Est. / Act.	ETE / ATE	ETA / ATA	Fuel / Rem.
JEFFERSON COUNTY										159				GPH 8.5
		D→	2500	230 20 / 18°	120	342 / -7	335	-9 / 326	+5 / 331°	9 / 150	127	9 (4+5)		
ISLAND IN RIVER														
			↓	↓	↓	↓	↓	↓	↓	20 / 130	↓	9		
SILSBEE														
			↓	↓	↓	↓	↓	↓	↓	33 / 97	↓	16		
WOODVILLE														
			↓	↓	↓	↓	↓	↓	↓	35 / 62	↓	17		
HUNTINGTON														
			↓	↓	↓	↓	↓	↓	↓	22 / 40	↓	10		
NACOGDOCHES														
			↓	↓	↓	↓	↓	↓	↓	40 / 0	↓	24 (19+5)		
RUSK COUNTY														
												(85.0)		
Totals ▶										159		(1:25)		12 GALS

Airport Frequencies

	Departure		Destination
	JEFFERSON CO		RUSK CO.
ATIS	——	ATIS	——
Grnd	121.9	Apch	——
Tower	119.5	Tower	——
Dep.	——	Grnd	——
FSS	122.2	FSS	
UNICOM	123.0	UNICOM	122.8
Field Elev.	16′	Field Elev.	440′

Airport & ATIS Advisories

Departure		Destination
	ATIS Code	
	Ceiling & Visibility	
	Wind	
	Altimeter	
	Approach	
	Runway	

Time Check		
Block In		Aircraft Number
Block Out		
Log Time		

Flight Plan and Weather Log on Reverse Side

SCALE 1:500,000 SECTIONAL AERONAUTICAL CHARTS

Nautical Miles 10 0 10 20 30 40 50
Statute Miles 10 0 10 20 30 40 50 60

Fig. 12-1

message to the FSS nearest the destination. The pilot may elect to *close* his flight plan with a flight service station other than the destination station, but this information should be relayed to the departure FSS when the flight plan is filed. At the destination airport, the flight plan must be canceled. An aircraft is considered overdue on a VFR flight plan when no communication has been received or when it fails to arrive at the destination within 30 minutes after the ETA. If a flight plan is not closed, ATC begins a telephone search to locate the overdue aircraft. If ATC is unable to locate the aircraft, costly search and rescue procedures are started.

It is important for a pilot to extend his flight plan with any flight service station within radio range when he knows his estimated time of arrival has changed significantly. It is also extremely important for him to close his flight plan by contacting the destination flight service station and letting them know he has arrived. This can be done by radio or by telephone after landing. In any event, it is always important for pilots to remember to CLOSE FLIGHT PLANS.

The FAA Flight Plan is to be filled in as follows: (See Fig. 12-2.)

1. TYPE OF FLIGHT PLAN: The VFR box will be checked for most flights. DVFR (Defense VFR Flight Plan) is used in accordance with the information found in Part 3 of the AIM. IFR flight plans can only be filed by qualified instrument pilots flying IFR equipped aircraft.
2. AIRCRAFT IDENTIFICATION: The full identification number of the aircraft should be used; e.g., N17356.
3. AIRCRAFT TYPE AND SPECIAL EQUIPMENT: The specific make and model is included; e.g., Clipwing 23. The special equipment code as listed in the bottom of the flight plan should be entered. For example, if Clipwing 23 was equipped

with a 4096-code transponder and DME, this information would be entered as "Clipwing 23/A." This letter suffix is *not* to be added to the aircraft identification during normal radio voice communications.

4. TRUE AIRSPEED: This figure is to be given in *knots* based on the estimated TAS at the planned flight altitude.
5. POINT OF DEPARTURE: The name of the departure airport is filled in. If the flight plan is filed in the air, the point of departure would be the position from which the flight plan was filed.
6. DEPARTURE TIME: This time is to be stated in Greenwich Mean Time, also referred to as Zulu (Z) time. The use of Zulu time eliminates any confusion resulting from differences in local time along the route of flight. For example, 1400Z is the same time any place in the world. After departure, the pilot will activate his VFR flight plan and give the FSS his actual time of departure. When the flight plan is filed in the air, the actual departure time is given rather than the proposed departure time.
7. INITIAL CRUISING ALTITUDE: The pilot's choice of the best altitude based on winds aloft, clouds, etc. is entered. The altitude should conform with the VFR cruising altitudes as specified in the regulations.
8. ROUTE OF FLIGHT: If the flight is direct, enter the word "direct." Otherwise, enter navigational aids and airways that will be used, or changes of course en route.
9. DESTINATION: The name of the airport and city should be entered. Quite often the destination city will have numerous airports.
10. ESTIMATED TIME EN ROUTE: The total time from takeoff to landing should be entered in hours and minutes and should include time for any enroute stops.

DEPARTMENT OF TRANSPORTATION— FEDERAL AVIATION ADMINISTRATION **FLIGHT PLAN**					Form Approved OMB No. 04-R0072		
1. TYPE VFR ☒ IFR DVFR	2. AIRCRAFT IDENTIFICATION N17356	3. AIRCRAFT TYPE/ SPECIAL EQUIPMENT CLIPWING 23 NONE	4. TRUE AIRSPEED 104 KTS	5. DEPARTURE POINT JEFFERSON COUNTY	6. DEPARTURE TIME PROPOSED (Z): 1500 / ACTUAL (Z): (1456) 0956		7. CRUISING ALTITUDE 2500'

8. ROUTE OF FLIGHT

DIRECT

9. DESTINATION (Name of airport and city) RUSK COUNTY HENDERSON, TEXAS	10. EST. TIME ENROUTE HOURS: 1 / MINUTES: 25	11. REMARKS NONE		
12. FUEL ON BOARD HOURS: 4 / MINUTES: 30	13. ALTERNATE AIRPORT (S) NONE	14. PILOT'S NAME, ADDRESS & TELEPHONE NUMBER & AIRCRAFT HOME BASE L.T. MONTANO 373 AIRE DRIVE BEAUMONT, TEXAS 388-5301		15. NUMBER ABOARD 1
16. COLOR OF AIRCRAFT WHITE & BLUE	CLOSE VFR FLIGHT PLAN WITH_____FSS ON ARRIVAL			

Fig. 12-2. Flight Plan Form

11. REMARKS: Stop-over airports can be listed here. Passengers' names can be entered, and the term "ADCUS" (meaning advise customs) may be listed if on an international flight.

12. FUEL ON BOARD: Total usable fuel on board the aircraft in hours and minutes should be entered.

13. ALTERNATE AIRPORT(S): Although not a requirement, if the weather is marginal, an alternate airport can be named. This entry is primarily an IFR requirement.

14. PILOT'S NAME, ADDRESS AND TELEPHONE NUMBER AND AIRCRAFT HOME BASE.

15. NUMBER OF PERSONS ABOARD.

16. COLOR OF AIRCRAFT: The major color should be listed first, followed by the trim color.

PRETAXI

After the flight plan has been filed, the pilot is ready to fly the trip. Some items to consider during the flight are listed below:

1. Note time off the ground at takeoff.
2. Open the flight plan as soon as practicable after leaving the traffic pattern at the point of departure.
3. Climb to altitude, pick up first checkpoint, and establish compass heading that is listed in navigation log.

4. Determine ETA (estimated time of arrival) by adding time en route to time off.
5. Determine time between checkpoints and, using distance between checkpoints listed in navigation log, determine groundspeed with computer.
6. With actual groundspeed known, readjust ETA as necessary.

EN ROUTE

After takeoff and when the climb is established, the pilot contacts Beaumont Flight Service Station to activate his flight plan. The time off the ground is 9:56 Central Daylight Time. The pilot can see the smoke from the oil refinery next to the island identified as the first checkpoint. After leaving the traffic pattern, the pilot heads directly for this smoke.

Now that the pilot is on his way, he can complete the navigation log by estimating the time of arrival at Rusk County Airport. Since the computed time en route is 1:25 and the departure time was 9:56, the estimated time of arrival is (9:56 plus 1:25) 11:21 Central Daylight Time. This time is entered in the navigation log. (See Fig. 12-3, item 1.)

The first circled checkpoint is the island located near the oil refinery. (See Fig.

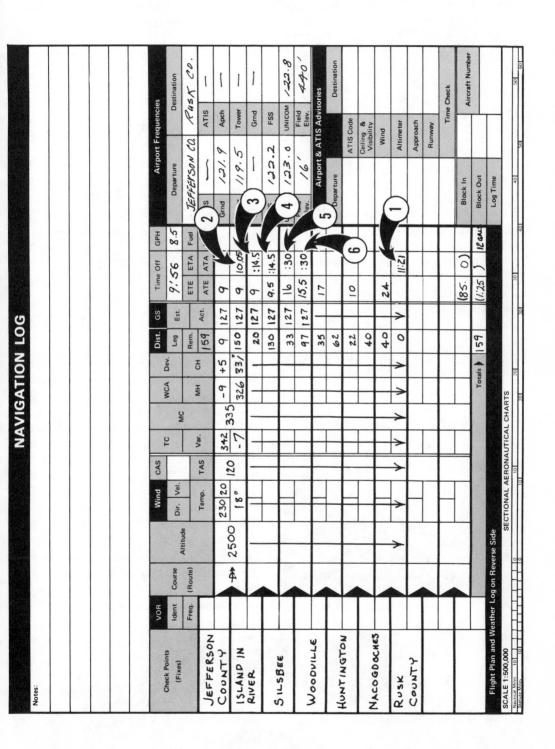

Fig. 12-3

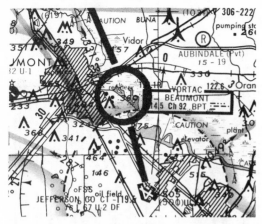

Fig. 12-4

12-4.) The time is 10:05 as the pilot flies over the island. (See Fig. 12-5.) He records the time. It is always good practice to record the time over checkpoints on the navigation log (See Fig. 12-3, item 2), or on the chart near the checkpoint. These times are used to make ground-speed checks. Then, if the groundspeed varies from the computed value recorded in the navigation log, the estimated time of arrival can be changed.

Fig. 12-5

Now the pilot turns the aircraft to a heading of 331°, which in his pre-trip planning, he found would be the compass heading for the trip.

The next checkpoint is Silsbee, 20 miles from the first checkpoint. If the compass heading is correct, he should pass just to the east of the city. (See Fig. 12-6.)

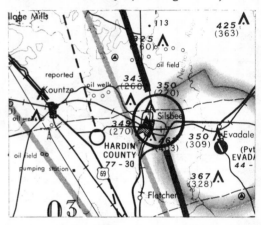

Fig. 12-6

By using the calculator side of the computer, the pilot can find the time required to fly 20 miles at 127 m.p.h. In figure 12-7, arrow "1" shows the index is set at 127, the predicted groundspeed. Arrow "2" points to 20, the distance. Below "20" the pilot reads approximately 9.5, representing 9-1/2 minutes. This time added to 10:05, the time over the first checkpoint, is 10:14-1/2, the estimated time over Silsbee. (See Fig. 12-3, item 3.)

The chart shows two tower symbols with the word "CAUTION" printed next to them. (See Fig. 12-8.) The symbol represents tall towers supporting powerlines

Fig. 12-7

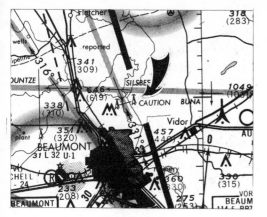

Fig. 12-8

that cross the stream. Figure 12-9 is a view of the powerlines crossing the stream. The path cleared of trees marks the powerline location and is easily visible from the air.

However, the wires and towers are almost invisible from the air. It can be seen that this powerline crossing is a hazard to a float-equipped airplane that may attempt to land or take off on this stream.

Fig. 12-9

At Silsbee, the pilot will check for the railroad shown on the chart in figure 12-10, going through the city from the west to the east.

Then, looking to the left as he passes over the tracks on the east side of Silsbee, the pilot is able to see the railroad track as it approaches Silsbee from the west. (See Fig. 12-11.) It makes a bend in the city and then proceeds in an easterly direction.

Railroads provide excellent means of identifying communities. Very often,

their patterns are visible for miles. However, the pilot must be aware that railroad tracks as well as roads can be easily hidden by trees and buildings.

The time, as the aircraft passes over the railroad tracks, is 10:14-1/2, as estimated. This time is recorded for the next ground-speed check. (See Fig. 12-3, item 2.)

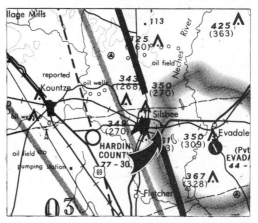

Fig. 12-10

Checking the chart, the pilot notices the dashed line running east and west labeled "underground pipeline." (See Fig. 12-12.) Naturally, he does not expect to see the pipeline, but, like powerlines, the right-of-way should be clearly visible. Looking east as he passes over the area marked with the dashed line, the pilot

Fig. 12-11

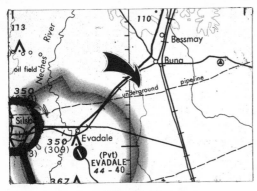

Fig. 12-12

can see the cleared right-of-way that marks the location of the underground pipeline. (See Fig. 12-13.) This is a *good* landmark; however, the pilot should be careful not to confuse this landmark with the powerline right-of-way, since powerlines and their supporting towers are very difficult to see from the air.

Fig. 12-13

Referring to the chart, the pilot estimates he should be almost over the tower represented by the symbol shown in figure 12-14. Figure 12-15 shows a close-up view

of tank towers and a radio tower that are shown on the chart with an obstruction symbol. The arrow points out the almost invisible radio tower, taller than tank towers. The guy-wires supporting this tower are completely invisible.

Very often, radio towers are located in areas lacking prominent landmarks, making them even more difficult to find. Because radio towers and the guy-wires are difficult to see, pilots are urged to use caution when flying low, such as during takeoffs or landings near radio towers. At night these towers are lighted with flashing red lights.

Progress of the flight to this point has been as planned. Now, the pilot is on the way to the third checkpoint, Woodville. It is important to hold a good compass heading since the course does not pass directly over Woodville, but passes over a primary road.

At this point, he decides to find the estimated time of arrival at Woodville using the known groundspeed of 127 m.p.h. The time to fly 33 miles, the distance between Silsbee and Woodville, is approximately 15-1/2 minutes. Adding this time to the time over the last checkpoint, Silsbee, 10:14-1/2, the estimated time of arrival over the primary road east of Woodville is 10:30. (See Fig. 12-3, item 5.)

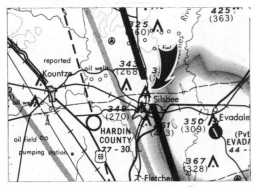

Fig. 12-14

Fig. 12-15

The symbol shown in figure 12-16, to the left of the course, represents a lookout tower. It is 1-1/2 miles from the pilot's course. As he passes this point, he is aware that it is difficult to see. Figure 12-17 shows the lookout tower from the closest point along the course. This tower would be a poor choice as a checkpoint. When choosing checkpoints, it is well for the pilot to remember that all features shown on a sectional chart are not always easily seen from the air. Whenever possible, a pilot must choose visual references that are easily seen and identified from information on the chart.

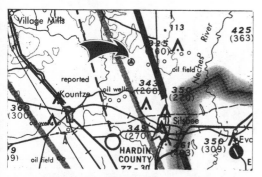

Fig. 12-16

Fig. 12-17

WEATHER DETOUR

As the flight approaches Woodville, the next checkpoint, the pilot notices that the thunderstorm activity reported in the preflight weather briefing covers a wide area ahead and to the left of the course. After confirming his position at Woodville, the pilot will call the Lufkin Flight Service Station for a check on the current weather. The time is now 10:30 and he should be over the road crossing his course line with Woodville to the west

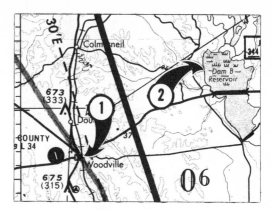

Fig. 12-18

(see Fig. 12-18, item 1), and the reservoir to the east (item 2). The sighting of these landmarks confirms his position and he records the time, 10:30. (See Fig. 12-3, item 6.)

From this point in the flight to its termination, additional references will be made regarding the *recording of time*, *groundspeed*, etc. This is good operational practice; however, for *clarity*, these entries will not be shown on the flight log for the remainder of the trip.

Now, the pilot calls Lufkin Flight Service Station on 122.6 MHz. Lufkin reports a band of thunderstorm activity from the east through the southeast quadrant, with the heaviest activity centered over Manning. Movement of the storm activity is reported to be northeast at about 10 knots. Lufkin also advises that the wind from the surface to 3,000 feet is from 230° at 17 knots.

After studying the chart, the pilot wisely decides to take a route behind the storm by way of Corrigan and north via the dual-lane highway to Del Rentzel Airport near Nacogdoches. The dual-lane highway will provide a continuous checkpoint. (See Fig. 12-19.)

A check with the plotter shows that a line from the Woodville checkpoint to Corrigan is 50° to the left of the present true course line of 342°. (See Fig. 12-20.) The pilot changes course to head for the place where the powerlines make

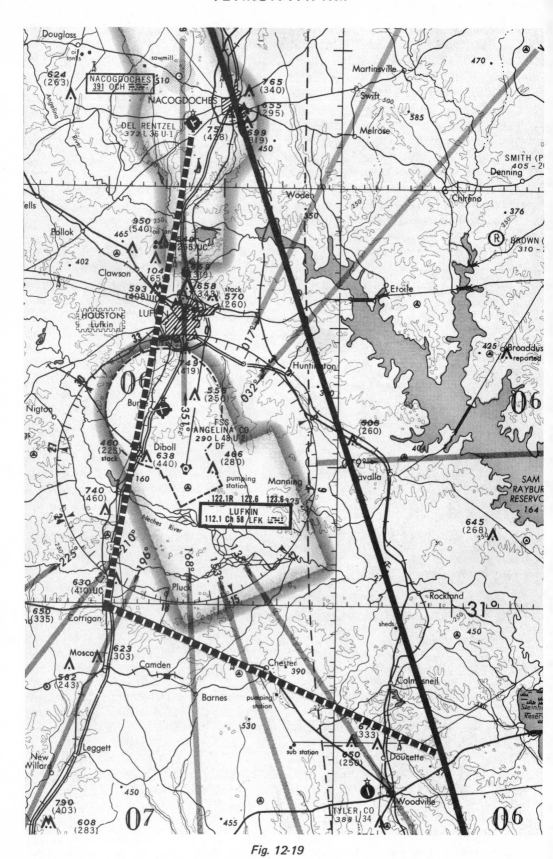

Fig. 12-19

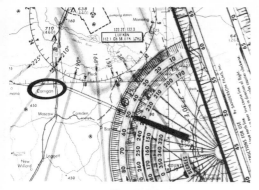

Fig. 12-20

an abrupt change in direction. (See Fig. 12-21.) The magnetic compass reads 280° as he passes over the powerlines. After holding the airplane on a heading of 280°, he arrives over the road and railroad. (See Fig. 12-22.) Notice how the railroad curves away from the road. The pilot notes and records the time, 10:33. This will be important at Chester when checking groundspeed.

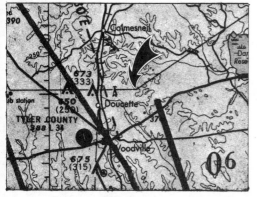

Fig. 12-21

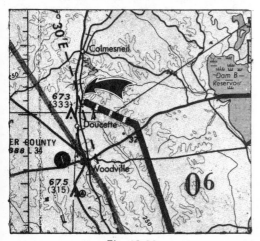

Fig. 12-22

As he approaches Chester, the pilot can plainly see the track made by the road and powerline right-of-way leading to Corrigan. (See Fig. 12-23.) He also notices that he is slightly to the right of his expected position. (See Fig. 12-24.) By correcting the heading of the aircraft to the left, he arrives directly over Chester at 10:39, 6 minutes and 11 miles from the last checkpoint.

Fig. 12-23

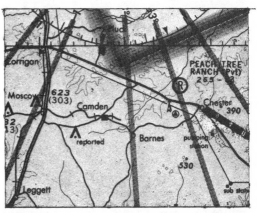

Fig. 12-24

He is careful to record the time over Chester so that he can again check groundspeed at Corrigan. The pilot can easily find the groundspeed now. He knows that six minutes is *one-tenth* of an hour. Multiplying the distance flown in one-tenth of an hour by 10 gives him the groundspeed in miles per hour. In this case, 11 times 10 is equal to 110 m.p.h. (See Fig. 12-25.) Now he can find the estimated time of arrival over Corrigan, which is 15 miles from Chester.

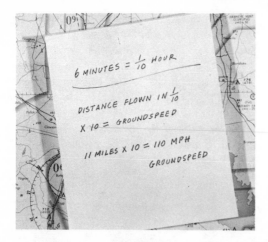

Fig. 12-25

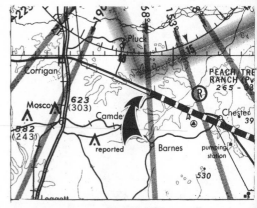

Fig. 12-27

Using the computer, he sets the speed index at 11 (see Fig. 12-26, item 1) representing 110 m.p.h., then proceeds to the right until he finds 15 (item 2), the distance between Chester and Corrigan. Under 15 on the B scale, he finds 8.2, or 8-2/10 minutes. The 2/10 of a minute is dropped. By adding 8 minutes to the time over Chester, 10:39, the pilot finds that the ETA for Corrigan is 10:47.

Fig. 12-28

line right-of-way going north. This is the place on the chart where the powerline changes direction. (See Fig. 12-29.)

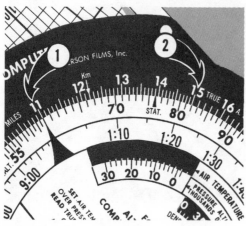

Fig. 12-26

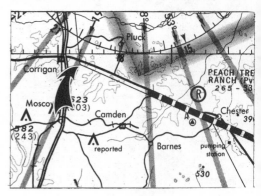

Fig. 12-29

As shown on the chart, the powerline right-of-way and the highway provide a prominent, continuous checkpoint to Corrigan. At this time, the flight is eight miles from Corrigan. (See Fig. 12-27.) The highway and the powerline as they appear from the air can be seen in figure 12-28. By looking ahead, he can see that the powerline right-of-way has been extended and continues ahead to Corrigan. However, there is evidence of a power-

The road and the cleared right-of-way marking the location of the powerline between Chester and Corrigan has provided a continuous checkpoint to follow directly to Corrigan. (See Fig. 12-30.) The chart shows another such continuous checkpoint; the dual-lane highway leading from Corrigan (item 1) north to Nacogdoches (item 2).

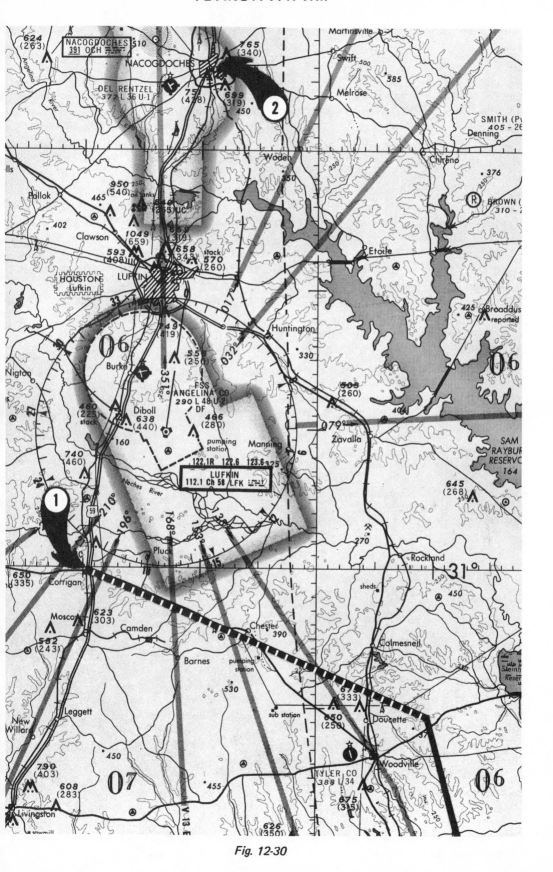

Fig. 12-30

Now, as he approaches from the east, the pilot can see that the dual-lane highway running north is barely visible because of the trees and buildings that hide it effectively. This is the road he will follow after he turns north. As he passes over Corrigan and turns north to follow the dual-lane highway, he notes the time, 10:48, and records it. The records show the times over Chester as 10:39 and Corrigan as 10:48. The time en route was nine minutes. Also, from the records, the pilot knows the distance between these two points is 15 miles.

Using the computer, he can compute the groundspeed between Chester and Corrigan. As shown by item 1 in figure 12-31, the time en route, nine minutes, represented by the 90 on the B scale, is set under the distance traveled, 15, on the A scale. Then, over the speed index pointed out by item 2 in figure 12-31, the groundspeed is indicated by the number 10, or 100 m.p.h.

Fig. 12-31

The groundspeed has dropped 10 m.p.h. from the previous check. A change in wind direction or velocity could be a contributing factor. The pilot decides to check this possibility with the Lufkin Flight Service Station by calling Lufkin radio on 122.6 MHz and request winds aloft information.

Lufkin reports winds from the surface to 3,000 feet are from 330° at 20 knots. The pilot carefully records the wind data for later reference.

Proceeding north, the pilot can see Diboll in the distance. By making minor heading adjustments, he can keep Diboll in a steady position relative to the nose of the airplane. Using this trial and error method will correct for wind drift. The pilot must refer to the magnetic compass from time to time to maintain a straight course. The magnetic compass now reads 355°.

The pilot checks his chart and notices the pattern of the dual-lane highway and the railroad going through the city of Diboll. (See Fig. 12-32.) To the west of the dual-lane highway, he should see evidence of the powerlines running almost parallel to the highway.

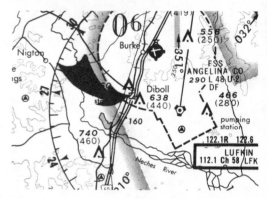

Fig. 12-32

As he is about to pass over Diboll, the pattern of the dual-lane highway and railroad are easily recognized, as in figure 12-33, item 1. To the west, in the upper left corner, the cleared area of the powerline track is seen (item 2).

Fig. 12-33

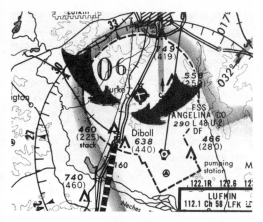

Fig. 12-34

From this position almost over Diboll, the pilot can identify the curves in the highway shown in figure 12-34. He sees the curves as sharp bends. (See Fig. 12-35.) The reason for this is that on the chart, the view is looking down on the highway from above, but by looking at the curves in the highway from a distance at a slant, the gentle curves appear as sharp angles.

Fig. 12-35

In a few minutes, the pilot sees an airport. The airport symbol on the chart helps him recognize the runway pattern of Angelina County Airport. (See Figs. 12-36 and 12-37.) The time as he passes Angelina County Airport is 10:58.

Fig. 12-36

It has taken 10 minutes to cover 17 miles, the distance between Corrigan and Angelina County Airport. The groundspeed is computed to be 102 m.p.h. This is a good time to compute a new ETA at Rusk County Airport. With a groundspeed calculated at 102 m.p.h., the pilot can expect to cover the remaining distance to Rusk County Airport in 38 minutes. When 38 minutes are added to 10:58, the time over Angelina County Airport, the pilot finds that the new estimated time of arrival at Rusk County Airport is 11:36. This is well within the airplane's fuel range. The pilot advises Lufkin radio of his revised ETA.

Fig. 12-37

The pilot's next major visual reference point is Del Rentzel Airport, southwest of Nacogdoches. Just north of Lufkin, he passes over oil tanks. (See Fig. 12-38.) They are valuable in helping a pilot establish his position. The notation on the chart identifies the solid round dots. (See Fig. 12-39.)

Fig. 12-38

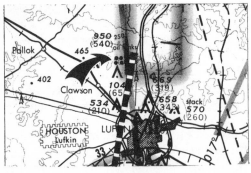

Fig. 12-39

lent path to follow. He estimates six minutes to Del Rentzel Airport. During this time, he plans a course from Del Rentzel to Rusk County Airport.

The course from Del Rentzel County Airport to Rusk County Airport has few good checkpoints. (See Fig. 12-40.) However, careful planning at this time will result in a magnetic compass heading that will take the pilot to his destination.

The road and powerline leading to Nacogdoches provides the pilot with an excel-

By using the wind information received from Lufkin Flight Service Station, he

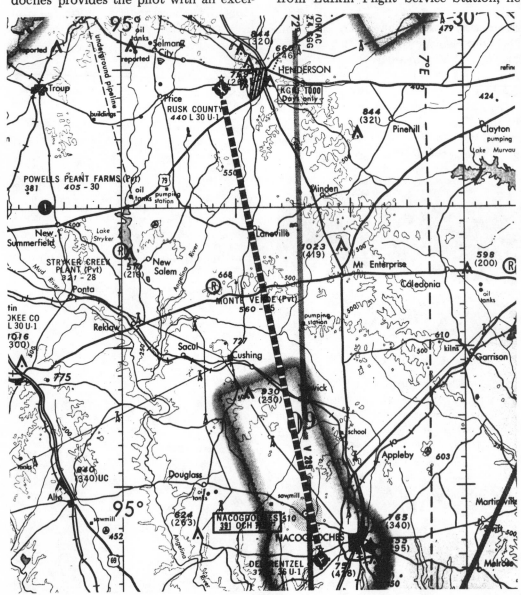

Fig. 12-40

finds the magnetic compass heading will be 335°. Because the new course is almost directly into the wind, the predicted groundspeed is 98 m.p.h. The estimated time en route is 25 minutes.

As he passes Del Rentzel Airport, the pilot records the time, 11:12. By adding the en route time of 25 minutes, he obtains the Rusk County ETA, 11:37. He then turns the airplane to 335°, the computed magnetic compass heading that will take him from Del Rentzel Airport to Rusk County Airport.

In eight minutes, he has a road and railroad in sight. (See Fig. 12-41.) To the right of his course, the pilot sees the small town of Trawick. The chart excerpt in figure 12-42 shows that the railroad makes a turn just east of Trawick (see item 1). The secondary road joins a primary road, as shown by item 2.

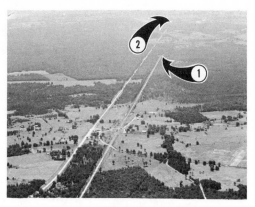

Fig. 12-43

From directly over the railroad tracks, the pilot can see the tracks disappear behind the trees as they make the first turn east of Trawick. (See Fig. 12-43, item 1.) The road continues into the distance where it seems to stop abruptly (see item 2), but actually, as indicated on the chart, joins a primary road running north and south. That distance from the pilot's present position on course is five miles.

To the northwest, the pilot identifies Cushing with the road and railroad almost converging. (See Fig. 12-44, item 1.) The pilot also notices the powerline right-of-way in the distance, as pointed out in figure 12-44, item 2. The pilot notes that the time is 11:20. He should be at his destination in 17 minutes.

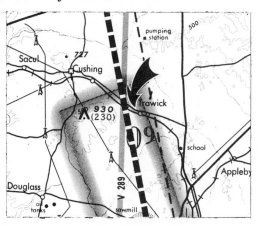

Fig. 12-41

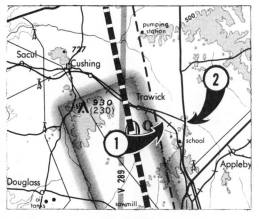

Fig. 12-42

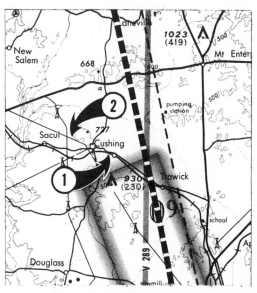

Fig. 12-44

LANDING

Proceeding north, he notices a road going north to a small town. (See Fig. 12-45.) Laneville should be directly ahead. From the east side of Laneville, the pilot can see how the road going west curves north and then disappears as it curves south. On the lower right side of figure 12-46 is the road going east.

The next landmark the pilot will see is the highway, and destination — Rusk County Airport. (See Fig. 12-47.) As he sights his destination, Rusk County Airport (see Fig. 12-48), the time is 11:36. The pilot lands and closes his flight plan by telephone with the Gregg County Flight Service Station.

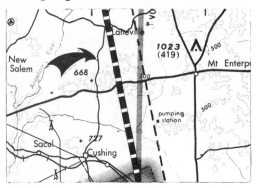

Fig. 12-45

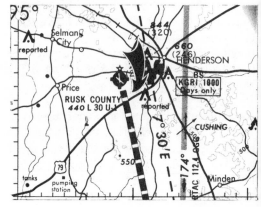

Fig. 12-47

Fig. 12-46

Fig. 12-48

chapter 13

AVIATION HISTORY

INTRODUCTION

Man's quest to raise himself from the surface of the earth has taken him from a dream, as he envied soaring birds, to successful space travel in a time span of less than 100 years. Volumes have been written on this subject, and a library could be filled to adequately cover all of man's exploits as he overcame the obstacles to flight.

It is the intent of this chapter to acquaint the reader with the highlights and key events of aviation history. This historic coverage concentrates on man's efforts to fly within the earth's atmosphere; rocket and spacecraft developments have purposely been omitted since they have a separate historical significance.

IMAGINATION, INVENTION, AND TECHNOLOGY

IDEAS WITHOUT TOOLS

In the case of aeronautics, the Greeks had the imagination to actually visualize man with the ability to fly, but they lacked the science as well as the technology to carry out their ideas. (See Fig. 13-1.) Leonardo da Vinci (see Fig. 13-2), gifted with the imagination which led to some excellent scientific speculation upon the construction of an apparatus to allow man to soar, lacked the technology to produce such a machine. (See Fig. 13-3.)

In the later 19th Century, another approach to flight was one which proved, to a degree, to be a "dead-end street" — man

Fig. 13-1. Greek Mythology

Fig. 13-2. Leonardo da Vinci

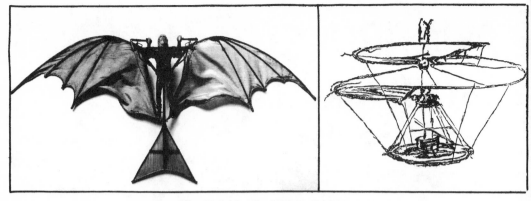

Fig. 13-3. da Vinci Flight Designs

being lifted from the ground by a lighter-than-air body. Although the technology that existed at the end of the 18th Century allowed construction of envelopes for balloons large enough to lift a man off the surface of the earth, ballooning reached a plateau of technological development in the late 19th Century. (See Fig. 13-4.) The imagination was there, but a lightweight engine of suitable power was the lacking component necessary to give the balloon maneuverability and control of altitude. It would have indeed been an unusual sight in the mid 19th Century to see a balloon or glider attempting to fly with the only type of engine available at the time — a stationary steam engine developed for agricultural use that developed one horsepower to each 3,000 pounds of weight.

In 1876, a man named Nikolaus A. Otto made the important discovery that led to the final component needed for the powered airship and powered heavier-than-air flight. This discovery was a fairly reliable internal combustion engine. (See Fig. 13-5.) In a few years, this engine was made more reliable as the power-to-weight ratio was improved, and the device was adapted for use in aircraft. With the completion of this device and its relationship to aeronautics, man now had the imagination, the basic scientific knowledge, and the technology for all the tremendous developments that have taken place up to the present.

Fig. 13-4. 19th Century Balloon

Fig. 13-5. Otto Internal Combustion Engine

To graphically display such an idea, the development of technology might be likened to a series of pyramids. Once all the stones of the pyramid are completed and in place, a base is laid for the construction of a new pyramid. Hence, a new technology was developed — aeronautics.

AEROSTATICS TO AERONAUTICS

FREE BALLOON

Two categories of lighter-than-air aircraft exist: the free balloon and the dirigible. Among free balloons, type differences exist based mainly on what lifts them into the atmosphere. Hot air was used for small paper toy balloons in China before the 6th Century, B.C., and for early ballooning attempts in France near the close of the 18th Century. Paper was used for the envelopes.

As the art of aerostatics progressed into the 19th Century, hydrogen was used in place of heated air for lift, and the balloon envelope was made of cloth. Since hydrogen tended to leak through the weave of the cloth, more advanced models carried a small gas generator to replenish the supply of gas in the envelope. Many ideas for propelling these devices with steam, horses, or human power were developed but never actually perfected.

One of the first successful balloon ascents was made by the Montgolfier brothers of France. (See Fig. 13-6.) They constructed a silk envelope to hold the heated air which came from a large brick oven underneath the aperture of the balloon. Quantities of wood shavings and paper were fed into the oven, the balloon was inflated and ascended. The balloon was a success even though the Montgolfiers thought that the lift was provided by smoke rather than by the heated air.

The Academie des Sciences became interested in the experiments of the Montgolfiers. J. A. C. Charles added an important contribution when he substituted hydrogen to obtain lift in the balloons he constructed.

Fig. 13-6. Montgolfier Balloon

Fig. 13-7. Pierre Blanchard

Fig. 13-8. John Wise

Fig. 13-9. T. S. C. Lowe

The expense for the construction of the envelope of paper, silk, or other cloth was great. The number of men necessary to handle the envelope at the time of inflation also cost a great deal, not to mention the cost of acid and other chemicals and equipment needed to produce the gas. The chance of making a balloon ascension a paying proposition was small since it was impossible to find an area for the ascension that was large enough and still prevent non-paying observers from seeing the balloon rise. At the end of the 18th Century, most people saw no practical purpose for the aerostat..

An American who made a series of major contributions to aerostatics, Dr. John Jeffries, crossed the English Channel by balloon in the company of Pierre Blanchard. (See Fig. 13-7.) As a result of his ascensions and experiments, Jeffries determined that a number of problems needed to be solved. Among these was the one discovered by John Haddock, a newspaper editor, and John La Montain, a balloonist. Intended only as a short demonstration flight, the two aeronauts left the ground at Watertown, New York, at 5:33 p.m. Because of unusual atmospheric conditions, they ascended to about

15,000 feet where they evidently experienced the unknown phenomenon of the jet stream. In a little over three hours, they descended some distance north of Ottawa, Canada, in the middle of the Canadian woods. It required four days of hiking for them to reach civilization. When the pair returned to Watertown after their four hour, three-hundred mile journey, they received a hero's welcome.

No history of ballooning could be complete without the mention of John Wise, who by 1859 had made hundreds of balloon ascensions. (See Fig. 13-8.) In that year he startled the world by riding in a balloon, cross-country, from St. Louis, Missouri, to Henderson, New York, in a little more than 19 hours. The balloon used for the trip was 60 feet in diameter and 100 feet high.

After unsuccessfully attempting to use the balloon for military purposes during the Civil War, Wise became interested in ballooning across the Atlantic. Although this expedition was probably one of the best scientifically equipped endeavors ever conceived, it ended in failure, as had all similar attempts at an Atlantic crossing.

Fig. 13-10. Portable Hydrogen Gas Generator

The free balloon was still at the whim of the wind with no method yet developed to control its direction. Altitude could not be accurately controlled since the only way a balloon could be made to ascend was by dropping off ballast to make the whole device lighter; and to descend, gas had to be released from the envelope. If, during an ascent, enough gas was lost through the envelope material or other leaks, the supply of hydrogen usually could not be replaced in flight. Experiments were made with the use of airborne gas generators, but for the most part, these were unsatisfactory.

Another individual that deserves more than passing notice is T. S. C. Lowe. (See Fig. 13-9.) Before the Civil War, Lowe had dreamed of crossing the Atlantic by balloon. His expedition was equipped with an aerostat large enough to carry a small crew and a steam-powered lifeboat. The lifeboat had both a screw to propel it through water and an aerial propeller to provide locomotion in the air. Before Lowe's balloon could attempt the Atlantic crossing, it was destroyed while being inflated for the journey.

Lowe made an additional attempt with a small aerostat, but he picked a rather inopportune time. The Civil War, in essence, had begun when Lowe started out once again to cross the Atlantic. Also,

the winds proved less favorable than he had expected, and the balloon was forced down in South Carolina. There he was taken prisoner, and it was only with great difficulty that he secured his release before open hostilities began.

After returning to the North, Lowe was instrumental in the development of a balloon corps for the Union Army. The work done by Lowe was mainly observation of a nearby Confederate Army during several battles. A reasonably efficient mode of field operation was established by the reconnaissance group. A portable gas generator was set up in the field behind Union lines to inflate the envelope with hydrogen. (See Fig. 13-10.) The balloon was kept captive by a strong rope and was allowed to go high enough to observe the Confederate troop movements and yet be out of the range of enemy weapons' fire. (See Fig. 13-11.) A telegraph was set up so that Lowe or one of the observers in the team could relay information to the Union Army headquarters below. This technique proved to be successful, but in spite of the advantages of this method of observation, the Balloon Corps was disbanded because of bad management.

THE AIRSHIP

The culmination of progress in lighter-than-air craft came with the technological

Fig. 13-11. Civil War Observation Balloon

Fig. 13-12. Count Ferdinand von Zeppelin

developments of Count Ferdinand Von Zeppelin. (See Fig. 13-12.) Zeppelin is included at this point in history because, although he was a German citizen, he served as a volunteer in the Union Army during the Civil War in the United States and was assigned to an observation balloon. It is not definitely known if the 23-year-old Zeppelin was acquainted with T. S. C. Lowe, but it is highly probable. It is known that Zeppelin's interest and imagination were stirred to a great enough degree that when he returned home to Germany after the war, he began work on aerostatic designs of his own.

The design Zeppelin completed in 1894 incorporated some important features,

none of which were actually original. The device he produced was a rigid airship; it had a rigid framework inside a cigar-shaped gas bag. It was controllable with a rudder and the equivalent of elevators and had a powerplant which allowed control of flight direction, providing the winds were not too unfavorable.

The rigid airship design achieved recognition in 1899 when the first Zeppelin was completed and test flown. The first model was not successful, but after some financial set-backs, Zeppelin was able to produce a machine which was in regularly scheduled service over specific routes by 1911. (See Fig. 13-13.) If the outbreak of World War I had not occurred, the civilian

Fig. 13-13. 1911 Zeppelin

Courtesy of Echelon Publishing Co., Minneapolis, Minnesota

Fig. 13-14. Spad Flames Observation Balloon

use of these airships would likely have continued to grow.

During World War I, Germany pressed all the Zeppelins in existence into military service. The airships were used mainly for long-range strategic bombing missions. Actually, the Zeppelin was not a success. Since its gas bag was filled with hydrogen and its size was so great, concealment in the air or on the ground was impossible. It was easy prey for Allied fighter-bomber pilots who needed only to machine-gun the gas bag and set the ship on fire in order to destroy it. (See Fig. 13-14.) The Zeppelin was particularly vulnerable on its top side as machine guns and bombs had to be carried under the gas bag and structure for balance.

The development of the Zeppelin reached its apex during the 1930s with the construction of the Hindenburg. (See Fig. 13-15.) This huge airship was also filled with hydrogen. Following one of its many transatlantic crossings, it was destroyed

Fig. 13-15. Hindenburg Zeppelin

while attempting to land at Lakehurst, New Jersey. The actual cause of the explosion of this magnificent machine is still a matter of conjecture.

If the Hindenburg tragedy had not occurred, it is possible that the Zeppelin might have had a great future. Also, if the relationship between the United States and Germany during the 1930s had been more favorable, the United States Government might have seen fit to sell natural helium to Germany, and the tragedy that occurred would have been avoided.

The only place in the world where helium is found in quantities sufficient to be extracted from the earth is in the southwest United States. (Helium is an inert gas that will not combine with other elements or substances to form new compounds. It is also odorless, tasteless, and will not ignite or support combustion.)

THE GLIDER

Verification of the many claims for glider flights are difficult to substantiate. For example, Sir George Cayley, in the late 19th Century, engraved on a plaque the major functions of control for a glider. This advance, which closely paralleled early ballooning attempts in France, was a significant contribution to aeronautics. It seems unusual that gliding did not continue to develop at a rate comparable to that of aerostatics.

On further examination, however, the problem of technology appears once again. A balloon could stay in the air for a considerable length of time without motive power. After the first quarter of the 19th Century, an individual with enough funds for components necessary to build an aerostat could have a balloon constructed that would carry him aloft in a public exhibition even though he had little control over the device. A glider was extremely difficult to handle since little was known about controlling a heavier-than-air device. A glider flight of ten seconds was considered successful.

Ballooning was inherently more popular and was participated in by a greater number of people as compared to gliding. Newspapers and fiction writers of the 19th Century saw greater possibilities for future development of the aerostat than for gliders because of the success the aerostat had already enjoyed. Balloonists were considered rather flamboyant characters and were regarded with some awe, whereas men who claimed to be able to glide through the air imitating the flight of birds were not considered flamboyant, but insane. No self-respecting scientist would risk his reputation by associating his name with such a device.

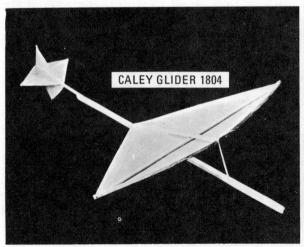

CALEY GLIDER 1804

Fig. 13-16. Caley Glider Design

Communications presented a problem also. George Cayley, who did some early speculation at the end of the 18th Century and the beginning of the 19th Century, was even reluctant to publish the results of his work until he discovered others had been working with some success in the field also. Figure 13-16 shows one of the Cayley glider designs.

During the 19th Century in France, a great deal of experimentation was done with small models of gliders; however, the general public regarded these models as nothing more than toys. The balloonist still had the edge since he could produce something large and tangible for public exhibition.

The problem of locomotion was one which plagued the builders of heavier-than-air machines as much as the balloonists. Cayley made some reasonably accurate estimates of needed power-to-weight ratios and devoted some thought to propellers, but generally during this period the approach was an attempt to imitate the flight of birds. Since manpower was considered for propelling the glider rather than an engine, it seemed logical that such movement would be generated by flapping the wings. This idea required complex engineering since the ornithopter had not been practical.

Fig. 13-17. Lilienthal Glider

Otto Lilienthal was probably one of the greatest experimenters in the field of gliding. By the time Lilienthal and his brother Gustav succeeded in making short flights in 1891, heavier-than-air machines had gained some respect. Despite this fact, both the Lilienthal brothers risked considerable ridicule for their activities. Since the flight of the Lilienthal brothers received favorable recognition in the German press, their activities became known over much of the Western world. They provided a focal point for gliding activities up to that time. Adequate money was available for their experiments and for financing the construction of a hill near their home.

The patent applied for by Otto Lilienthal on February 28, 1894, was for an experimental glider similar in many respects to the one he used for glides of as much as 300 feet. The wings were generally shaped like those of a bird at rest. These features aided considerably in control of the device. (See Fig. 13-17.)

Lilienthal's intention was not only to promote experimentation, but also to establish, in the future, teams of athletes who could compete with one another in gliding competitions with this or a similar device. Lilienthal believed that a tremendous amount of physical skill was necessary to make successful glides. He believed that when teams eventually began competing with one another in national and international competitions, the distance of glides would be extended greatly, and new techniques for the aviator-athlete would develop rapidly as a result of this team competition. The relationship between skill, weight and balance, and control was carefully studied by Lilienthal. As a result, much useful data was relayed to other experimenters.

Octave Chanute, a French-born engineer who was well-known as an expert on railroad construction in the United States, made some essential contributions to gliding. (See Fig. 13-18.)

Fig. 13-18. Octave Chanute

Chanute's approach to the problem of gliding was similar in some ways to that of Lilienthal and others before him. He carefully studied the flight of birds and categorized the flight of different species as to their respective methods of flight.

Chanute, who began with actual gliding rather late in life, constructed a series of hanging gliders which were controlled mainly by the pilot's action of shifting his weight from side to side and forward and backward. (See Figs. 13-19 and 13-20.)

Chanute was on reasonably good terms with most of the individuals in the United States and abroad who were experimenting in the field. He never held back any of the information that he obtained and did his best to convince others in the field to share their knowledge so that progress could move more rapidly.

By the late 19th and early 20th Centuries, heavier-than-air flight was being taken much more seriously by the experimenters, if not by the general public, and keen rivalry existed in many cases. Often experimenters were forced to make the same mistakes that someone else had made because, if information were exchanged, a rival might receive a clue to solving a problem that was plaguing him.

Samuel Langley, who held positions with several universities in the East and was the secretary of the Smithsonian Institution, was participating in numerous

Fig. 13-19. Chanute Glider

Fig. 13-20. Chanute Gliders in Flight

gliding experiments during this period. (See Fig. 13-21.) He used the scaling method. He maintained that by building a small working model in advance of constructing a larger machine, one could conduct inexpensive tests before risking a full-sized version. Langley painstakingly labored with small models of hundreds of designs but actually achieved little success with his work.

In gliding, the work of the Wright brothers cannot be discounted. They built a series of gliders, attempting to devise some type of control system. One method they used was flying a glider as a kite in the stable and steady winds of the North Carolina coast. Cables were attached to the different experimental

Fig. 13-21. Samuel Langley

Fig. 13-22. Wright Brothers' Glider

control systems and could be actuated from the ground so that no danger of injury existed for the pilot in the untried machine. (See Fig. 13-22.)

TRANSITION TO POWERED FLIGHT

As mentioned earlier, Nikolaus Otto was responsible for the development of an internal combustion engine. The first model, although it did function, left much to be desired. It took the work of Daimler to construct an engine powered by a petroleum-based fuel for an aeronautics application. By the late 1890s, petroleum-based fuel was available at a suitable price, and new oil discoveries were being made almost daily.

Fig. 13-23. Sir Hiram Maxim

While engine progress was occurring, some of the first experiments with powered heavier-than-air flight were made by Sir Hiram Maxim, the inventor of an early machine gun. (See Fig. 13-23.) Since ample funds were available, this experimenter approached the problem on a grand scale. He constructed several steam engines that developed over 300 horsepower to be used in a heavier-than-air flying machine.

By 1894, Maxim had designed and built a huge three and one-half ton aircraft that was 110 feet long. It was powered with a Maxim 350 horsepower engine. (See Fig. 13-24.) Rather than risk destroying the engine and the vehicle which it powered, Maxim designed a long track for the engine-driven flying machine to move on. The tracks were constructed so that the huge machine was captive to the track. As the machine gained momentum and attained speed enough to lift it from the ground, the control surfaces, consisting of two large elevator-like controls at each end of the machine, were moved to a lifting position.

One day when Maxim had attained considerable speed, he pulled back on the controls, and the machine developed so much lift that it tore itself loose from the tracks. Afraid that the machine might get away from him, Maxim immediately reduced power to prevent the machine from going further. In doing so, the

Fig. 13-24. Maxim Flying Machine (1894)

machine crashed to the ground destroying it. But luckily, the inventor's life was spared in the accident.

Following the mishap, his funds for experimentation were nearly exhausted, restricting additional experiments. Could this be considered the first powered heavier-than-air flight? Probably not, because the important element of being able to land the machine intact was not achieved; but results of the experiments were published, and this in itself was a contribution. Anyone who read the account of the experiments would be able to avoid some of the problems that beset Maxim.

Samuel Langley, previously mentioned in relation to gliding activities, was another who worked diligently to build a powered heavier-than-air machine. He used the same method of experimentation that he had used with gliders — that of scaling. Small steam-driven models were constructed and flown to determine good flying characteristics for a man-carrying model. Some of the models built were actually rather large with wingspans of up to twelve feet.

The problem that beset Langley in the development of a man-carrying machine was the problem of finding a suitable engine. Since he was the secretary of the Smithsonian Institution at the time, he was more fortunate than many experimenters and succeeded in obtaining a government grant of $50,000 for the construction of a flying machine, provided that it met certain criteria. The government required that the machine include an engine fueled by gasoline which could develop 12 horsepower and not exceed 100 pounds in weight.

Attempts were made to have an engine constructed to the government specifications, but no one would accept the job, arguing that it could not be done. Often in the development of aviation, pioneers in the field have heard the phrase, "That's impossible." In this case, however, it proved to be quite possible for Langley's chief engineer, Charles Manly. (See Fig. 13-25.) He not only suc-

Fig. 13-25. Charles Manly

Fig. 13-26. Manly Engine

Manly's engine was bench-run three times for ten hours each time without a major mechanical failure; however, the airframe of the Langley "aerodrome" did not measure up in quality to the engine produced by Manly. The tail assembly was fixed, and two wings were mounted on the open fuselage in tandem rather than in biplane configuration. Directional control was provided by a rudder located behind the two large pusher propellers. (See Fig. 13-27.) The "aerodrome" was to be launched by a catapult from the top of a houseboat anchored in the Potomac River.

Several launching attempts, with Charles Manly at the controls, ended in failure. Finally, with the original $50,000 appropriated by the United States gone and another $20,000 funded by the Smithsonian Institution expended, a final attempt was made on October 7, 1903. This effort ended in disaster for Langley. The "aerodrome" barely left the catapult before it descended sharply down to the

ceeded in constructing such an engine, but also he produced a radial design engine that weighed 207.45 pounds and developed 52.4 horsepower. (See Fig. 13-26 above.)

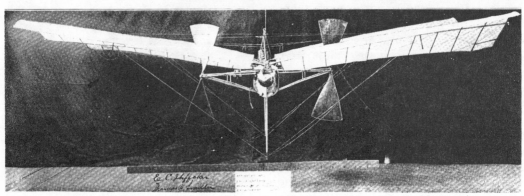

Fig. 13-27. Langley Aerodrome

bottom of the Potomac River. Newspaper reporters observing the accident used the opportunity to discredit Langley and the United States Government for wasting money and for getting involved in such a foolhardy scheme in the first place. The failure of this project did much to retard the growth of aviation under government support in the United States.

NOTE: This model was later flown by Glenn Curtiss after a few modifications in 1914.

Two of history's more renowned experimenters in heavier-than-air flight differed in many ways from their predecessors since they were not engineers and held no academic degrees. In fact, neither one had a high school diploma. They did not possess fortunes to follow their aeronautical pursuits. They were, however, among the most successful in applying technology to all the theoretical speculations of men supposedly more knowledgeable than they. These two men were Orville and Wilbur Wright. (See Figs. 13-28 and 13-29.) Sons of a United Brethren bishop, they owned a small bicycle manufacturing shop in Dayton, Ohio. Heavier-than-air flight had excited the Wright brothers since mid-1890.

The one man who had a tremendous influence on their career was Octave Chanute. They wrote to Chanute before they succeeded in doing much in the field of gliding, and characteristically, he aided

Fig. 13-28. Orville Wright

Fig. 13-29. Wilbur Wright

Fig. 13-30. Wright Brothers' First Flight, December 17, 1903

them in every way he could. After considerable correspondence, Chanute even went to Kitty Hawk, North Carolina to conduct gliding experiments with the Wrights. Kitty Hawk was selected by the Wrights after correspondence with the Weather Bureau in Washington as to the place with the most suitable winds for gliding experiments.

The method used by the Wrights in their gliders was subsequently used in their powered aircraft as well. The wings of the Wright planes were light and fragile with wires or cables attached to the tips so that one or the other of the wings could be warped. Coordinated turns could be made with manual warping of the wing in conjunction with rudder deflection.

The configuration of the first powered model that maintained sustained flight on December 17, 1903, was unusual by aircraft standards of today. The elevator was ahead of the aircraft; the rudder, behind. (See Fig. 13-30.)

Two large pusher propellers were driven by chains from one centrally located position. The engine itself was built in the Wrights' own bicycle shop in Dayton with the help of one of their machinists. One propeller was driven by a cross-over chain to help eliminate adverse torque by counter-rotation. The propellers were unique because they were manufactured by the Wrights with a profile like that of a wing. Weight and balance of their fragile machine, together with the efficiency of the propellers, made flight possible in spite of the poor power-to-weight ratio of the machine.

The Wrights had a problem since even though they were confident in themselves, they disliked news coverage of their activities. They lived very frugally. They did not drink or smoke and were considered rather colorless personalities. Perhaps the bad publicity given the Langley failures made them shy away from the press even more.

On the day of their first powered controlled flight (December 17, 1903) no professional newsman saw the event. Several people were at the remote sites of a nearby lifesaving station and a nearby village. The reports that came from this group were eventually telegraphed to the press and were so exaggerated that even when printed, their credibility was questionable.

The takeoff was effective without the aid of a catapult or other booster device. These aids were not incorporated by the Wright brothers until their flight in 1904. It is interesting to note that the first flight of the Wright brothers was actually shorter than the wingspan of the Boeing 747.

Fig. 13-31. 1908 Wright Airplane

With the Wrights' successful flight, the reader would expect the future to be happy for them, but unfortunately, this did not happen. Their flying activities in 1904 were moved to a pasture near their Dayton home. At this stage, the press was still dubious of their success. The Wrights applied for a patent of the unique features of their machine, particularly the wing-warping device. The United States Patent Office refused them the patent until 1906 when they demonstrated their achievement.

The patent office finally recognized the accomplishments of the Wrights, but the United States Government still rejected the thought of purchasing the Wright airplane. After exhausting every possibility to sell their ideas in the United States, the Wrights concentrated on France. Wilbur Wright went to France in 1907 taking one airplane with him while Orville took care of matters at home. Negotiations dragged on in France since some of the European experimenters were having limited success.

In late 1907, the United States finally became interested in purchasing an airplane. Considering the state of technological development and the time involved in filing a bid, the required specifications were very rigid. The aircraft was to have a minimum speed of 36 miles per hour, and it was to carry two persons and enough fuel to fly 125 miles.

Subsequent exhibition flights in France, Britain and Germany in 1908 and 1909 proved to be rewarding to the Wrights financially and in honors received. In March 1908, the Wright brothers received $100,000 plus half of the shares of stock in a French company formed to build airplanes using their patented ideas. (See Fig. 13-31.) In September 1908, Orville made a series of public demonstration flights at Fort Meyer, near Washington, D. C. To climax a generally successful year, Wilbur, in France, won the 20,000 franc Michaelin Prize for long distance flying (2 hours and 20 minutes).

After 1910, the Wrights had little to offer in the way of new ideas and designs. Their major concern was to justify themselves to the world by proving the authenticity of their patents. They succeeded in winning a series of legal battles involving their patents; however, their rights in Germany could not be substantiated. There, it was proven that information regarding the wing-warping technique had been related in a speech given by Octave Chanute before the Wrights' first powered flight. The Wrights believed that their inventions for control allowed all other experimenters in the field to have success. They contended that without these inventions no one would have been successful.

The question of the uniqueness of technological development came again with the experimenter, Glenn Curtiss. (See Fig. 13-32.) Curtiss, before becoming interested in aviation, had been in the bicycle business as had the Wrights. He, however, carried the bicycle one step further by adding an engine to it and, thus, was responsible for some of the first important developments in motorcycles in the United States.

Fig. 13-32. Glen Curtiss

Because motorcycle engines produce high power for weight, they have excellent power-to-weight ratios; therefore, aviation experimenters began showing interest in them. The first engine that Curtiss produced was actually for a lighter-than-air ship. The powerplants he designed were based, to some extent, on his motorcycle-building experience. They were powerful, light, and dependable.

Alexander Graham Bell, at this same time, was interested in aeronautical experiments. He first designed various types of kites to study problems of control. Other members of an organization known as the Aerial Experiment Association were at the same time flying a Chanute biplane glider. Curtiss built most of the engines for the various airships and airplanes designed and built by this experimental group. While working as an engine builder for the group, Curtiss learned a considerable amount about the building of airframes. He was a person who could visualize a completed project and was extremely good at making improvements to existing models and methods of construction.

The first airplane he built and designed for the association contained many important improvements. One of the most significant was the addition of ailerons as separate control surfaces located between the wings of his biplane design.

This innovation eliminated the need for wing-warping and gave better control stability and greater rigidity to the wings. (See Fig. 13-33.) With construction of the third of his machines, Curtiss, now independent of the Aerial Experiment Association, became a rival of the Wright brothers in aircraft development.

After competing in several air meets in France with considerable success, Curtiss returned to the United States to accomplish a feat which had great impact upon many American experimenters. Curtiss flew from Albany to New York City in one of his own designs to win a prize of $10,000.

Technological improvements in aviation are quite evident in an examination of the aircraft used for exhibition flights from 1911 to 1917. One of the major changes was moving the location of the ailerons from a position between the wings to the trailing edge of the wings. Wing-warping as developed by the Wrights was virtually abandoned by 1918. By 1914, another significant change had appeared. Engines were being mounted in a tractor position rather than a pusher position. (See Fig. 13-34.) This arrangement provided for simpler fuselage construction, better protection for the pilot, and improved weight and balance.

A major problem for aeronautical designers was eliminating the number of com-

Fig. 13-33. Curtiss Pusher

Fig. 13-34. Cessna Airplane

ponents of the airframe which produced drag, while at the same time improving structural integrity through use of stronger materials. A tandem seat for the passenger was a feature of many of the exhibition planes touring after 1914. About the same time that the tractor planes were introduced and tandem seating for a passenger was provided in an open cockpit, dual wing surfaces were employed and elevators in the front of the aircraft were eliminated. With the empennage, fuselage, and other changes, the airplane began to take on the appearance recognized today. (See Fig. 13-35.)

LITIGATION HINDERS AVIATION DEVELOPMENT

After 1909, the Wright brothers abandoned much of their developmental work to become involved in a series of lawsuits regarding the patent rights. The Wrights could not patent the idea of heavier-than-air controlled flight but did succeed in using their patent for a wing-warping device to demand royalties and licensing fees from manufacturers of other airplanes. It is difficult to determine whether Orville and Wilbur wanted the rights to their patents for financial gain or to gain for themselves the honor of being first with the original ideas behind controlled

Fig. 13-35. Curtiss Jenny JN-4

flight. It is known that they violently disapproved of using airplanes for exhibition flying which they considered to be a "circus performance." A problem arose when the Wrights were not willing to give adequate demonstration flights to satisfy the curiosity of the public. Perhaps misunderstanding the motives of the Wrights, the public began to turn against them.

Glenn Curtiss, who was their only really important competitor, presented a different public image. Friendly and outgoing, Curtiss made arrangements for many exhibition flights and entered numerous competitions with the airplanes his company constructed. He, with the aid of others in his camp, was responsible for starting one of the first schools of flying in the United States. Curtiss worked to see that others had an opportunity to purchase airplanes for exhibition purposes.

The Wright brothers were successful in getting some injunctions from the court demanding that other manufacturers, such as Curtiss, pay royalties on each airplane manufactured and also fees for using airplanes in exhibition flights. The effect of these injunctions did not last long, as more and more individuals were building and exhibiting their own aircraft. As a result of the limited success of the Wrights in their suits, the statistics of the United States Government show that no airplanes were manufactured in 1911. Actually,

many were produced on a limited basis, but the manufacturers were reluctant to admit the fact. If they had, they might also have been designated as co-defendants with Glenn Curtiss in the patent infringement suits.

Gradually, the scientific community of this country was turning away from the Wright brothers. The last scientific award received by the Wrights was the Collier Trophy Award in 1913, the year after Wilbur died of typhoid fever. The Smithsonian Institution refused to give the Wright brothers adequate recognition and went so far as to hire Glenn Curtiss, the Wrights' competitor, to fly the Langley "aerodrome" which had been carefully stored in Hammonds Port, New York. Curtiss equipped the airplane with floats, made a series of changes to provide for better structural integrity, and succeeded in flying the machine. (See Fig. 13-36.) The Smithsonian contended that as a result of this demonstration, the Langley "aerodrome" was capable of sustained flight, and it was only due to the unfavorable circumstances that Langley encountered that the Wright brothers' model succeeded before the Langley model.

The controversy between Orville Wright and Glenn Curtiss continued. Orville was unwilling to have the Wright airplane of 1903 exhibited in this country and made arrangements to send the plane to London to be exhibited in the South Kensington

Fig. 13-36. Langley Aerodrome in Flight

Museum. It was not until after Orville's death in 1948 and the publication of his will that the Wright airplane was returned to the United States and exhibited in the Smithsonian Institution, where to this day, the label of this airplane reads, "The World's First Power-Driven Heavier-Than-Air Machine in Which Man Made Free, Controlled, and Sustained Flight."

As the growth rate of the aviation industry increased rapidly due to public curiosity and World War I, these preliminary deterrents to aviation development were overcome. The United States Government, with active support through appropriations from Congress, was ready to become a major producer of aircraft in a very short time.

WORLD WAR I AND AMERICAN DEVELOPMENT

An informative report, issued by the Assistant Secretary of War and compiled in 1919 by Col. B. W. Mixter of the U. S. Army and Lt. H. H. Emmons of the U. S. Navy, brings some interesting information to light concerning the effect of the war effort upon aviation. When war was declared in Europe in 1914, the total number of military and civilian personnel connected with military aviation in the U. S. was 194 men.

France entered the war with 300 airplanes, Britain with 250, and Germany with an estimated 1,000 planes. In contrast, the United States, in the eight years prior to 1916, had ordered only 59 airplanes. In 1916, 3,666 planes were ordered but only 83 were delivered. When the United States entered the war in 1917, the total number of planes that had been delivered to the military subsequent to the flight of the Wright brothers was 142. Considering that some of these aircraft had been destroyed and that most of the models were definitely obsolete, a person can visualize our complete lack of preparedness.

No support industries had been developed to complement the production of aircraft. Machine guns for use on military planes, instrument production, and camera equipment had not been developed in this country. In addition to these plaguing problems, no facilities existed for the mass production of airplanes for training or for war. The airplanes built before this time were manufactured by hand as individual units. The advantage of interchangeable parts, already a practice at this time by the automobile industry, had not been implemented by the aircraft industry.

On May 12, 1917, there were standing orders for 334 aircraft designated for military use to be delivered from sixteen different manufacturers. The aircraft were of ten entirely different types and 32 different designs. The largest order placed with a single manufacturer, the Curtiss Airplane and Motor Company, was for 126 airplanes.

The United States Congress felt that something had to be done and soon. Large sums of money were appropriated to build the military aviation industry, a thing which had been advocated many years earlier by several farsighted individuals. It was an idea which had already been made a reality in France and Germany prior to the outbreak of World War I.

Now that funds were available, there was the question of what to do with the money. Mechanical designers and engineers were sent to Europe to study the latest types of aircraft. Models of these aircraft were purchased and brought to McCook Field at Dayton, Ohio, to be studied in detail.

Licensing agreements were made with various European manufacturers to produce proven designs rather than to begin immediate production of untried American types. A joint military group, known as the Joint Army and Navy Technical Board, recommended a program calling for the production of 22,000 airplanes by July 1, 1918. In view of the underdeveloped state of manufacturing in the United States, this task was impossible.

GENTLEMEN OF THE AIR IN COMBAT

At the beginning of World War I in Europe, the strategic value of the airplane was definitely questioned. Prior to this conflict, balloons had been used in the United States in the Civil War and in Europe in the Franco-Prussian War. Heavier-than-air aircraft had not been used in combat. World War I provided a chance to experiment with an entirely new medium of warfare — the airplane.

Since the previous application of an aircraft was for observation purposes, it was natural that heavier-than-air aircraft were initially used for the same purpose. When the war began, small light airplanes flew over the trenches of the enemy to study fortifications and troop movements. The only contact that the pilot of one side had with that of the other was an occasional wave of the hand if either flew near enough for visual contact.

The passive attitude did not last long, however. In the two-place observation planes or "ships" as they were sometimes called in imitation of the naval term, the observers began arming themselves with pistols. The first armed confrontation between heavier-than-air aircraft amounted to no more than an exchange of pistol shots; but this encounter marked the beginning of much bigger armed conflicts to come.

After the initial assaults of World War I, the lines of battle became relatively static until the American entry into the war in 1917. Ground mobility of troops was limited because of road conditions and the lack of fast motorized vehicles. This was an excellent opportunity for aircraft to demonstrate their ability to attack troop concentrations.

At the beginning of the war, armament for use against the trenches did not exist. Occasionally, a particularly nasty pilot or observer threw rocks down on the troops. The troops replied first with small arms fire, whereupon planes reciprocated with the observer dropping grenades and firing pistols at the troops. Many of these early attempts to attack the trenches were met with reprimands since the intended purpose of the early military aircraft was construed to be strictly observation. This attitude on the part of the military authorities on both sides was short-lived. Pilots were beginning to drop hand-armed bombs over the sides of their cockpits when particularly troublesome ground fire was encountered.

Contact between aircraft rapidly became more sophisticated, and light machine guns were mounted at the rear of the observer's cockpit to fire at airborne enemies. Bomb racks were installed when the pilots found that detonation of loose bombs carried on board sometimes occurred during aerobatic maneuvers used to avoid enemy machine gun bullets. A group of French fighters received a shock one morning while out on a mission when a group of German fighters attacked and fired head-on at the Allied planes. This was a tremendous breakthrough — a machine gun which was synchronized to fire through a propeller. The Allies could reciprocate only by getting into position for the rear-seat observer-gunman to return fire.

Anthony Fokker (see Fig. 13-37) was responsible for the Germans' success in perfecting a method of shooting through the propeller. He developed the idea of

Fig. 13-37. Anthony Fokker

piece, let alone fly against a relatively well-trained enemy in combat. The desire for the United States to get men into air combat was as great as the previously mentioned effort to have 22,000 aircraft and spares completed within one year.

Training was not the only problem. Pilots were also plagued with aircraft of poor structural integrity. The airplanes of World War I were, for the most part, constructed hurriedly by poorly-trained factory workers. Wings warped with a change in humidity, and fabric ripped from wings and control surfaces during aerobatic maneuvers in combat. Occasionally, wings and control surfaces came off entirely. It was not unusual for an entire squadron to go out on a mission and none return, all having made forced landings because of mechanical failures.

firing the gun each time the propeller blades were parallel to the wings. This meant that two rounds of ammunition could be fired for each revolution of the propeller, and that at a maximum RPM of 1800, a machine gun could get off 3600 rounds per minute. If the weapon was not capable of firing that many rounds of ammunition, it could be adjusted to fire once each time the engine turned over, thus reducing the number of rounds per minute by half. The triggering mechanism could be fired by either of two methods: electrically by the magneto when a cylinder was fired, thereby giving propeller clearance, or by a lobe on the camshaft that would actuate the trigger mechanism when it revolved. In a very short time, the Allies had the same method of firepower, and the age of the dogfight began in earnest.

In spite of the tremendous prestige enjoyed at home by the pilots in the "Great War," they came to the front lines in most cases without adequate training and were expected to fly equipment that was poorly designed. Many pilots came to the front with less than ten hours total flying time, half of which was dual instruction. They were fortunate to be able to get the airplane off the ground and back in one

Engines presented another problem. Accurate machining and the measurement of parts to 0.0010 of an inch was becoming a reality; however, 80 hours between major overhauls was considered nearly a phenomenon. The OX-5 engine, produced by Curtiss, was known for its stamina with an overall life of 60 hours. (See Fig. 13-38.)

The Gnome and LeRhone rotary engines (see Fig. 13-39) had overhaul lives of 80 hours each. In contrast to the types of engines used in airplanes today, the OX-5 was an inverted V-8 which was water-cooled and produced 80 horsepower. The

Fig. 13-38. Curtiss OX-5 Engine

Fig. 13-39. LeRohne Rotary Engine

Fig. 13-40. Anzani Engine

rotary engines were air-cooled but were very different from today's engines in that the cylinders and crankcase revolved around a crankshaft that was bolted to the firewall. (In today's engines the crankshaft revolves, while the cylinders and crankcase are stationary.)

The OX-5 engine had variable throttle settings, but the rotaries had only two throttle positions — wide open and off. This limitation was because of carburetor

designs. In many newsreel films of the First World War, a viewer can watch closely and see that when the engine is started, the propeller and the crankcase of the rotary engines move together.

Another popular engine in use by the Allies was the Anzani engine. (See Fig. 13-40.) Both this engine and the rotary engine used castor oil as a lubricant, just as fuel and oil are mixed together in two-cycle outboard marine and lawn mower engines today. These engines threw out a tremendous quantity of castor oil over the cockpit and into the face of the pilot.

POSTWAR AVIATION

A vast aviation industry was created in the United States in one year through large expenditures appropriated by Congress and through the untiring efforts of many dedicated military men and civilians. At the close of the war, 195,024 men in the military were associated in some way with aircraft, and a total of 16,952 airplanes had been delivered. In spite of this "maximum effort," not a single American-designed aircraft saw combat in Europe.

Due to the extensive expenditures from Congressional appropriations, a tremendous quantity of aircraft and aircraft components became available after World

War I. So many aircraft engines were manufactured that after the war, they were put on the market as surplus and sold for ridiculously low prices. The Curtiss OX-5 engine, for example, sold for as little as $50, brand new and still packed in preservative cosmoline. Since this particular engine had excellent application in medium-weight aircraft, it was possible for an airplane to be constructed practically for the cost of the airframe. If an airplane owner had trouble with the engine installed in his plane, he simply had it removed and installed another one for $50. The old engine which had amounted to more than half the cost of the airplane was then disposed of as junk.

Fig. 13-41. Lindbergh, The Barnstormer

Several airframes, such as the JN-4 "Jenny," were also available from the government as surplus. This airplane was manufactured in quantity by several manufacturers for training purposes during the war. Many in good flying condition were sold after the war for as little as $200. A number of pilots who were to become famous during the 1920s took advantage of these low-priced airplanes to learn to fly. Learning to fly often consisted of a few minutes dual instruction with someone who had some flying time. The next step was to buy an airplane and learn to fly on one's own. The "Jenny" represented the most popular do-it-yourself, learn-to-fly kit.

After World War I, the flying circus became one of the biggest disposal areas for surplus. With the beginning of the 1920s, what had consisted of nothing more than public exhibition flights for a few underwent a dramatic change. Observing an airplane in flight was no longer enough to excite the crowds. The day of the specta-

tor sport had come. Since baseball and football drew huge crowds, promoters decided that air shows could do equally well if enough showmanship was included to hold crowd interest.

At first, these flying circuses could draw crowds by giving extended aerobatic shows. In a short time, this imitation of World War I dogfights could no longer fill the stands. Next, acrobats were suspended from a trapeze under an airplane. Women and children were used to draw even larger crowds who hoped to see blood spilled. Wing-walking became the vogue, and finally the wing-walker was suspended from the top of a biplane where he stayed in the same position while the airplane went through a series of aerobatic maneuvers. Plane-to-plane transfers became popular, and plane-to-automobile transfers followed. Barnstorming provided a training ground for many famous pilots of the period. Among these were Art Gobel, Roscoe Turner and Charles Lindbergh. (See Fig. 13-41.)

ERA OF BILLY MITCHELL

With the Warren Harding administration and its "return to normalcy," interest in government spending for military aviation waned. "Why should the government spend money on military aviation in time

of peace?" people asked. A few dedicated and farsighted individuals did work to further military aviation, but unfortunately with little success.

Another factor which retarded the further development of military aviation in the 1920s was the beginning of some serious intra-service rivalries. Following the impact of the book, *The Influence of Sea Power Upon History, 1660 - 1783*, by Captain Alfred Mahan, the naval forces in the United States were given priority over the other military services. The basic point of the book was that the greatest, most powerful nations in the world, historically, had risen largely because of their great sea power. In order to support such a navy on a worldwide basis, Mahan stated that it was necessary to have colonies to use as naval bases all over the world. Even though this book was published in 1890, its influence was particularly strong after the close of World War I.

Brigadier General William "Billy" Mitchell (see Fig. 13-42), one of the foremost proponents of a powerful military Air Force, challenged Mahan's basic idea of an invincible Navy. At the time of his crossing swords with the Navy, Mitchell was Assistant Chief of the Air Services and as the result of his battle with the Navy and members of Congress, his appointment was not renewed. He reverted to his permanent rank of Colonel and was assigned to Fort Sam Houston, Texas. This got him far away from Washington but did not quiet his tongue. This "remote outpost" actually was one of the most important in the Air Service; however, it was from Fort Sam Houston that he issued a blistering 6,000-word statement in response to the Navy's abortive California to Hawaii flight and the crash of a Naval Dirigible, the *Shenandoah*. It was this statement made in September of 1925 that eventually resulted in his court martial.

This furor actually began much earlier. Mitchell's support of air power as a separate military service received tremendous support when Mitchell's bombers sunk the German battle ship, *Ostfriesland* on July 21, 1921. It was after three hits by 1,100-pound bombs that Mitchell's pilots care-

Fig. 13-42. Brigadier General "Billy" Mitchell

fully laid two additional specially-prepared 2,000-pound bombs alongside the battle ship and the resulting underwater reverberations ruptured the hull. The "unsinkable" vessel turned over and went down by the stern 25 minutes after the first bomb exploded. (See Fig. 13-43.)

The dramatic sinking of the *Ostfriesland* greatly sped up America's aircraft carrier program and reshaped the thinking of many military experts regarding the role of air power.

After the controversial court martial, President Coolidge attempted to lighten the punishment by restoring a portion of Mitchell's pay and allowances. Mitchell replied by resigning from the Army; was free of the military shackles on his speech, and carried on his crusade as a private citizen. He set out on a lecture tour and pointed out that the major mission of military aviation was first to destroy the military air forces of the enemy country. When this was accomplished, the next step was to bring an attack to the seat of enemy war-making capability after the control of the skies had been assured. The forerunner to the concept of strategic

Fig. 13-43. Billy Mitchell's Sinking of the Ostfriesland Battleship

bombing was carried one step further when he insisted that the most important and vulnerable area of the United States extended from the Chesapeake Bay to Boston and from Chicago to New York. If an enemy with a well-equipped air force could destroy communication within this "T" formation, industry, agriculture, population, and production could be disrupted and destroyed. He continued to speak and write in this manner for as long as he lived, at the same time enjoying the life of a country gentleman on his Virginia estate.

At the age of 56, he experienced heart trouble and died on February 19, 1936. Coincidentally, the newly created Army Air Corps had just placed its order for a dozen new B-17 Flying Fortresses, soon to become the focal point of the air power for which he had long struggled.

CROSSING THE OCEANS

One of the most significant problems throughout United States history has been the "space barrier." When referring to this term, it must be realized that during the early development of this country, few settlers lived in the more remote areas. Settlement was encouraged by an extensive railroad system that was essentially completed by the time of the United States' entry into the First World War.

By the 1920s, the automobile presented a possibility for an extensive new transportation system. Was there a need for another system to be developed? Many farsighted individuals believed that there was a need for an air transportation network. Before such a system could become a reality, the reliability of this new mode of transportation had to be proven.

Various attempts at long distance, speed, and endurance records helped this new concept in transportation to gain acceptance by stirring the imagination of the public. Fostering a future transportation system, the most significant records were those involving long distance flights.

One of the earliest long distance flights after World War I occurred in May of 1919 when three Navy NC-4 flying boats started across the Atlantic to Europe by way of the Azores. (See Fig. 13-44.) Of the three that began the trip, only one actually made it and then only by taxiing through rough seas over part of the route.

The next significant flight was in a British R-34 lighter-than-air dirigible which flew from Britain to the United States and returned without mishap. (See Fig. 13-45.) Alcock and Brown made an at-

Fig. 13-44. Navy NC-4 Flying Boat

tempt to fly the Atlantic during the same summer and would have enjoyed complete success had it not been for a crash landing in Ireland. They fortunately escaped injury.

Many other attempts to fly the Atlantic were made, but often money could not be raised for such a venture, or if finances were available, many of the takeoffs ended in disaster. The failure of many of these unsuccessful early attempts was due to overloading of the aircraft in an effort to guarantee enough fuel for the lengthy journey. An offer that renewed interest in transatlantic flight was the Orteig Prize of $25,000 offered to the pilot who successfully completed a flight from New York to Paris.

Three American attempts were planned by Commander Richard Byrd, Clarence Chamberlin, and Charles Lindbergh. The first two entries in the contest had almost unlimited funds at their disposal for the attempt; Charles Lindbergh (see Fig. 13-46) was limited in his expenditure Lindbergh originally proposed the idea of

Fig. 13-45. British R-34 Dirigible Crossing the Atlantic

Fig. 13-46. Charles Lindbergh

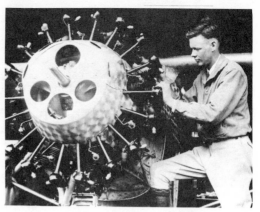

Fig. 13-47. Charles Lindbergh
and the Spirit of St. Louis

flying over the Atlantic to Paris to several airplane manufacturers. They agreed that it was an excellent idea and offered to allow him to take part in preparation for such a flight. They would not, however, agree to allow him to be the pilot-in-command on such a venture.

Disappointed, Lindbergh approached a group of businessmen in St. Louis to sponsor such a flight. They agreed on the condition that the name of the airplane contain the name of the sponsoring city, thus the name, *Spirit of St. Louis*. Lindbergh personally negotiated for the airplane which was built by Ryan Aircraft of California. The airplane was constructed in 60 days. (See Fig. 13-47.) A few test hops were flown, and then the airplane was flown to the east coast with a stop in St. Louis.

On May 20, 1927, Lindbergh, armed with a favorable weather forecast, took off with a tremendous load of gasoline and was on his way to France. He experienced bad weather on the initial leg of his trip, but as the trip continued, the weather improved. In just over 33 hours, Lindbergh reached Paris with enough fuel left on board to continue on to Rome.

Without modern radio navigation aids, alone, and with only one very reliable engine, he succeeded in doing what at the time was thought to be impossible. The flight from New York to Paris was completed, and the Orteig Prize of $25,000 was his. This exploit in 1927 stirred the imagination of the public throughout the United States and Europe. If this young man was able to make such a trip with a minimum of equipment, could it be that transatlantic air transportation for the public could become a reality?

Byrd and Chamberlin decided to continue with their respective plans for an Atlantic crossing, but neither was able to do so without damaging his airplane. Byrd's trimotor ditched into the English Channel without injury to the crew, and Chamberlin made a forced landing in Germany, damaging his airplane. Lindbergh had succeeded where, in a sense, others had failed.

After the successful completion of Lindbergh's flight to Paris, another space became evident as a challenge to be conquered. James Dole, president of the Hawaiian Pineapple Company, offered a prize of $25,000 to the first pilot who flew non-stop from the mainland of the United States to Honolulu. The second successful completion would net the pilot $10,000. The overwater stretch to be covered by this flight was about 2,400 miles, while the longest overwater stretch that Lindbergh had to face was 1,800 miles.

Fig. 13-48. Art Gobel's Woolaroc

Dole's offer sparked other offers of prize money for various attempts at distance records. For example, challenges included a round-trip flight from the mainland to Hawaii and a flight to Tokyo from Dallas with a maximum of three fuel stops. None of the newly announced prizes was great enough to attract Lindbergh. They did, however, attract many others.

Art Gobel, flying a Travel Air named *Woolaroc* (after the ranch in Oklahoma owned by the sponsor, Frank Phillips) (see Fig. 13-48) won the race to Honolulu and was followed on landing by another monoplane, the *Aloha*, piloted by Martin

Jensen. After the two had landed, news of the disasters of the race reached them. When the final toll was taken, ten of the thirteen final entrants had lost their lives. This tragedy occurred even though the United States Government licensed the pilots specifically to take part in this particular race. The public attitude at this time toward the feasibility of aviation for transportation on such long hauls was unfavorable. Once again, although brave attempts were made to forge ahead in the development of new aviation routes, public opinion regarded the whole affair as nothing more than dangerous, impractical stunting.

AN AERONAUTICS BRANCH FOR THE GOVERNMENT

Prior to these record-setting attempts, the United States Government, in 1926, passed an Air Commerce Act. After the bill was signed into law by President Coolidge, the Secretary of Commerce was given authority to "foster air commerce, designate and establish airways, establish, operate, and maintain aids to navigation (except airports), arrange for research and development to improve such aids, license pilots, issue airworthiness certificates for aircraft and major components, and investigate accidents."

As 1926 ended, the Department of Commerce had taken on new regulatory functions connected with civil aviation. By

this time, regulations required that airplanes engaged in private and commercial flying be registered and carry proper identification markings, and that all pilots in interstate commerce be required to have the proper license. Mechanics were also required to be licensed under the same provisions.

Government appropriation for the new Aeronautics Branch of the Department of Commerce increased rapidly until the middle years of the depression. In 1927, the new agency received an appropriation of $550,000, and in 1928, $3,791,000. The high appropriation for the period was $10,362,300 in 1931. By 1936, the depression budget for this new office was reduced to a little over $5,000,000. With the small appropriations received by this agency during the middle '30s, it is amazing how it was so successful in enforcing so many of its programs.

AIRMAIL SERVICE

Even before the government revealed interest in appropriating funds for the regulation of civilian aircraft transportation, manufacturing, piloting, and repair, many farsighted individuals were thinking in terms of providing a new transportation system to the public. However, technology had not yet reached the level necessary to make such an idea as airline passenger service practical, profitable, and acceptable to the public.

Although the governments of France, Germany, and the United Kingdom were willing to subsidize an air transportation system, the idea of subsidy was not yet acceptable in the United States because the more conservative legislators in this country saw this type of government help as a deterrent to free enterprise. Socialist institutions had made some inroads into the governments of the most conservative states in Europe; however, socialistic ideas in this country after World War I and into the 1920s carried the stigma of being anarchistic. As a result of this attitude, Europe preceded the United States in scheduled air carrier service by as much as ten years.

One program that did receive the support of the government in the United States was the development of an airmail system. The first airmail route between New York and Washington was put into service on May 15, 1918, but in a short time the United States Post Office Department took over the job of handling the mail, using hired civilian pilots. Most of these mail routes were originally equipped under direct supervision of the Post Office Department with surplus airplanes left over from the First World War.

A daylight airmail system spanning the United States was in operation by 1920,

and a transcontinental route to be used at night as well as in the daytime was also set up by the Post Office Department. This route was not set up on a radio navigation aids system as we know it today. Instead, a series of lights, spaced about 25 miles apart, extended from one town to another along the route running all the way from the east coast to the west coast of the United States.

By 1924, a regularly scheduled service was completed whereby an airplane with mail would leave the west coast of the United States the morning of one day and arrive at its destination on the east coast the evening of the following day. With the passage of the Kelly Air Mail Act of 1925, the Post Office Department was relieved of the responsibility of supplying equipment and pilots for the mail service. Under this act, private companies were to compete on a bid basis for the right to carry air mail across the country.

With this act came several important results. The new operators were not compelled to use war surplus airplanes to carry the mail. This allowed designers in the field to have a new market for an airplane to be used for a specific purpose. The second important result of the Kelly Act was that finally a method had been initiated whereby a passenger service could be practically subsidized by the government. Since a company could be paid for carrying mail, why then would it not be possible to carry passengers for additional revenue?

The mail-passenger concept led to a number of designs which could be used for this dual purpose. Usually, the combination type airplane had a single engine, an open pilot's cockpit, and a small cabin for the mail and two or three passengers.

Fig. 13-49. Early Airmail/Passenger Airplane

Airmail carriers were paid on the basis of the number of pounds of mail they carried, and often, when this mail load was heavy, passengers would be asked to stay in preference for the mail. (See Fig. 13-49.)

By 1929, an attempt was made to put a regular passenger service into operation. The unique thing about this service was its planned operation as a transcontinental airline-railroad combination. Passengers could leave New York City's Pennsylvania Station by train and travel to Columbus, Ohio. On arrival in Columbus, the train was switched to a siding near the Columbus airport, and the passengers were chauffeured a short distance to board Ford Tri-Motor transports for a flight to Waynoka, Oklahoma. There, the passengers were driven to a train equipped with Pullman cars. At Clovis, New Mexico, the passengers were taken to more waiting Ford Tri-Motors to complete the trip to the west coast by air. This service reduced the travel time from the east coast to the west coast from 72 hours to 48 hours. By this arrangement, the passengers traveled only two of the four legs of the trip by airplane. Even though this combination system did operate, it lost millions of dollars in its first year of operation.

Without a series of developments in the aircraft industry, the experiment mentioned above would have been impossible. This series of developments led to dependable multi-engine transportation.

Fig. 13-50. Curtiss Twin-Engine Passenger Airplane

Fig. 13-51. Fokker Tri-Motor

Glenn Curtiss had designed a successful twin-engine airplane that saw limited service for short hops to various places in the United States. This airplane had a major drawback — it was built by outdated construction methods using old type materials. It was a large, twin-engine biplane constructed of wood, metal tubing, and fabric. Its single-engine performance was not adequate, and its appearance did not inspire prospective passengers with feelings of security. (See Fig. 13-50.)

Something new needed to come from the designers. One breakthrough that made many new designs possible was the manufacture of quantities of the metal alloy, Duralumin. When this metal became available, the Ford Motor Company of Dearborn, Michigan, was just beginning work in research and development of a transport plane. It is unclear if the famous World War I Dutch airplane designer, Anthony Fokker, came up with the idea of installing three engines on a large, high-winged monoplane, or if the idea was original with the Ford designers. It may be

possible that technological development reached a stage that the ideas and designs came about simultaneously.

In any case, the tri-motor airplane produced by Fokker (see Fig. 13-51) and the Ford Tri-Motor (see Fig. 13-52) were very similar in appearance and performance; however, construction techniques were different. Fokker held to the idea of using plywood skins over a wood structure for the wings of his airplane for a lighter, smoother surface. Ford, on the other hand, constructed his tri-motor models entirely of metal. Humorous names were given by some pilots and members of the flying public to the Ford Tri-Motor whose corrugated Duralumin was suggestive of the walls of chicken houses. The name "Tin Goose" became firmly attached to the aircraft during this period.

Luckily, before the stock market crash in the fall of 1929, the aircraft industry underwent considerable financial consolidation. Had it not been for the interest of such large companies as General

Fig. 13-52. Ford Tri-Motor

Fig. 13-53. Boeing 247

Motors, Ford, Curtiss Wright, and others, the aircraft manufacturers would not have approached the stock market crash with sufficient financial health to weather such a storm. Revenues in this growing transportation industry declined during the early 1930s, but not to the same degree as revenues in other industries.

Boeing Aircraft, a company which had concentrated mainly on building single-engine planes, came upon the aviation scene in the early 1930s with something new to offer to the airlines — the Boeing 247. (See Fig. 13-53.) The airplane was of all-metal construction with twin engines, carrying ten passengers and 400 pounds of mail, with a cruising speed of 155 MPH. The difference in comfort and performance between this Boeing model and the Ford Tri-Motor was extraordinary.

During this same period, Douglas Aircraft was given specifications by Trans World Airlines to build what they believed to be the real need of the airlines at the time. Douglas met the challenge by surpassing the specifications by a considerable margin and produced the first of the DC series. The DC-1 was built as the prototype for the series, and the Douglas factory immediately came out with an improved version, the DC-2. This airplane was similar in many aspects to the airplane that succeeded in putting the world on wings, the famous and still active DC-3. (See Fig. 13-54.) This new airplane was far enough ahead in design that it is still in extensive use around the world today.

Although most of these models have been phased out for airline work in the United States, over half of all the 10,000 DC-3's built are still flying. This all-metal twin-engine airplane achieved one of the most enviable records of reliability in the aircraft industry. More than 90% of all airline traffic by 1940 was carried in this famous airplane.

Fig. 13-54. Douglas DC-3

Fig. 13-55. 1929 Light Airplane Factory

THE LIGHT AIRPLANE

By the mid-1920s, Henry Ford had succeeded, through mass production techniques, in producing automobiles at a cost low enough to "put America on wheels." With aviation developing in the technological sense, some designers and manufacturers saw the day when they would have a part in putting America "on wings." On the surface, such an assumption would appear to have been a safe one but in the final analysis, it was wrong.

One basic problem in any manufacturing concern involves the existence of a market adequate to absorb production. With the automobile, the populace was ready and waiting to buy. What would an airplane be replacing? Are there any skills which can be immediately transferred to learning to fly?

The market for light airplanes was much smaller than that of the automobile. It was, however, large enough to make profits for the manufacturer. After the First World War, the airplane had taken on its familiar appearance as we know it today. It had one engine and was either a biplane or a monoplane. The engine was located in front of the pilot and was equipped with a tractor propeller. Provision was usually made for one or more passengers. Similarities end there, for designs were as varied as the materials used for construction.

It is not surprising that construction of the early, post World War I airplanes was of wood and fabric, for wood was a material that was easily worked by hand and technology of woodworking skills and tools had developed to a very high level. (See Fig. 13-55.) Fabric could be applied with ease and shrunk on the wood structure for the fuselage assembly and installed in a similar manner over the wooden ribs of the wings. The material used for this shrinking process was a dope made of a nitrate or butyrate base.

This dope was applied over balloon cloth, linen, or cotton fabric. If constructed properly with a sufficient number of coats of dope, the airplane, with proper hangaring, would last for years. The cloth, as it absorbed and acquired coat after coat of dope, became almost the equal of a smooth metal surface. The major advantage of this method of construction was that no sophisticated tools were necessary. Most of the airplanes of World War I were built in this way, just as the light planes of the 1920s and 1930s which followed.

A substitute for fabric and dope on wings for the lifting surface that came into limited use during this period was borrowed from small boat construction techniques. Large pieces of plywood could be steamed to form skins for wings and fuselage structures. When the skins were installed properly on formers, much of the performance-robbing weight of the inner structural members was eliminated. (See

Fig. 13-56. Plywood Construction (Wiley Post's "Winnie Mae")

Fig. 13-56.) Fabric was then installed as an outer covering in much the same manner as it was before. The fabric was employed in this method of construction to seal out moisture. It was found that as long as the moisture content of a properly selected aircraft grade wood was sealed, it lasted indefinitely. Additional moisture or excessive drying destroyed the structural integrity of the wood.

With the advent of metal for aircraft structures during the 1920s, it is interesting to note that many pilots and manufacturers feared that if metal tubing (steel) were used for structure under fabric, it would rust without the knowledge of the operator of the aircraft. To combat this problem, the steel tubing in many of the early airframes was filled with oil so that rusting could not occur. It was soon discovered that this method was not necessary since the tubing structure had no more inherent weaknesses than did wood structure. (See Fig. 13-57.)

Through the efforts of many individuals working in aviation research, methods were devised to replace plywood skin over wooden formers with metal skins (aluminum alloy) over metal tubing and formers. Techniques in welding and machining of metals moved forward at a rapid pace in the 1930s, making metal airplanes far more desirable to the public than aircraft made of wood and fabric. (See Fig. 13-58.) The only factor that prevented low-cost

Fig. 13-57. Welded Steel Tube Fuselage Construction (Cessna AW)

Fig. 13-58. All Metal Construction (Beech Model 18)

airplanes like the Taylor, Aeronca, and early Piper airplanes from being constructed in all-metal designs was the cost. Employees could be hired at lower rates to build airplanes using a combination of metal, wood, and fabric, than for constructing aircraft of all-metal designs. Metal work required training and skills not readily available in the labor market. (It is interesting to note that today aircraft metal workers are available in quantity, but few aircraft woodworkers are available.)

A major consideration of design for light airplanes at the end of the 1920s was whether or not a plane should be constructed with single or double wings. The biplane had been by far the more popular design up to that time, but the monoplane was gaining in popularity. The proponents of the biplane claimed that its extra lifting surface made it safer to fly and easier to handle with a shorter take-off and landing roll. The proponents of the monoplane stated that a biplane had too much drag and an unnecessarily large area of lifting surface.

In Wichita, Kansas, the argument between the two designs actually caused old companies to break up and new ones to be formed. Walter Beech was a proponent of the biplane and probably carried the design to the ultimate in light airplanes with the advent of his model 17 series that was first introduced in 1933. (See Fig. 13-59.) Clyde Cessna, on the other

Fig. 13-59. Beech Model 17

Fig. 13-60. Cessna Airmaster

hand, remained a proponent of the monoplane design and did an extraordinary job in designing a high-wing, light plane called Airmaster in the middle and late 1930s. (See Fig. 13-60.) The Cessna Airmaster achieved the almost impossible efficiency for the day by providing one mile per hour for each horsepower the engine developed. The Beechcraft Model 17 Staggerwing, with the addition of a more powerful engine, matches the performance of some of the medium-priced, single-engine airplanes being offered today. In its final model, it could cruise at 200 MPH.

Both of the planes were excellent performers. Neither, however, could be considered an "economy" airplane suitable for training and light personal use, since they were expensive for their day. What, then, did the industry have to offer in the area of light, economical trainers and personal airplanes?

Before the advent of the Aeronautical Branch of the Department of Commerce, anyone who had the skill could build a light airplane and fly it. No inspection was necessary and no license was needed. If an airframe could be built to accommodate a Model "T" automobile engine or some other left-over powerplant, and if the builder had the courage to fly it, it was his business.

After the Aeronautics Branch was initiated, fortunately, much of this type of activity was curtailed. Engines were required to have a dual ignition system and airframes had to meet certain construction requirements of safety. Components for aircraft were manufactured under government supervision. As a result, many dreams of building cheap airplanes were abandoned.

The major problem in building a low-cost airplane was not the problem of manufacturing an airframe that was suitable, but of finding a good, reliable and licensable engine to power it. During the 1920s, several suitable powerplants were

Fig. 13-61. Air-Cooled Radial Engine

Fig. 13-62. Horizontally Opposed Engine

produced, but none in sufficient quantity to insure delivery to airframe manufacturers.

Larger engines had been perfected by the time Lindbergh made his famous flight to Paris, but unfortunately, they were bulky, air-cooled radials which required a large, heavy airframe for installation. (See Fig. 13-61) It was not until the designs produced by Continental and Lycoming in the late 1930s that the light-plane power-plant problem was solved.

These new engines were air-cooled, but they were not radials. They were small, four-cylinder opposed engines developing from 40 to 75 horsepower, were inexpensive to purchase and maintain, and had small appetites for fuel and oil. (See Fig. 13-62.) Through a combination of

circumstances and with the right men behind the ideas and designs, a light plane at moderate cost became a reality by the end of the 1930s.

In 1929, a man named G. C. Taylor designed a lightweight airplane which was eventually to be one of the most famous and popular light airplanes in history. Having failed at the time of the depression to get his small craft into production, an oil man named William T. Piper bought Taylor's share in the company. In 1937, the factory was moved to Lock Haven, Pennsylvania, where the 40 horsepower J-2 Piper Cub was built.

In a short time the famous J-3 Cub was designed and put into limited production. (See Fig. 13-63.) Then, in 1939, a $4,000,000 appropriation was voted by Congress which established the Civilian Pilot Training Program. This new program was for training 10,000 pilots in 460 colleges throughout the United States. The Piper J-3 Cub accounted for 75% of all the airplanes used in this program. As a result, thousands of the Piper Cub models were built and sold in the United States and abroad. This one act of Congress put Piper on the map of the aviation industry. Other companies shared in this appropriation to some extent, but the Piper company reaped most of the

Fig. 13-63. Piper J-3 Cub

Fig. 13-64. Aeronca C-3 ("Flying Bathtub")

benefit and, at the same time, built an airplane that became a standard training craft for many years thereafter.

The development of lightweight opposed engines by Lycoming and Continental made it possible for manufacturers to design excellent lightweight airframes. Aeronca, which came out with its original "flying bathtub" (see Fig. 13-64) featuring a two-cylinder engine of its own design, started manufacturing a series of heavier and more popular models. Among these was the Aeronca Champion. (See Fig. 13-65.)

Simplicity was an important feature of these new light airplanes. They had a very dependable starting system that consisted of the strong arm of the pilot or line boy. They had no electrical system, no hy-

draulic system, or anything else that required a sophisticated mechanic to maintain. The engines, with proper care, could run as long as 1,000 hours between major overhauls.

At this stage of light plane development in aviation, the coming of the Second World War was to bring mixed blessings. Money was to be available for extensive research, but the idea of manufacturing light planes for public consumption had to be curtailed until the end of the war.

Light plane manufacturers were compelled to change their mode of thinking also. Aircraft required by the government were to be of heavier and more sophisticated types so that the transition of military pilots from the training airplane to a heavy single-engine fighter or

Fig. 13-65. Aeronca Champion

multi-engine bomber would not be so great. At least, that represented the thinking of the military at the time.

Orders from foreign countries already engaged in hostilities also dictated a transition of interest on the part of many light plane manufacturers. Many did not manufacture complete airplanes but instead found it more profitable to manufacture components for some of the larger companies.

PRELUDE TO WAR

While the United States Government was in the process of recovering from a severe economic depression, Japan, Germany, and Italy, under militaristic governments, were building arsenals of weapons. The depression that the United States endured was the same one that allowed these militaristic governments to gain power. One of the promises of the leaders of these powers was that if they achieved power, there would be full employment. Full employment was achieved through employment in the industries building military airplanes and other tools of war.

By the mid 1930s, Germany had broken the Treaty of Versailles, which previously had placed tight limitations on the production of weapons of war. Germany's government-subsidized airline, Lufthansa, was in the process of training bomber pilots even while scheduled airline activities were being conducted. Powered training airplanes were not available in quantity, so many of the future fighter pilots for the Third Reich had their first taste of flying in gliders. Since flying a sailplane requires unique skill, this may be a partial explanation for the excellent quality of German pilots at the beginning of World War II.

The major advantage to an aggressor nation is that, to a degree, it has the advantage of selecting where, when, and how hostilities will begin. Germany, in this case, exhibited very careful advance planning for a supreme war effort. Long before German military aircraft production became a threat, detailed plans had been made outlining which type airplane was to be manufactured and in what plant. In many cases the factories had not yet been constructed. Few people could believe the boast that in six months a factory could be built and begin production of advanced fighter planes. This boast turned out to be a very accurate projection.

The Spanish Civil War of the mid-to-late 1930s offered an opportunity for testing much of Germany's new equipment in actual combat. Germany and Italy went to the aid of General Franco in his armed bid to take over Spain and establish a Fascist state. It is ironic that this is the only Fascist state left in Western Europe today. The Fascist governments of Germany and Italy were completely destroyed as a result of World War II.

Italy, under the Fascist rule of Benito Mussolini since the 1920s, had begun warlike actions. Generally, the countries that Italy sought to conquer were backward, agrarian states with little in the way of material for mechanized warfare. Italian successes were mainly in Africa where Mussolini hoped to re-establish "the grandeur that once was Rome." Most of the airplanes used by the Italian forces were not as unique and advanced in design as those of Hitler's Germany.

With a foreign policy of isolationism in the United States, it was difficult for this country, still reeling under the effects of the depression, to become interested in major defense appropriations. Only after September 1, 1939, when the attack of Poland had begun and Czechoslovakia had fallen, did the citizens of the United States realize that the possibility existed that they might become involved in a war of survival.

Fig. 13-66. Japanese Zero

Hitler's Luftwaffe succeeded in destroying the Polish air force in two days and then began a concentrated bombing attack of the major Polish cities. Billy Mitchell, in testimony before a Congressional committee in 1921, had stated that modern air warfare would be conducted in exactly this manner. The major difference was that Mitchell's example involved attack upon the United States, and Hitler's objective was Poland. With air superiority established, only a few weeks passed before all of Poland was in German hands. By June 20, 1940, France had fallen, and Western Europe was in the hands of Nazi Germany.

Conditions in the Far East were no more encouraging. Japan proved that industrialization of a country was possible in areas of the world other than the West. The Japanese proved their might in a war with Russia at the turn of the century, in which they emerged victorious. Korea was invaded and occupied, and then Manchuria was taken in the early 1930s by the Japanese.

The next step in the Japanese plan for domination of the Far East was an attack on China. Because of the huge population and land area of the Chinese mainland, Japan only partially succeeded by gaining control of the Chinese coastal areas. The successes that were enjoyed by Japan would have been impossible without a large, powerful air force.

Some of the designs originated by the Japanese for military aircraft were impressive. The famous Japanese Zero was designed by a Japanese who gained much of his training while studying and working in a number of aircraft manufacturing plants in the United States. (See Fig. 13-66.)

After France fell in 1940 and the collaborationist government was established at Vichy, the door was open for the Japanese to occupy French Indo-China. (Viet Nam is a part of this old French colonial holding.) On December 7, 1941, the confrontation between Japan and the United States finally came in the form of a devastating surprise air attack on the major United States military base in the Pacific, Pearl Harbor, Hawaii. The attack came at 7:55 a.m. on a Sunday morning. In one hour and five minutes, 353 Japanese carrier-based airplanes had succeeded in killing 2,000 Americans, destroying most of the United States Pacific Fleet, and in smashing a number of military aircraft that never got off the ground. (See Fig. 13-67.) The Japanese lost 29 airplanes and 55 men in this attack. Once again the prophecy of Billy Mitchell was realized.

In 1938, the limited number of orders for aircraft was in no way indicative of what was to come. By 1940, our allies, mainly Great Britain, were desperate for military aircraft, but expenditures were so great

Fig. 13-67. Bombing of Pearl Harbor

that the British economy was not able to stand the strain of purchasing them. In the late summer of 1940, Germany, in control of the continent of Europe, began a massive air assault on Britain.

The blitz had begun. With it came the destruction of many industries which manufactured goods for export to earn Britain's livelihood in the world economy. Britain could no longer pay cash.

It was to the advantage of the United States for military security and economic well-being to assure that a German invasion of Britain did not take place. President Roosevelt's plan, the Lend-Lease, was put into operation. After a stormy debate in Congress, seven billion dollars was set aside for war material for our allies, including the Soviet Union, following the German attack of June, 1941.

How directly did this affect the aircraft industry? The output of aircraft in the United States in 1940 totaled 6,000. By 1941, this production total was increased to 19,000. By the time the United States was in the war in 1942, 48,000 airplanes were produced, and by 1944, the United States was producing 96,000 airplanes a year for the war effort. The aviation industry has not seen production like this at any time — before the war, or since.

AIRCRAFT FOR COMBAT: ALLIES VS AXIS

In time of war and with unlimited funds, many ideas had an opportunity to leave the drawing board and become a reality. The Allied state of technological development, including American, was well behind that of Germany. In 1934, Will Messerschmitt produced a four-place, all-metal, low-winged monoplane with retractable gear that was a sensation. It could fly as slow as 38 MPH and as fast as 189 MPH. The model designation for this remarkable airplane was BF-108, Taifon. (See Fig. 13-68.)

With experience only in building the Taifon, Messerschmitt entered a design competition for a fighter plane for the Luftwaffe. The result of his effort was the famous ME-109 series. This basic design started with a 695 horsepower

Fig. 13-68. Messerschmitt BF-108 Taifon

Rolls-Royce engine. By the time the war in Europe came to a close, the same basic airframe carried a Mercedes Benz, liquid-cooled, inverted V-12 engine of 1800 horsepower. During its production in Germany from 1936 to the end of World War II, 33,000 of these excellent little fighters were produced.

Another famous and effective fighter used by Germany in World War II was the Focke-Wulf 190. Its preliminary design work was started in 1928, and in 1939, the first prototype was completed. The production model, an all-metal fighter airplane, was powered by a 1600 horsepower BMW radial air-cooled engine. It first saw combat against the RAF in 1941 and was a successful match for the British Spitfire. More than 20,000 of this model were built. (See Fig. 13-69.)

Fig. 13-69. Focke-Wulf 190

Germany also produced an airplane design that, had it been put into service early enough and in the proper role, might have turned the tide of the war in favor of Germany. It was the first operational military jet in history, the ME-262 twin jet fighter-bomber. This airplane was capable of over 500 MPH in level flight.

At the end of hostilities, the Allies had nothing that could compare with this airplane. The proponents of this model were unable to get approval for its extensive manufacture and use in time to be of any value. Its mission was also clouded, as Hitler himself would have preferred to see this airplane used as an attack bomber rather than as a fighter. At the end of the war, hundreds of these airplanes were

Fig. 13-70. German ME-262 Twin Jet Fighter-Bomber

found in Germany fresh from the factory and ready for combat. (See Fig. 13-70.)

The fighter designs of the German aviation industry were impressive, but the bomber designs left much to be desired. A few four-engine models were produced, but most German bomber aircraft were of the twin-engine variety. They were not particularly fast and not capable of great accuracy at the high altitudes from which the Allied bombers operated.

At the outbreak of the war in Europe, Britain was fortunate to have in its arsenal one excellent fighter, the Spitfire. (See Fig. 13-71.) This airplane was based on a design used to win the Schneider Trophy for England in 1931 with a speed of 340.08 MPH. The Spitfire was a more-than-even match for the ME-109 and an even match for the FW-190. At the time of the Battle of Britain, a small number of Spitfires succeeded in breaking the back of the Luftwaffe bombing raid over Britain. This remarkable achievement probably · was responsible for the failure of Germany to carry out an invasion of Britain.

Fig. 13-71. Supermarine Spitfire

Fig. 13-72. P-47 Thunderbolt

Fig. 13-73. P-51 Mustang

The fighters produced in the United States were of a heavier type than those produced in Britain or by the Axis powers. The P-47 is an example of a fighter that was built for a specific mission, that of protecting bombers all the way to the target and back to their home base. (See Fig. 13-72.) Before the manufacture of this type of fighter, the Allies had nothing in their arsenal with the range to stay with the bombers to their destination. As a result, bomber losses were high, and raids within Germany were infrequent.

Following introduction of the P-47, the United States introduced another fighter of considerable merit. This machine was the P-51. Unlike the P-47, it was powered by a twelve-cylinder, inverted V-type engine that was liquid-cooled. This airplane made an excellent record for itself in Europe and the Far East. (See Fig. 13-73.)

The arsenal of bombers manufactured by the United States was most impressive.

The concept of the military at that time included extensive strategic bombing, and designs produced to serve this mission were, in some cases, an outgrowth of transport designs. The B-17 bomber, which was the terror of Germany during the war, was the outgrowth of research on the Boeing 247 transport. (See Fig. 13-74.)

The B-24 Liberator came from the United States earlier in the war, making raids against such Central European targets as the oil fields at Polesti, Rumania. A bomber that did not see service in the European theater of war but was the scourge of Japan was the famous B-29 Superfortress. In contrast to the first raids made on Tokyo with 13 twin-engined B-25 bombers early in the Pacific War, B-29s flew in formations numbering in the hundreds and dumped millions of tons of bombs on the Japanese homeland.

The two final blows against the Japanese cities of Hiroshima and Nagasaki were de-

Fig. 13-74. Boeing B-17 Bomber

Fig. 13-75. Boeing B-29 Bomber

livered by the B-29's, *Enola Gay* and *Bock's Car*. (See Fig. 13-75.) These acts marked the end of the Second World War. Along with the atomic bomb, it was the superior ability of the Allies in the use of strategic bombing which brought the war to a conclusion in 1945. This factor, coupled with the fantastic ability of the United States to produce excellent aircraft, made it possible for the Allies to dictate the terms of unconditional surrender to the Axis powers.

POSTWAR EFFECT ON THE LIGHT AIRPLANE

With the end of the war and the immediate release of most of the flying personnel in the United States military, market analysts believed the time had come for the aviation industry to begin producing small, light airplanes of the

Cessna 140

Cessna 190

Cessna 170

Fig. 13-76. Cessna Postwar Models

Piper Vagabond

Piper Super Cruiser

Fig. 13-77. Piper Postwar Models

two and four-seat type on a large scale. Cessna, which had built thousands of T-50 and UC-78 models, came out with the 120 and 140 series in 1946. These were two-place, and each was powered with an 85 horsepower Continental engine. In 1947, Cessna added the five-place 190 and 195 models powered by radial engines ranging from 220 to 230 horsepower. By 1948, Cessna had re-entered the four-place market with the 170 model, which was similar in appearance to the 140 and powered by a six-cylinder, 145 horsepower Continental engine. (See Fig. 13-76.)

Piper continued production of its Cub series with several power options, ranging from 65 horsepower up to an eventual 150 horsepower. Other models were added to the Piper line such as the Vagabond, a side-by-side two-seater with a 65 horsepower engine, and the Super Cruiser powered with the new 108 horsepower Lycoming engine. (See Fig. 13-77.)

Swift came out with an all-metal, two-place airplane with retractable gear and aeromatic propeller. This airplane was powered by a variety of engines, ranging from 85 to 145 horsepower. (See Fig. 13-78.)

Aeronca, which had been involved in military production, began producing the Aeronca Chief and Super Chief with 65 and 85 horsepower respectively. (See Fig. 13-79.) Mooney probed the low-cost, single-place airplane market with the Mooney Mite which was powered with a 40-50 horsepower Crosley engine. (See Fig. 13-80.)

Ercoupe, with its non-spin design and no-rudder-pedal control system, was finally on the market. (See Fig. 13-81.) Beechcraft continued to build, for a short time, its famous Model 17 Staggerwing biplane with 450 horsepower, and a new butterfly-tail model, the Bonanza. (See Fig. 13-82.) North American built the four-place Navion, a single-engine airplane based to some degree on its successful fighter, the P-51. (See Fig. 13-83.)

One of the most successful of the immediate postwar airplanes was the Stinson

Fig. 13-78. Globe Swift

Fig. 13-79. Aeronca Chief

Fig. 13-80. Mooney Mite

Fig. 13-82. Beech Bonanza

Voyager, a four-place airplane. (See Fig. 13-84.) It was fabric-covered and was powered with a 150 horsepower Franklin engine. Its four-place capacity appealed to businessmen and started the airplane market trend for business usage.

Another postwar airplane that proved to be a trend-setter was the Luscombe Silvair, a two-place airplane powered with a 65 horsepower Continental engine. (See Fig. 13-85.) This airplane was the first all-metal light airplane in the two-place market. The all-metal technology, developed extensively during World War II, soon was adopted by other light aircraft manufacturers.

Fig. 13-83. North American Navion

The number and variety of airplanes manufactured and those conceived on the drawing board are too numerous to mention. Many were built, but the manufacturers, much to their disappointment, discovered that they had drastically overestimated the market. Some of the companies went under financially within a few years. Some began manufacturing farm machinery and furniture, hoping that some time in the future a cautious approach to airplane manufacturing might prove financially rewarding.

Fig. 13-84. Stinson Voyager

Fig. 13-81. Ercoupe

Fig. 13-85. Luscombe Silvair

A NEW CONCEPT FOR THE LIGHT AIRPLANE

Fig. 13-86. Piper Cherokee

During the 1950s, many businesses found that, by utilizing airplanes for their salesmen and executives, key personnel could cover territory in a fraction of the travel time required by automobile or public transportation. More and more people began to learn to fly and to utilize the airplanes they purchased for both business and pleasure. Some of the airplanes manufactured during the false boom after World War II were available on the used market so that it was possible for a person of more limited means to learn to fly and own his own airplane.

The light planes that were manufactured at the end of the Second World War were usually constructed of fabric over metal airframes. By 1950, most of the fabric airplanes were on the way out. Cessna, which had been constructing fuselages of all-metal design since the end of the war, changed their wings from fabric to metal.

Later, Piper replaced its Tri-Pacer series with the new all-metal Cherokee series. (See Fig. 13-86.) By the mid 1960s, only a few fabric-covered models were still manufactured. Among these were the Piper Super Cub, the Maule airplanes, and the Bellanca which retained a wooden, fabric-covered wing.

Light airplanes manufactured today in the United States may not appear to have changed a great deal in appearance over the last 15 years, but this appearance is deceptive. Marvelous changes have taken place. Engines which once were good for 1,000 hours between overhauls are now rated for 2,000 hours. New avionics, once available only to the airlines, are now available for use in light, private airplanes because of reduced cost. Light airplanes are more rugged, quiet, and efficient than ever before.

The light plane manufacturers are engaged in active market development projects and are attempting to interest people from all walks of life in learning to fly. Light airplanes are no longer spartan. They now have every comfort and convenience to be found in the best the automobile industry has to offer the public. The number of general aviation aircraft has increased at a steady, if not spectacular rate, as have the number of people learning to fly.

THE AIRLINES IN POSTWAR AMERICA

Immediately before the United States' involvement in the war, two important designs for airline application were developed, the Boeing 307 Stratoliner, the first pressurized commerical airliner (see Fig. 13-87), and the Douglas DC-4. (See Fig. 13-88.) They were a departure in land-based transports. They carried more passengers than previous land-based airplanes and had the added safety and reliability advantage of four engines.

Fig. 13-87. Boeing 307

Fig. 13-88. Douglas DC-4

Fig. 13-89. Boeing Stratocruiser

The race to develop pressurized transport aircraft which would fly faster and carry more passengers continued during the war. In 1943, the first of the 300 MPH Lockheed Constellation series flew. By 1950, it was one of the mainstays of airlines in the United States and was used on both international and domestic flights. (See Fig. 13-91.) Boeing came out with a transport in 1949 based on the B-29 and B-50 bomber series that carried 75 passengers in luxury. This two-level airplane called the Stratocruiser had a lounge and a bar on the lower level. (See Fig. 13-89.) The DC-4 four-engine series was re-powered and extensively re-designed until it became the DC-6, which is one of the last four-engine, piston-driven airplanes developed. (See Fig. 13-90.)

A revolution was taking place in the development of engines for transport airplanes. The airlines and industry were constantly searching for a quieter, more vibration-free engine with a greater over-

Fig. 13-90. Douglas DC-6

haul life. The British thought they had the answer when they designed and began delivery of the De Havilland Comet series in 1953. (See Fig. 13-92.) With this pure jet, BOAC was the envy of all the other manufacturers. This airplane, well ahead of its time, suffered from lack of technical knowledge in metallurgy and was retired from service due to metal fatigue problems.

A transitional step between the pure jet and the piston engine airplane was the

Fig. 13-91. Lockheed Constellation

Fig. 13-92. DeHavilland Comet

Fig. 13-93. Vickers Viscount

sidering the state of jet engine development at the time. (See Fig. 13-93.)

In 1954, an event took place that was eventually to change airline history. The first Boeing 707 flew. By 1958, this airplane was in airline service, and this same model is doing a commendable job on international and domestic flights around the world today. (See Fig. 13-94.) With this airplane, scheduled flights across the continent in five hours for less than $200 became a reality.

turbine-driven prop jet. The airplanes resembled the piston-driven transport airplanes, but instead of the old conventional engines driving the propellers, they were driven by a propeller shaft geared to a jet engine. These airplanes proved to be especially efficient on short hauls in domestic service. They were practically vibration free and had maintenance records far superior to the piston-driven models. It might be said that they incorporated the best features of both the pure jet and the piston-engine airplane, con-

In 1959, Douglas made its debut in pure jet transports with the DC-8. (See Fig. 13-95.) During the 1960s, the British designed several new jet airplanes for airline use. Among these models were an improved Comet and the BAC-111 jet. (See Fig. 13-96.) The French designed the Caravelle, a medium jet which met with great success around the world. (See Fig. 13-97.) The Soviets equipped their airline, Aeroflot, with new turbo-props and the pure jet TU-104. (See Fig. 13-98.)

Fig. 13-94. Boeing 707

Fig. 13-95. Douglas DC-8

Fig. 13-96. BAC-111

The use of turbo-props and jets for business aviation came into being also. Lear Jet produced an excellent airplane (see Fig. 13-99) which could outclimb some of the operational fighters in the United States Air Force. Sud Aviation of France produced a slightly larger executive jet, the Fan Jet Falcon. (See Fig. 13-100.) North American Aviation marketed a successful business jet, the Sabreliner, based on a military aircraft they had produced. (See Fig. 13-101.)

Experience gained in the Korean conflict, several near-East confrontations, and Viet Nam revealed the need for transports that could move large groups of men and outsized equipment to distant points in the world quickly and efficiently.

These military needs and the ever-growing airline passenger projections fostered the development of a new familiy of so-called "jumbo jets." Although not new in terms of their flight environment, speed range, and type of propulsion, the very immensity of these second-generation jet transports required advanced fabrication techniques and manufacturing facilities, innovative techniques for final assembly of the aircraft, redesigned ground support equipment, and even new passenger terminal facilities.

These large aircraft benefited from advanced technology in metals, improved engine combustion chambers designed to reduce air pollution, and new navigation systems. One example is the enor-

Fig. 13-97. French Caravelle

Fig. 13-99. Lear Jet

Fig. 13-98. Russian TU-104

Fig. 13-100. Fan Jet Falcon

Fig. 13-101. North American Sabreliner

Fig. 13-102. Lockheed C-5

mous Lockheed C-5 developed for the U. S. Air Force. Its size may be determined by the fact that six Greyhound buses could be parked in the C-5 cargo compartment with room to spare. (See Fig. 13-102.)

Passenger travel has also benefited from the introduction of the mammoth Boeing 747. (See Fig. 13-103.) For the first time the cabin has a "living room" look with a width of twenty feet and a ceiling height of eight feet, four inches. Side walls are six feet high allowing passengers to stand erect even at the window seats as seen in figure 13-104.

The rear of this first-class cabin contains a spiral staircase leading to an upper level lounge. Tourist and economy passengers have also benefited from increased spaciousness. The interior of the aircraft is divided into "salons" to prevent any long auditorium-like appearance.

Hundreds of passengers on long domestic or international flights are presently enjoying the benefits and luxury of this giant aircraft. When the new tri-jets, the Lockheed L-1011 and the Douglas DC-10 join regular commercial service, the benefits will be brought to passengers on shorter domestic flights.

If the airlines can succeed in carrying hundreds more passengers per single flight at greater speeds with less down time for maintenance, the overall cost of operating the business can be reduced.

Fig. 13-103. Boeing 747

Fig. 13-104. Boeing 747 Interior

Fig. 13-105. The Concorde

THE FUTURE

What does the future hold for aviation? After the wide-bodied jets, the trend seems to be for faster and higher flying aircraft. The SST (supersonic transport) is representative of this trend. However, it does present many problems. The speed is advantageous, but the passenger carrying ability of present and proposed models is not greater than subsonic jets such as the 707. The noise problem from sonic booms has not as yet been solved. This problem may relegate the SST to flight only over sparsely populated areas and across the oceans. With the combined efforts of France and Britain, the Concorde has been successfully flown and is now available for airline service. (See Fig. 13-105.)

The Soviet Union has also successfully developed an SST of its own which is available to the world airline market. The United States has done some developmental studies on the supersonic transport and has the advantage of experience learned through the B-70 supersonic bomber program, which actually served as a flight test bed for similar passenger-carrying aircraft of the future.

However, high developmental costs coupled with environmental problems, such as air and noise pollution, have hampered further U.S. development efforts. As illustrated in figure 13-106, projected levels of world atmospheric pollution show alarming trends. In addition, world fuel shortages and general ecological concerns over pollution and resource allocation have raised serious questions about the future of the SST.

In spite of these developments, refinements in space technology have placed U.S. astronauts on the moon and allowed extended operations of manned orbiting space stations, such as Skylab. In addition, both the United States and the Soviet Union are concentrating space efforts in exploration of the solar system. These sophisticated space probes are designed to gather scientific data about the sun and neighboring planets.

PROJECTED LEVELS OF WORLD ATMOSPHERIC POLLUTION

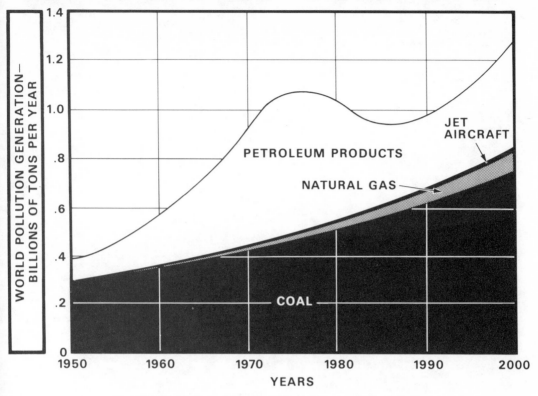

Fig. 13-106. Projected Levels of World Atmospheric Pollution

In the more immediate future, transportation of goods and people by air will continue to become less expensive and more commonplace as payloads and speeds increase rapidly. Increasingly, the business of the world is being geared to the air age. Air transportation will be a dominant factor economically and socially in world development in the coming decade, influencing attitudes and shaping the quality of life.

chapter 14

THE AIR AND SPACE AGE

THE IMPACT OF AEROSPACE

The requirements of aerospace missions insure the generation of new knowledge and technology. New technology comes into being in the course of the research and development necessary to insure the success of a particular mission. In addition to the obvious development of new "hardware," this technology includes the use of materials and the development of new techniques. As will be pointed out in succeeding paragraphs, the new knowledge spans the complete spectrum of the sciences and many areas of modern day technology.

"Spin-off" is a term often applied to the use of technology for a purpose other than that for which it was originally developed. Many people are unaware of the vast technological spin-off from aerospace efforts. This spin-off is so vast that an examination of only two areas — the national space effort and the mission of the U. S. Air Force will be developed here.

MINIATURIZATION

Miniaturization, as it applies to aerospace, can be defined simply as reducing the size of component parts. The use of aviation-related electronic equipment underwent a rapid expansion at the beginning of World War II. Limitations imposed by available space and weight became quickly apparent and a large-scale program to reduce the size, bulk, and weight of component parts was instituted.

Solid state electronics was an outgrowth of the radar detection equipment employed during the conflict when it became evident that bulky vacuum tubes were incapable of handling the radio frequencies effectively. In later years, the trend toward smallness continued. In booster rockets and spacecraft, small components are an absolute necessity because of the limited payload of rocket vehicles and the high amount of fuel required to put a small payload in orbit.

Miniaturization is especially apparent in the area of electronic computers. One of the early computers had vacuum tube components that filled several rooms. The functions performed by this array of equipment can now be performed by solid state electronic components no larger than a small shoebox.

An example of a direct benefit to the consumer is the development of transistors, and solid-state circuits which replace vacuum tubes, and other electronic components. (See Fig. 14-1.) Most people are familiar with the small pocket radios, television sets, tape recorders and pocket size computers made possible largely through the technology of miniaturization.

COMMERCIAL USES OF AEROSPACE TECHNOLOGY

The remotely controlled walking device conceived as a means of transport on the

Fig. 14-1. Size Comparison of Ladybug with Miniaturized Solid-State Circuits.

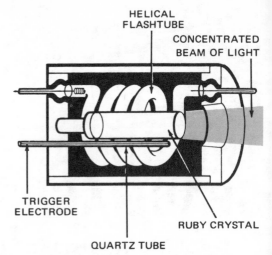

HELICAL FLASHTUBE

CONCENTRATED BEAM OF LIGHT

TRIGGER ELECTRODE

RUBY CRYSTAL

QUARTZ TUBE

Fig. 14-2. Typical Laser Components

lunar surface has been adapted as a walking chair for crippled persons. Unlike the wheelchair, this device moves easily across a sandy beach, gravel, or uneven road, and can even climb over curbs.

Examples of improved medical techniques and equipment stemming from Air Force research include lasers used in "welding" a detached eye retina to the underlying tissues (see Fig. 14-2), and sensors to monitor a patient's heart action, respiration, and blood pressure by remote control. These sensors were derived from those first used to detect changes in the physical condition of Air Force pilots flying high-speed experimental aircraft. Medicine has further benefited from the miniaturization of devices for artificial heart and kidney machines that reduced these machines to a more practical size.

Because it was once thought that an astronaut would be unable to manipulate controls during the high "G" conditions of blastoff, a device was developed which would enable the astronaut to maintain control of the space vehicle by simply moving his eyes. (See Fig. 14-3) This sight switch can be used to permit a paralyzed person to dial a telephone and even to raise and lower a hospital bed.

An improved diagnostic tool has evolved which can detect and record the slight

postural reflexes, or muscle tremors associated with certain illnesses such as Parkinson's disease. This diagnostic tool, known as a muscle accelerometer, is based on technology originally developed for the detection of micrometeorites in space.

Pinpoint-sized ball bearings for satellite equipment are now used in ultra high speed, painless dental drills. Tiny mercury batteries, developed from those used in satellites, have been implanted in the bodies of heart patients to control the rhythm of their heartbeats. Miniature transistors, designed for spacecraft equipment, are now used in tiny, lightweight hearing aids. The use of space age cryogenics, or low temperature liquids, has lead to a new cancer surgery technique in which cancer cells are literally frozen to death.

The gigantic astrodome in Houston, Texas, is a descendent of the radome pioneered by Air Force research to protect radar equipment from the weather.

The NASA Langley Research Center was assigned the task of learning why airplanes skid on wet runways and of seeking a means of preventing this hazard. The researchers found that at

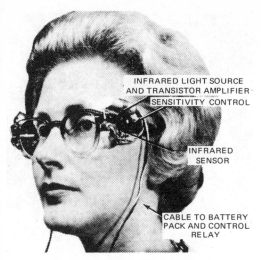

INFRARED LIGHT SOURCE
AND TRANSISTOR AMPLIFIER
SENSITIVITY CONTROL

INFRARED
SENSOR

CABLE TO BATTERY
PACK AND CONTROL
RELAY

Fig. 14-3. Control of Devices by Sight

high speeds water was forced between the runway and tires, literally lifting the tire from the runway, or "hydroplaning" on the surface of the water. Closely spaced grooves in the runway, perpendicular to the runway centerline, were the result of this research effort. Further spin-off resulted when several states undertook pilot projects, grooving stretches of automobile highways where accidents were frequent during rainy weather. As a result of grooving five stretches of highway in Los Angeles, California, such accidents were reduced by 92%.

MANUFACTURING SPIN-OFFS

The plastics industry, the food industry, and the metal-casting industry, to cite three examples, use linear accelerators to make better plastic products for the home, to sterilize and preserve foods, and to detect flaws in metal castings. The accelerators are the result of a military-sponsored research program.

High resolution radar, originally developed for the national defense effort, is now used by oil companies searching for oil deposits, by agricultural experts to make crop and soil surveys, and by geologists studying earth formations in earthquake research.

Manufacturers of railway tank cars and storage tanks for corrosive chemicals are using lightweight, corrosion-resistant materials originally developed through Air Force research for solid fuel rocket cases. Infrared sensor systems, used in intelligence missions, have been adapted by the steel manufacturers to control the thickness of rolled steel.

OTHER SPIN-OFFS

Digital computer techniques, developed to clarify photographs telemetered across 140 million miles of space between Earth and Mars are now being used to make medical X-rays more revealing. Ceramic substances, developed during research on missile nose cone materials, have been adapted to high temperature oven cookware. Non-stick cookware is also a product of the space age.

Portable color TV cameras were made possible through the use of integrated circuits and miniaturization of parts. Space age solar cell research has evolved into solar cells that power the emergency telephone system along the Los Angeles freeways. Electric wristwatches operate on tiny nickel cadmium batteries developed for space use, and research rocket sleds have contributed to auto safety standards.

Sunglasses that become more opaque as sunlight brightens stem from Air Force research on glasses to protect aircraft crews exposed to sudden nuclear flashes. Stainproof, fire-resistant fabrics and carpeting owe their existence to basic research for space requirements; and even the TV dinner resulted from the military's need to provide aircraft crews with frozen meals that could be heated quickly and easily in flight.

To list all of the specific items that were derived from space age related research would be a time-consuming task. The examples given, however, will serve to illustrate that the space age has far-reaching effects on the daily life of Mr. Average Citizen.

Fig. 14-4. TIROS Weather Observation Satellite

Fig. 14-5. ECHO Satellite Balloon

IMPROVED WEATHER FORECASTING

It is very advantageous to be able to observe and collect data about weather patterns over large areas of the world. It is for this reason that the first series of weather satellites were put into an earth orbit. These early weather satellites were nicknamed TIROS.

TIROS I operated for 77 days. During the following five years, a total of nine other TIROS satellites were launched. From these research satellites, over half a million useable pictures were processed. These photographs and the heat data gathered permitted a "giant step" forward in the science of weather observation and forecasting.

In late 1961, for example, TIROS III sent, from space, television pictures of a storm brewing in the Atlantic Ocean. Within days the largest evacuation in U.S. history had saved over 350,000 Gulf Coast inhabitants from hurricane Carla. These pictures, plus information that TIROS provided on heat radiation from clouds (see Fig. 14-4), provided this early warning of a hurricane hazard.

The capability of weather satellites has been improved tremendously since the TIROS series. Pictures taken from camera-equipped weather satellites are now commonplace in the daily weather forecasts seen on many television stations.

The benefits of long-range accurate weather forecasts are immeasurable. Farmers will be able to take advantage of the best days for planting and harvesting. They can even determine the best crops to plant. It may be possible to dissipate hurricanes and tornadoes before they become destructive. Vacations can be better planned. Floods and other natural disasters can be foreseen and necessary countermeasures prepared.

Recently, the significance of correct weather reporting and forecasting was presented in a report to Congress. The report stated, "an improvement of only 10% in accuracy could result in savings totaling hundreds of millions of dollars annually to farmers, builders, airlines, shipping, the tourist trade, and many other enterprises."

As the many complex and interrelated factors responsible for weather systems are better understood, projects for actual weather control may become more commonplace. Rainfall might be increased on certain arid areas of the world as well as modifications to undesirably warm or cool areas. However, the political, social, and psychological implications of such capabilities are enormous. A master plan, agreed upon by the nations of the world, would seem to be an intregal part of any large-scale weather modification program.

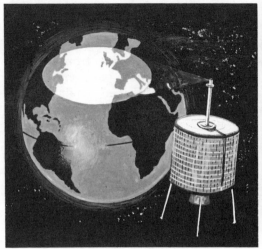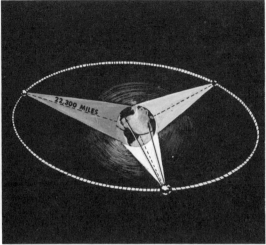

Fig. 14-6. "Early Bird" Satellite

EXPANSION OF COMMUNICATION CAPABILITY

Long distance communication capability has been rapidly expanded with the advent of the space age. Early efforts in applying satellite technology to communication involved the use of passive reflector-type satellite balloons such as Echo I and II. (See Fig. 14-5.) Active repeater satellites such as "Early Bird" were another giant step forward in communication. Three of these satellites were placed in orbits synchronized with the speed of the earth's rotation at an altitude of 22,300 miles. (See Fig. 14-6.) Their antennae linked North America and Europe with 240 high-quality voice circuits and made live television across the Atlantic possible for the first time. Communications satellites are now capable of simultaneously transmitting 12 television channels or 20,000 point-to-point messages or any combination thereof.

The benefit of this rapidly expanding field of communication to the average American is enormous. Direct dialing for international phone calls is presently available between several countries on a trial basis. Communication satellites will soon extend the use of this technique to inter-continental distances.

Although international transmission of live television programs was unheard of a few short years ago, rapid advances in satellite technology have made live TV programming from foreign shores routine. On a worldwide basis there are almost as many TV sets as telephones. Since television appeals to both the senses of sight and sound, satellite television programming is expected to have a more profound effect on education, cultural exchange, and worldwide news coverage than any other communication's medium.

Educational television via satellite offers many interesting possibilities. For example, while many people in under-developed nations may not be able to write or read, they can see and hear and can, therefore, be given valuable training in agricultural techniques, basic health and sanitary practices, and vocational skills via television.

Medicine has also benefited from international TV, as witnessed by a live telecast of open heart surgery performed in Houston, Texas, and observed by doctors in Geneva, Switzerland. The medical personnel involved in the operation and the observers were able to discuss the techniques being witnessed.

An outgrowth of this type of communication might be the formation of a worldwide diagnostic communication center.

Libraries in several different countries may become readily accessible to students, scholars, and scientists in other countries of the world. They would be able to obtain facsimile reproductions of articles, pictures, or entire books via satellite. Computer correspondence through satellites would make it possible for companies with offices around the world to maintain centralized inventory control and quality control.

In the not-so-distant future, a type of communications utility will be possible in metropolitan areas. From the home or office, a person could receive a wide assortment of communication services. Personal banking, shopping, and billing services as well as picturephone, library information, television, and information on favorite books, magazines and newspapers would be possible. Contacts outside the local area would be routed via satellite to other more distant areas. Considerable work in development and implementation of some of these services still remains to be accomplished while others are approaching reality. When developed to its full potential, communication via satellite is expected to have far-reaching influence on the lifestyle of modern man.

SOCIOLOGICAL INVOLVEMENT OF AEROSPACE COMPANIES

Certainly in a discussion such as the preceeding in which the emphasis has been on the technological effects of aerospace, there are elements of social importance. It may be enlightening to examine the more direct involvement of several aerospace companies in today's social problems. The companies cited are only selected examples from a rather extensive listing.

TRAINING THE UNEMPLOYABLE

As part of its national commitment to train and provide jobs for persons previously considered unemployable, the Fort Worth Division of General Dynamics opened a manufacturing plant in San Antonio, Texas. Several groups of fifty persons, each drawn from the ranks of unemployed, have been enrolled in a program of skilled training and on-the-job instruction.

This training was largely in the production of parts and services for an advanced Air Force fighter. The results have exceeded anticipation, the rate of retention and the number of employees advanced to higher positions has been similar to that of long established departments.

A number of other companies which are basically aerospace in nature have initiated similar programs with encouraging results.

LOW COST HOUSING

Another aerospace firm, ROHR Corporation, is contributing equipment and technology to a government-industry program to provide low cost housing for migrant farm workers. The program also teaches mechanical and carpentry skills to unemployed farm workers.

COMMUNITY DEVELOPMENT

Another example of an aerospace company directly applying its pool of technical skill to a social problem is TRW Incorporated. This company is assisting the city of Fresno, California, in the preparation of a community renewal plan which identifies and analyzes Fresno's renewal needs and the resources it has available. The program also includes preparing an economic development plan which takes into account the relationship of the community to its surrounding region and design of a management system to direct and implement the projects. Constant review and updating of the program is also provided.

CRIME CONTROL

For several years, General Electric's electronics laboratory has been working with the Syracuse Police Department on the problem of crime in the city. What was originally intended to be a cooperative program from which ideas for new police equipment would come, has evolved into a whole new concept of police organization.

Two experimental crime control teams have been deployed in Syracuse, and data collected during their deployment indicated a marked reduction in criminal activity in the areas served.

CRIME DETECTION TECHNIQUES

Improved methods of fighting crime through the use of modern scientific developments have been outlined by Aerojet General Corporation during national and local meetings on crime prevention. "Nuclear fingerprints" have been advanced as the most useful method of comparing clues found at the scene of a crime with similar samples taken from a suspect. This "neutron activation analysis" can be used, for example, in comparing tire marks, hair samples, and marijuana. Matching probabilities are said to be 99.999%.

The use of chemistry has been cited as a method for marking and tracking. For example, it could be possible to "sniff" around a house to determine if narcotics are inside without ever entering the house. Special markings on vehicles would enable police officers to follow a car chemically.

SCHOOL-COMMUNITY RELATIONS

Attacks on some of the problems posed by urban poverty and cultural gaps, particularly the barriers to education, have been mounted by Lockheed Aircraft Corporation. One project employed aerospace skills to design questions for more than 250 teachers in poverty areas and later to analyze their answers. Findings showed the success or failure of schools to educate youngsters with ethnic, cultural, or economic handicaps can be a major contributor to either contentment or unrest in the modern city.

In another effort, Lockheed systems experts are helping educators make great gains in motivating junior high school students who previously had "underachieved" because of social and economic factors. Both of these projects have drawn heavily on a major aerospace company's experience in the analysis of enormously complex problems and the design of workable solutions to them.

SOCIOLOGICAL IMPACT OF AVIATION IN EVERY DAY LIFE

The discussion thus far in this chapter has centered about the technological influence of aerospace and the direct involvement of aerospace in solving sociological problems. There are other sociological implications of a less direct nature that should be examined. Some concept of the magnitude of social changes to be expected during the remainder of this twentieth century can be gained by a review of some of the sociological effects of aviation.

The development of aviation has had a far reaching influence on every segment of man's life. It is difficult to imagine the changes that would occur in the mode of living if suddenly aviation ceased to exist. An awareness of the dependence on air travel was dramatically illustrated during the week long strike of airline employees in 1966. Entire patterns of travel underwent drastic changes which, though temporary, seriously hampered many kinds of businesses.

Aviation has played a number of sociological roles since its inception. It is necessary to examine some of the most important of these roles for better understanding of the world lived in today.

POPULATION DISTRIBUTION

Air transportation has helped to re-distribute populations and to change population trends. The cities in which we live developed primarily in response to the transportation system that was in existence when the city was founded.

For example, when the dominant form of transportation was the ship, cities were founded along ports and waterways. When railroads became more dominant as a mode of transportation, those cities located near the railroad lines became important industrial centers. Today the development of the airlines as a dominant means of transportation has again produced a change in population trend, and those cities which are developing as important air centers are among those cities experiencing the greatest population growth. Such air centers, however, will differ from the cities of the past.

Since people are able to travel by air at extremely high speeds, the effective size of the world has become considerably smaller. It is no longer necessary to crowd large masses of people into a particular area. The use of helicopters, STOL (short takeoff and landing) aircraft and air cushion vehicles may help to accelerate the trend toward de-centralization. Individuals may then commute distances which were formerly considered impractical. Therefore, our new "air centers" will not be the congested cities of the past, but will trend toward less highly populated suburban complexes.

The state of Alaska is presently experiencing the effects of population redistribution. Aviation is perhaps the dominant form of transportation as evidenced by the wide use of general aviation type aircraft for the movement of products and persons. Alaska is largely unfettered by earlier population distribution centering around water and rail transportation and may therefore become a model for future development. (See Fig. 14-7.)

ECONOMIC PROGRESS

Aviation has had the effect of eliminating the restricting influence of physical

Fig. 14-7. Alaska and Light Aircraft

barriers. Jungles, mountains, oceans, and frozen waste lands no longer render an area inaccessible. The ability to move goods and people and to exchange ideas is essential to economic progress as well as to population distribution. Prior to the development of air transportation, many areas of the world which were rich in natural resources remained undeveloped because they were surrounded by formidable natural barriers. In large regions of Africa and Asia, for example, people lived as they did centuries ago. Aviation, however, has provided a great stimulus to the development of these regions into important, modern trade and cultural centers. Four important contributing factors are: (1) air transportation routes do not require the large investment in time and money necessary for the development of highway and rail transportation; (2) it is possible to deliver equipment and supplies essential to a region much more quickly by air than is possible by surface transportation; (3) the flexibility of the air transportation system makes it quickly adaptable to changing needs of the region; (4) the disadvantage of great distance has decreased thus allowing people to travel quickly from one region to another.

RESPONSE TO EMERGENCIES AND LARGE SCALE DISASTERS

The ability to respond to emergencies has been greatly affected by air transportation. In times of natural disaster and sickness, supplies, medicines, and trained

Fig. 14-8. Flood Survivor Being Hoisted Aboard Helicopter

personnel can be sent to devastated areas. Rescue and emergency operations can be carried out quickly and efficiently. The ability to evacuate, by air, persons who are stranded in an area ravaged by floods and to provide rapid medical attention to those injured by hurricanes, earthquakes, tornadoes, fires, and floods has saved the lives of many persons. (See Fig. 14-8.)

INDIVIDUAL AND GROUP LIFE STYLE

Air transportation has increased mobility. Today, individuals are more willing

Fig. 14-9. Visiting Relatives and Friends

to accept jobs and relocate in other parts of the country. The speed of air transportation makes it possible to visit relatives and friends in distant places with ease and comfort. (See Fig. 14-9.)

Air transportation has had a distinct effect on eating habits. The use of air cargo has made perishable goods from distant areas more readily available to consumers throughout the country and around the world. It is now possible for an individual in the midwest to enjoy fresh fish from the seacoast, fresh strawberries from Mexico, and fresh pineapple from Hawaii without great expense. Furthermore, the speed of the airplane has made it possible to introduce new products to consumers quickly. If a new product becomes available it can be shipped by air immediately and be on the shelves ready for purchase the next day.

Air transportation has also influenced vacation travel. Prior to the modern air-

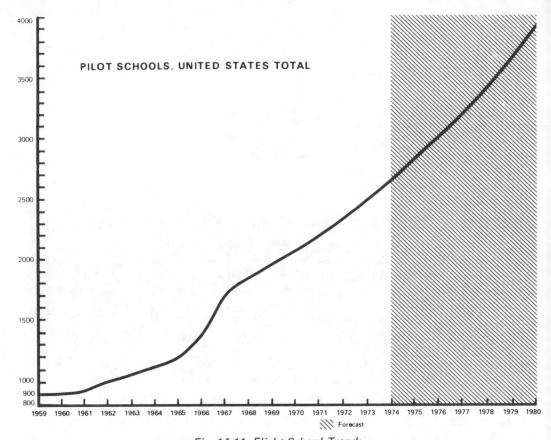

PILOT SCHOOLS, UNITED STATES TOTAL

Forecast

Fig. 14-11. Flight School Trends

Fig. 14-10. Visiting Distant Lands

EDUCATION

To provide the expanding field of aviation with competent mechanics and pilots, there are approximately 265 airframe and aircraft powerplant mechanics schools and approximately 2,600 pilot schools. By 1980 the number of flight schools is predicted to reach 3,900, an indication of the accelerating education requirements of this field. (See Fig. 14-11.) In addition, many aircraft plants and airlines conduct schools to train employees.

plane, man was limited visiting places of interest that were distant. The speed of air travel has made it possible to visit and explore countries and to meet people many thousands of miles away. (See Fig. 14-10.) The ability to see how the rest of the country (and, in fact, the rest of the world) lives and to find out how other people think has increased man's understanding of his neighbors and his world.

Aviation has also influenced the education system. In recognition of the increasing opportunities afforded by aviation, a number of colleges, especially community colleges, offer courses in some phase of aviation education. The aviation programs provide students with the following opportunities:

1. Personal enrichment.
2. Increased knowledge of the role of aviation in society.
3. Flight training for business and pleasure.
4. Development of aviation job skills.
5. Upgrading aviation job skills.

These objectives are accomplished by means of individual credit and non-credit courses, seminars and workshops, certificate programs less than two years in duration, career-oriented associate degree programs, and transfer associate degree programs.

GEOGRAPHY

Aviation has influenced everyday lives by changing the concepts of the size of the world. People formerly thought of distance as the number of miles they had to travel on the Earth's surface. Today, in many cases, the concern is the number of "flying hours" necessary to get to the destination.

Air transportation has changed global geography concepts. For example, a great circle is actually the shortest distance between two points on the Earth's surface. On most commonly used charts a great circle would appear as a curved line. However, when viewed on the globe itself, as shown in Figure 14-12, the great circle route is actually shorter than the

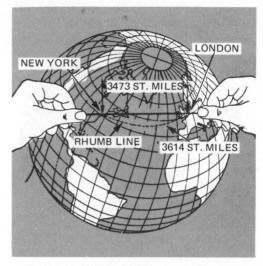

Fig. 14-12. Comparison of Great Circle and Rhumb Line Routes

straight line (also called a rhumb line) route drawn on a chart. Charles Lindbergh was one of the early pioneers using the great circle concept of navigation and he flew to the Orient by heading Northwest rather than West. Presently, non-stop daily commercial flights between American cities and European population centers go over the Polar region.

INTERDEPENDENCE

Perhaps the greatest influence of aviation has been the role it has played in increasing the interdependence among people. Specialization in agriculture, business, services, the growth of great urban centers, the increasing rate of population

Fig. 14-13. Executive Jet Aircraft

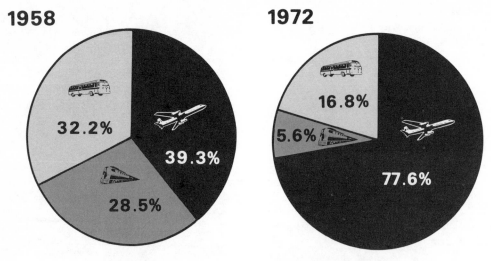

Fig. 14-14. Distribution of Inter-City Passenger Travel

growth, and increasing technology are all directly related to the interdependence of people that are influenced directly by aviation.

AIR PASSENGER TRANSPORTATION

SPEED AND TIME SAVINGS

Aviation's primary advantage is the reduction of non-productive time involving getting from one place to another. For example, assume the executive jet aircraft, shown in figure 14-13, was based in Des Moines, Iowa. Any point in the U.S. would be within four hours flying time. The speed of air transportation conserves a most valuable resource, that of time. Therefore, aviation enables society to produce more, to raise standards of living, and to enjoy more leisure time.

Aviation has produced a revolutionary shift in passenger travel from surface to air transportation. This is evidenced by the fact that, in 1958, less than 40% of all the public inter-city transportation was by air. This figure had risen to 77.6% in 1972. (See Fig. 14-14.) If the present growth rate continues, more than 90% of all inter-city travel will be by air in the near future. In international travel, the importance of air transportation is even more pronounced. In 1947,

about half of all international travel was by air. At present, that figure is nearing 100%.

Modern commercial airlines travel at speeds up to 640 miles per hour. Many flights are direct, non-stop. A businessman can leave his east coast office in the morning and be in the west coast office of an associate by mid-morning. Without air travel, similar trips would take not four or five hours, but four or five days. Since the time of a businessman is usually very valuable, the time that he saves by utilizing air travel represents a substantial cost savings to his company.

Many businessmen have found greater utility by owning and operating their own aircraft. (See Fig. 14-15.) The freedom to utilize his own schedule offers both speed and flexibility to business travel.

COMFORT

The aircraft industry has actively promoted air travel by striving to provide travelers with numerous comforts and with the most up-to-date planes available. Keen competition within the industry has caused manufacturers to design the most comfortable planes possible. In both general aviation and in air transport aircraft, noise and vibration are so minimal that they are virtually non-existent.

Fig. 14-15. Owner–Operated Business Aircraft

Fig. 14-16. A Modern Time-Saver

Pressurized cabins are finding greater usage so that passengers will have no physical discomfort at extreme altitudes.

Despite the increased comforts and time savings provided by air travel, the cost of this form of transportation is being constantly lowered. This is true for both general and commercial aviation. For example, although the cost of consumers goods has risen over 25% since 1962, the cost of airline travel has decreased at approximately the same rate. Translated into the traveler's point of view, in 1947 the minimum round trip fare between Los Angeles and New York was $300 including tax, and took nine hours each way. This same trip today costs $307 (excursion fare) and takes slightly less than five hours each way.

The airlines have been able to reduce fares because the greater speeds of the modern jet airplanes permit a single plane to fly many more miles during a single day, thus producing more passenger miles. Also, modern jets seat many more passengers.

Since the airlines receive no revenue for empty seats, they have made increased use of promotional fare plans such as family fares, youth fares, excursion fares, and of coach and economy rates. Two major airlines report that more than 50% of all their passengers now fly a discount or promotional fare. The "jumbo jets" are expected to continue the trend toward lower fares.

Light general aviation aircraft, of the type normally associated with travel, can be operated at costs comparable to automobiles and at the same time retain the aircraft's decided speed advantage. Airplanes of the type shown in figure 14-16 are capable of rapid point-to-point travel times and are comparable to airliners in terms of time savings on shorter flights.

SAFETY

The unfavorable safety record of aviation in its early development was one of the major disadvantages of air travel. Today, these records constitute one of the air industry's primary advantages. Although accidents do occasionally occur and are given extensive publicity, comparative figures emphasize the safety of air travel.

Figure 14-17 shows that during the 1960's the miles flown by general aviation aircraft more than doubled while the fatal accident rates showed a decrease. For the airlines the number of total accidents per million miles dropped from .24 in 1960 to .17 in 1969.

A comparison of the fatal accident rate for passenger automobiles and scheduled

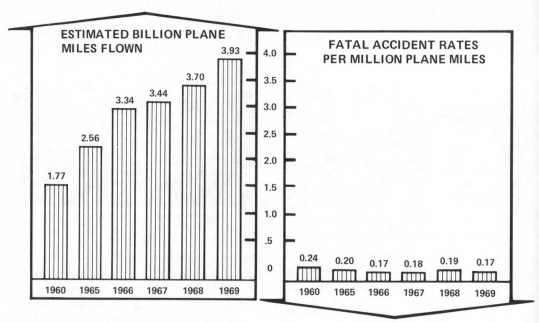

Fig. 14-17. Comparison of General Aviation Accident Rate to Miles Flown

airlines shows a decided advantage in favor of the scheduled airlines with a generally improving safety record during the years studied. (See Fig. 14-18.)

RELIABILITY

Both mechanical and systems reliability are of great importance to the aircraft industry because of the great emphasis on speed and time savings. Poor weather conditions formerly made it necessary to delay or cancel many flights. The continuous improvements that have been made in radar facilities and instrument landing systems, for instance, have greatly increased the schedule reliability of all types of air travel. This trend can be

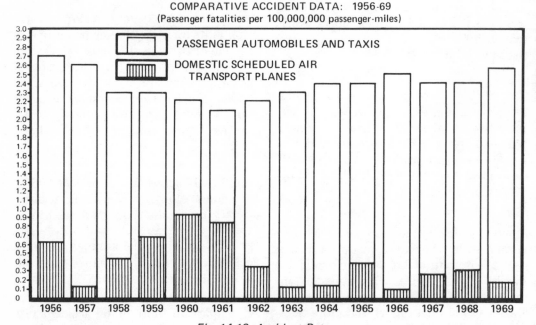

Fig. 14-18. Accident Rate

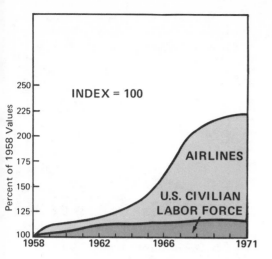

Fig. 14-19. Airline Employment vs. U.S.
Labor Force Growth

expected to show continued improvement in the years ahead.

UTILITY AND ACCESSIBILITY TO REMOTE AREAS

As pointed out earlier, formerly remote or inaccessible areas of the world have been opened to trade and cultural exchange by the airplane. In the U.S. where surface transportation has been the best in the world, air travel has made more accessible those areas where limited population or underdevelopment formerly produced a minimum of travel. Such areas were often by-passed by good highways, railroads, and other conventional surface travel means. An excellent example is Alaska where use of the airplane has become an indispensable way of life.

EMPLOYMENT

Although the specific area of employment and careers is treated more fully in Chapter 15, an example of the sociological impact of new jobs may be seen by considering future airline employment trends. The rapid expansion of airline traffic during the past years has dramatically increased the number of persons employed in air transportation and related fields. The rapid increase in jobs in these fields has been much greater than for the rest of the civilian work force.

Today, the air transportation payroll is a significant factor in the nation's economy.

Since the introduction of commercial jet aircraft in 1958, employment by the airlines has more than doubled. The civilian work force increased by about 20% during the same period of time. (See Fig. 14-19). By the end of 1972 over 301,000 employees were working for scheduled airlines in the United States.

Airline payrolls also experienced corresponding increases. During 1972 this payroll amounted to 4.2 billion dollars, four times higher than in 1960 and eleven times higher than in 1950. The average annual airlines salary in 1972 was $13,921. This figure is 213% higher than in 1960. These salaries tend to be higher than for workers in private industry and other transportation industries. (See Fig. 14-20.)

There have been other important influences besides the number of jobs and the increasing payrolls. Today, one can see that some airports have become virtual small cities in themselves. Other businesses related to air travel, such as hotels, restaurants, and car rental agencies as well as barber shops, gift shops, and other service-oriented businesses, have arisen within the airport complex. The purchase of new aircraft has given rise to many thousands of new jobs in

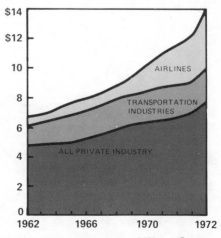

Fig. 14-20. Average Annual Wages Compared

aircraft manufacturing, electronics, and maintenance industries. It has been estimated that for every airline employee there are three other employees in related fields. The related fields spoken of here are not intended to include general aviation, for it generates its own "employment spin-off" in fields related to general aviation.

A prediction has been made that the nation's airlines may add 200,000 new jobs during the decade of the 1970's bringing the total airline work force to about 500,000 people. Assuming the three to one ratio previously cited holds true, by 1980 some 1,500,000 persons will hold jobs in work related to air travel.

AIR CARGO TRANSPORTATION

Not only has the airplane revolutionized passenger travel, it has likewise had a dramatic impact on cargo transportation. As shown in figure 14-21, cargo ton miles (freight, mail and express) reached 5.5 billion in 1972 — over five times the 1958 level and 7.6% more than 1971.

The great stimulus to the growth of air freight has been the jet freighter which was brought into service in 1963. (See Fig. 14-22.) With more than 300 jet freighters now in service, the airlines

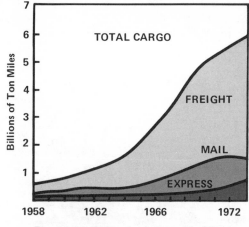

Fig. 14-21. Growth of Cargo Traffic

schedule cargo aircraft into approximately 70 U.S. cities.

One benefit of the jet freighter is that it has made possible the containerization of air freight. (See Fig. 14-23.) An analysis of the program of containerization shows a dramatic increase in container traffic and also the revenue obtained from containerized traffic.

There has been an increasing effort made by the airlines and the post office department to move greater volumes of mail by air. This cooperation has fostered a number of innovations. They are: (1) construction of modern mail handling facilities at airports. (2) greater utilization of plane-side pickup and delivery of mail; (3) closer coordination of

Fig. 14-22. Jet Air Freighter

Fig. 14-23. Containerization of Air Freight

late evening airline flights with post office close-out times; (4) computerization of airline scheduling data by the post office department in order to make use of available airlift; and (5) stepped up research and design of aircraft containers specifically designed for mail.

The "jumbo jets" now used in airline service will continue the growth of air cargo. Entire units of bulky items may be handled with ease. A major producer of luxury custom cars already plans to supply distant customers by air cargo, promising delivery within hours and eliminating long and wearying overland trips customary in the past. The capacities of these huge jets, coupled with the reduced per mile costs of goods carried, offers stimulating competition to present surface methods of cargo transportation. The general acceptance of this new form of air cargo transportation is primarily due to the following factors:

1. Broader area of distribution.
2. Reduction of inventories.
3. Reduction in warehouse space.
4. Reduction of loss, theft, damage, and spoilage.
5. Lower crating costs.

BROADER AREA OF DISTRIBUTION

The geographical area over which a particular type of goods is distributed has often been limited by the perishable nature or timely importance of such goods. The use of air cargo has greatly expanded the distribution area of these goods because of the speed with which shipment can be accomplished.

For example, new styles and new products can be introduced simultaneously throughout the country due to the speed of air travel. National advertising campaigns require that a new product be available throughout the entire country at the time the campaign begins. Often this can be accomplished only by the use of air cargo. (See Fig. 14-24.)

Air cargo also enables consumers in distant places to enjoy fresh fish, flowers and other perishable goods which would otherwise be restricted by the distance that they could be safely shipped. The value of newspapers and magazines is also enhanced by air transportation. Time renders these products valueless rather quickly. Air cargo, however, enables these publications to be widely

Fig. 14-24. Air Cargo Promotes Simultaneous Introduction of Products Throughout the Country

distributed while retaining their timely importance.

REDUCTION OF INVENTORIES

The speed with which goods can be supplied to the businessman enables him to reduce the quantity of goods which he must keep on hand. This is a distinct advantage when a particular item is extremely costly or not in great demand.

For example, many of the component parts used in construction machinery are very costly. Replacements must seldom be made but it is important that parts be available if needed. With the use of air freight the replacement part can be obtained quickly from the manufacturer so that the businessman or operator does not have to tie up large amounts of capital in non-productive inventory. The ability to reduce inventories also may eliminate losses attributable to mark-downs which occur as a result of overbuying at the beginning of a season.

REDUCTION IN WAREHOUSE SPACE

The ability of businessmen to reduce inventories with the use of air freight means that warehousing costs can be reduced. Smaller inventories mean that less space is needed to store the goods that are on hand.

Furthermore, air freight makes it possible to deliver finished goods directly from the factory to the retail store. (See Fig. 14-25.) This technique, for example, is employed by a west coast company when a customer places an order for carpet. The firm telephones the order to the mill and the order is shipped by air freight to the company. Thus, the firm is able to deliver the goods to the consumer rapidly and at the same time eliminate the need to warehouse the goods.

REDUCTION OF LOSS, THEFT, DAMAGE AND SPOILAGE

Air freight emphasizes careful handling and speed. The reliability of this form of transportation means that goods are transported and delivered with the minimum amount of delay. Thus, there is a reduction of losses attributable to spoilage, theft and damage and the result is an overall reduction in the cost of shipping. Insurance companies readily

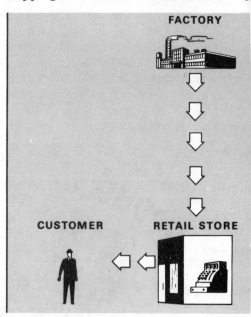

Distribution Without Air Freight

Distribution With Air Freight

Fig. 14-25. Advantages of Air Freight

YEAR ENDING DECEMBER 31	TOTAL GROSS NATIONAL PRODUCT*	SALES OF AEROSPACE INDUSTRY*	AEROSPACE SALES AS PERCENT OF GNP
1960	$503.7	$17.3	3.4
1961	520.1	18.0	3.5
1962	560.3	19.2	3.4
1963	590.5	20.1	3.4
1964	632.4	20.6	3.3
1965	684.9	20.7	3.0
1966	747.6	24.6	3.3
1967	793.5	27.3	3.4
1968	865.7	29.8	3.4
1969	930.3	26.1	2.8
1970	976.4	24.9	2.6
1971	1,050.4	22.2	2.1
1972	1,152.1	22.3	1.9

*Billions of dollars

Fig. 14-26. Aerospace Sales and the National Economy

recognize these advantages of air freight, and insurance rates on goods transported by air are lower than those shipped by other forms of transportation.

LOWER CRATING COSTS

As touched upon previously, standardized shipping containers have recently been developed for the air freight industry. Use of these containers for shipping results in substantial rate reductions for the shipper and enables the planes to be loaded and unloaded rapidly. This means that goods can be transported with maximum speed and at minimum cost.

THE EFFECT OF AEROSPACE ON THE ECONOMY

Aerospace has made major contributions to the economic growth of this country. For example, figure 14-26 indicates that aerospace sales averaged slightly more than 3% of the gross national product during the period 1960-1972. The dollar amount ranged from slightly over 17 billion in 1960 to over 22 billion in 1972. Additional insight may be gained from figure 14-27. During the time period 1949-72, total aerospace sales showed a dramatic increase. An analysis of these sales by product group reveals that missiles and space vehicles became a larger percentage of the total sales.

Another aspect of the effect of aerospace on the economy may be recog-

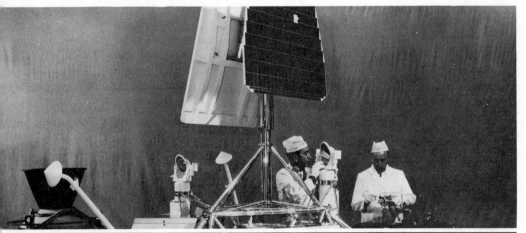

YEAR ENDING DECEMBER 31	TOTAL SALES (Millions of Dollars)	PRODUCT GROUP			
		AIRCRAFT	MISSILES	SPACE VEHICLES	NON-AEROSPACE
1949	$2,232	$2,032			$ 200
1954	12,807	10,460	$1,194		1,153
1959	16,640	9,714	5,042	$ 386	1,498
1964	20,594	8,911	5,242	4,720	1,721
1969	26,882	14,816	5,083	4,308	2,705
1972	22,313	11,883	4,764	3,048	2,621

Fig. 14-27. Estimated Sales of Aerospace Industry by Product Group   1949-1969

YEAR ENDING DECEMBER 31	ANNUAL AVERAGE AEROSPACE EMPLOYMENT TOTAL (THOUSANDS OF EMPLOYEES)	AEROSPACE PAYROLL TOTAL (MILLIONS OF DOLLARS)	AEROSPACE AS % OF TOTAL MANUFACTURING PAYROLL
1959	1,128	$7,427	8.5
1960	1,074	7,317	8.2
1961	1,096	7,809	8.7
1962	1,177	8,889	9.2
1963	1,174	9,102	9.0
1964	1,117	8,897	8.3
1965	1,133	9,502	8.2
1966	1,298	11,394	8.9
1967	1,392	12,659	9.4
1968	1,418	13,748	9.5
1969	1,354	14,150	9.0
1970	1,199	12,687	8.0
1971	969	10,746	6.7
1972	922	11,249	6.4

Fig. 14-28. Estimated Employment and Payroll in the Aerospace Industry: 1959-1969

nized by examining the employment opportunities offered by this largest industry in the United States. (See Fig. 14-28.) The number of persons employed in aerospace showed an increase from 1959 to 1969 with a decrease through 1972. However, the average individual payroll check increased 14% between 1969 and 1972. An examination of the column "aerospace as percent of total" indicates that among manufacturers, aerospace employees earned approximately 6.4% of the total manufacturing payroll.

The total assets of the aerospace industry represents a substantial proportion of the nation's total wealth. These assets increased from over 18 billion in 1968 to more than 23 billion in 1972. The aerospace industry for new plants and equipment in 1972 spent approximately 92 billion and current plans call for an annual increase of 9%. The continued high level of investment by this industry added to the stability of the U.S. economy and provided support for a high level of economic growth.

The aerospace industry also made a major contribution to the volume of goods which this country exported to other countries. Figure 14-29 reveals that from 1960 to 1969 aerospace exports experienced substantial growth. The minor role played by aerospace imports indicates that the aerospace industry was a major factor in reducing the overall deficit balance of trade. Another facet of aerospace in the world economy is indicated by Figure 14-30. Airliners of U.S. manufacture remained over three-fourths of the total in use throughout the world.

YEAR	TOTAL U.S. TRADE BALANCE*	AEROSPACE			AEROSPACE TRADE BALANCE AS PERCENT OF U.S. TOTAL
		TRADE BALANCE*	EXPORTS*	IMPORTS*	
1960	$5,369	$1,665	$1,726	$ 61	31.0
1961	6,096	1,501	1,653	152	24.6
1962	5,178	1,795	1,923	128	34.7
1963	6,060	1,532	1,627	95	25.3
1964	7,556	1,518	1,608	90	20.1
1965	5,852	1,459	1,618	159	24.9
1966	4,524	1,370	1,673	303	30.3
1967	4,409	1,961	2,248	287	44.4
1968	1,133	2,661	2,994	333	234.9
1969	1,289	2,831	3,138	307	219.6
1970	2,708	3,089	3,397	308	114.6
1971	2,014	3,823	4,196	373	
1972	6,439	3,258	3,823	565	

*In millions of dollars

Fig. 14-29. Aerospace Imports and Exports

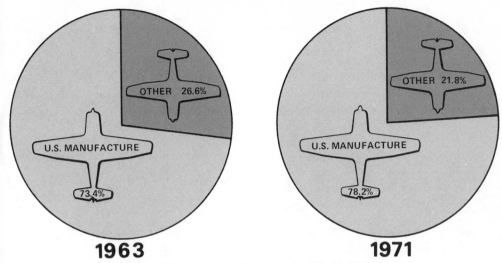

1963   **1971**

Fig. 14-30. Aircraft in Operation by World Airlines

Therefore, the economy is much different than it would be if there was no aerospace industry. Certainly it has been a stimulating force in the development of a high level of prosperity.

THE IMPORTANCE OF THE AIRPORT TO COMMUNITY GROWTH AND DEVELOPMENT

Aviation has played an important role in the economic growth of many cities. Just how important this role is to a particular community is often determined by the airport facilities which that community provides.

It is vital to recognize that adequate airports are as important to a community's economic well-being in the age of aviation as was railroad linkage or highway linkage with other communities in earlier years. An airport puts a community on the doorstep of the world. It is a means of facilitating and encouraging trade.

Due to the importance of air travel, businesses and industries are attracted to those communities which maintain efficient airports. Thus, if a community's airport facilities are deficient, economic growth will be impaired. But if they are efficient, economic growth will be facilitated and enhanced.

Some concept of the growth of the air transportation system may be comprehended from figure 14-31. Although this

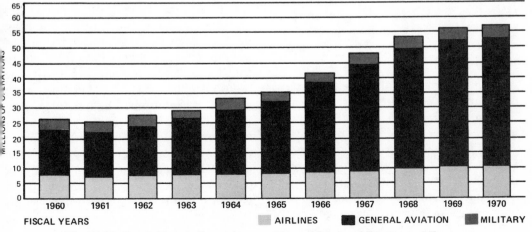

Fig. 14-31. Total Aircraft Operations at Airports Having FAA Control Towers

Fig. 14-32. Airport in Harmony with the
Environment

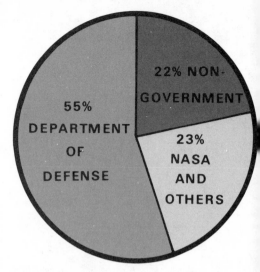

Fig. 14-33. Primary Customers of Aerospace
Products

graph covers only airports having an
FAA control tower, it does portray the
rapid growth of flying and, in particular,
general aviation. This growth could not
have been achieved if the growth of the
airport system had not kept pace.

Based upon FAA statistics, aircraft
operations at airports with FAA traffic
control service are expected to show a
rather dramatic increase during the years
1970-1980. In order for these operations
to be handled effectively, the airport
system must be expanded sufficiently to
handle the increased traffic. This can
only be accomplished by the concerted
effort of industry, government, and private individuals. Since the inception of
the Federal Aid to Airports Program
(FAAP) in 1947, the federal government
has spent nearly $2 billion for the im-
provement and expansion of airport
service.

As the advantages of an airport to a
community's economic growth have
been recognized, the ecological aspects
have also received greater recognition.
(See Fig. 14-32.) For example, one
eastern airport has a bird sanctuary at
the end of the runway. Another was
built and covered over a refuse dump. In
the south, one airport's drainage system

prevents erosion, thus, preserving the
existing water table. As seen in this
photograph, general aviation airports can
harmonize with the aesthetics of the
environment.

One of the major problems of aviation
growth at present is the airport congestion at larger terminals during peak
hours. This congestion has resulted in
airlines being delayed thousands of hours
at estimated costs amounting to millions
of dollars. Of course, no dollar figure can
be placed on the ill-will created by the
inconvenience experienced during the
delays.

Financing of airports also presents a
major problem. Traditionally, much of
the construction costs of airports, terminals, taxiways, and other facilities
directly related to ground handling of
aircraft and passengers has been left to
the local levels as has much of the
required planning to keep up with
growth expectations. The federal government has assisted some communities
with grants of money and matching
funds to help overcome some of the
immediate needs. The Airway User Act
passed by Congress was intended to
channel additional funds into airport
financing.

The construction of new runways, landing aids, and air traffic control facilities are only a part of the solution. The number of persons using the facilities of major air terminals has skyrocketed in the past few years and has created problems in baggage handling, passenger processing, ticket purchases, and lobby facilities.

Imaginative solutions to the problems cited such as monorail transportation to and from terminals, construction of satellite airports around major air terminals, better access to the terminal, computerized passenger processing, and innovations in baggage handling, are being considered. New concepts in terminal air traffic control, the use of short takeoff and landing aircraft, and vertical takeoff and landing aircraft may also be part of the solution. Whatever form the solutions take, it is important to recognize the problems that exist and to deal with the deficiencies effectively so that economic progress does not suffer from an inability to plan for the future.

THE AEROSPACE INDUSTRY

The aerospace industry has become an extremely broad and encompassing industrial effort. It can, however, be broken down into three basic areas. These are:
1. The aerospace manufacturing industry (civilian and military).
2. The air transport industry.
3. The general aviation industry.

Certain aspects of the aerospace manufacturing industry and the air transport industry have been touched upon elsewhere, therefore the major expansion in this section will be upon general aviation.

THE AEROSPACE MANUFACTURING INDUSTRY

The aerospace manufacturing industry is composed of all research, development, fabrication, assembly, and sales related to aircraft, missiles, and spacecraft and their parts and accessories. It also includes any maintenance and modification facilities.

The primary customer for aerospace products is the federal government. For example, in 1972, approximately 78% the sales of the aerospace industry were made to government agencies. Fifty-five percent of the total was in support of national defense. (See Fig. 14-33.) Despite the fact that such a large portion of the industry's products are sold to the government, the aerospace manufacturing industry remains a highly competitive, free enterprise.

THE AIR TRANSPORT INDUSTRY

The air transport industry includes the flight activities of the commercial airlines and all cargo carriers. These carriers can be classified according to the type of operation they perform:

1. Domestic Operations.
   a) Trunk carriers.
   b) Local service carriers.
   c) Helicopters.
   d) Intra-Alaska carriers.
   e) Intra-Hawaii carriers.
   f) All-cargo carriers.

2. International-Territorial Operations.
   a) Passenger-cargo carriers.
   b) All-cargo carriers.

Passenger enplanements as shown in Figure 14-34 indicate a steady growth from 1948 to 1973. The shaded forecast zone indicates an acceleration of this growth rate through 1980 with total passenger enplanements nearing 500 million. Further evidence of the growth of the air transport industry is depicted in figure 14-35.

DOMESTIC FREIGHT TON MILES

The growth rate of air freight through 1980 shows an even sharper increase than passenger enplanements. The growth rate and forecasts pictured in the preceding graphs are further evidence that aviation

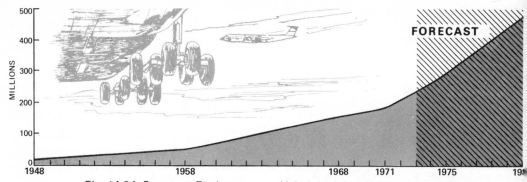

Fig. 14-34. Passenger Enplanement – U.S. Scheduled Airlines 1948-1980

is becoming the dominant form of transportation.

GENERAL AVIATION

General aviation is a heterogenous category which includes all civilian flying except that performed by the air transport industry. It includes a wide range of aircraft uses which can be classified into five categories. These include:

1. Business flying.
2. Personal flying.
3. Commercial flying.
4. Instructional flying.
5. Other flying.

The use of the airplane by private operators is the most rapidly growing portion of air transportation and the importance of general aviation can be seen by considering several statistics. The general aviation fleet of aircraft num-

bered well above 125,000 active aircraft at the beginning of 1971. This figure may be readily appreciated by comparing it with the number of air transport aircraft, reported as 2,340 in mid-1971. The general aviation fleet is expected to increase to over 225,000 by 1980. (See (See Fig. 14-36.)

It is interesting to note that general aviation aircraft flew approximately 4 times as many hours as the air transport industry in 1973. (See Fig. 14-37). The FAA forecasts almost 10 million hours will be flown in 1980 by air transport aircraft. However, the hours flown by general aviation are expected to total 42.8 million, some 4.2 times greater than air transport aircraft.

During 1972 aircraft manufacturers produced nearly 10,000 aircraft for general aviation having a total value of over 558

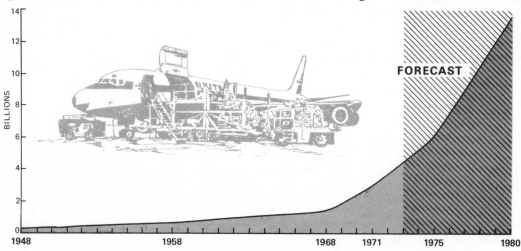

Fig. 14-35. Domestic Freight Ton Miles U.S. Scheduled Airlines 1948-1980

AS OF JANUARY 1	TOTAL AIRCRAFT
1964	85,088
1965	88,742
1966	95,442
1967	104,706
1968	114,186
1969	130,806
1970	142,000
1971	151,000
1972	160,000
1973	169,000
1974*	178,000
1975*	187,000
1980*	225,700

*Forecast

Fig. 14-36. FAA General Aviation Forecast of Aircraft

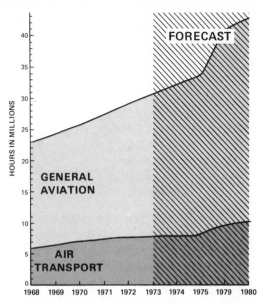

Fig. 14-37. Forecast of Hours Flown by General Aviation vs. Air Transport Aircraft

million dollars. This was a significant contribution to the economy.

The number of pilot certificates issued by the FAA is another indicator of the growth of aviation. During the period 1964-1970, issuance of pilot certificates showed marked growth with the exception of 1968 and 1969. The upward trend has resumed in the 1970's.

A rather universal expansion of general aviation activities is expected to take place in the future. It may be helpful to examine the part played by various segments of general aviation within the total realm of general aviation flying.

BUSINESS FLYING

Business flying includes those aircraft that are owned or leased by a company or an individual to transport persons or property required by the business. It does not include those used for a special purpose such as crop dusting or for air

Fig. 14-38. Business Aircraft is Most Rapidly Growing Segment

taxi. However, every type and size of general aviation aircraft is represented in this business classification except gliders and balloons.

The business flying segment of general aviation is the most rapidly growing area of civil aviation. Businesses have found private air transportation to be an extremely efficient and effective form of transportation. (See Fig. 14-38.)

Business aviation is not restricted solely to the transportation of executives. The business activities of most firms are becoming extremely diversified geographically. Many corporations have found that they are able to save valuable time by using aircraft to transport specialists and technical personnel to branch plants for special or emergency assignments. Company aircraft are also widely used in sales programs to broaden marketing areas and to maintain closer customer contact, thus generating good will between the firm and its customers.

In 1970 an estimated 7.4 million flight hours were flown for business purposes. This represents approximately 31% of the hours flown by general aviation aircraft. By 1980 business aircraft are expected to fly 12.5 million hours, a 40% increase over 1970.

PERSONAL FLYING

Personal flying includes any use of an aircraft for personal purposes not associated with a business or profession and not for hire. It also includes the flying required to maintain pilot proficiency.

At the end of World War II, thousands of servicemen who had learned to fly during the war returned home, and as a result of the training provisions of the Servicemen's Readjustment Act of 1944, many former servicemen continued to fly. Thus, personal flying increased dramatically. This boom lasted only until 1949 when the provisions of the act were changed and from 1949 to 1953 personal flying experienced a period of decline. After 1953 it began to increase once more. By 1980 the hours flown are expected to nearly double from the present 6.5 million hours per year to nearly 12 million hours per year.

There are a number of factors that could substantially boost these figures. The nation's population is rising, disposable incomes are increasing, and individuals

Fig. 14-39. Air Taxi Passengers Boarding Aircraft

Fig. 14-40. Aerial Application

are enjoying greater amounts of leisure time. This means that people should have both increased time and money to spend on air travel. Furthermore, people have become extremely mobile and have a strong desire to travel. Therefore, they are interested in utilizing the advantages of air travel.

Also, flying clubs and other forms of group or partner ownership have increased at a rapid pace and, as cited earlier, airport facilities and accommodations are being expanded and improved. All of these factors have a significant influence on personal flying and account for the rapid growth in the number of individuals who are beginning to think of the airplane as a practical travel medium.

COMMERCIAL FLYING

General aviation commercial flying encompasses three classes of service: scheduled and non-scheduled taxi service, aerial application, and instructional flying.

Air Taxi

The air taxi service has grown rapidly in the past several years. Air carriers recognize that the air taxi is an important part of the transportation system and that they are filling a void in the service to small communities that do not have enough traffic for scheduled airline service. The airlines also recognize that the air taxi provides an important link between air carrier airports and outlying areas. (See Fig. 14-39.)

Furthermore, some businesses are finding the air taxi more flexible, economical, and convenient that maintaining their own fleet of aircraft. Therefore, a number of firms are reducing the size of their own fleets and are beginning to rely on air taxi service.

Aerial Applications

The airplane has found application as a working vehicle and has replaced a number of farm implements and ground vehicles in certain phases of food and fiber production and health control. Airplanes have been used in the distribution of chemicals and seeds for agricultural purposes and even in reforestation and insect control. (See Fig. 14-40.)

Among the ·uses of general aviation aircraft in agriculture and forestry are insect and plant disease control, weed and brush control, animal pest control, application of fertilizer, seeding defoliation, restocking of fish and wild life, cloud seeding, and the production of air turbulence for frost prevention, drying fruit and athletic fields, and harvesting fruit and nuts.

One of the first uses of aircraft for aerial application was crop dusting in 1918. However, due to the high cost of application and the inefficient methods and chemicals used, the growth of this field was slow. After World War II many techniques developed for military use were found to be applicable to civilian needs. Also, new and more effective chemicals were developed and a large number of wartime trained pilots were available for hire.

Therefore, after World War II, the use of aerial spraying grew rapidly and the entire field of aerial application expanded. With the increasing industrialization of agriculture, it is expected that the demand for aerial application will sustain continued growth.

Flight Instruction

As the number of pilots has grown through recent years, so has the requirement for formal instruction to develop piloting skills. Training hours flown have shown a steady increase since a low point in 1953. In 1969 and 1970, however, flight training reflected a downward trend in response to a nationwide economic decline. Rigorous stimulation from within the industry and an improving economic forecast is expected to cause an increase in the number of hours devoted to flight training throughout the 1970's. (See Fig. 14-41.)

OTHER USES

Other activities include research and development, demonstration, sport parachuting, and ferrying flights. Also included are pipeline and highway patrol, aerial surveying, emergency and rescue operations, advertising, photography, and fire fighting. The number of planes and the number of flight hours recorded for this segment of general aviation are not significant in comparison to the other categories.

History has witnessed the growth of the air and space age as a dominant influence in modern life. A realization of this vast impact of aerospace on modern life is a realistic goal for the well-educated person living in the dynamic and exciting years ahead.

Fig. 14-41. Modern Flight Instruction

chapter 15

CAREERS IN AVIATION AND AEROSPACE

INTRODUCTION

The aviation industry is an exciting and dynamic world, surrounded by a magnetism that captivates and challenges the people who are attracted to it. The aerospace industry is the nation's number one industrial employer, and while everyone is familiar with the role of the pilot in aviation, it is important to remember that the greatest majority of job classifications and opportunities are available in ground-based operations which support the flight activities. *The first section of this chapter will be devoted to presenting a brief introduction to non-pilot aviation careers available in the industry. The next section will survey the job opportunities available to pilots, and then the chapter will conclude with the steps that an aerospace student can follow to prepare for an aviation career.*

NON-PILOT AVIATION CAREERS

MANUFACTURING AND PRODUCTION CAREERS

The aerospace industry is engaged in the design, development, manufacture and operation of aircraft, missiles, spacecraft, propulsion and navigational guidance systems, and other aeronautical and astronautical systems. (See Fig. 15-1.)

Job opportunities in this industry can be divided into four major groups: scientists and engineers, administrative and support personnel, technicians, and production workers.

Fig. 15-1. Aircraft Manufacturing Plant

SCIENTISTS AND ENGINEERS

Scientists and engineers include all persons engaged in technical areas such as aerodynamics, avionics, ceramics, chemistry, cryogenics, mathematics, meteorology, metallurgy, physics, physiology, and psychology. For example, a scientist might be involved in the development of special metals for a spacecraft development project.

At least half of the scientists and engineers in the aerospace industry are directly involved in research and development work on aircraft designs, propulsion systems, instrumentation, flight testing, and aircraft structural reliability. (See Fig. 15-2 and Fig. 15-3.) The remainder are involved in production planning, quality control, tool design, and similar fields.

AVIATION MANAGEMENT

The roll of a manager in the aviation industry is basically the same as that of a manager in any other industry. He must administer, establish, and improve company policy, strategy, objectives, and organization and operation techniques. Aviation management fields include: fixed-base operations, airport management, aircraft manufacturing, and airline management. (See Fig. 15-4.)

A great need for aviation managerial and administrative personnel is projected in order to keep pace with the anticipated growth of all sectors of the industry. The forecast increase in the demand for professionally-trained managers also stems from the need to upgrade those managers and management personnel who have extensive knowledge of the aviation aspects of their job, but who lack a thorough understanding of basic business administration principles and procedures.

Managerial positions, especially in the area of fixed-base operations and aircraft manufacturing, will increase in the future as general aviation production and development increases. In addition, air taxi, charter operations, and aviation maintenance facilities will offer an increasing number of management positions.

TECHNICIANS

Science and engineering technicians work in all of the fields previously mentioned, in addition to serving the aerospace industry as draftsmen, technical writers, and illustrators. Electronic technicians, on the other hand, install and maintain the indispensable and complex components of the federal airways network, including

Fig. 15-2. Aircraft Engineering Departments Utilize a Variety of Engineering Specialists

Fig. 15-3. Engineers Reviewing Aircraft Design Layout

Fig. 15-4. Airport Manager

radar, radios, computers, and teletypewriters. At busy airports, an electronic technician may be called upon at unusual hours or during adverse weather conditions to maintain the airport's electronic aids which are important to the safe flow of air traffic. Other electronic technicians specialize in the design, development, and evaluation of new types of equipment for the federal airways, and some are even selected for airborne instrumentation duty in flight test aircraft to collect data for engineering purposes. (See Fig. 15-5.)

Several factors indicate that the long range demands for technicians will significantly exceed the supply. In addition, the growth of air transportation, the increasing complexity of modern aircraft, and the trend toward automatic and semi-automatic electronic systems for air traffic control, collision avoidance, all-weather landings, and area navigation will probably result in an increased number of employment opportunities for technically oriented personnel.

Fig. 15-5. Electronic Technician

PRODUCTION WORKERS

Production workers, including working foremen and all non-supervisory workers engaged in fabricating, processing, assembling, inspecting, receiving, storing, handling, packing, warehousing, shipping, and many other similar occupations, comprise the greatest majority of aerospace industry employees. (See Fig. 15-6.)

Fig. 15-6. Inspector on Production Line

AIRCRAFT MAINTENANCE CAREERS

AVIATION MECHANICS

The preceding paragraphs introduced the occupations which are primarily concerned with the design and production of aerospace products. The following paragraphs concern jobs that are involved with the maintenance of these products. (See Fig. 15-7.)

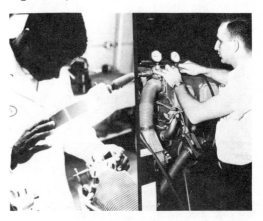

Fig. 15-7. Mechanics Working on Small Aircraft Engine

In this group are mechanics who have the important responsibility of keeping aircraft airworthy and in safe and efficient flight condition. They service, repair, overhaul, and test aircraft, airframes, engines, propellers, aircraft systems, electronic equipment, and aircraft instruments. Thousands of mechanics are employed by air taxi operators, aerial applicators, aircraft manufacturers, and the Federal Aviation Administration. Likewise, mechanics employed by the airlines perform either line maintenance work at terminals or major repairs, overhauls, and inspections at an airline's overhaul base. (See Fig. 15-8.)

Line mechanics work under the pressure of keeping the airlines on schedule and have the responsibility of deciding whether an aircraft is airworthy. In addition, since the maximum time between overhauls and mandatory retirement life cycles for all airliner components are care-

Fig. 15-8. Line Mechanic

fully calculated by engineers, a certain amount of "built in" work is constantly demanding attention. (See Fig. 15-9.)

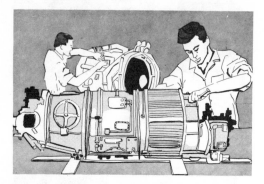

Fig. 15-9. Airline Engine Overhaul Base

Many large corporations utilize extensive fleets of aircraft to assure rapid mobility for their key personnel, and thus they require the services of many mechanics who are also pilots. These mechanics/pilots are charged with the responsibility of supervising the maintenance work performed on the corporation's aircraft, even though the actual work is usually performed at a fixed-base operation or specialty repair shop.

Aviation mechanics can be either licensed or unlicensed. A licensed mechanic holds either an Airframe, Powerplant, Airplane and Powerplant (A & P) Mechanic's Certificate, or a Repairman's Certificate from the Federal Aviation Administration (FAA). The A & P Certificate allows a mechanic to work on any part of an aircraft's engine, systems, or airframe. (See Fig. 15-10.)

FAA Repairman's Certificates are issued on the basis of specific work experience and allow a mechanic to work on those

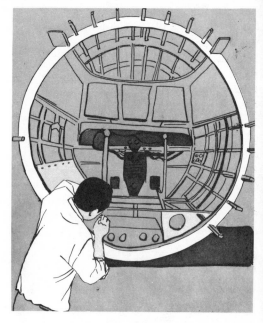

Fig. 15-10. Airframe Mechanic

parts of the aircraft that his certificate specifically allows, such as radios or instruments. (See Fig. 15-11.) If a repairman or A & P mechanic works on transmitting equipment aboard an aircraft, he must hold at least an FCC Second Class Radiotelephone License. However, repairmen who work only on autopilots, navigation equipment or receivers do not necessarily have to have an FCC license.

Fig. 15-11. Radio Technician

Unlicensed mechanics and apprentice mechanics may work on various parts of the airplane under the supervision of a licensed mechanic who must sign his approval of the work before the aircraft or its equipment are considered airworthy. The military services employ many aviation mechanics to work on military aircraft at Army, Navy, and Air Force aviation installations all over the world. These positions do not normally require an FAA mechanic certificate and may not even require enlistment in the military services.

Mechanics represent the largest job classification in civil aviation. The future demand for trained mechanics will experience short-range fluctuations corresponding to those of the economy. However, long-range forecasts anticipate a substantial growth pattern.

The following principle factors will contribute to an increasing demand for aviation mechanics: first, the demand for air transportation will necessitate a growing civilian aviation fleet consisting of aircraft of increasing complexity; second, the relative ages of mechanics now in the occupation point toward high annual retirement rates in the 1970's and 1980's. Of the FAA certificates held by mechanics in 1965, 78% were held by individuals 35 years of age or older. Annual retirements are expected to increase from about 1,200 in 1964 to about 3,300 in 1980.

AIRLINE FLIGHT SERVICE CAREERS

Air transportation has mushroomed into a highly competitive and diversified 24-hour operation. Within this world-wide network, there are thousands of job classifications requiring all levels of education and training from a high school education to a college degree and several thousand hours of flight experience. The following job classifications are only a partial list of the available positions: flight dispatcher, meteorologist, schedule coordinator, station manager, reservationist, teletypist, ground hostess, skycap, air cargo agent, passenger service agent, representative, sales manager, serviceman, maintenance mechanic, food service em-

Fig. 15-12. Airline Ticketing

ployee, planner, engineer, and clerical and administrative personnel. (See Fig. 15-12.)

The demand for flight service personnel is geared directly to the growth of the airline industry. In the past, the industry's expansion has exceeded the most optimistic expectations. This has been largely a result of the jet revolution. The growth trend is forecast to continue well into the future, although fluctuations with the economy are anticipated. The underlying growth factors are twofold — society's increasing mobility and the production of the next generation of jet aircraft.

AVIATION CAREERS IN GOVERNMENT

A major source of aviation careers lies in jobs with federal, state, and local government agencies. Civil aviation and federal government careers for both men and women are found within the Department of Transportation, Federal Aviation Administration, and a growing number of other federal departments and agencies. Civilians are also employed in aviation jobs by the military services. All of these aviation jobs come under the federal civil service.

The largest number of aviation jobs within the federal government is found in the Federal Aviation Administration. The FAA is charged with the administration and enforcement of all federal aviation regulations to insure the safety of air transportation. The FAA also promotes, guides, and assists in the development of a national system of civil airports. In addition, the FAA provides pilots with flight information ranging from flight planning to takeoff and landing procedures. Another function of this administration is to develop, staff, and maintain air navigation aids and air traffic control facilities. Major career opportunities within the FAA are: air traffic control specialists, flight service specialists, aviation safety officers, pilots, and engineers.

AIR TRAFFIC CONTROL SPECIALISTS

Air traffic control specialists work at air traffic control towers and air route traffic control centers across the country. The air traffic control specialist is responsible for the smooth and orderly flow of air traffic. For example, a pilot departing an airport that has a control tower communicates with the tower for taxi and takeoff instructions and clearances. On large airports, the tower operator is invaluable in assisting the pilot through the complex systems of runways and taxiways. The tower operator also directs the flow of air traffic departing from and arriving at the airport. (See Fig. 15-13.) This is accomplished partially by visual observation from the control tower cab, referred to as *local control*, and also by radar, which is usually located in a room below the control tower cab.

Fig. 15-13. FAA Tower Controllers

Air traffic control specialists in the radar room handle incoming and departing air traffic. They also operate runway lighting systems, airport traffic direction indicators, and prepare reports on air traffic communications. They must be able to remember the types and speeds of all aircraft under their control, as well as the different airplanes' positions in relation to other aircraft or navigational aids.

Air traffic control specialists at FAA air route traffic control centers (ARTCC), are primarily responsible for the control of aircraft that are flying on instrument flight plans (a flight plan designed for use in poor weather conditions). When an airplane that is operating on an instrument flight plan departs the area controlled by a tower, the departure controller in the tower relays the control of the airplane to the air route traffic control center (ARTCC). (See Fig. 15-14.) The ARTCC operator communicates with the pilot and identifies the location of the airplane on his radar scope. (See Fig. 15-15.) The controller then gives the pilot instructions, air traffic clearances, and advice regarding flight conditions along his flight path. On a long cross-country flight, the control of an airplane may be handed over from one ARTCC operator to another as the airplane flies through the various sectors of the air route traffic control system. Finally, as the airplane nears its destination, its control is transferred from the air route traffic control center to the approach controller at the destination tower.

Air traffic control specialists are usually trained at the Federal Aviation Administration Academy at Oklahoma City, and then later they receive on-the-job training. There is a shortage of air traffic control specialists and this manpower gap is expected to grow in the 1970's.

It is estimated that during the 1970's an average of over 3,000 additional air traffic control specialists will be needed annually to compensate for growth, replacement, and attrition. Additionally, congested airways are threatening to

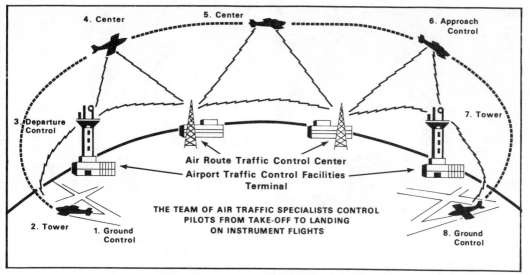

Fig. 15-14. Air Traffic Control Sequence

seriously curtail the continued needed growth and services of the nation's air transportation system. In response to this dilemma, the Airport and Airway Development Act of 1970 was enacted by Congress with the objective of substantially expanding airport and air traffic control facilities and capabilities.

FLIGHT SERVICE SPECIALISTS

The specialist at FAA flight service stations provides preflight, inflight, and emergency assistance upon request to all pilots. Unlike the air route traffic control center, the FSS does not control air traffic, but does act as a primary source of flight information to all types of aviation activities. Flight service stations are strategically located at airports as well as many remote areas across the country.

Fig. 15-15. Air Traffic Controller

The system is engineered so that pilots on flights anywhere in the United States will be within communication range of a flight service station. At some remote locations, the flight service specialist is responsible for weather observations. Actual forecasting, however, remains the responsibility of the Weather Bureau.

Flight service stations are interconnected by a network of teletype communications. The teletype system is used to relay information regarding weather forecasts and reports, advisories for pilots, and flight plan information. The majority of the non-airline and non-military weather briefings and flight plans is handled by a flight service station. A pilot may take advantage of weather briefings and flight plan service by calling the FSS on the telephone or, once airborne, by using the aircraft radio. Pilots may also contact a flight service station en route to update weather information or to make amendments to their flight plans. (See Fig. 15-16.)

U.S. WEATHER BUREAU

Aviation is one of the largest consumers of weather information. Flight and weather are so interrelated that many people in aviation look upon the weatherman as a member of the aviation team. (See Fig. 15-17.) Thus, the meteorologist deserves

Fig. 15-16. Flight Service Station Specialist
at Radio Console

Fig. 15-17. Weatherman at Work

at least some mention in any discussion of vocations in aviation, even though his functions are not *entirely* for the benefit of the aviation community.

AVIATION SAFETY OFFICERS

The Federal Aviation Administration employs aviation safety officers for manufacturing inspection, aviation electronics inspection, maintenance inspection, and flight operations inspection.

The manufacturing inspector checks to see that the design, materials, and methods of producing civil aircraft, aircraft engines, systems, and equipment are according to Federal Aviation Regulation safety standards and that the finished products are airworthy. (See Fig. 15-18.) He works with aircraft manufacturing engineers from the blueprint stage through the entire manufacturing process, supervises activities of FAA designated inspection representatives, and participates in civil aircraft accident investigations.

The aviation electronics inspector assures compliance with air safety rules relating to airworthiness, adequacy, and proper operation of communication and electronic navigation equipment aboard the aircraft. The maintenance inspector checks for compliance with air safety rules relating to maintenance of aircraft,

inspecting maintenance procedures, methods, spare parts stock, employee training programs, instructional materials, and maintenance manuals. He inspects airworthiness of airframes, engines, propellers, and components of the aircraft as well as repairs, alterations, and general maintenance.

NATIONAL TRANSPORTATION SAFETY BOARD

The National Transportation Safety Board (NTSB) is charged with the dual aviation responsibility of safety and accident investigation. NTSB personnel conduct interviews, examine aircraft parts, instruments, engines, maintenance records, and flight records to determine

Fig. 15-18. FAA Safety Officer At Work

probable causes of airplane accidents. Thus, this organization is in an ideal position to recommend the use of new safety procedures and equipment.

An on-duty accident investigator must be prepared to be dispatched, with short notice, to very remote areas of the country. In some cases, this requires the use of helicopters or rough terrain vehicles to finally reach the scene of the investigation.

CIVIL AERONAUTICS BOARD

The Civil Aeronautics Board is another federal agency concerned with aviation, particularly commercial air transportation. It employs lawyers to develop regulations, to make decisions about airline routes and fares, and to represent the U.S. in international airline legal discussions. The CAB also employs economists and statisticians who gather and interpret statistics so decisions can be made about changes in airline fares, routes, and services.

ENGINEERS

The FAA, as well as the National Aeronautics and Space Administration and the Department of Defense, employs engineers of all specialities to work on research and development problems in aviation, such as V/STOL (very short takeoff and landing) aircraft, aircraft noise, the sonic boom, hypersonic aircraft, and new equipment to increase aviation safety. Engineers also provide guidance in the design, construction, operation, and maintenance of airports.

OTHER FAA EMPLOYEES

A wide scope of additional professions are represented in the FAA. These include airport safety specialists, urban planners, economists, mathematicians, physicians, program officers, statisticians, psychologists, management analysts, and budget analysts. The FAA also employs lawyers to write Federal Aviation Regulations, to interpret them, and to represent the FAA in legal controversies.

AIRPORT CAREERS

Airports may be privately owned by a single operator or owned and operated by a city, county, regional, or interstate government authority. Airports are usually operated by a director or manager responsible either to the owners of the airport or the government authorities. Depending on the size of the airport, the airport director may have assistants such as assistant director, engineer, controller, personnel officer, maintenance superintendent, or supporting office workers such as secretaries, typists, and clerks.

FIXED-BASE OPERATIONS

The airport-based businessman who provides aviation services for pilots and aircraft is usually referred to as a fixed-base operator (FBO).

The services he offers may include aircraft repair, aviation fuel sales, flight school operations, new and used aircraft sales, and air taxi or charter flights. (See Fig. 15-19.) The fixed-base operator is a typical small businessman in the aviation industry and, depending on the size of his operation, may employ aviation mechanics, flight instructors, and aircraft salesmen.

The fixed-base operator also employs linemen, or ramp servicemen, who greet and service arriving aircraft. (See Fig. 15-20.) Normal servicing includes filling the tanks with fuel, checking the oil, emp-

Fig. 15-19. Fixed-Base Operation

Fig. 15-20. Lineman Directing Pilot

OTHER AVIATION – RELATED CAREERS

The aviation industry has many associated organizations requiring physicians, lawyers, salesmen, writers, photographers, researchers, publishers, and administrators who have specialized knowledge backgrounds plus some experience in aviation.

Their services are utilized by aviation trade journals, training manual publishers, flight safety research organizations, insurance companies, trade or professional and industrial aviation organizations, such as the Air Transport Association (ATA), American Institute of Aeronautics and Astronautics (AIAA), the Aerospace Industries Association (AIA), National Business Aircraft Association (NBAA), Airline Pilots Association (ALPA), Aircraft Owners and Pilots Association (AOPA), General Aircraft Manufacturers Association (GAMA), and many others.

tying ashtrays, vacuuming the interiors, washing the windshield, and reporting to the owner any signs of incipient trouble such as oil leaks. Linemen are usually young men who are interested in aviation. Working as a lineman, they can begin their aviation careers by building up experience with aircraft under the guidance of the fixed-base operator.

PILOT CAREERS

In the early days of aviation, it appeared as though the pilot's objective was to demonstrate his own abilities, or to prove the capacity of the airplane to do a given job. That job might have been to deliver the air mail or stage air shows in city after city. The early pilot's personal requirements were similar to those of a daredevil. In contrast, however, today's career pilot is considered to be a professional who has years of specialized training. (See Fig. 15-21.) His responsibilities include the utmost regard for safety in the conduct of flight operations. The successful pilot is concerned with accomplishing his job as efficiently and safely as possible.

A pilot's duties include the responsibility for determining the airworthiness of his aircraft, the preparation of his aircraft for flight, the safe and efficient operation of the aircraft; and, in some cases, the management of the crew and complex air-

Fig. 15-21. Today's Professional Pilot

borne systems. The pilot in command is the final authority in regard to the operation of his aircraft.

The modern airplane performs a wide variety of tasks, thus providing many different types of flying opportunities. Flight activities are divided into the following categories:

1. general aviation,
2. airline,
3. military, and
4. aerospace.

There are many reasons for learning to fly, only one of which is to obtain employment as a pilot. In fact, statistics indicate that less than 10% of all active pilots are engaged in flying as a career. Opportunities for those who seek to attain non-career goals are outstanding and are limited only by desire and personal resourcefulness.

On the other hand, opportunities for a student seeking a career as a professional pilot are dependent upon many variable factors. Most current information regarding long-range employment forecasts for civil aviation represents a favorable picture in terms of pilot demand in general aviation and a somewhat more modest outlook with regard to the airlines. A summary of the manpower projection made by the Bureau of Labor Statistics is given in figure 15-22.

The overview of demand, however, must be tempered by the existence of a potential which equals or exceeds the projected demand. Most employers, especially the airlines, agree that there is no shortage of available pilots to meet the short term growth requirements of the industry. Pilots with specialized skills such as agricultural flying, of course, would be an exception.

Part of the adequate pilot supply comes from the number of students enrolled in flight training. This number is expected to more than double from the present level of 200,000 over the next decade. Of course, only a small percentage of these students will have flight career aspirations; and, of these, only a portion will extend their aspirations and training to the point that flying will be their principal profession.

GENERAL AVIATION PILOTS

INSTRUCTIONAL FLYING

General aviation activities include instructional, commercial, business, and pleasure flying. Many pilots are employed as flight instructors due to the ever-increasing interest in training for aviation careers and pleasure flying. (See Fig. 15-23.)

	PROJECTED MANPOWER REQUIREMENTS 1968-77		
TYPE OF PILOT	Growth	Retirements and Deaths	Total Need
AIR CARRIER	18,000	4,800	22,800
GENERAL AVIATION	30,700	5,500	36,200
EXECUTIVE	16,500	2,700	19,300
AIR TAXI	9,600	1,500	11,100
AGRICULTURAL	700	300	900
INDUSTRIAL	600	200	800
INSTRUCTIONAL	3,000	700	3,700
OTHER*	300	100	400

* LESS THAN 50
Note: Individual parts may not add to totals due to rounding.

Fig. 15-22. Projected Requirements for Civilian Professional Pilots

Fig. 15-23. Flight Instructor and Student

Some instructors make their careers in aviation education, while others move into commercial, business, or airline flying positions as experience is gained. The type of training an instructor may give varies from teaching a beginning student to fly in a light, two-place trainer to providing advanced instruction to other professional pilots in jets or helicopters.

COMMERCIAL FLYING

Commercial flying in general aviation includes a wide variety of aircraft and types of flying. Some of these include:

1. *Air taxi operations* — flying passengers or cargo in light to medium weight aircraft on scheduled commuter or charter flights.
2. *Aerial patrol* — powerline, pipeline, highway, forest, and border patrol;
3. *Aircraft sales* — the demonstration and sales of the many types of aircraft;
4. *Ferry* — transport of aircraft from factory to dealer, or from dealer to customer;
5. *Aerial survey and photography* — of commercial property, highway right-of-ways, mapping, etc;
6. *Advertising* — towing banners and lighted signs;
7. *Sightseeing* — over nature's wonders and metropolitan areas of interest;
8. *Ambulance* — mercy flights for the ill or injured;
9. *Construction support* — usually with helicopters for powerline or pipeline construction and delivery of critical parts and crews;
10. *Agricultural support* — aerial seeding, pollination, insect and weed control;
11. *Test* — experimental and production test flying of aircraft and systems;
12. *Atmospheric research* — investigation of the Earth's atmosphere; and
13. *Missionary support* — flying missionaries and supplies in the remote areas of the world.

A helicopter pilot might be employed to fly a regular schedule carrying workers and supplies to offshore oil rigs or to fly accident victims to the hospital heliport. He might, at some time, be called upon to lift heavy loads to tops of buildings or remote mountain sites, rescue people stranded by floods, carry smoke jumpers to fight forest fires, or even deliver Santa Claus to a shopping center parking lot. (See Fig. 15-24.)

BUSINESS-RELATED FLYING

Corporate pilots fly the aircraft owned by a particular business or industrial firm on a full-time basis with no additional responsibilities except the supervision of the aircraft's maintenance. These pilots transport company executives to branch plants for business conferences, to sales meetings, and occasionally to secluded vacation spots. Many large corporations have large fleets of executive aircraft and employ many pilots as well as other flight crew members. (See Fig. 15-25.)

Many pilots are employed by smaller companies as business pilots, with accompanying additional responsibilities such as advertising, research, or sales management. Also, salesmen, lecturers, musicians, free-lance writers, and photographers, as well as many other people in similar vocations who must cover a wide territory in minimum time, have found that owning their own plane is the most efficient means of transportation, especially when compared to other surface-based modes.

Fig. 15-24. Helicopter Carrying Heavy Load

Fig. 15-25. Corporate Pilot

PLEASURE FLYING

Thus far, our discussion has been devoted to flying in connection with a vocation. By far, the largest segment of the aviation community is comprised of people who have turned to flying as a means of recreation. An airplane allows a family, group, or even a single individual to stretch a weekend into a vacation. For instance, imagine a quick trip to another city to visit relatives, then off to the mountains for some fishing or skiing. With an airplane available to provide fast, efficient transportation, you can still make it back to the city in time for the opening of a new play. (See Fig. 15-26.)

AIRLINE PILOTS

One of the most popular pilot vocations is that of the airline pilot. The airline pilot usually has excellent working conditions, good pay, broad benefits, and flies some of the best equipment available. (See Fig. 15-27.) In a three-man pilot crew, the positions are Captain, First Officer (Co-Pilot), and Second Officer (Flight Engineer).

In days past, the Second Officer was referred to as the Flight Engineer, and often was not a rated pilot. Today, however, the Second Officer is normally a rated pilot. After a prescribed period of training and experience, he advances to the Co-Pilot or First Officer position, and eventually to the Captain's seat.

The airline pilot plans each flight with the airline's flight dispatcher and meteorologist and checks the aircraft weight, fuel supply, alternate destination, weather, and route. The captain briefs his crew, checks takeoff procedures, and satisfies himself that the airplane is airworthy. He is in complete command of the airplane, and, therefore, responsible for the safety of the airplane, its passengers, crew, and cargo. The airplane he flies may range from a twin-engine jet, with a few passengers on a 100-mile hop, to a four-engine jumbo jet crossing the ocean with more than 350 passengers.

The increase in volume of airline passengers and the explosive growth of air cargo

Fig. 15-26. Pleasure Flying

Fig. 15-27. Airline Pilots on Flight Deck

service will require a larger air transport fleet and a rise in the number of additional flight officers required.

MILITARY PILOTS

The military pilot is a specialist in management of his aircraft, its systems, and weaponry. (See Fig. 15-28.) A military pilot receives years of valuable training in the most sophisticated aircraft available. Ever increasing earnings and rapid advancement make military flying a very attractive career field. Because of the broad experience and quality of training, the former military pilot is in demand by the airlines, providing at least 50% of the new-hire airline pilots. In addition, military flying offers the widest variety of equipment — everything from a two-place, single-engine trainer to a massive cargo or transport aircraft, not to mention helicopters, single-seat fighters, and research aircraft.

Fig. 15-28. Military Pilot in Training

SPACE PILOTS

Space travel has captured the imagination of the world. A very limited number of pilot positions are found in this area, the most specialized type of flying in aviation. (See Fig. 15-29.) Many of today's space pilots have an extensive background in military aviation, although some pilots have a general aviation background prior to their military career. Space pilots are employed by the National Aeronautics and Space Administration and have very attractive pay and career opportunities.

Fig. 15-29. Astronaut Walking in Space

PREPARING FOR A CAREER IN AVIATION

ACQUIRING A BROAD EDUCATIONAL BACKGROUND

Up to this point, this chapter has described the various jobs available in the aviation industry. The next logical question is: "How does a student prepare for an aerospace career?" Opportunities are waiting for those who begin their preparations during their high school and college years. The aerospace industry is a dynamic and growing enterprise which will require many employees in the future. The personnel chosen, however, will be those who planned ahead during the years they were still in school and adequately prepared themselves to meet the demands of the industry. Suggested school subjects for various careers are shown in figure 15-30. In addition, figure 15-31 shows aptitudes required in the different aerospace occupations.

FLIGHT TRAINING

Flight instruction may be provided by an individual flight instructor or a fixed-base

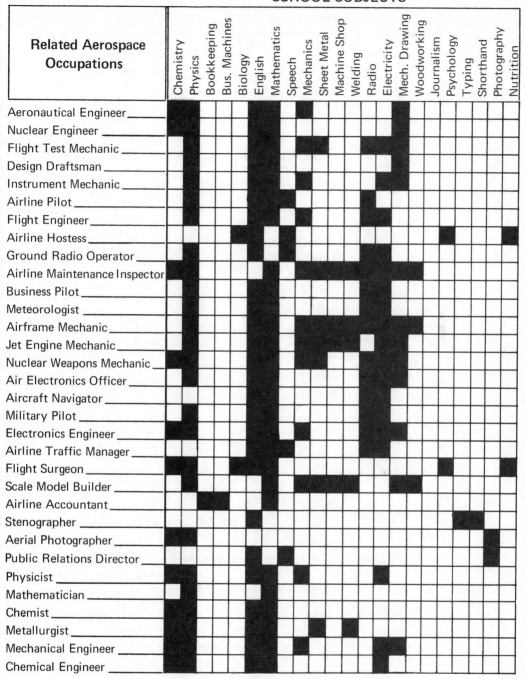

Fig. 15-30. School Subjects and Aerospace Careers

operator. In addition, several colleges are now offering flight training as part of a professional pilot or aerospace career program.

Flight schools provide two types of instruction — ground school and flight training. These forms of instruction usually may be taken separately or simultaneously and involve such areas as aerodynamics, meteorology, navigation theory and techniques, radio communication principles, and FAA regulations. Flight instruction is offered at various levels of

APTITUDES	RELATED VOCATIONAL ACTIVITIES	SELECTED AEROSPACE AGE CAREERS
Mechanical	Equipment Development Aircraft Maintenance Machinery Repair	Aeromechanical Engineer Astronautical Engineer Production Technician Powerplant Mechanic Instrument Repairman Design Engineer
Verbal	Speaking and Writing Giving Instructions Persuasive Activities	Flight Instructor Public Relations Director Air Traffic Controller Military Information Specialist Airline Sales Representative
Scientific	Research and Invention Experimentation Scientific Investigation	Aeronautical Engineer Physical Chemist Research Metallurgist Astrophysicist Aeromedical Lab Technician Process Engineer
Manipulative	Equipment Operation Machinery Control Instrument Supervision	Aircraft Pilot Flight Engineer Radar Specialist Machine Tool Operator Production Expediter
Numerical	Mathematical Calculations Arithmetic Reasoning Computational Activities	Data Processing Engineer Aircraft Navigator Research Mathematician Industrial Accountant Airline Statistician
Administrative	Managerial Activities Supervisory Responsibility Secretarial Duties	Research Project Director Management Engineer Airport Operator Military Administrative Officer Stenographer Administrative Engineer
Social	Service, Advice, and Assistance to Individuals and Groups	Aviation Psychologist Personnel Manager Flight Nurse Training Director Airline Hostess
Artistic	Self-expression Through Design, Drawing, and Other Creative Skills	Design Engineer Draftsman Airline Architect Photographic Technician Technical Illustrator Scale Model Builder

Fig. 15-31. Aptitudes and Aerospace Careers

proficiency until the desired licenses and ratings are obtained. The regulations which control the licensing of pilots are contained in Federal Aviation Regulation Part 61, found in Chapter 6 of this textbook. Pilot certificates and ratings, along with accompanying privileges, are described in figure 15-32.

CERTIFICATE OR RATING	
PRIVATE PILOT CERTIFICATE	**COMMERCIAL PILOT CERTIFICATE**
ELIGIBILITY Age: 17 years (minimum) Other: Ability to read, speak, and understand the English language. (FAR 61.103)	**ELIGIBILITY** Age: 18 years (minimum) Other: Must be able to read, speak, and understand the English language. (FAR 61.121)

AERONAUTICAL KNOWLEDGE, EXPERIENCE, AND SKILL REQUIREMENTS
 	
Knowledge:
 	
Applicant must be able to pass FAA Written Examination covering the following:
    Federal Aviation Regulations
    Cross-Country Flying
    Meteorology
    General Safety Practices
    (FAR 61.105)
 	
Experience:
 	
40 hours including flight instruction and solo time. (FAR 61.109)
 	
Skill:
 	
Applicant must pass practical test on applicable procedures and maneuvers. (FAR 61.43)
	AERONAUTICAL KNOWLEDGE, EXPERIENCE, AND SKILL REQUIREMENTS
	
	Knowledge:
	
	Applicant must be able to pass FAA Written Examination covering:
	    Federal Aviation Regulations
	    Basic Aerodynamics
	    Airplane Operations
	    Principles of Safe Flight Operations
	    (FAR 61.125)
	
	Experience:
	
	250 hours of flight time. (FAR 61.129)
	
	Skill:
	
	Applicant must pass practical test on applicable procedures and maneuvers. (FAR 61.43)

MEDICAL STANDARDS*
 	
Third Class Medical Certificate which includes the following criteria:
    Vision must be 20/50 in each eye without correction or if vision is less than 20/50, then each eye must be correctable to 20/30 or better.
 	
All other First Class Medical Certificate requirements apply except a Third Class Applicant must be able to hear a whisper from 3 feet. (See ATR certificate.
	MEDICAL STANDARDS*
	
	Second Class Medical Certificate which includes the following criteria:
	    Vision must be 20/100 in each eye, separately correctable to 20/20 or better with the ability to distinguish aviation red, green, and white.
	
	All other First Class Medical Certificate requirements apply except Second Class Applicant must be able to hear a whisper from 8 feet. (See ATR certificate.)

PRIVILEGES AND LIMITATIONS Privileges: Private Pilot may act as pilot in command of an aircraft in connection with any business if passengers or property are not carried for compensation or hire. Private Pilot may share the operating expenses of a flight with his passengers. If he is an aircraft salesman and has logged at least 200 hours of flight time, he can demonstrate an aircraft in flight to a prospective buyer. Limitations: Private Pilot may not act as pilot in command of an aircraft that is carrying passengers or property for compensation or hire; nor may he, for compensation or hire, act as pilot in command of an aircraft. A Private Pilot may fly only by visual reference unless instrument rated.	**PRIVILEGES AND LIMITATIONS** Privileges: Commercial Pilot may act as pilot in command of an aircraft for compensation or hire (FAR 61.139) Commercial Pilot may act as pilot in command of an aircraft that is carrying passengers or property for compensation or hire. (FAR 61.139) Limitations: An applicant must hold an instrument rating (airplane), or the commercial pilot certificate that is issued is endorsed with a limitation prohibiting the carriage of passengers for hire in airplanes on cross-country flights of more than 50 nautical miles, or at night. (FAR 61.129)
* *If an applicant for a medical certificate is unable to meet some of the requirements listed above, it is possible, in some cases, to receive a waiver from the FAA Aeromedical Branch.*	

Fig. 15-32. Pilot Certificates, Ratings, Privileges, and Limitations (Sheet 1 of 3)

CERTIFICATE OR RATING

FLIGHT INSTRUCTOR CERTIFICATE	AIRLINE TRANSPORT PILOT (ATP)
ELIGIBILITY Age: 18 years (minimum) Other: Must be able to read, write, and converse fluently in English. (FAR 61.183) To obtain a Flight Instructor's Certificate, an applicant must hold a commercial or airline transport pilot certificate with an aircraft rating appropriate to the flight instructor rating sought, and an instrument rating, if the person is applying for an airplane or an instrument instructor rating. (FAR 61.183)	**ELIGIBILITY** Age: 23 years (minimum) Other: Must be able to read, write, and understand the English language and speak it without accent or impediment of speech that would interfere with a two-way radio conversation. Applicant must be a high school graduate or the equivalent. Applicant must be of good moral character. (FAR 61.151)

AERONAUTICAL KNOWLEDGE, EXPERIENCE, AND SKILL REQUIREMENTS
 	
Knowledge:
 	
Applicant must be able to pass FAA Written Examination covering:
    The learning process.
    Elements of effective teaching.
    Student evaluation, quizzing, and testing.
    Course development.
    Lesson planning.
    Classroom instructing techniques. (FAR 61.185)
 	
Experience:
 	
Appropriate to accompanying certificate(s).
 	
Skill:
 	
Performance and analysis of standard flight training procedures and maneuvers appropriate to the flight instructor rating sought. (FAR 61.187)
 	
MEDICAL STANDARDS*
 	
Appropriate to accompanying certificate.
	AERONAUTICAL KNOWLEDGE, EXPERIENCE, AND SKILL REQUIREMENTS
	Knowledge:
	
	ATP applicant must be able to pass FAA Written Examination covering: Federal Aviation Regulations, Fundamentals of Air Navigation, Meteorology, Air Navigation Facilities Used on Federal Airways, Radio Communication Procedures, Basic Principles of Loading and Weight Distribution and their Effects on Flight Characteristics (FAR 61.153)
	
	Experience:
	
	Commercial Pilot Certificate with at least 1,500 hours of flight time. (FAR 61.155)
	Skill:
	
	Applicant must pass a practical test (FAR 61.147)
	
	MEDICAL STANDARDS*
	First Class Medical Certificate which includes the following criteria:
	    Vision must be 20/100 in each eye, separately correctable to 20/20 or better with normal color vision.
	    Applicant must have normal ears, nose, throat, equilibrium, nervous system, and ability to hear a whisper at 20 feet.
	    No history of heart disease that might lead to myocardial infarction. After age 35, an electrocardiograph exam is required.
	    Normal blood pressure according to age bracket.
	    No medical history of diabetes.
	    No history of other diseases that would make applicant unable to perform the duties of a pilot.

PRIVILEGES AND LIMITATIONS Privileges: Flight Instructor may give the flight instruction required and endorse a Student Pilot's Certificate and Logbook for solo or cross-country flights. Flight Instructor may give flight instruction. (FAR 61.193) Limitations: Flight Instructor may not give more than 8 hours of flight instruction a day. (FAR 61.195)	**PRIVILEGES AND LIMITATIONS** Privileges: ATP has all of the privileges of a Commercial Pilot with an Instrument Rating. (FAR 61.171) An airline transport pilot may instruct other pilots in air transportation service in aircraft of the category, class, and type for which he is rated. (FAR 61.169) Limitations: The holder of a commercial pilot certificate who qualified for an airline transport pilot certificate retains the ratings on his commercial pilot certificate, but he may exercise only the privileges of a commercial pilot with respect to them. (FAR 61.171)
If an applicant is unable to meet some of the requirements it may be possible, to receive a waiver from the FAA Aeromedical Branch.	

Fig. 15-32. (Sheet 2 of 3)

CERTIFICATE OR RATING
INSTRUMENT RATING
ELIGIBILITY
Applicant must hold a current private or commercial pilot certificate with an aircraft rating appropriate to the instrument rating sought. (FAR 61.65)
AERONAUTICAL KNOWLEDGE, EXPERIENCE, AND SKILL REQUIREMENTS

Knowledge:
Applicant must be able to pass FAA Written Examination covering:
    Federal Aviation Regulations
    Radio Navigation Systems and Procedures
    Meteorology
    (Safe operations of aircraft under IFR weather conditions)
    (FAR 61.65)
Experience:
200 hours of pilot flight time including 40 hours of Instrument Time. (FAR 61.65)
Skill:
Applicant must be able to pass a practical test on appropriate procedures and maneuvers. (FAR 61.65)

MEDICAL STANDARDS*
Appropriate to accompanying certificate.
PRIVILEGES AND LIMITATIONS
Privileges: Instrument rated pilot may exercise privileges of his valid Private Pilot or Commercial Pilot license without using outside visual references. (FAR 61.3) Limitations: Appropriate to accompanying certificate.(FAR 61.3)
*If an applicant for a medical certificate is unable to meet some of the requirements listed above, it is possible, in some cases, to receive a waiver from the FAA Aeromedical Branch.

Fig. 15-32. (Sheet 3 of 3)

Types of training required after graduation from high school will vary with the type of piloting job an individual desires. To become a flight instructor, agricultural pilot, or to enter most fields in general aviation, an average of two years of college is very helpful. To become an airline pilot, however, most airlines require four years of college. In addition, the type of college courses pursued may provide an advantage when applying for a job. A degree in aeronautical engineering, or some other aviation-related degree, would be viewed favorably by airline personnel offices.

FLIGHT SERVICE

The large airlines have elaborate college-type training centers for their service personnel, such as reservation agents, flight line personnel, and flight stewardesses. Classes are normally six weeks in duration and the students study airline equipment, emergency procedures, safety procedures, first aid, route scheduling, serving methods, company regulations, and a host of other details related to work in the airline industry. Flight service personnel must learn rigid FAA emergency procedures if they plan to work on board an aircraft. A six-month probationary period usually follows the completion of these classes.

MANAGEMENT

Management jobs exist at many levels in the aviation industry and represent various degrees of responsibility and authority. A college degree in business administration, accounting, or economics will be beneficial when applying for most managerial positions. Although a college degree is not normally required for managers of fixed-base operations, college courses in business, accounting, and finance are valuable.

Management personnel in most fields go through a certain period of on-the-job training which involves several different positions. Examples of positions which require on-the-job management training for airline positions include airline station manager, sales manager, and customer services manager.

Certain colleges now offer programs that prepare students for future roles in aviation management. These courses of instruction range from two to four years and, when combined with flight training, provide excellent preparation for fixed-base managers. In addition, such programs prepare students for a variety of positions with the aircraft manufacturing industry or airlines. Figure 15-33 shows a representative two-year college curriculum for aviation management.

AIR TRAFFIC CONTROL

The responsibility for the training of air traffic control specialists lies with the Federal Aviation Administration. Technical training is highly sophisticated and requires instruction at the FAA Academy in Oklahoma City as well as on-the-job training at the various air traffic control facilities. It is estimated that a minimum of three and one-half years of training and/or experience are required to become a professional controller.

The Federal Aviation Regulations specify that a person acting as an air traffic controller must hold an appropriate certificate and rating issued according to Part 65. Specialists employed in air route traffic control centers, flight service stations, as well as control towers, must pass a written exam which covers flight rules, airport traffic control procedures, communications, operating procedures, flight assistance service, air navigation procedures, aids to air navigation, and aviation weather.

Generally, applicants applying for an air traffic control position need a minimum of two years of college, with preferential

consideration given to those applicants who hold a commercial pilot certificate. Air traffic controllers are all civil service employees, and therefore, must pass the appropriate civil service exam prior to consideration of their application.

AVIATION MECHANICS

Important sources of mechanics for the aviation industry are the FAA certificated maintenance technician schools across the United States. These schools are educational or training institutions with programs which meet the specific requirements established by the FAA. They are administered by high schools, technical institutes, vocational schools, two-year community colleges, and four-year colleges or universities.

In order to provide the maximum assurances that the aircraft and its components are maintained in an airworthy condition, the FAA prescribes and enforces certification criteria for aviation maintenance technician schools and for aviation mechanic training programs. Aircraft mechanics must hold the appropriate certificate or work under the supervision of a mechanic who is certificated. The regulations which control the certification of aviation mechanics are contained in Part 65 of the Federal Aviation Regulations.

The applicant for a mechanic certificate or rating must successfully complete a written examination, an oral examination,

---

A REPRESENTATIVE ASSOCIATE DEGREE AVIATION MANAGEMENT CURRICULUM*			
Freshman Year	**Credit Hours**	**Sophomore Year**	**Credit Hours**
Communications	6	Economics	6
Mathematics	6	Business Law	3
Accounting	6	Introduction to Data Processing	3
Business Administration	3	Aviation Law	3
Social Science	8	Aviation Administration Electives**	6
Introduction to Aviation	3	Humanities	6
	—	Electives	3
	32		—
			30

**Aviation Administration electives: Basic Flight, Basic Ground, Airport Management, Airline Management, Principles of Transportation, Aviation Problems, Aviation Marketing, Public Administration.

Fig. 15-33. Typical Two-Year Curriculum in Aviation Management

A REPRESENTATIVE ASSOCIATE DEGREE AVIATION MECHANIC CURRICULUM*

Freshman Year (including summer session)	Credit Hours	Sophomore Year (including summer session)	Credit Hours
Communications	6	Social Science	8
Mathematics	3	Welding	2
Natural Science	4	Aircraft Propellers	2
Drafting	2	Powerplant Repair Stations	4
Machine Shop	1	Aircraft Electrical Systems	6
Introduction to Aviation	3	Aircraft Hydraulic and Pneumatic Systems	2
Aircraft Fuel and Oil Systems	3	Sheet Metal Structures	3
Aircraft Powerplants	3	Aircraft Finishing	3
Aircraft Powerplant Maintenance	10	Aircraft Repair Stations	5
	—	Aircraft Auxiliary Systems	2
	35		—
			37

Fig. 15-34. Sample Two-Year Curriculum for Aviation Mechanics

and a practical test of his skills. The written test covers the construction and maintenance of the aircraft as it pertains to the appropriate rating and must be successfully completed before applying for the oral and practical tests.

A representative aviation mechanic curriculum for a typical certificated mechanics school is shown in figure 15-34. Also, a typical avionics technician curriculum is shown in figure 15-35.

A REPRESENTATIVE ASSOCIATE DEGREE AVIONICS TECHNICIAN CURRICULUM*

Freshman Year	Credit Hours	Sophomore Year	Credit Hours
Communications	6	Natural Science	8
Mathematics	6	Aviation Electronics	6
Basic Electricity and Electronics	6	Avionics Laboratory	4
Drafting	2	Aircraft Navigation Systems	3
Introduction to Aviation	3	Microwave Aviation Applications	3
Avionics Laboratory	4	Electives	6
Radio Aids and Communications	3		30
	30		

Fig. 15-35. Typical Two-Year Avionics Technician Curriculum

GLOSSARY

ABSOLUTE ALTITUDE. (See altitude.)

ACROBATIC FLIGHT. Maneuvers intentionally performed by an aircraft involving an abrupt change in its altitude, an abnormal attitude, or abnormal acceleration. NOTE: The term "acrobatic flight" is not intended to include turns or maneuvers necessary to normal flight.

ADVECTION FOG. Fog resulting from the movement of warm, humid air over a cold surface, especially a cold ocean surface; sometimes applied to steam fog which results from movement of cold air over a relatively warm ocean surface.

AERONAUTICAL (LIGHT) BEACONS. A visual navaid displaying flashes of white and/or colored light, which is used to indicate an airport or landing area, a route leading to an airport or landing area, or a point from which bearings can be taken to an airport or landing area, an outstanding landmark, or an object or area presenting a hazard to flying. The principal light used is a rotating beacon of relatively high intensity, which may be supplemented by non-rotating flashing lights of lesser intensity.

AGL. Above ground level.

AGONIC LINE. Line along which no magnetic variation occurs.

AILERONS. Control surfaces on the wings which rotate the aircraft around the roll axis.

AIR DEFENSE IDENTIFICATION ZONE (ADIZ). The area of airspace over land or water within which the ready identification, the location, and the control of aircraft are required in the interest of national security. For operating details see ADIZ procedures.

AIR DENSITY. The mass density of the air in terms of weight per unit volume.

AIR MASS. An extensive body of air having the same properties of temperature and moisture in a horizontal plane.

AIR NAVIGATION FACILITY (NAVAID). Any facility used in, available for use in, or designed for use in aid of air navigation, including landing areas, lights, any apparatus or equipment for disseminating weather information, for signaling, for radio direction-finding, or for radio or other electronic communication, any other structure or mechanism having a similar purpose for guiding or controlling flight in the air or the landing or takeoff of aircraft.

AIR ROUTE TRAFFIC CONTROL CENTER (ARTCC) A facility established to provide air traffic control service to IFR flights operating within controlled airspace and principally during the en route phase of flight.

AIR TRAFFIC CONTROL (ATC). A service operated by appropriate authority to promote the safe, orderly, and expeditious flow of air traffic.

AIRCRAFT. An aircraft is any contrivance now known or hereafter invented, used, or designed for navigation of or flight in the air.

AIRFOIL. A surface or body, such as a wing or propeller blade, designed to obtain a reaction, as lift or thrust, from the air through which it moves.

AIRFRAME. The fuselage, booms, nacelles, cowlings, fairings, airfoil surfaces (including rotors but excluding propellers and rotating airfoils of engines), and landing gear of an aircraft and their accessories and controls.

AIRMAN'S INFORMATION MANUAL (AIM). A pilot's operational manual containing information needed for the planning and conduct of flight in the National Airspace System (conterminous U.S. only).

AIRMET (contraction for airmen's meteorological information). An in-flight weather advisory.

AIRPLANE. An airplane is a power-driven fixed-wing aircraft, heavier-than-air, which is supported by the dynamic reaction of the air against its wings.

AIRPORT. A defined area on land or water, including any buildings and installations, normally used for the takeoff and landing of aircraft.

AIRPORT ADVISORY AREA. The area within five statute miles of an uncontrolled airport on which is located a flight service station so depicted on the appropriate sectional chart.

AIRPORT ADVISORY SERVICE. A service provided by flight service stations at airports not served a the control tower. This service consists of providing information to landing and departing aircraft concerning wind direction and velocity, favored runway, altimeter setting pertinent known traffic, pertient known field conditions, airport taxi routes and traffic patterns, and authorized instrument approach procedures.

AIRPORT SURVEILLANCE RADAR (ASR). See radar.

AIRPORT TRAFFIC AREA. The airspace within a horizontal radius of 5 statute miles from the geographical center of any airport at which a control tower is operating extending from the surface up to, but not including, 3000 feet above the airport elevation.

AIRSHIP. A mechanically propelled aircraft whose support is derived from lighter-than-air gas.

AIRSPEED. The speed of an aircraft relative to the air.

AIRSPEED INDICATOR (ASI). An instrument which gives a measure of the rate of motion of an aircraft relative to the surrounding air.

AIRWAY. An air corridor established for the control of traffic.

AIRWORTHY. The status of being in condition suitable for safe flight.

ALTERNATE AIRPORT. An airport specified in the flight plan to which a flight may proceed when a landing at the point of first intended landing becomes inadvisable.

ALTERNATOR. An alternating current generator.

ALTIMETER. An instrument that measures altitude using the change in air pressure with height, utilizing an aneroid as its sensitive element.

ALTIMETER SETTING. Local station pressure corrected to sea level — an altimeter set to local station pressure which will provide the indicated height above sea level, whether airborne or on the ground (without such correction made for temperature as may be necessary, this is not a true altitude height).

ALTITUDE. Height expressed in units of distance above a reference plane, usually above mean sea level or above ground. See below:

> **Absolute Altitude.** True altitude corrected for terrain elevation; the vertical distance of the aircraft above the terrain.

> **Calibrated Altitude.** Indicated altitude of an aircraft altimeter corrected for the temperature of the column of air

below the aircraft, the correction being based on the estimated departure of the existing temperature of the column from the standard atmospheric temperature; an approximation of true altitude.

Density Altitude. The altitude in the standard atmosphere at which the air has the same density as the air at the point in question. An aircraft will have the same performance characteristics as it would have in a standard atmosphere at this altitude.

Indicated Altitude. The altitude indicated on a standard aircraft altimeter (uncorrected for temperature); above mean sea level when the altimeter is set at the prescribed altimeter setting. When the aircraft is on the ground and the altimeter is set at the local altimeter setting, it will read field elevation or true altitude of the field.

Pressure Altitude. The altitude in the standard atmosphere at which the pressure is the same as at the point in question. Since an altimeter operates solely on pressure, this is the uncorrected altitude indicated by an altimeter set at standard sea level pressure (29.92) inches, i.e., when an altimeter is set at 29.92 inches, indicated altitude and pressure altitude will be identical.

True Altitude. The exact distance above mean sea level.

ALTOCUMULUS (abbreviated Ac). Gray or white layers or patches of cloud, often with a waved appearance; cloud elements appear as rounded masses or rolls; composed mostly of liquid water droplets which may be supercooled; may contain ice crystals at subfreezing temperatures.

ALTOSTRATUS. A form of middle cloud.

AMMETER. An instrument for measuring electric current in amperes.

ANEMOMETER. An instrument for measuring the speed of the wind.

ANEROID BAROMETER. A barometer which operates on the principle of having changing atmospheric pressure bend a metallic surface which, in turn, moves a pointer across a scale graduated in units of pressure. In U.S. usually graduated in inches of mercury. Usually graduated in millibars outside of U.S.

ANGLE OF ATTACK. The acute angle between the chord of an airfoil and a line representing the relative wind.

ANNUAL INSPECTION. An annual inspection is an inspection of an aircraft required once each 12 calendar months and is a complete airworthiness inspection of such aircraft and its various components and systems in accordance with procedures prescribed by the Administrator.

ANVIL CLOUD. Popular name given to the top portion of a cumulonimbus cloud having an anvil-like form.

APPROACH CONTROL (App). A service established to control IFR flights arriving at, departing from, and operating in the vicinity of airports by means of direct and instantaneous communication between approach control personnel and all aircraft operating under their control. Available to VFR traffic on a workload permitting basis.

ATA. Actual time of arrival.

ATC. An abbreviation used in radiotelephone, interphone, or other conversation to mean the air traffic control service of the FAA.

ATMOSPHERIC PRESSURE. The force exerted by the weight of the atmosphere per unit area.

ATTITUDE. The position of an airplane considering the inclination of its axes in relation to the horizon.

AUTOGYRO. An airplane that derives lift from airfoils that rotate without engine power.

AUTOMATIC DIRECTION FINDER (ADF). A radio device composed of a radio receiver, sense and directional (loop) antennas and a bearing indicator; uses radio transmissions from ground stations to automatically indicate the bearing of an aircraft relative to the ground transmitter.

AUTOMATIC TERMINAL INFORMATION SERVICE (ATIS). A continuous broadcast of recorded noncontrol information in selected areas of high terminal activity. Information such as: ceiling, visibility, wind, altimeter, instrument approach and runways in use. This data is broadcast over the voice feature of VOR, VOT, or ILS located at or near the airport, or by a discrete VHF tower frequency.

AUTOPILOT. Those units and components which furnish a means of automatically controlling the aircraft.

AXIS. The theoretical line extending through the center of gravity of an airplane in each major plane; fore and aft, crosswise, and up and down. These are the longitudinal, lateral, and vertical axes.

AZIMUTH. The angle, starting from north and moving eastward, graduated into 360 degrees.

BALANCE. The condition of the aircraft load relative to the aircraft's center-of-gravity. The loading of passengers and

baggage must be within approved limits of balance as well as total weight.

BALANCED CONTROL SURFACE. (1) A control surface which is aerodynamically balanced to reduce the force necessary to displace it by providing some area ahead of the hinge lines, or (2) control surface which is balanced by weights ahead of its hinge line to prevent possible flutter.

BALLOON. An aircraft, excluding moored ballons, without mechanical means of propulsion, the support of which is derived from lighter-than-air gas.

BALLOON CEILING. Surface aviation weather observation, the ceiling classification applied when the height of the ceiling layer has been determined by the timing of the ascension of a ceiling balloon or pilot balloon and precipitation is not occurring or precipitation is of a type and/or intensity considered to have no significant effect on the ascension rate of a balloon. Identified by the ceiling designator B.

BANK. To tip, or roll about the longitudinal axis of the airplane. Banks are incident to all properly executed turns.

BAROMETER. An instrument for measuring atmospheric pressure. The two main types of barometers are mercury barometer and aneroid barometer.

BAROMETRIC TENDENCY. The changes of atmospheric pressure within a specific time (usually 3 hours) before the observation.

BASE LEG. A flight path at right angles to the landing runway off its approach end and extending from the downwind leg to the intersection of the extended runway centerline.

BEARING. The horizontal angle at a given point, measured from a specific reference datum, to a second point. The direction of one point relative to another, as measured from a specific reference datum.

BEST ANGLE-OF-CLIMB AIRSPEED. The airspeed which results in the greatest increase in altitude (in a unit of time) with respect to the distance over the ground.

BEST RATE-OF-CLIMB AIRSPEED. The airspeed which results in the greatest increase in altitude in a unit of time.

BI-METALLIC STRIP. A strip composed of two unlike pieces of metal which, when heated, will bend.

BIPLANE. An airplane having two main lifting surfaces, one above the other.

BLIMP. A lighter-than-air craft normally powered by aircraft engines, having no rigid framework.

BRAKE HORSEPOWER. The power delivered at the propeller shaft (main drive or main output) of an aircraft engine.

BROKEN. (1) In surface aviation weather reports, descriptive of a sky cover of 0.6 to 0.9 coverage; applicable only when clouds or other obscuring phenomena are present aloft; designated by the symbol " ⓪ ;" may be identified as thin (predominantly transparent), otherwise a predominantly opaque status is implied. (2) In radar reporting, a classification of *echo* coverage of 0.6 to 0.9 of the area to which applied; designated by the symbol " ⓪ ."

BUFFETING. The beating effect of the disturbed airstream on an airplane's structure during flight.

BURBLE. Separation in the boundary layer of air over the upper surface of an airfoil resulting in loss of lift and increase of drag.

—— C ——

CALIBRATED AIRSPEED (CAS). Indicated airspeed of an aircraft, corrected for position and instrument error. Calibrated airspeed is equal to true airspeed in standard atmosphere at sea level.

CALIBRATED ALTITUDE. (See altitude.)

CALM. The absence of apparent motion of the air.

CAMBER. The curve of an airfoil section from the leading edge to the trailing edge.

CANTILEVER. A beam, wing, or other member supported at one end only. Most commonly used to describe a wing with no external supports.

CARBURETOR. A device that vaporizes the fuel and mixes it with air in the proper proportions for combustion.

CATEGORY. (1) As used with respect to the certification, ratings, privileges, and limitations of airmen, means a broad classification of aircraft. Examples include: airplane; rotorcraft; glider; and lighter-than-air; and (2) as used with respect to the certification of aircraft, means a grouping of aircraft upon intended use or operating limitations. Examples include: transport; normal; utility; acrobatic; limited; restricted; and provisional.

CAUTION AREA. An area of defined dimensions within which the military training activities conducted, though not hazardous, are of interest to nonparticipating pilots.

CEILING. The height above the ground or water of the lowest layer of clouds or obscuring phenomena that is reported as "broken," "overcast," or "obscuration" and not classified as "thin" or "partial."

CEILING, ABSOLUTE. The maximum height at which a particular airplane is capable of operating.

CEILING, SERVICE. The altitude beyond which a particular airplane is incapable of maintaining a rate of climb exceeding 100 feet per minute.

CEILOMETER. An electronic device which measures the height of cloud bases.

CELSIUS TEMPERATURE SCALE (abbreviated C). By recent convention, the same as the centigrade temperature scale with zero degrees as the melting point of ice and 100 degrees as the boiling point of water at standard sea level atmospheric pressure.

CENTER OF GRAVITY (CG). The point at which all of the mass of an object appears to be concentrated. An aircraft, suspended from this point, would remain level with no tendency to pitch in nose-up or nose-down attitudes.

CENTER-OF-GRAVITY RANGE. The distance between the most forward and most aft center-of-gravity limits specified by the manufacturer.

CHART. A graphic representation of a section of the earth's surface specifically designed for navigational purposes. A chart may also be referred to as a Map. Although a chart is usually specifically designed as a plotting medium for marine or aerial navigation, it may be devoid of cultural or topographical data.

CHECKLIST. A list, usually carried in the pilot's compartment, of items requiring the airman's attention for various flight operations.

CHECKPOINT. A geographical reference point used for checking the position of an aircraft in flight. As generally used, it is a well-defined reference point easily

discernible from the air. Its exact position is known or plotted on the navigational chart, and was selected in preflight planning for use in checking aircraft position in flight.

CHINOOK. A warm, dry foehn wind blowing down the eastern slopes of the Rocky Mountains over the adjacent plains in the U.S. and Canada.

CHORD. An assumed straight line tangent to the lower surface of an airfoil section at two points, or a straight line between the leading and trailing edges of an airfoil section; the distance between the leading and trailing edges of an airfoil section.

CIRCUIT BREAKER. A device which takes the place of a fuse in breaking an electrical circuit in case of an overload. Most aircraft circuit breakers can be reset by pushing a button, in case the overload was temporary.

CIRRIFORM. All species and varieties of *cirrus, cirrocumulus,* and *cirrostratus* clouds; descriptive of clouds composed mostly or entirely of small ice crystals usually transparant and white; often producing halo phenomena not observed with other cloud forms. Average height ranges upward from 20,000 feet in middle latitude.

CIRROCUMULUS (abbreviated Cc). A principal cirriform cloud type appearing as a whitish veil, usually fibrous, sometimes smooth; often produces halo phenomena; may totally cover the sky.

CIRRUS (abbreviated Ci). A principal cirriform cloud type in the form of thin, white feather-like clouds in patches or narrow bands; have a fibrous and/or silky sheen, large ice crystals often trail down a considerable vertical extent in fibrous, slanted, or irregularly curved wisps called "mare's tails."

CLASS (of aircraft). A class is a classification of aircraft within a category differentiating between single-engine and multi-engine and land and water configurations.

CLEAR. In U.S. aviation meteorology, the state of the sky when it is cloudless or when the sky cover is less than 0.1 (to the nearest tenth).

CLEAR AIR TURBULENCE (CAT). Turbulence that occurs in clear air and not associated with cloud formation, such as that associated with winds at low altitudes and with the jet stream at high altitudes.

CLEAR ICE. Generally, the formation of a layer or mass of ice which is relatively transparent because of its homogeneous structure and small number and size of airspaces; used commonly as synonymous with glaze, particularly with respect to aircraft icing. Compare with rime icing. Factors which favor clear icing are large drop size, such as those found in cumuliform clouds, rapid accretion of supercooled water, and slow dissipation of latent heat of fusion.

CLEARANCE (IFR). Authorization to follow a specified flight outline. Clearances are issued by the control agency appropriate for the controlled area within which the flight will operiate, and are used to prevent collisions between aircraft.

CLOUD. A visible cluster of minute water and/or ice particles in the atmosphere above the earth's surface.

COASTAL AIR DEFENSE IDENTIFICATION ZONE. An ADIZ over the coastal waters of the United States. See Air Defense Identification Zone.

COCKPIT. The space in the fuselage with seats for the pilot and passengers; also used to denote the pilot's compartment in a large airplane.

COLD AIR MASS. (1) An air mass that is colder than the underlying surface. (2) A mass of air that is cold relative to adjacent or surrounding air.

COLD FRONT. The discontinuity at the forward edge of an advancing cold air mass which is displacing warmer air in its path.

COMPASS. An instrument which indicates direction measured clockwise from magnetic north.

COMPASS CORRECTION CARD. A card mounted upon the instrument panel listing the amount of deviation to be expected on various headings.

COMPASS DEVIATION. The difference between magnetic heading and compass heading.

COMPRESSIBILITY. The effect encountered at extremely high speeds, near the speed of sound, when air ceases to flow smoothly over the wings, and "piles up" against the leading edge, causing extreme buffeting and other effects.

CONDENSATION LEVEL. The height at which a rising column of air reaches saturation, and clouds form.

CONES. Tapered nerve cells in the retina important in color vision.

CONSOLAN. A low frequency, long-distance navaid used principally for transoceanic navigation.

CONTERMINOUS U.S. Forty-eight states and the District of Columbia.

CONTINENTAL CONTROL AREA. The airspace of the 48 contiguous states and the District of Columbia at and above 14,500 feet MSL, but does not include (1) the airspace less than 1500 feet above the surface of the earth; or (2) prohibited and restricted areas, other than restricted area military climb corridors and the restricted areas listed in sub-part E of FAR 71.

CONTOUR LINES. Lines drawn on maps and charts joining points of equal elevation; also, a line connecting points of equal altitude on a constant-pressure chart.

CONTROL ZONE. Controlled airspace from ground to base of continental control area. A control zone may include one or more airports and is normally a circular area with a radius of five miles and any extensions necessary to include instrument approach and departure paths.

CONTROLLED AIRSPACE. Airspace, designated as continental control area, control area, control zone, or transition area, within which some or all aircraft may be subject to air traffic control.

CONTROLS. The devices used by a pilot in operating an airplane.

CONVECTION. (1) In general, mass motions within a fluid resulting in transport and mixing of the properties of that fluid. (2) In meteorology, atmospheric motions that are predominantly vertical, resulting in vertical transport and mixing of atmospheric properties; distinguished from advection.

CONVENTIONAL LANDING GEAR. Landing gear which consists of two main wheels and a tail-wheel.

COORDINATED TURN. A turn utilizing both ailerons and rudder controls so that neither slipping or skidding is experienced.

COORDINATES. Description of a geographical location by latitude and longitude. Latitude is measured north or south from the "zero" line of the equator. Longitude is measured east or west from the "zero" line of the Greenwich

Meridian. Points of equal latitude determine lines called "parallels" which are parallel to the equator and to each other. Points of equal longitude determine lines called "meridians" which extend from pole to pole, crossing the equator and the parallels at right angles.

COPILOT. A copilot is a pilot serving in any capacity other than as pilot in command on aircraft requiring two pilots for normal operations, but excluding a pilot who is on board the aircraft for the sole purpose of receiving dual instruction.

CORIOLIS FORCE. An apparent deflecting force resulting from the earth's rotation; it acts normal to the velocity of a particle, to the right of motion in the Northern Hemisphere and to the left in the Southern Hemisphere. It cannot alter the speed of the particle. It tends to balance the pressure gradient force between highs and lows and to cause the air to move parallel to the isobars. In the lower levels, winds do not quite parallel the isobars due to friction but are deflected toward the low pressure area.

COURSE. (1) The direction toward the destination as charted, described in degrees of deviation from north. True course is measured from true north; magnetic course is measured from true north; magnetic course is measured from magnetic north. All courses formed by VORs on sectional charts are magnetic. (See track.) (2) In connection with an ILS, the visually indicated beam which provides directional guidance to the runway.

COWL FLAP. One of several shutters in an aircraft engine cowling, used to regulate the flow of cooling air around the engine.

COWLING. A covering placed over or around an aircraft component for streamlining, protection, or to direct the flow of air for cooling purposes. The term is usually used in conjunction with the component such as "engine cowling."

CRAB ANGLE. A wind correction angle. The angular difference between the course and the heading of an aircraft due to the effects of a crosswind.

CRASH LOCATOR BEACON. An electronic device attached to the aircraft structure as far aft as practicable in the fuselage, or in the tail surface, in such a manner that damage to the beacon will be minimized in the event of crash impact. It may be automatically ejectable or be permanently mounted. If it is automatically ejectable it will also have provision for manual removal and operation. The beacon operates from its own power source on 121.5 MHz and/or 243 MHz, preferably on both emergency frequencies, transmitting a distinctive downward swept audio tone for homing purposes, and is designed to function without human action after an accident.

CROSSWIND. A wind blowing across the line of flight of an aircraft.

CROSSWIND COMPONENT. A wind component which is at a right angle to the longitudinal axis of the runway or the aircraft's flight path.

CROSSWIND LEG. A flight path at right angles to the landing runway off its upwind leg.

CUMULIFORM. A general term descriptive of all clouds exhibiting vertical development in contrast to the horizontally extended stratiform clouds.

CUMULONIMBUS (abbreviated Cb). A principal cumuliform cloud type; it is heavy and dense, with considerable vertical extent, in the forms of massive towers; often with tops in the shape of an anvil or massive plume, under the base of cumulonimbus, which often is very dark, there frequently exists virga (see precipitation), and low ragged clouds (scud), either merged with it or not; frequently accompanied by lightning, thunder, and sometimes hail; occasionally produces a tornado or a waterspout, the ultimate manifestation of the growth of a cumulus cloud, occasionally extending well into the stratosphere.

CUMULUS (abbreviated Cu). A principal cloud type in the form of individual detached domes or towers which are usually dense and well defined. Develops vertically in the form of rising mounds of which the bulging upper part often resembles a cauliflower. The sunlit parts of these clouds are mostly brilliant white; their bases are relatively dark and nearly horizontal. If precipitation occurs, it is usually of a showery nature; composed of a great density of small water droplets, frequently supercooled. Larger drops often develop within the cloud and fall from the bases as trails of evaporating rain (virga). Ice crystals sometimes form in the upper portion and grow larger by taking water from the water droplets. Maximum frequency and development is in the afternoon over land and during the night over water.

—— d ——

DANGER AREA. A specified area within or over which there may exist activities constituting a potential danger to aircraft

DATUM. (1) Refers to a direction, level, or position from which angles, heights, depths, speeds, or distances are conventionally measured. (2) A vertical reference line from which horizontal measurements are made to calculate weight and balance.

DEAD RECKONING. The directing of an aircraft and determining of its position by the application of direction and speed data to a previous position.

DEFENSE VISUAL FLIGHT RULES (DVFR). A flight within an ADIZ conducted under the fisual flight rules in Part 91.

DENSITY. The mass of a substance per unit volume. The density of a gas is particularly sensitive to changes in temperature (and pressure). The weight of a substance varies directly with its density.

DEPARTURE CONTROL. A function of approach control providing service for departing IFR aircraft and on occasion, VFR aircraft.

DEVIATION. A compass error caused by the magnetism within an aircraft; the angle measured from magnetic north eastward or westward to the direction of the earth's lines of magnetic force as deflected by the aircraft's magnetism.

DEWPOINT (or dewpoint temperature). The temperature to which a sample of air must be cooled; while the mixing ratio and barometric pressure remain constant, in order to attain saturation with respect to water. In U.S. surface aviation weather observations, dewpoint is expressed to the nearest whole degree Fahrenheit.

DIHEDRAL. The angle formed between the plane of the wing and horizontal plane at the root of the wing. This tilt is usually upward and is designed to increase stability.

DIP, MAGNETIC. The vertical displacement of the compass needle from the horizontal caused by the earth's magnetic field.

DIRECTION FINDER (DF). A radio receiver with a direction finding antenna (commonly a loop) which takes bearings on two or more fixed transmitters of known position. The intersection of two bearing lines establishes the observer's position (a fix).

DISPLACED THRESHOLD. A threshold located on the runway at a point other than at the end and/or beginning of the full strength pavement.

DISTANCE MEASURING EQUIPMENT (DME). An interrogation-reply system which provides the pilot with a continuous presentation of distance in nautical miles to the DME site.

DOWNDRAFT. A relatively small scale downward current of air; often observed on the lee side of large objects restricting the smooth flow of air or in precipitation areas in or near cumuliform clouds.

DOWNWASH. The downward thrust imparted on the air to provide lift for the airplane.

DRAG. The force opposing the movement of the airplane through the air.

  Induced. The part of the total drag on an airplane produced by the flow of air over lifting surfaces.

  Parasite. Drag produced by attachments to the aircraft and no-lift devices such as landing gear and struts.

DRIFT. Deflection of an airplane from its intended course by action of the wind.

DRY ADIABATIC LAPSE RATE. The rate of decrease of temperature with height of unsaturated air lifted adiabatically (due to expansion as it is lifted to lower pressure). It is equal to approximately 5.4°F per 1000 feet.

DUPLEX. Transmitting on a frequency other than the receiving frequency. See simplex.

— e —

ELEVATOR. A hinged, horizontal control surface used to control movement around the pitch or lateral axis.

EMPENNAGE (am-pa-nazh). The assembly at the rear end of an aircraft comprised usually of horizontal and vertical stabilizers and their associated control surfaces.

EMPTY WEIGHT. The empty weight of an aircraft includes the following:

(1) Basic aircraft structure.
(2) All fixed equipment.
(3) Unuseable fuel supply.
(4) Undrainable oil.
(5) Full engine coolant.
(6) Hydraulic fluid.

ESTIMATED CEILING. In aviation weather observations, the ceiling classification applied when the ceiling layer of clouds or obscuring phenomena aloft has been estimated by the observer or has been determined by some other method, but, because of the specified limits of time, distance or precipitation conditions, a more descriptive classification cannot be applied. Identified by the ceiling designator "E" for non-cirriform ceiling layers and "D" for cirriform ceiling layers.

ETA. Estimated time of arrival.

ETD. Estimated time of departure.

— f —

FAHRENHEIT. A temperature scale on which 32° denotes the temperature of melting ice, and 212° the temperature of boiling water, both under standard atmospheric pressure.

FAIRING. A member or structure the primary function of which is to produce a smooth outline and to reduce drag.

FIN. A fixed airfoil to increase the stability of an airplane. Usually applied to the vertical surface to which the rudder is hinged.

FINAL APPROACH VFR. A flight path of a landing aircraft in the direction of landing along the extended runway centerline from the base leg to the runway.

FIX. The geographic position of an aircraft for a specified time, established by navigational aids.

FLAPS. (See wing flap.)

FLARE OUT. To round out a landing by decreasing the rate of descent and airspeed by slowly raising the nose.

FLIGHT LEVEL. A level of constant atmospheric pressure related to a reference datum of 29.92 inches of mercury. Each is stated in three digits that represent hundreds of feet. For example, flight level 250 represents a barometric altimeter indication of 25,000 feet; flight level 255, an indication of 25,500 feet.

FLIGHT PLAN. Specified information relating to the intended flight of an aircraft that is filed orally or in writing with an air traffic control facility.

FLIGHT SERVICE STATION (FSS). An FAA-operated air-ground voice communication station which relays clearances, requests for clearances, and position reports between en route aircraft and the Air Route Traffic Control Center. In addition, an FSS provides pre-flight briefing for either IFR or VFR flights, gives in-flight assistance, broadcasts weather once each hour, monitors radio navigational facilities, accepts VFR flight plans and provides for notification of arrival, and broadcasts notices to airmen concerning local navigational aids, airfields and other flight data.

FLIGHT TIME. The time from the moment the aircraft first moves under its own power for the purpose of flight until the moment it comes to rest at the next point of landing ("block-to-block" time.)

FLIGHT VISIBILITY. The average forward horizontal distance, from the cockpit of an aircraft in flight, at which prominent unlighted objects may be seen and identified by day and prominent lighted objects may be seen and identified by night.

FOG. A cloud at or near the earth's surface. Fog consists of numerous droplets of water, which individually are so small that they can not readily be distinguished by the naked eye.

FRACTUS. Clouds in the form of irregular shreds, appearing as if torn; having a clearly ragged appearance; applies only to stratus and cumulus, i.e., cumulus fractus and stratus fractus.

FREEZING LEVEL. A level in the atmosphere at which the temperature is $32^\circ F$ ($0^\circ C$).

FRONT. A surface, interface, or transition zone of discontinuity between two adjacent air masses of different densities; more simply *the boundary between two different air masses.*

FULCRUM. The point about which a level will rotate; the pivot point of a teeter-totter.

FUSELAGE. The main structure of central section of an airplane, which houses or contains the crew, passengers, cargo, etc.

—— g ——

"G." In this text, "G" represents the force of gravity. One "G" is equal to the actual weight of an object at rest. A force of two "G's" is equal to two times the weight of the object upon which the force is being exerted. (In physics "G" represents the downward acceleration due to gravity.)

GENERAL AVIATION. That segment of aviation which includes all aircraft except the military and the commercial airlines.

GENERATOR. A machine, run by the engine, which converts mechanical energy into electrical energy.

GLAZE ICE. A coating of ice, generally clear and smooth, but containing some air pockets. Synonymous with clear ice, it forms on exposed surfaces of the aircraft.

GLIDE. Sustained forward flight in which speed is maintained only by the loss of altitude.

GLIDE SLOPE. An ILS navigation facility in the terminal area electronic navigation system providing vertical guidance for aircraft during approach and landing by radiating a directional pattern of UHF radio waves modulated by two signals which, when received with equal intensity, are displayed by compatible airborne equipment as an "on-path" indication.

GLIDER. A glider is a heavier-than-air aircraft, the free flight of which does not depend upon a power-generating unit.

GRAVITY. The force of attraction between two bodies; the force that makes a body, if free to move, accelerate toward the center of the earth.

GREENWICH MEAN TIME (GMT). Local time at the Greenwich meridian measured by reference to the mean sun. It is the angle measured at the pole or along the equator or equinoctial (and converted to time) from the lower branch of the Greenwich meridian westward through 360° to the upper branch of the hour circle through the mean sun (formerly called Greenwich civil time).

GROSS WEIGHT. The total weight of the aircraft ready for flight. This weight consists of the following: (1) Aircraft empty weight. (2) Oil. (3) Fuel. (4) Pilot. (5) Passengers. (6) Baggage. (7) Cargo. (8) Removable equipment.

GROUND CONTROLLER APPROACH (GCA). (See radar.)

GROUND EFFECT. The influence, usually beneficial, upon aircraft performance, while flying close to the earth's surface, caused by the rebounding from the surface of the air disturbed by the movement of the aircraft and/or its propulsion system.

GROUND FOG. In the United States, a fog that conceals less than 0.6 of the sky and is not contiguous with the base of clouds. In surface aviation weather observations, it is reported as an obstruction to vision, encoded "GF."

GROUND SPEED. The speed of the aircraft relative to the ground.

GROUND VISIBILITY. Prevailing horizontal visibility near the earth's surface as reported by the National Weather Service or an accredited observer.

GYRO INSTRUMENTS. Any instruments actuated or controlled by a gyroscope.

GYROSCOPE. A device that has a wheel mounted so that it is free to rotate in two axes perpendicular to itself and to each other.

─── h ───

HAIL. Precipitation consisting of balls of irregular lumps of ice often of considerable size; a single unit of hail is called a hailstone. Large hailstone usually have a center surrounded by alternating layers of clear and cloudy ice. Hail falls almost exclusively in connection with thunderstorms. The largest hailstone observed in the United States was 17 inches in circumference and weighed 1½ pounds.

HAZE. Fine dust or salt particles scattered through a protion of the atmosphere. Particles are so small that they cannot be seen individually, but they diminish horizontal visibility.

HEADING. The direction in which the nose of the airplane points during flight. Corrections made to compensate for wind will cause differences to arise between track and heading. If no change is made in heading to compensate for wind, differences will arise between track and course as the aircraft drifts.

HELICOPTER. A type of rotorcraft, the support of which, in the air, is normally derived from airfoils mechanically rotated about an approximately vertical axis.

HERTZ (Hz). Cycle per second.

HIGH. An area of high barometric pressure; an anticyclone; also high pressure system.

HOME. To fly toward a radiation-emitting source, especially a radio transmitter, using the radiated waves as a guide; to bring an aircraft to a given spot by radio transmissions, light beacons, or other signals.

HORIZON:

Artificial. A gyroscopic instrument which indicates the attitude of the aircraft with respect to the true horizon.

Natural or Visible. The circle around the observer where earth and sky appear to meet. Also called natural horizon or sea horizon.

HORSEPOWER. A unit for measurement of power output of an engine. It is the power required to raise 550 pounds one foot in one second.

HUMIDITY. The measure of water vapor content in the air.

HURRICANE. A tropical cyclone in the Western Hemisphere with winds in excess of 65 knots (U.S.) or in excess of 64 knots International.

HYDRAULIC. Designating devices which operate on fluids under pressure such as brakes, and control surfaces on high performance aircraft.

HYPERVENTILATION. A physiological condition in which too much carbon dioxide has been removed from the bloodstream by over-breathing.

HYPOXIA. Oxygen deficiency in blood cells or tissue.

———— i ————

ICE FOG. A type of fog composed of minute suspended particles of ice; occurs at very low temperatures and may cause a halo phenomena. Denoted in aviation weather observations as obstruction to vision, encoded "IF."

ICING. In general, any deposit of ice on an object. See clear icing, rime icing, glaze.

IDENT FEATURE. The special feature in the ATCRBS equipment used to distinguish one aircraft from other aircraft on the same code.

IFR CONDITIONS. Weather conditions below the minimum prescribed for flight under VFR.

INCIDENCE, ANGLE OF. The angle between the mean chord of the wing and the longitudinal axis of the airplane.

INDEFINITE CEILING. In aviation weather observations, a ceiling classification applied to the ceiling height value used to denote vertical visibility into a surface-based total obscuration unless the vertical visibility is reported as an aircraft ceiling.

INDICATED AIRSPEED (IAS). The reading obtained from the airspeed indicator uncorrected for density altitude or instrument error.

INDICATED ALTITUDE. Altitude as shown by any altimeter. With a pressure or barometric altimeter it is altitude as shown by the reading uncorrected for instrument error and uncompensated for variations from standard atmospheric conditions.

INDUCED DRAG. The drag produced indirectly by the effect of the induced lift.

INERTIA. The tendency of a body to remain as it is, either at rest or in motion, until acted upon by some outside force.

INSTRUMENT FLIGHT RULES (IFR). When weather conditions are below the minimums prescribed for visual meteorological conditions, pilots must fly in accordance with IFR. Pilots may elect to fly an IFR flight plan during VFR conditions.

INSTRUMENT LANDING SYSTEM (ILS). A system which provides in the aircraft, the lateral, longitudinal, and vertical guidance necessary for a landing.

INSTRUMENT TIME. Instrument time is that time during which a pilot is operating an aircraft under actual or simulated instrument flight conditions solely by reference to instruments, or time acquired in a synthetic trainer approved for instrument flight training.

INTERNATIONAL CIVIL AVIATION ORGANIZATION (ICAO). An international body in the field of aeronautics. ICAO standards and recommended practices are not binding; final decision rests with the sovereign state.

INTERROGATOR. The ground-based surveillance radar beacon transmitter-receiver which scans in synchronism with a primary radar, transmitting discrete radio signals which repetitiously request all transponders, on the mode being used, to reply. The replies received are then mixed with the primary radar video to be displayed on the same plan position indicators.

INTERSECTION. A fix established by the intersection of specified radials emanating from two VOR stations, or one VOR station and a localizer.

INVERSION. An increase in temperature with height — a reversal of the normal decrease with height in the troposphere; may also be applied to other meteorological properties.

IONOSHPERE. The outer region of the atmosphere which contains layers of ionized air particles.

ISOBAR. A line of equal barometric pressure; their pattern is a main feature of surface chart analysis.

ISOGONIC LINES. Lines determined by points at which the amount of magnetic variation is equal.

----- j -----

JET ROUTES. A high altitude route system, at or above 18,000 feet MSL, predicated on a network of designated high altitude VHF/UHF facilities.

JET STREAM. A narrow meandering stream of winds with speeds of 50 knots and greater, embedded in the normal wind flow aloft.

----- k -----

KNOT. A unit of speed equal to one nautical mile per hour.

----- l -----

LAG. The failure of the output of an instrument to respond instantly to variations of input; in meteorological instruments, it usually refers to time lag.

LAMBERT CONFORMAL PROJECTION. A conformal projection in which the meridians are first described upon the surface of a cone, and in which the parallels are spaced mathematically to impart the property of true shape.

LAND PLANE. An airplane designed to rise from and alight on the ground.

LANDING. The act of terminating flight and bringing the airplane to rest, used both for land and seaplanes.

LANDING GEAR. The under structure which supports the weight of the airplane while at rest.

LAPSE RATE. The rate of decrease of an atmosphere variable with height, the variable being temperature unless otherwise specified.

LARGE AIRCRAFT. Aircraft of more than 12,500 pounds maximum certificated takeoff weight.

LATERAL AXIS. An imaginary line running through the center of gravity of an aircraft, parallel to the straight line through both wing tips. The pitch axis.

LATITUDE. Angular distance on the surface as viewed from the earth's center, measuring north or south from 0° through 90°, beginning at the equator.

LEADING EDGE. The forward edge of any airfoil.

LENTICULAR CLOUD (or lenticularis). A species of cloud the elements of which have the form of more or less isolated, generally smooth lenses or almonds. These clouds appear most often in formations of orographic origin, the result of lee waves, in which case they remain nearly stationary with respect to the terrain (standing cloud), but they also occur in regions without marked orography. When this species is present in the altocumulus form (altocumulus lenticularis, abbreviated ACSL) it is generally associated with moderate to severe turbulence, hence its presence requires a mandatory remark in surface weather observations.

LIFT. The force on an airfoil, perpendicular to the relative wind, exerted normally upward, opposing the pull of gravity.

LIGHT GUN. An intense, narrowly focused spotlight with which a green, red, or white signal may be directed at any selected airplane in the traffic on or about an airport. Usually used in control towers.

L/MF. Low/Medium frequency.

LOAD FACTOR. (See Chapter 12, Section C.)

LOCALIZER. An ILS navigation facility in the terminal area electronic navigation system, providing horizontal guidance to the runway centerline for aircraft during approach and landing by radiating a directional pattern of VFR radio waves modulated by two signals which, when received with equal intensity, are displayed by compatible airborne equipment as an "on-course" indication, and when received in unequal intensity are displayed as an "off-course" indication.

LOG. To make a flight-by-flight record of all operations of an airplane, engine, or pilot, listing flight time, area of operation, and other pertinent information.

LONGERON. A relatively heavy longitudinal structural member in the airplane, usually running continuously across a number of forms.

LONGITUDE. The angular distance east or west of the Greenwich meridian, measured in the plane of the equator or of a parallel from 0° to 180°.

LONGITUDINAL AXIS. An imaginary line running fore and aft through the center of gravity of an aircraft parallel to the axis of the propeller or thrust line. The roll axis.

LORAN. An electronic aid to navigation utilizing pulse techniques for long range navigation (usually over water). Use of this navigational aid requires a loran receiver-indicator in conjunction with suitable charts and tables. Accuracy is approximately within one percent of the distance from the transmitters.

LOW. An area of low barometric pressure, with its attendant system of winds. Also called a barometric depression or cyclone.

LOW FREQUENCY (LF). A frequency in the 30-300 kHz band. When used in comparison of the colored airway system with the VHF or Victor airway system, the term "ILF" is applied to all components in the LF and MF frequency bands.

—— m ——

MACH NUMBER. (See Chapter 12, Section C.)

MAGNETIC COURSE. The true course or track, corrected for magnetic variation, between two points on the surface of the earth.

MAGNETO. A generator using permanent magnets to generate an electric current.

MANEUVERING SPEED. The maneuvering speed is the maximum speed at which the flight controls can be fully deflected without damage to the aircraft structure. It may be found in the airplane flight manual and is useful for guidance in performing flight maneuvers, or normal operations in severe turbulence.

MANIFOLD PRESSURE GAGE. A gage which measures the pressure in the intake manifold.

MAXIMUM CONTINUOUS POWER. With respect to reciprocating engines, means the brake horsepower that is developed (1) in standard atmosphere at a specified altitude, and (2) under the maximum conditions of crankshaft rotational speed and engine manifold pressure that are approved for use of unrestricted duration.

MAXIMUM GROSS WEIGHT. The maximum weight authorized by the FAA for operation of the aircraft involved.

MEAN AERODYNAMIC CHORD (MAC). The mean, or average, chord of the wing.

MEAN SEA LEVEL (MSL). The average level of the sea, used to compute barometric pressure.

MEASURED CEILING. In aviation weather observations, the ceiling classification that is applied when the ceiling value has been determined by means of: (a) a ceiling light or ceilometer: (b) the known heights of unobscured portions of objects, other than natural landmarks, within 1½ nautical miles of any runway of the airport. It applies only to clouds and obscuring phenomena aloft and is identified by the ceiling designator "M".

MERIDIAN. A great circle on the earth that passes through the poles.

MILLIBAR. A unit of pressure equal to a force of 1000 dynes per square centimeter. The standard sea level pressure of 29.92 inches of mercury is equal to 1013.2 millibars.

MOMENT. The tendency for an object to rotate; equal to the weight of an object multiplied by its "arm."

MONOCOQUE. A type of aircraft construction in which the external skin constitutes the primary structure. (An egg is of monocoque construction.)

MONOPLANE. An airplane having one lifting surface.

MSL. See mean see level.

—— n ——

NACELLE. A streamlined structure, housing, or compartment on an aircraft such as: (1) a housing for an engine; a power car or gondola on an airship; (2) a crew compartment or cabin or an airplane that does not have a conventional fuselage; (3) as housing for some external components such as a loop antenna.

NAUTICAL MILE (NM). A unit of distance used in navigation, 6080 feet; the minimum length of one minute of longitude on the equator; approximately one minute of latitude; 1.15 statute miles.

NAVIGABLE AIRSPACE. Airspace at and above the minimum flight altitudes prescribed by or under this chapter, including airspace needed for safe takeoff and landing.

NAVIGATION LIGHT. Any one of a group of lights mounted on an aircraft to make its dimensions, position, and direction of motion visible at night or under other conditions of poor visibility.

NAVIGATIONAL AID (NAVAID). Any visual or electronic device located on the ground or on shipboard which provides point-to-point guidance information and/or partial or complete position data to aircraft in flight.

NIMBOSTRATUS (abbreviated Ns). A principal cloud type, gray colored, often dark, the appearance of which is rendered diffuse by more or less continuously falling rain or snow, which in most cases reaches the ground. It is thick enough throughout to blot out the sun. Low, ragged clouds frequently occur below the layer, with which they may or may not merge.

NONDIRECTIONAL RADIO BEACON (NDB). A surface-based radio facility which broadcasts a carrier wave in all directions simultaneously.

NOSE-HEAVY. A condition of trim in an airplane in which the nose tends to sink when the elevator control is released.

NOSE WHEEL. A swiveling or steerable wheel mounted forward in tricycle-geared airplanes.

NOTICE TO AIRMEN (NOTAM). A notice to Airmen in message form requiring expeditious and wide dissemination by tele-communication means.

——— O ———

OBSCURATION (also called obscured sky cover). (1) In weather observing practice, the designation for sky cover when more than 0.9 of the sky is hidden by surface-based obscuring phenomena and vertical visibility is restricted overhead. When an obscuration is reported, a ceiling also must be reported. Ceiling height ascribed to an obscuration is vertical visibility into the obscuration. Compare partial obscuration. (2) Surface-based obscuration phenomena.

OBSTRUCTION LIGHT. A light, or one of a group of lights, usually red, mounted on a surface structure or natural terrain to warn pilots of the presence of a flight hazard; usually either an incandescent lamp with a red globe or a red gaseous discharge lamp.

OCCLUDED FRONT (commonly called occlusion, also called frontal occlusion). A composite of two fronts as a cold front overtakes a warm front or quasi-stationary front.

OIL BURNER. A military operation simulating jet aircraft bombing runs.

OLEO. A shock-absorbing strut in which the spring action is dampened by oil.

OMNI BEARING SELECTOR (OBS). An instrument capable of being set manually to any desired bearing of an omnirange station and which controls a course deviation indicator.

OROGRAPHIC. Of, pertaining to, or caused by mountains as in orographic clouds, orographic lift, or orographic precipitation.

OUTER MARKER BEACON (OM). An ILS navigation facility in the terminal area electronic navigation system located four to seven miles from the runway edge of the extended centerline transmitting a 75 MHz fan-shaped radiation pattern, modulated at 400 Hz, keyed at two dashes per second, received by compatible airborn equipment, indicating to the pilot, both aurally and visually, that he is passing over the facility and can begin his final approach.

OUTER MARKER COMPASS LOCATOR. A facility in the terminal area electronic navigation system, located at the outer marker beacon, which transmits a continuous carrier, L/MF radio wave in an omni-directional pattern, enabling the pilot of an aircraft equipped with a direction finder to determine his bearing relative to the outer marker.

OVERCAST. In surface aviation weather observations, descriptive of sky cover of 1.0 (95 percent or more) when at least a portion of this amount is attributable to clouds or obscuring phenomena aloft; that is, when the total sky cover is not due entirely to surface-based obscuring phenomena (see obscuration); denoted by the symbol " ⊕ "; may be identified as thin "- ⊕ " when predominantly transparent, otherwise a predominant opaque cover is implied.

—— P ——

PARALLEL. A true east-west line of points of equal latitude extending around the world by which position (latitude) is measured north or south from the equator. All parallels are parallel to the equator and to each other. See coordinates.

PARTIAL OBSCURATION. In surface aviation weather observations, a designation of sky cover when (1) part (0.1 to 0.9) of the sky is completely hidden by surface-based phenomena, or (2) ten tenths (to the nearest tenth) of the sky is hidden by surface-based phenomena and vertical visibility is not restricted; compare obscuration.

PILOT. A pilot is a person holding a valid pilot certificate issued by the Administrator.

PILOT IN COMMAND. The pilot responsible for the operation and safety of an aircraft during flight time.

PILOT WEATHER REPORT (commonly contracted (PIREP). A report of in-flight weather by an aircraft pilot or crew member.

PILOTAGE. Navigation by visual reference to landmarks.

PITCH. (1) The blade angle of a propeller. (2) The movement of an aircraft about its lateral axis.

PITOT TUBE. A cylindrical tube with an open end pointed upstream, used in measuring impact pressure, particularly in an airspeed indicator.

PLOTTER. An instrument or device with scales calibrated in linear measure and in degrees, used in drawing and measuring courses and in measuring distances on aeronautical charts.

POLAR FRONT. The semi-permanent, semi-continuous front separating air masses of tropical and polar origin.

POSITIVE CONTROL. Control of all air traffic, within designated airspace, by air traffic control.

PRECESSION, APPARANT. The apparent deflection of the gyro axis, relative to the earth, due to the rotating effect of the earth and not due to any applied forces.

PRECESSION, INDUCED. The movement of the axis of a spinning gyro when a force is applied. The gyro precesses 90° from the point of applied pressure in the direction of rotation.

PRECIPITATION. The collective name for moisture in liquid and solid form large enough to fall from the atmosphere.

PRECISION APPROACH RADAR (PAR). (See radar.)

PRESSURE ALTIMETER. An aneroid barometer calibrated to convert atmospheric pressure to altitude using standard atmospheric pressure-height relationships; shows indicated altitude (not necessarily true altitude); may be set to measure altitude (indicated) from any arbitrarily chosen level. (See altimeter setting, altitude.)

PRESSURE ALTITUDE. (1) generically, altitude above the standard datum plane, as determined by applying the standard pressure lapse rate to the atmospheric pressure at altitude; (2) either indicated pressure altitude or calibrated pressure altitude; (3) a simulated pressure altitude inside a low-pressure chamber; (4) can be obtained by setting altimeter to 29.92. See altitude.

PRESSURE GRADIENT. The rate of decrease (gradient) of pressure per unit distance at a fixed time.

PREVAILING WIND. The wind direction most frequently observed during a given period.

PROGNOSIS. A presentation of forecast weather conditions as they are expected to exist at some future time or interval of time.

PROGNOSTIC CHART. A forecast of weather versus an analysis chart which reports existing weather.

PROHIBITED AREA. Airspace of defined dimensions identified by an area on the surface of the earth within which flight is prohibited.

PROPELLER. A device for propelling an aircraft that has blades on an engine-driven shaft and that, when rotated, produces by its action on the air, a thrust approximately perpendicular to its plane of rotation. It includes control components normally supplied by its manufacturer, but does not include main and auxiliary rotors or rotating airfoils of engines.

PUSHER. An airplane in which the propeller is mounted aft of the engine, and pushes the air away from it.

——— ⌐ ———

RADAR.

Air Route Surveillance Radar (ARSR). A long-range radar (approximately 150 miles) used by Air Route Traffic Control Center personnel to control air traffic along the airways between terminals.

Airport Surveillance Radar (ASR). A relatively short-range radar used by control tower personnel for arrival and departure control within a 30-mile area or for radar instrument approaches. Range and azimuth information only available.

Air Traffic Control Radar Beacon System (ATCRBS). Radar transponders, installed in the aircraft, used with secondary radar beacon ground equipment to provide a reinforced target pip for air traffic control use at higher altitudes in areas of high traffic density.

Ground Controlled Approach (GCA). Instrument approach directed by ground personnel who observe the position of the aircraft by the use of radar. Either Precision Approach Radar (PAR) or Airport Surveillance Radar (ASR) may be available.

Precision Approach Radar (PAR). Radar providing positive position of aircraft in range, azimuth, and altitude, permitting the controller to accurately guide aircraft to touchdown on runway.

RADAR ADVISORY SERVICE. The provision of advice or information based on radar observation.

RADAR ALTITUDE. (See altitude.)

RADAR BEACON (secondary radar). A radar system in which the object to be detected is fitted with cooperative equipment in the form of a radio receiver transmitter (transponder). Radio pulses transmitted from the searching transmitter/receiver (interrogator) site are received in the cooperative equipment and used to trigger a distinctive transmission from the transponder. This latter transmission, rather than a reflected signal, is then received back at the transmitter/receiver site.

RADAR IDENTIFICATION. The process of ascertaining that a radar target is the radar return from a particular aircraft.

RADAR SERVICE. A term which encompasses one or more of the following services based on the use of radar which can be provided by a controller to a pilot of a radar-identified aircraft.

Radar Monitoring. The radar flight following of aircraft, whose primary navigation is being performed by the pilot to observe and note deviations from its authorized flight path airway, or route. This includes noting aircraft position relative to approach fixes and major obstructions.

Radar Navigation Guidance. Vectoring aircraft to provide course guidance.

Radar Separation. Radar spacing of aircraft in accordance with established minima.

RADAR SUMMARY CHART. A chart presenting a synoptic analysis of weather radar observations.

RADAR SURVEILLANCE. The radar observation of a given geographical area for the purpose of performing some radar function.

RADAR VECTOR. A heading issued to an aircraft to provide navigational guidance by radar.

RADIAL. A navigational signal generated by a VOR or VORTAC, measured as a magnetic bearing from the station.

RADIAL ENGINE. An internal combustion engine with cylinders mounted in a circle around the crankshaft.

RADIATION. The emission of energy by a medium and transferred, either through free space or another medium, in the form of electromagnetic waves. The emitting body is cooled as heat is transformed into radiant energy, while any medium absorbing radiation energy converts it to heat raising the temperature of the absorbing body.

RADIO ALTIMETER. An electronic instrument used for determining the height of an aircraft above the earth's surface. Three basic types operate on the principles of (1) pulsed radar, (2) continuous-wave radar, and (3) measurement of variations in electrical potential between the aircraft and the surface.

RADIO COMPASS (ADF). A radio receiver equipped with a rotatable loop antenna which is used to measure the bearing to a radio transmitter.

RADIOSONDE (CONTRACTION FOR RADIO SOUNDING). An instrument carried aloft by a radiosonde balloon for the simultaneous measurement and transmission of meteorological data, i.e., pressure, temperature, and humidity.

RANGE MAXIMUM. The maximum distance a given aircraft can cover under given conditions by flying at the economical speed and altitude at all stages of the flight.

RAPCON. Radar Approach Control facility associated with the United States Air Force.

RATE-OF-CLIMB INDICATOR. An instrument which indicates the rate of ascent or descent of an airplane.

RBN. Radio Beacon. See nondirectional radio beacon.

RELATIVE HUMIDITY. The ratio of the amount of moisture in the air to the amount which the air could hold at the same temperature if it were saturated; usually expressed in percent.

RELATIVE WIND. The direction of an airflow with respect to an airfoil.

RESTRICTED AREA. Airspace indentified by an area on the surface of the earth within which the flight of aircraft, while not wholly prohibited, is subject to restrictions. A restricted area may be established by the President of the United States or by any State of the United States pursuant to the Air Commerce Act of 1926, or it may be established pursuant to the City Aeronautics Act of 1938, as amended, or it may be established by the Administrator of the Federal Aviation Administration

RHUMB LINE. The line drawn on a Lambert chart between points for navigational purposes. In practice it is the line on the map which the pilot attempts to follow.

RIDGE (also called ridge line). In meteorology, an elongated area of relatively high atmospheric pressure; usually associated with and most clearly identified as an area of maximum anticyclonic curvature of the wind flow (isobars, contours, or streamlines).

RIME ICE. The formation of a white or milky and opaque granular deposit of ice formed by the rapid freezing of super cooled water drops as they impinge upon an exposed aircraft; formation is favored by small drop size, slow accretion, a high degree of supercooling and rapid dissipation of latent heat of fusion, i.e., one particle freezes before the next one strikes; white appearance is the result of numerous, relatively large air pockets, weighs less than clear ice, but may seriously distort airfoil shape and therefore diminish aerodynamic efficiency.

RODS. Cylindrically shaped nerve cells in the retina that are especially sensitive to low levels of illumination.

ROLL. Movement of an aircraft about its longitudinal axis.

ROTOR CLOUD (sometimes improperly called roll cloud). A turbulent altocumulus type cloud formation found in the lee of some large mountain barriers, the air in the cloud rotates around an axis parallel to the range; indicative of severe to extreme turbulence and, therefore, requires a mandatory remark in surface aviation weather observations when observed.

ROTORCRAFT. A rotorcraft is a power-driven aircraft, heavier-than-air which is supported during flight by one or more rotors.

ROUTE. A defined path, consisting of one or more courses, which an aircraft traverses in a horizontal plane over the surface of the earth.

RUDDER. A movable control surface, usually attached to a vertical stabilizer, by which an air vehicle is guided in the horizontal plane.

RUNWAY. A strip, either paved or improved, on which takeoffs and landings are effected.

RUNWAY THRESHOLD. The beginning of the landing area of the runway.

RUNWAY VISIBILITY VALUE. (See visibility.)

RUNWAY VISUAL RANGE. (See visibility.)

—— S ——

SATURATED AIR. Air that contains the maximum amount of water vapor it can hold at a given pressure and temperature (relative humidity of 100 per cent).

SCATTERED. (1) in surface aviation weather reports, descriptive of a sky cover of from 0.1 to 0.5 coverage; applicable only when clouds or other obscuring phenomena are present aloft; designated by the symbol " ⊕ "; may be identified as thin (predominately transparent), otherwise a predominately opaque status is implied; never constitutes a ceiling. (2) In a thunderstorm forecast, denotes area coverage of thunderstorms of 16% to 45%. (3) In radar reporting, a classification of *echo* coverage denoting a coverage of 1/10 to 5/10 of the area to which applied; designated by the symbol " ⊕ ".

SCUD. Small detached masses of stratus fractus clouds below a layer of higher clouds, usually nimbostratus.

SEA LEVEL PRESSURE. The atmospheric pressure at mean sea level, either directly measured by stations at sea level or empirically determined from the station pressure and temperature by stations not at sea level; used as a common reference for analyses of surface pressure patterns; standard practice is to reduce the observed station pressure to the value it would have at a point at sea level directly below if air of a corresponding temperature that is actually present at the surface were present all the way down to sea level.

SEAPLANE. An airplane equipped to rise from and alight on the water. Usually used to denote an airplane with detachable floats, as contrasted with a flying boat.

SEARCH AND RESCUE (SAR). A service to seek missing aircraft and to assist those found to be in need of assistance.

SEGMENTED CIRCLE. A basic marking device located on an airport which indicates direction of traffic. The circle may encompass a wind direction indicator.

SEPARATION. In air traffic control, the spacing of aircraft to achieve their safe and orderly movement in flight and while landing and taking off.

SEQUENCE FLASHING LIGHTS (SFL). A line of very high intensity lights centered on a runway approach course flashing in rapid sequence to show direction; provides pilot with a "tracer bullet" effect.

SERVICE CEILING. The height above sea level, under normal air conditions, at which a given airplane is unable to climb faster than 100 feet per minute.

SIGMET (contraction for significant meteorological information). (1) A message containing information of particular interest to pilots of *all* aircraft. (2) In the U.S. an inflight weather advisory issued to include tornadoes, squall lines, damaging hail, severe and extreme turbulence, heavy icing, and widespread duststorms.

SIGNIFICANT WEATHER PROG. A prognostic weather chart prepared specifically for aviation depicting some combination of precipitation types and coverage, icing, and turbulence, cloud heights, areas and amount of cloud cover, and freezing level.

SIMPLEX. Transmitting and receiving on same frequency.

SKID. The sidewise sliding by an airborne aircraft away from the center of the curve while turning.

SKY COVER. In weather observations, a term used to denote one or more of the following:

(1) the amount of sky covered, but not necessarily concealed, by clouds or by obscuring phenomena aloft;

(2) relative to the surface, the amount of sky obscured by surface based obscuring pheomena; or

(3) the amount of sky covered or concealed by a combination of (1) and (2).

SLANT RANGE. The line-of-sight distance between two points not at the same elevation.

SLEET. Generally transparent, globular, solid grains of ice which have formed from the freezing of raindrops, or the refreezing of largely melted snowflakes when falling through a below-freezing layer of air near the Earth's surface.

SLIP. The motion of an aircraft in which a relative flow of air moves along the lateral axis; the sidewise motion of an aircraft while turning.

SLIPSTREAM. The current of air driven astern by the propeller.

SMALL AIRCRAFT. Aircraft with a maximum certified takeoff weight of 12,500 lbs. or less.

SMOG. A natural fog contaminated by industrial pollutants; a mixture of smoke and fog.

SNOW. Precipitation in the form of white or translucent ice crystals, chiefly in complex branched hexagonal form and often clustered into snowflakes.

SOLO. A flight during which a pilot is the only occupant of the airplane.

SOLO FLIGHT TIME. Solo flight time is flight time during which the pilot is the sole occupant of the aircraft.

SONIC BOOM. A noise caused by a shock wave that emanates from an aircraft or other object traveling at or above the speed of sound. A shock wave is a pressure disturbance, and is received by the ear as a noise or clap.

SOUNDING BALLOON. A free, unmanned balloon instrumented and/or observed for obtaining a sounding of the atmosphere.

SPAR. A principal longitudinal structural member in an airfoil.

SPECIAL VFR CONDITIONS (special VFR minimum weather conditions). Weather conditions which are less than basic VFR weather conditions and which permit flight under Visual Flight Rules in a control zone. Not authorized in 33 airports listed in FAR Part 93.

SPIN. A prolonged stall in which an airplane rotates about its center of gravity while it descends, usually with its nose well down.

SPIRAL. A prolonged gliding or climbing turn during which at least 360° change of direction is affected.

SQUALL. A wind that increases suddenly in speed, maintains a peak speed of 16 knots or more for a period of two or more minutes, and then decreases in speed; the increases and decreases recur intermittently. Essential difference between squall and gust is the duration and speed of the peak wind.

SQUALL LINE. Any nonfrontal line or narrow band of active thunderstorms (with or without squalls); a mature instability line.

SQUELCH. A control on a radio receiver to surpress the background noise.

STABILITY. The tendency of an airplane in flight to remain in straight, level, upright flight, or to return to this attitude if displaced, without attention of the pilot.

STABILIZER. The fixed airfoil of an airplane used to increase stability; usually, the aft fixed horizontal surface to which the elevators are hinged (vertical stabilizer).

STALL. The flight maneuver or condition in which the air passing over and under the wings stops providing sufficient lift to hold the aircraft's attitude, caused by an excessive angle of attack.

STANDARD ALTIMETER SETTING (QNE). An altimeter set to the standard pressure of 29.92" Hg or 1013.2 Mb. Enroute altitudes based on this setting will provide accurate separation of all traffic using QNE, but will be proportionately higher or lower than the indicated altitude as the ambient barometric pressure is higher or lower than standard. Cruising altitudes will be called "flight levels" and abbreviated as "FL" followed by the indicated altitude in hundreds of feet.

STANDARD ATMOSPHERE. A hypothetical atmosphere based on climatological averages comprised of numerous physical constants of which the most important are (1) a surface temperature of 59° F (15° C) and a surface pressure of 29.92 inches of mercury (1013.2 millibars) at sea level; (2) a lapse rate in the troposphere of approximately 3.5° F (2° C) per 1000 feet; (3) a tropopause of approximately 36,000 feet with a temperature of -56.5° C; and (4) an isothermal lapse rate in the stratosphere to an altitude of approximately 80,000 feet.

STANDARD PRESSURE LAPSE RATE. A decrease in pressure of approximately 2° Celsius for each 1,000 feet.

STANDARD TEMPERATURE LAPSE RATE. A temperature decrease of approximately 2° Celsius for each 1,000 feet increase in altitude.

STATIC SYSTEM. A vent to transmit atmospheric pressure to the vertical speed indicator, altimeter, and airspeed indicator.

STATUTE MILE. 5,280 feet or 0.867 nautical miles.

STRAIGHT-IN APPROACH VFR. Entry of the traffic pattern by interception of the extended runway centerline without executing any other portion of the traffic pattern.

STRATIFORM. Descriptive of clouds of extensive horizontal development, as contrasted to vertically developed cumuliform clouds; characteristic of stable air and, therefore, composed of small water droplets.

STRATOCUMULUS (abbreviated Sc). A principal type of low cloud, predominantly stratiform in gray and/or whitish patches or layers, may or may not merge; elements are tessellated, rounded or roll-shaped with relatively flat tops.

STRATOSPHERE. The atmospheric layer above the tropopause, average altitude of base and top, seven and 22 miles respectively; characterized by a slight average increase of temperature from base to top and is very stable; also characterized by low moisture content and absence of clouds.

STRATUS (abbreviated St). A principal type of low stratiform cloud; a low, gray cloud layer or sheet with a fairly uniform base; sometimes appears in ragged patches; usually does not produce precipitation, but occasionally may give drizzle or snow grains; often forms by evaporation or lifting of the lower layers of a fog bank; composed of minute water droplets or if the temperature is low enough, partly of ice crystals; usually widespread horizontally.

STRUT. A compression or tension member in a truss structure. In airplanes, usually applied to an external major structural member.

SUPERCHARGER. A pump or compressor for forcing more air or fuel-air mixture into an internal combustion, reciprocating engine than it would normally induct at the prevailing density altitude.

SUPERCOOLED WATER. Water that has been cooled below the freezing point but still is in a liquid state.

SURVEILLANCE APPROACH. (See radar.)

—— t ——

TAB. A small auxiliary airfoil usually attached to a movable control surface to aid in its movement, or to effect a slight displacement of it for the purpose of trimming the airplane for varying conditions of power, load, or airspeed.

TACHOMETER. An instrument which registers in revolutions per minute (r.p.m.) the speed of the engine.

TACTICAL AIR NAVIGATION (TACAN). A radio transmitter facility in the en route electronic navigation system, transmitting a pulse train UHF modulated radio wave, utilized by compatible airborne receiver/interrogator equipment to derive bearing relative to the facility in terms of reference pulse/modulation coincidence and distance in terms of time delay between interrogation and receipt of reply.

TAIL GROUP. The airfoil members of the assembly located at the rear of an airplane. (See empennage).

TAIL HEAVY. A condition of trim in an airplane in which the tail tends to sink when the elevator control is released.

TAIL WHEEL. A wheel located at the aft end of the fuselage. Airplanes with such wheels are referred to as tailwheel-type airplanes.

TARGET. The indication shown on a radar display resulting from a primary radar return or a radar beacon reply.

TAXI. To operate an airplane under its own power on the ground, except that movement incident to actual takeoff and landing.

TERMINAL FORECASTS. Weather forecasts available each six hours at all FAA flight service stations, covering the airways weather.

TERMINAL VELOCITY. The hypothetical maximum speed which could be obtained in a prolonged vertical dive.

TETRAHEDRON. As pertaining to airport terminology, a device with four triangular sides which indicates wind direction.

THERMOCOUPLE. A connection or junction of two pieces of unlike metals which produces an electrical current when heated.

THRUST. The forward force on an airplane in the air, provided by the engine.

TIME ZONE. A bank on the earth approximately 15° of the longitude wide, the central meridian of each zone generally being 15° or a multiple removed from the Greenwich meridian so that the standard time of successive zones differs by one hour.

TORNADO. A violently rotating column of air attended by a funnel-shaped or tubular cloud hanging beneath a cumulonimbus cloud.

TORQUE. Any turning, or twisting force. Applied to the rolling force imposed on an airplane by the engine in turning the propeller.

TRACK. The flight path made good over the ground by an aircraft. A track may be called a course when in reference to the charted route and is described in terms of a magnetic bearing.

TRAFFIC PATTERN. The traffic flow that is prescribed for aircraft landing at, taxiing on, and taking off from an airport. The usual components of a traffic pattern are upwind leg, crosswind leg, downwind leg, base leg, and final approach.

TRAFFIC PATTERN INDICATORS. Markers associated with the segmented circle system for the purpose of controlling the direction of the traffic pattern when there is any variation from the normal left hand pattern.

TRANSCRIBED WEATHER BROADCAST (TWEB). Broadcasts provided at selected flight service stations which include meteorological and NOTAM data recorded on tapes and broadcast continuously over the low frequency navigational aids and certain VOR and VORTAC stations.

TRANSITION AREA. Transition areas extend upward from 700 feet or higher above the surface when designated in conjunction with an airport for which an instrument approach procedure has been prescribed or from 1,200 feet or higher above the surface when designated in conjunction with airway route structures or segments. Unless otherwise limited, transition areas terminate at the base of the overlying controlled airspace.

TRANSPONDER. The airborne radar beacon receiver-transmitter which receives radio signals from all interrogators on the ground and which selectively replies with a specific pulse reply pulse or pulse group only to those interrogations being received on the mode to which it is set to respond.

TRICYCLE LANDING GEAR. Landing gear that consists of two main wheels and a single nose wheel. Because the CG is ahead of the main wheels, the aircraft is directionally stable on the ground, and the tendency to ground loop is eliminated.

TROPOPAUSE. The boundary between the troposphere and stratosphere, usually characterized by an abrupt change of lapse rate. The change is in the direction of increased atmospheric stability from regions below to regions above the tropopause. Its height varies from 49,000 feet to 65,000 feet in the tropics to about 32,000 feet in the polar regions. In polar regions in winter it is often difficult or impossible to determine just where the tropopause lies, since under some conditions there is no abrupt change in lapse rate at any height. It has become apparent that the tropopause consists of several discrete, overlapping "leaves," a multiple tropopause, rather than a single continuous surface. In general, the leaves descend step-wise, from the equator to the poles.

TROPOSPHERE. That portion of the atmosphere from the earth's surface to the tropopause; that is, the lowest 25,000 to 65,000 feet of the atmosphere. The troposphere is characterized by decreasing temperature with height, appreciable vertical wind motion, appreciable water vapor content and weather.

TROUGH. In meteorology, an elongated area of relatively low atmospheric pressure; usually associated with and most clearly identified as an area of maximum cyclonic curvature of the wind flow (isobars, contours, or streamlines). Compare with ridge.

TRUE ALTITUDE. The altitude above mean sea level. (See altitude.)

TURBOJET. An engine that derives power from a vaned wheel spinning in reaction to burning gases escaping from a combustion chamber. The turbine in turn drives a compressor and other accessories.

TURBOPROP. A turbine engine in which the rotating turbine turns a propeller.

TURBULENCE. Irregular motion of the atmosphere produced when air flows over a comparatively uneven surface, such as the surface of the earth, or when two currents of air flow past or over each other in different directions or at different speeds.

TURN-AND-SLIP INDICATOR. A gyroscopic instrument for indicating the rate of turning and the quality of coordination.

TURN COORDINATOR. An instrument with a 35° canted gyro which utilizes an airplane silhouette to indicate rate of turn.

TWO-MINUTE TURN. Turning at a rate in which the aircraft will turn 180° in one minute or 360° in two minutes.

TYPE (Of Aircraft). Type is a specific classification of aircraft having the same basic design including all modifications which result in a change in handling or flight characteristics.

——— U ———

ULTRA-HIGH FREQUENCY (UHF). A frequency band from 300 to 3,000 MHz.

UNICOM. Frequencies authorized for aeronautical advisory services to private aircraft. Only one such station is authorized at any landing area. The frequency 123.0 MHz is used at airports served by an airport traffic control tower or a flight service station and 122.8 MHz is used for other landing areas. Services available are advisory in nature, primarily concerning the airport services and airport utilization.

UPSLOPE FOG. A type of fog formed when air flows upward over rising terrain and is, consequently, adiabatically cooled to or below its initial dewpoint.

UPWIND LEG. A flight path parallel to the landing runway in the direction of landing.

USEFUL LOAD. In airplanes, the difference, in pounds, between the empty weight and the maximum authorized gross weight.

—— V ——

VARIATION (VAR). The angle difference at a given point between true north and magnetic north expressed as the number of degrees which magnetic north is displaced east or west from true north. The angle to be added algebraically to true directions to obtain magnetic directions.

VECTOR. The resultant of two quantities (forces, speeds, or deflection); used in aviation to compute load factors, headings, or drift.

VENTURI (or venturi tube). A tube with a restriction, used to provide suction to operate flight instruments by allowing the slipstream to pass through it.

VERTICAL AXIS. An imaginary line running vertically through the center of gravity and about which the aircraft yaws left or right. The yaw axis.

VERTIGO. Sensory confusion, usually associated with dizziness and inability to tell which way is up.

VERY HIGH FREQUENCY (VHF). Frequency band from 30 to 300 MHz.

VFR. The symbol used to designate visual flight rules.

VHF OMNIDIRECTIONAL RANGE (VOR). The VHF omnidirectional range provides positive guidance on any selected magnetic course to or from the station.

VICTOR AIRWAY. An airway system based on the use of VOR facilities. The north-south airways have odd numbers, i.e., Victor 11, and the east-west airways have even numbers.

VISIBILITY. In weather observing practice, the greatest distance in a given direction at which it is possible to see and identify with the unaided eye (or the instrumentally determined equivalent) specified objects under specified conditions at specified locations. For weather observing purposes, visibility is categorized as (1) meteorological visibility, (2) vertical visibility, or (3) runway visual range.

VISIBILITY, RUNWAY VISUAL RANGE (RVR). An instrumentally derived value, based on standard calibrations, that represents the horizontal distance a pilot will see down the runway from the approach end; it is based on the sighting of either high intensity runway lights or on the visual contrast of other targets — whichever yields the greater visual range. RVR, in contrast to prevailing or runway visibility, is based on what a pilot in a moving aircraft should see looking down the runway. RVR is horizontal, *and not slant*, visual range. It is based on the measurement of a transmissometer made near the touchdown point of the instrument runway and is *reported in hundreds of feet.*

VISUAL APPROACH SLOPE INDICATOR (VASI). An airport lighting facility providing vertical guidance to aircraft during approach and landing, by radiating a directional pattern of high intensity red and white focused light beams which indicate to the pilot that he is "on path "if red/white, "below path" if red/red, and "above path" if white /white.

VISUAL FLIGHT RULES (VFR). When weather conditions are above the minimums prescribed for visual meteorological conditions pilots may fly with visual reference to the ground and without reference to radio navigational aids.

VOR. Very high frequency omnirange station. (See VHF omnidirectional range.)

VORTICES. As pertaining to aircraft, circular patterns of air created by the movement of an airfoil through the atmosphere. As an airfoil moves through the atmosphere in sustained flight, an area of high pressure is created beneath it and an area of low pressure is created above it. The air flowing from the high pressure area to the low pressure area around and about the tips of the airfoil tends to roll up into two rapidly rotating vortices, cylindrical in shape. These vortices are the most predominant parts of aircraft wake turbulence and their rotational force is dependent upon the wing loading, gross weight and speed of the generating aircraft.

VOT (VOR test signal). A ground facility which emits a test signal to check VOR receiver accuracy.

——— W ———

WAKE TURBULENCE. Turbulence found to the rear of a solid body in motion relative to a fluid. In aviation terminology, the turbulence caused by a moving aircraft.

WARM AIR MASS. (1) An air mass that is warmer than the underlying surface (see air mass classification). (2) A mass of air that is warm relative to adjacent or surrounding air.

WARM FRONT. The discontinuity at the forward edge of an advancing current of relatively warm air which is displacing a retreating colder air mass.

WARNING AREA. Airspace which may contain hazards to nonparticipating aircraft in international airspace.

WEATHER DEPICTION CHART. A weather analysis portraying areas of precipitation and obstructions to vision, cloud cover and cloud heights.

WIND CORRECTION ANGLE (WCA). (See crab angle.)

WIND SHEAR. The change of either wind speed or direction or both, in any direction, conventionally expressed as vertical wind or horizontal wind shear.

WIND SOCK. A cloth sleeve, mounted aloft at an airport to use for estimating wind direction and velocity.

WIND TEE. An indicator of wind direction for takeoff and landing.

WINDS ALOFT. Winds at high altitudes, unaffected by surface features.

WING. An airfoil whose major function is to provide lift by the dynamic reaction of the mass of air swept downward.

WING FLAP. A movable section of an airfoil used to change the effect of air flow over the airfoil. Wing flaps are located along the trailing edge of the wing and are lowered during takeoff and landing in order to increase the effective lift of the wing.

WING ROOT. The end of a wing which joins the fuselage, or the opposite wing.

WING TIP. The end of the wing farthest from the fuselage, or cabin.

WINGSPAN. The distance from one wingtip of an aircraft to the opposite wingtip.

——— Y ———

YAW. To turn about the vertical axis. (An airplane is said to yaw as the nose turns without the accompanying appropriate bank.)

——— Z ———

ZEPPELIN. A lighter-than-air aircraft with a rigid framework, generally made of light metal.

ZULU TIME. Formerly known as Zebra time, and also known as Universal time; it is the local civil time in Greenwich, England, and is used throughout the world in navigation.

ALPHABETICAL INDEX

(Note: The Federal Aviation Regulations are not indexed in this section.
See page 6-1 for the index to the FARs.)

—— d ——

—— e ——

—— m ——

—— n ——